AF815155

Nonlinear Wave Processes in Excitable Media

NATO ASI Series

Advanced Science Institutes Series

A series presenting the results of activities sponsored by the NATO Science Committee, which aims at the dissemination of advanced scientific and technological knowledge, with a view to strengthening links between scientific communities.

The series is published by an international board of publishers in conjunction with the NATO Scientific Affairs Division

A	**Life Sciences**	Plenum Publishing Corporation
B	**Physics**	New York and London
C	**Mathematical and Physical Sciences**	Kluwer Academic Publishers Dordrecht, Boston, and London
D	**Behavioral and Social Sciences**	
E	**Applied Sciences**	
F	**Computer and Systems Sciences**	Springer-Verlag
G	**Ecological Sciences**	Berlin, Heidelberg, New York, London,
H	**Cell Biology**	Paris, and Tokyo

Recent Volumes in this Series

Volume 238—Physics, Geometry, and Topology
edited by H. C. Lee

Volume 239—Kinetics of Ordering and Growth at Surfaces
edited by Max G. Lagally

Volume 240—Global Climate and Ecosystem Change
edited by Gordon J. MacDonald and Luigi Sertorio

Volume 241—Applied Laser Spectroscopy
edited by Wolfgang Demtröder and Massimo Inguscio

Volume 242—Light, Lasers, and Synchrotron Radiation: A Health Risk Assessment
edited by M. Grandolfo, A. Rindi, and D. H. Sliney

Volume 243—Davydov's Soliton Revisited: Self-Trapping of Vibrational Energy in Protein
edited by Peter L. Christiansen and Alwyn C. Scott

Volume 244—Nonlinear Wave Processes in Excitable Media
edited by Arun V. Holden, Mario Markus, and Hans G. Othmer

Volume 245—Differential Geometric Methods in Theoretical Physics: Physics and Geometry
edited by Ling-Lie Chau and Werner Nahm

Series B: Physics

Nonlinear Wave Processes in Excitable Media

Edited by

Arun V. Holden

University of Leeds
Leeds, United Kingdom

Mario Markus

Max-Planck-Institut für Ernährungsphysiologie
Dortmund, Federal Republic of Germany

and

Hans G. Othmer

University of Utah
Salt Lake City, Utah

Springer Science+Business Media, LLC

Proceedings of a NATO Advanced Research Workshop
on Nonlinear Wave Processes in Excitable Media,
held September 11-15, 1989,
in Leeds, United Kingdom

Library of Congress Cataloging-in-Publication Data

NATO Advanced Research Workshop on Nonlinear Wave Processes in
Excitable Media (1989 : Leeds, England)
Nonlinear wave processes in excitable media / edited by Arun
V. Holden, Mario Markus, and Hans G. Othmer.
p. cm. -- (NATO ASI series. Series B, Physics ; v. 244)
"Proceedings of a NATO Advanced Research Workshop on Nonlinear
Wave Processes in Excitable Media, held September 11-15, 1989, in
Leeds, United Kingdom"--T.p. verso.
"Published in cooperation with NATO Scientific Affairs Division."
Includes bibliographical references and index.
ISBN 978-1-4899-3685-1 ISBN 978-1-4899-3683-7 (eBook)
DOI 10.1007/978-1-4899-3683-7
1. Nonlinear waves--Congresses. 2. Excitation (Physiology)-
-Congresses. I. Holden, Arun V., 1947- . II. Markus, M.
(Mario), 1944- . III. Othmer, H. G. (Hans G.), 1943- .
IV. North Atlantic Treaty Organization. Scientific Affairs
Division. V. Title. VI. Series.
QA927.N27 1991
574.19'12--dc20 90-49843
CIP

Originally published by Plenum Press, New York in 1991.
Softcover reprint of the hardcover 1st edition 1991

SPECIAL PROGRAM ON CHAOS, ORDER, AND PATTERNS

This book contains the proceedings of a NATO Advanced Research Workshop held within the program of activities of the NATO Special Program on Chaos, Order, and Patterns.

Volume 208—MEASURES OF COMPLEXITY AND CHAOS
edited by Neal B. Abraham, Alfonso M. Albano, Anthony Passamante, and Paul E. Rapp

Volume 225—NONLINEAR EVOLUTION OF SPATIO-TEMPORAL STRUCTURES IN DISSIPATIVE CONTINUOUS SYSTEMS
edited by F. H. Busse and L. Kramer

Volume 235—DISORDER AND FRACTURE
edited by J. C. Charmet, S. Roux, and E. Guyon

Volume 236—MICROSCOPIC SIMULATIONS OF COMPLEX FLOWS
edited by Michel Mareschal

Volume 240—GLOBAL CLIMATE AND ECOSYSTEM CHANGE
edited by Gordon J. MacDonald and Luigi Sertorio

Volume 243—DAVYDOV'S SOLITON REVISITED: Self-Trapping of Vibrational Energy in Protein
edited by Peter L. Christiansen and Alwyn C. Scott

Volume 244—NONLINEAR WAVE PROCESSES IN EXCITABLE MEDIA
edited by Arun V. Holden, Mario Markus, and Hans G. Othmer

PREFACE

This volume contains the contributions, both the oral and poster presentations, to the NATO Advanced Research Workshop on "Nonlinear Wave Processes in Biological Media" held at Tetley Hall, University of Leeds, Leeds U.K., September 11-15, 1989. At this meeting 55 researchers from North America and Western Europe, and the Soviet Union and Eastern Europe presented mathematical and experimental results on nonlinear wave phenomena in excitable media and related nonlinear systems through lectures and posters. Tetley Hall is a small, self contained residence for students and is set in secluded gardens, and so the site favoured the small informal discussions that carried on into the late summer evenings. In spite of their involvement with excitable media for many years many of the participants met here for the first time, and so the main result of the workshop will only appear in the next few years, as these preliminary contacts lead to more extensive collaborations.

As organisers we feel that the outcome of the meeting exceeded our expectations, and that we were extremely fortunate in organising this meeting at a time when political developments in the Soviet Union enabled all the invited lecturers to attend. We would like to thank all participants for the enthusiasm they brought with them and for contributing to this volume. We also acknowledge the staff at Tetley Hall for their help with the organisation. Special thanks are due to Mrs Joanna Barugh, the Domestic Bursar of Tetley Hall, for her assistance before and during the meeting, and to Mrs Jackie Hill for typing the manuscripts.

We are grateful for the financial support of the NATO Scientific Affairs Division, Brussels, who funded the meeting under the Special Program Panel on Chaos, Order and Patterns: Aspects of Nonlinearity; the Commission of the European Communities, Directorate-General for Science Research and Development, Brussels; and the Wellcome Trust.

Arun V. Holden
Mario Markus
Hans G. Othmer

CONTENTS

1. EXCITING TIMES IN EXCITABLE MEDIA

A.V. Holden

Centre for Nonlinear Studies
The University
Leeds LS2 9JT, UK

"There is no better, there is no more open door by which you can enter into the study of natural philosophy than by considering the physical phenomena of a candle".

In spite of this early recognition by Michael Faraday [1] of the importance of nonlinear reaction-diffusion systems, most of the explosive interest in nonlinear systems that has resulted in the founding of Centers (and Centres), journals, book series and research inititiatives in the area of Nonlinear Science have been concerned with either solitons or chaos. It is only recently that spectacular results have begun to emerge in the area of nonlinear wave processes in excitable media.

An early example of the analysis of a nonlinear diffusion equation

$$p_{xx} - p_t = F(p)$$

is the paper by Kolmogorov, Petrovsy and Piskunov [2]. The problem was related to the diffusion of a dominant advantageous gene through a population, and $F(p)$ was taken of the form $p(1-p)$ which leads, in the one dimensional case, to the propagation of a wavefront through the medium.

To obtain a solitary travelling wave solution a cubic nonlinearity is required: with $F(p) = p(p - a)(p - 1)$ a solitary travelling wave solution

$$p^* = [1 + exp\{(x - ut)/\sqrt{2}\}]^{-1}$$

at a velocity $u = (1-2a)/\sqrt{2}$ results. This cubic was introduced by Nagumo [3] as a model for a propagating nerve impulse on an axon. Unlike in the theory of solitons, such prototype equations, with simple polynomial nonlinearities, are rare in the area of excitable media, and variants of the FitzHugh-Nagumo caricature for excitation are widely used as they form the only example of a simple, tractable excitation equation.

Both these examples come from biology, and this is one reason why progress in the study of excitable media has been slow. Partial differential equations that accurately describe reaction diffusion systems in biology, such as the Hodgkin-Huxley equations for squid giant axon [4], the Noble *et al.*

descriptions of electrical activity in cardiac muscle [5] and the Monk-Othmer equations for aggregation in the slime mold *Dictylostelium* (see chapter 23) are generally complicated, empirical and often have high order, stiff nonlinearities. Such equations lead to numerical methods rather than analyis, and mathematical approaches such as those of Carpenter [6] are rare. It is only recently that computational power has become readily available, so that investigations on important but complicated partial differential equations can be carried out by scientists and applied mathematicians (eg [7]), rather than by expert numerical analysts (eg [8]). Further, in general for

$$D\mathbf{p}_{\mathbf{xx}} - \mathbf{p}_t = F(\mathbf{p})$$

$\mathbf{p} \in \mathbb{R}^n$ of a high order state space and $\mathbf{x} \in \mathbb{R}^3$ of physical space, and so as well as computational problems there are also problems in how to display the spatio-temporal patterned and irregular results of computations. Here the graphics workstation is proving invaluable.

Glancing through this volume, it is obvious that computer generated figures far outweigh actual experimental observations. Although examples of excitable media occur throughout the sciences, from laboratory biophysics to galactic physics, the shortage of laboratory preparations where well controlled experimental investigations can be carried out has contributed to the slow growth of studies on excitable media. The best studied excitable medium is the Belousov-Zhabotinsky reaction [9] in a thin layer of medium, and when combined with two-dimensional microspectrophotometry and image processing techniques [10] quantitatively precise data can be obtained. It is such data that are required to drive the development of theory and stimulate numerical computations, and the recent development of methods for controlling the excitability of an excitable medium, by using a photosensitive catalyst [11], and for controlling the diffusion coefficients in a reaction diffusion reactor [12] promise a new range of phenomenology that will require explanation and quantitative simulation.

A great deal of modern research is linked to industrial applications, such as the use of optical solitons in telecommunications. Excitable media have not had such an obvious industrial application, with the major applications interest coming from biomedicine and the possibility of regulating abnormal wave processes in the heart. One promising application is the use of photosensitive excitable media to pre-process images: if this technology becomes viable then excitable media will move from being of little more than academic interest to being of economic interest.

Much of the pioneering work in excitable media, their mathematical analysis, experimental phenomenology and applications, has been carried out in the Soviet Union. Both because of language and political problems, work in the West and in the Soviet Union has followed parallel, relatively independent courses, with repeated rediscoveries of the same results. The political changes of the late eighties have altered the research environment, allowing rapid exchange of results and the beginings of collaborative projects. The first Euromech Colloquium to be held in the Soviet Union was in Tallinn, Estonia in October 1988, and on *Nonlinear waves in active media* [13]. This NATO ARW was attended by Soviet theoreticians and experimentalists, and in June 1990 there will be an international workshop on *Waves and patterns* at Pushchino, Moscow Region, that will concentrate on excitable media [14]. This rapid succession of meetings will foster collaborations and focus attention on the major problems in excitable media: spatio-temporal patterning and irregularity in two and three dimensional excitable media. Now that precisely controllable laboratory preparations and computational power are available one can anticipate a rapid growth in our understanding of nonlinear wave phenomena in excitable media. The next few years promise to be exciting times in the area of excitable media.

References

[1] Faraday, M. The chemical history of a candle. Harvard Classics **30** P.F.Collier & Son New York (1910)
[2] Kolmogorov, A.N., Petrovsky, I.G. and Piskunov, N.S. (1937) Investigation of the equation of diffusion associated with increase of the quantity of a substance and its application to a biological problem. *Byull. MGU,* sec. A, **1**(6)
[3] Nagumo, J., Arimoto, S and Yoshizawa, S. (1962) An active pulse transmission line simulating nerve axon. *Proc. IRE* **50** 2061
[4] Hodgkin, A.L. and Huxley, A.F. (1952) A quantitative description of membrane current and its application to conduction in nerve. *J.Phyiol (London)* **117** 500
[5] HEART. From Oxsoft, Oxford UK
[6] Carpenter, G.A. A geometric approach to singular perturbation problems with applications to nerve impulse equations. (1977) *J. Diff. Equations* **23** 335
[7] Coutemarche, M. and Winfree, A. (1990) Two-dimensional rotating depolarization waves in a modified Beeler-Reuter model of cardiac cell activity. Annual Research Peport for 1990.Consortium for Scientific Computing, Princeton New Jersey. also in *Pixel* **1 (2) 34**
[8] Cole, K.S., Antosiewicz, H.A. and Rabinowitz, P. (1955) Automatic computation of nerve excitation. *J. Soc. Indust. Appl. Math.* **3** 153
[9] Zhabotinsky, A.M. The first period 1961-1969 of systematic studies of oscillations and waves in Belousov chemcial systems. in; *Oscillations and waves in chemical systems*, ed. R.Field, M. Burger. New York: Wiley:
[10] Muller, S.C., Plesser, T and Hess, B. (1986) Two-dimensional spectrophotometry and pseudo-color representation of chemcial reaction patterns. *Naturwissenschaften* **73** 165
[11] Kuhnert, L., Agladze, K.I. and Krinsky, V. (1989) *Nature* **337** 244
[12] Noszticzius, Z., Horsthemke, W.,McCormick, W.D., Swinney, H.L. and Tam, W.Y. (1987) *Nature* **329** 619
[13] Engelbrecht, J., ed. (1989) *Nonlinear waves in active media.* Berlin: Springer Verlag.

2. A STUDY OF THE AUTOWAVE MECHANISMS OF CARDIAC ARRHYTHMIAS

V. Krinsky, A Pertsov, V. Fast and V. Biktashev*

Institute of Biological Physics
USSR Academy of Science
Pushchino Moscow Region 142292, USSR
*Research Computing Centre, USSR Academy of Science
Pushchino Moscow Region 142292, USSR

It has been shown by multielectrode electrophysiological mapping that many types of cardiac arrhythmias are based on re-entry, i.e. excitation wave circulation along a closed circuit. On the other hand, rotating waves of excitation have long been studied by physical methods. They constitute an important class of strongly nonlinear waves, the so-called autowave vortices.

Our aim is to apply the concepts, methods and results of studying the rotating waves in excitable media to cardiac arrhythmias and to give a short review of new results obtained. The properties of rotating vortices in chemical active medium (BZ reaction) and cardiac tissue were compared. The mechanisms were described for the origin of rotating autowave vortices, as well as their interation with different wave sources, particularly, with a focal-type wave source. The processes were considered that impose limitations on the vortex life-time. In particular the drift of autowave vortices and the relation of this phenomenon to the duration of cardiac arrhythmias are considered.

Experimental data are presented on three dimensional vortices in the myocardium. A new defibrillation mechanism, which is connected with the movement of leading cycles, is considered along with the ways of reducing discharge energy.

1. Introduction

The paper is concerned with arrhythmias based on re-entry, circulation of an excitation wave. Such arrhythmias have been studied for about a century but many problems still remain. The most important of them is why an extrasystole attack under similar conditions may result in quite different consequences, from a short-term, almost harmless attack of arrhythmia to sudden death. What causes an ectopic focus either to vanish after emitting a few pulses or to create a threatening picture of the progressive impairment of cardiac rhythm up to fibrillation?

The problem has been extensively studied in terms of general medical and specific cardiological factors such as the state of blood circulation and concomitant diseases. Lately approaches at cellular and molecular levels have been developed.

There is, however, an aspect of the problem that has received little attention, although it may turn out to be very essential. We mean the wave propagation processes during cardiac arrhythmias. The cardiac muscle is a fairly large organ as compared to the characteristic sizes of excitation processes therein. Therefore, the consideration of excitation waves along with excitation pulses is necessary.

2. Fibrillation and rotating waves

Wiener & Rosenbluth [1] were the first to attempt a mathematical modelling of the propagation of excitation waves in cardiac muscle. They have shown that an excitation wave can circulate around a hole whose effective perimeter exceeds a certain wave length. In this case the wave takes the form of a spiral, the involute of the hole. In his computational experiments, Moe [2] could observe what was very much like fibrillation.

From the standpoint of modern theory, the fibrillation type phenomena are not exclusively specific to the heart. They belong to the so-called chaotic processes characteristic of the broad class of systems with nonlinear dynamics. There are some typical scenarios of chaos development. The specificity of the heart, however, is in its being an excitable system whose elements are potential auto-oscillators. Waves propagating in such systems have been called autowaves [3].

Figure 1 shows one of the most spectacular examples of autowave systems, the BZ reaction. Many of the phenomena that underlie cardiac arrhythmias occur also in this case. In a Petri dish containing the BZ reagent (100 mM malonic acid, $CH_2(COOH)_2$, 100 mM sodium bromate, $NaBrO_3$, 300 mM sulphuric acid, H_2SO_4, and 1.3 mM ferroin), an electric pulse can evoke a propagating wave of excitation (Fig. 1a). It appears as a blue wave of the oxidized state of the ferric-phenanthroline complex (Fe^{3+}) against the bright red background of the reduced state (Fe^{2+}).

As in the case of cardiac tissue, the wave processes in the BZ reaction show a refractory period; two waves of excitation do not pass through each other, but annihilate when they collide; pacemakers and re-entries may occur. A pacemaker emits concentric waves (Fig. 1a,b, left) and re-entry gives rise to spiral waves (Fig. 1a,b, right) rotating with a period close to the refractory period. It is seen from Fig. 1.b that the pacemaker is suppressed by re-entry. Since the waves sent by re-entry have higher frequencies than those emitted by the pacemaker (shorter wave lengths are evident from the figure) and colliding waves annihilate pairwise, each next collision occurs increasingly closer to the pacemaker. Eventually, the re-entry synchronizes it. Figure 1b shows the last instances of a pacemaker's life: before it disappears it has time to emit only one wave.

The propagation of autowaves in different excitable media is governed by the same fundamental laws [3]. Chaos taking place in such systems has been termed autowave chaos. The specific scenario of its appearance is as follows [4,5]: 1) occurrence of wave breaks at inhomogeneities; 2) whirling of the break ends to give spiral waves (vortices); 3) multiplication of vortices; 4) their decay.

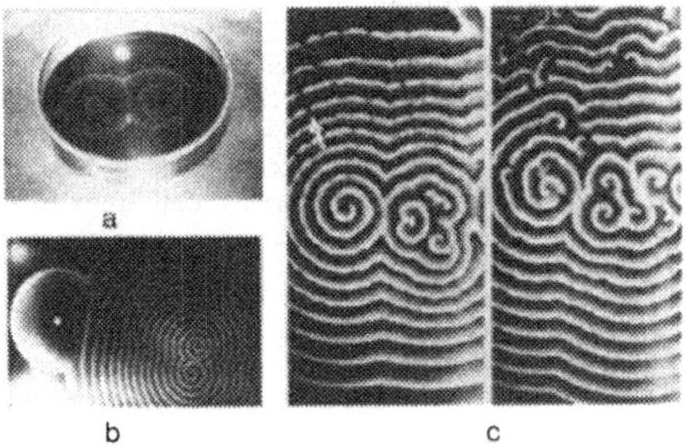

Figure 1. Waves of excitation in a chemical active medium. a) Propagating concentrational waves in a Petri dish with the BZ reagent. On the left, a pacemaker that has sent two concentric waves is seen. On the right, a broken wave is shown which starts twisting into two counter-rotating spirals.
b) Spiral waves (re-entry) suppress a lower-rate pacemaker.
c) Reproduction of spiral waves. Initially, vortices are located only in the central part of the frame (left). Five minutes later (right), multiple generation of vortices is seen in the upper part of the frame, where inhomogeneities were created. In the absence of inhomogeneities, no wave breaks are seen (bottom).

Figure 1c illustrates the appearance of wave breaks at inhomogeneities [6]. The breaks evolve into spirals (right) and produce a multiplicity of rotating vortices. In the absence of inhomogeneities vortices do not appear (cf. the behaviour of waves in the upper and lower parts of Fig. 1c).

It is at inhomogeneities that vortices decay. It follows from both theory [5,7,8] and experiment [9] that at an inhomogeneity a drift of vortices takes place: the higher the inhomogeneity gradient the greater the velocity of the drift. When encountering the medium border, vortices decay. They also decay upon collision.

Whether chaos will be self-sustained or will rapidly disappear is determined by the relationship between the rates of births and decays of vortices: if the number of vortices born in unit time (dN_b/dt) is greater than the vortices decaying in unit time (dN_d/dt), the wave pattern becomes complicated and fibrillation may occur:

$$dN_b/dt > dN_d/dt. \tag{1}$$

With the reverse relationship

$$dN_b/dt < dN_d/dt \tag{2}$$

the complexity of the pattern decreases and this may result in termination of fibrillation.

3. Rotating waves in three dimensions

Cardiac tissue is essentially three-dimensional. It was not until very recently that experimental investigation of three-dimensional rotating waves in the myocard became possible [10]. While in two dimensions, a wave rotates around a point, in three dimensions such points form a filament around which a scroll-like wave rotates (Fig. 2a). The filament can be closed into a ring, in which case a scroll ring occurs [11,12] (Fig. 2b). As seen from the figure, in a cross-section perpendicular to the scroll ring filament there are two spirals with opposite sense of rotation. The surface of the medium displays a specific wave pattern with a circular wave source (projection of the filament onto the surface). Inside the circle, waves propagate inwards to its centre. Outside the circle the wave fronts propagate outward in a concentric manner, as in the case of the pacemaker.

A wave pattern typical of the scroll ring was recorded on the surface of the heart wall [13], using the technique of multielectrode mapping, Fig. 2c. Such a pattern may occur if the filament plane is somewhat tilted with respect to the surface. As a result, waves emitted by a filament closed into a circle will emerge on the surface at different times. The wave pattern recorded at t=10 ms corresponds to the case where waves emitted by the left part of the circle come to the surface first, and a horseshoe shape of the wave results. Of course, this is only indirect evidence for the scroll ring. Direct evidence can be produced using intramural recordings.

Scrolls and scroll rings are expected to occur only in the case of an

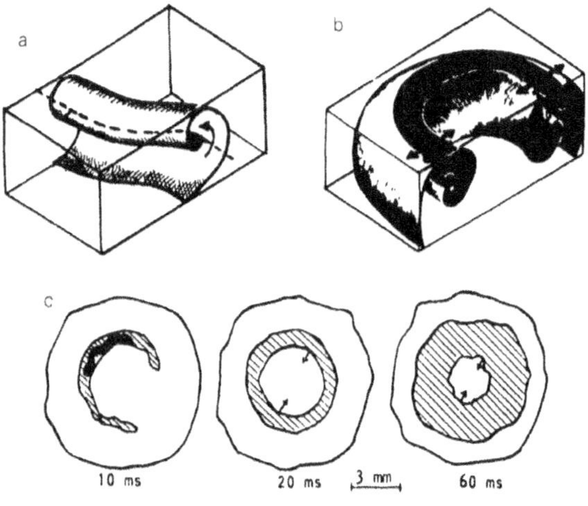

Figure 2. Three-dimensional vortex structures.
a) Rotating scroll. The dashed line is a filament around which the rotation occurs.
b) Scroll ring (rotating scroll whose filament is closed into a ring). A wave pattern on the myocardial surface: inside a small region, ring waves run towards its centre and outside it, they propagate outward (is indicated by arrows).
c) Excitation pattern in isolated rabbit atrium. One can see an annular region (hatched) whose outer border expands and inner border shrinks to the centre in a manner typical of the vortex ring.

insignificant damage of cardiac tissue, when the heart wall remains excitable throughout its entire thickness (early stage of infarction). When extended inexcitable areas appear in the myocard, quite different three-dimensional patterns determined by inhomogeneities may occur.

4. Three-dimensional circulation in superfused rabbit ventricle

During arrhythmias, jumpwise changes in the cycle length are known to occur. We also observed such changes in experiments with isolated rabbit ventricles during tachycardia [14]. Curiously, a focal pattern typical of the pacemaker has been shown by multielectrode mapping. It is difficult to conceive of a mechanism that would give rise to jumpwise changes in the pacemaker rhythm. By simultaneous endocardial and epicardial mapping we have shown [15] that such effects are based on three-dimensional circulation.

The studies were carried out with an experimental model of sustained ventricular tachycardias in isolated preparations of rabbit ventricle superfused with Tyrode's solution. Figure 3a presents endocardial and epicardial activation patterns recorded during experiments. It is seen that a focal type of activation is registered on the myocardial surfaces. An excitation arising periodically from a small region on the surface spreads in a radial fashion in all directions. It is interesting to note that such an activation pattern is often seen in patients undergoing surgical treatment [16] as well as in infarcted canine hearts [17].

Multielectrode mapping provides evidence for a circulatory mechanism of the tachycardia. The circulation involves the endocardial and epicardial sufaces and two pathways of transmural conduction. A wave spreads along the endocardium and enters the epicardium through the right transmural pathway. Further the wave moves along the epicardium retrogradely reaching the left transmural pathway and then enters the endocardium thus closing the cycle and giving rise to a new one.

To prove that tachycardia is caused by re-entry, Mine's third criterion should be satisfied, that is cutting the circuit must abolish abnormal rhythms believed to be due to re-entry [18]. For that reason, we cut the re-entrant circuit approximately midway between the transmural paths. In no case could arrhythmia be reinitiated.

We could show by mapping techniques that changes in the cycle length occurred when there were more than two transmural conduction pathways and they depended on where the pathways lay. The onset of tachycardia in a preparation with three transmural conduction pathways P_1, P_2 and P_3 is shown in Fig. 3. The initial re-entry circuit involved P_1 and P_2 (see map T_3). The change in the activation pattern occurred after the third beat of tachycardia. The wave was blocked in P_2 and circulation proceeded through P_3 (see map T_4). The earliest activation site on the endocardial surface shifted by 6 mm from the apex to the base. This resulted in an increase in the re-entry circuit length by 12 mm, which caused the tachycardia cycle length to increase from 115 to 200 ms.

5. Defibrillation

The development of implantable defibrillators raises an acute problem of decreasing the power of a defibrillating pulse. This has renewed interest in

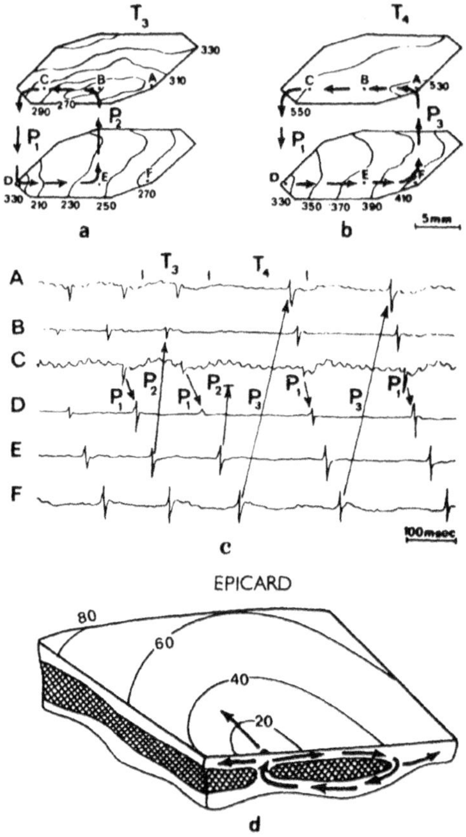

Figure 3. Three-dimensional circulation in isolated right rabbit ventricle resulting in a focal activation pattern on the myocardial surface.
a) Activation maps for epicardium (bottom) and endocardium (top). One can see a radial activation pattern typical of the pacemaker. Here it is due to excitation circulation through the epicardium, endocardium and two transmural pathways.
b) continuation of the record shown in (a).Abrupt displacement of the earliest activation point and increasing of the cycle length during the fourth tachycardia cycle due to the change in the circulation pathway.
c) Electrograms registered at different points on the epicardium and endocardium (cf. Fig. 3a,b).
d) A scheme illustrating the mechanism of tachycardia. An injured zone (hatched) occurs in the thickness of the heart wall because of the termination of coronary circulation.

studying in detail the defibrillation mechanism.

Fibrillation is generally considered as a process of multiple generation of functionally determined re-entries (leading circles, or reverberators). The highest frequency tachycardias requiring cardioversion and not amenable to other treatments are also related to leading circles.

The common defibrillation mechanism is in bringing all myocardial cells, owing to the large amplitude of the applied current, to the state of refractoriness, which makes impossible the occurrence of re-entries [19]. Analysis of mathematical models suggests a principally different mechanism of defibrillation: displacement of leading circles to the outside of the excitable tissue and pairwise annihilation of leading circles with opposite sense of rotation. Its realization, as in the case of common defibrillation, presupposes a uniform tissue depolarization, but does not require large current amplitudes. Currents exceeding the threshold amplitudes as little as several-fold may turn out to be sufficient. It has been found that a small-amplitude defibrillating pulse can shift leading circles. Figure 4a presents a

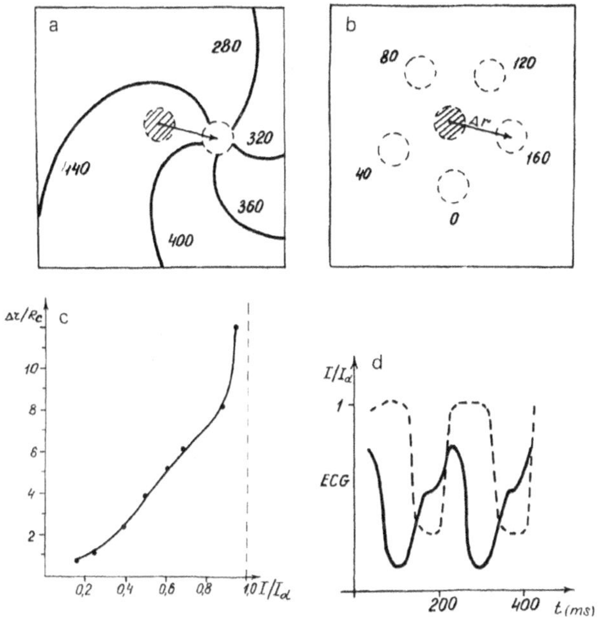

Figure 4. Investigation of the defibrillation mechanism in a mathematical model of cardiac tissue.
a) Displacement of a leading circle caused by current impulse below the defibrillation threshold. The initial position of the leading circle is hatched. Numbers near isochronal contours indicate time in ms.
b) Change in the direction of the displacement depends on the time of application of a stimulus.
c) Displacement as a function of the amplitude of applied current. R_c is the leading circle core radius and I_d is defibrillation current.
d) Dependence of the minimum defibrillation current (dashed line) on the time of application of a stimulus. Solid line - a model ECG.

typical isochronic map of a leading circle. The dashed line shows a central region, the core, around which an excitation wave rotates. On application of a defibrillating pulse, the core shifts. The direction of the shift depends on the phase at which the pulse was applied. By varying the phase, it is possible to obtain shifts in any direction, Fig. 4b.

The distance of the displacement, as a rule, grows with defibrillating current amplitude (Fig. 4c). Displacements become noticeable at amplitudes approaching the tissue excitation threshold. Large displacements (comparable to the core size) can be observed when currents significantly smaller than the defibrillation current are used.

If the core is displaced as far away as the border of an excitable tissue, circulation will terminate. Selecting a moment of application of a pulse so that the core was displaced in the direction of the nearest border it is possible to decrease the maximum current required. Figure 4d shows a change in the minimum current sufficient to interrupt circulation during tachycardia. In this case, the termination of arrhythmia can be achieved using an amplitude only one fourth that of the defibrillation current or less.

6. Conclusion

The autowave theory outlines the conditions of the transition from paroxysmal tachycardia to fibrillation, explains the critical mass phenomenon (requirement of the minimum heart tissue mass for fibrillation to occur), the phenomenon of short-lived fibrillation, its reversible transition to tachycardia and termination of an attack.

References

[1] Wiener, N. & Rosenbluth, A. (1946). The mathematical formulation of the problem of conduction of impulses in a network of connected excitable elements, specifically in cardiac muscle. *Arch. Inst. Cardio. Mex.* **16**, 205-265.

[2] Moe, G.K., Rheinboldt, W.C. & Abildskov, T.A. (1964). A computer model of atrial fibrillation. *Am. Heart J.* **67**, 200.

[3] Krinsky, V.I. (1984). Autowaves: Results, Problems, Outlooks. In *Self-Organization. Autowaves and Structures far from Equlibrium*, Krinsky, V.I. (ed.), pp. 9-18. Springer-Verlag: Berlin.

[4] Krinsky, V.I. (1981). Mathematical models of cardiac arrhythmias. *International Encyclopaedia of Pharmacology and Therapeutics*, pp. 105-121. Pergamon Press: London.

[5] Krinsky, V.I. (1968). Fibrillation in excitable media. *Problemy Kibernetiki N* **20**: 59-80.

[6] Agladze, K.I., Krinsky, V.I & Pertsov, A.M. (1984). Chaos in the non-stirred Belousov-Zhabotinsky reaction is induced by interaction of waves and stationary dissipative structures. *Nature* **308**, 834-836.

[7] Hramov, R.N., Rudenko, A.N., Panfilov, A.V. & Krinsky, V.I. (1984). Drift of vortices in heterogeneous active medium (simplified analysis). *Studia Biophys.* **102**, 69-74.

[8] Pertsov, A.M. & Ermakova, E.A. (1988). Mechanism of the drift of a spiral wave in an inhomogeneous medium. *Biofizika* **33**, 338-342.

[9] Fast, V.G. & Pertsov, A.M. (1989). Drift of vortices in myocardium. *Biofizika*, in press.

[10] Pogwizd, S.M. & Corr, P.B. (1987). Reentrant and Nonreentrant Mechanisms Contribute to Arrhythmogenesis During Early Myocardial Ischemia: Results

Using Three-Dimensional Mapping. *Circ. Res.* **61**, 352-371.
[11] Panfilov, A.V. & Pertsov, A.M. (1984). A vortex ring in a three-dimensional active medium described by the reaction- diffusion equation. *Doklady AN SSSR* **274**, 1500.
[12] Winfree, A.T. & Strogatz, S.J. (1984). Organizing centres for three-dimensional chemical waves. *Nature* **311**, 611-615.
[13] Medvinsky, A.B., Panfilov, A.V. & Pertsov, A.M. (1984). Properties of Rotating Waves in Three Dimensions. Scroll Rings in Myocard. In *Self-Organization. Autowaves and Structures far from Equilibrium*, Krinsky, V.I. (ed.), pp. 195-199. Springer-Verlag: Berlin.
[14] Pertsov, A.M., Fast, V.G. & Grenader, A.K. (1986). Effects of lidocaine on "leading circle" and focal activity in isolated rabbit heart tissue under hypothermia. *Kardiologia* **N4**, 83-86.
[15] Pertsov, A.M. & Fast, V.G. (1987). Threedimensional circulation during paroxysmal ventricular tachycardias: results of electrophysiologic mapping. *Kardiologia* **N5**, 75-78.
[16] De Bakker, J.M.T., van Capelle, F.J.L., Janse, M.J., Wilde, A.A.M., Coronel, R., Becker, A.E., Dingenmans, K.P., van Hemel, N.M. & Hauer, R.N.W. (1988). Reentry as a cause of ventricular tachycardia in patients with chronic ischemia heart disease: electrophysiologic and anatomic correlation. *Circulation* **77**, 589-606.
[17] Wit, A.L., Allessie, M.A., Bonke, F.I.M., Lammers, W., Smeets, J. & Fenoglio, J.J. (1982). Electrophysiological mapping to determine the mechanism of experimental ventricular tachycardia initiated by premature impulses. *Am. J. Cardiol.* **49**, 166-185.
[18] Mines, G.R. (1914). On circulating excitations in heart muscles and their possible relation to tachycardia and fibrillation. *Trans. R. Soc. Can.* **8**, 43-52.
[19] Zipes, D.P., Fischer, J., King, R.M., Nicoll, A.D. & Jolly, W.W. (1975). Termination of ventricular fibrillation in dogs by depolarizing a critical amount of myocardium. *Am. J. Cardiol.* **36**, 37-44.

3. DYNAMICS OF SPIRAL CENTRES IN THE FERROIN-CATALYZED BELOUSOV-ZHABOTINSKII REACTION

S.C. Müller and Th. Plesser

Max-Planck-Institut für Ernährungsphysiologie
Rheinlanddamm 201, D-4600 Dortmund 1, FRG

1. Chemical spirals as a function of excitability

Wave propagation under far from equilibrium conditions has been studied in some detail in chemical excitable media, mostly in the ferroin catalyzed Belousov-Zhabotinskii reaction [3,13]. In this reaction a chemical wave is detectable by its blue excited front. It indicates the presence of ferriin, the oxidized state of the catalyst, which travels through a quiescent, red solution layer where the catalyst remains in its reduced form ferroin. The spatial distribution of the catalyst concentration in such wave patterns can be determined quantitatively with devices for 1D and 2D spectrophotometry [6,11,15]. Progress is being made in applying these precise recording and evaluation techniques to the analysis of wave shapes and profiles under systematic variation of system parameters, in particular of the initial concentrations of the chemicals mixed into the solution [9,10].

In excitable systems, which are capable of transmitting a local pulse of excitation into the resting, non-excited neighbourhood, one frequently observes the dynamic evolution of spiral-shaped structures. The spiral is a widespread structural element, in the large variety of shapes, created by nature and this geometric form belongs to the archetypes of spatial organization. Therefore it is not surprising that since the discovery of travelling chemical waves there has been particular interest in studying the phenomenon of rotating spirals. Early research was concerned mainly with spirals of approximately Archimedian shape where the spiral tip moves steadily on a circular path around its rotation centre. It was soon noticed that the spiral geometry and the dynamics of the tip motion become more complex when the excitability of the solution is changed.

If the choice of initial chemical composition results in a sufficiently low excitability of the medium, distorted spirals appear and the tip starts to perform a looping action (often referred to as "meandering", although the authors think that this term is not appropriate). This was first pointed out by Winfree [14] and by Agladze [1]. Only recently more attention has been given to a systematic approach in terms of concentration dependence of characteristic parameters describing the complex dynamics of the distorted spirals [2,4,12].

Nonlinear Wave Processes in Excitable Media
Edited by A. V. Holden *et al.*, Plenum Press, New York

The goal of this contribution is to present observations and evaluations on non-Archmedian spiral dynamics which were obtained on the basis of quantitative measurements of the concentration distribution of the catalyst using space-resolved 2D spectrophotometry. The properties of spirals in media of low excitability are quantified by a series of experiments concerned with the dependence of spiral rotation on the initial concentration of sulphuric acid, i.e. the pH value. We report measurements of spiral tip motions and transient spiral cores which are formed for each of the loops of the tip.

2. Quantitative recording of the patterns

2.1. *Sample preparation*

Solutions were prepared with reagent grade chemicals and distilled water. The three initial reactant solutions were sodium bromate in sulphuric acid, sodium bromide in water, and malonic acid in water. In addition, a 25 mM solution of ferroin was used. All solutions were filtered through a 0.44 μm Millipore filter and stored in separate containers. Quiescent, excitable solutions of the BZ reaction were obtained by preparing mixtures of 60 mM sodium bromide, 330 mM sodium bromate, 120 mM malonic acid, and various concentrations of sulphuric acid in the range from 0.15 M to 0.70 M in a final volume of 5 ml. After the orange-brown colour of the solution, indicating the generation of molecular bromine, had completely disappeared, 0.6 ml of 25 mM ferroin solution was added resulting in a final concentration of 3.0 mM. After filtering, a volume of 3.63 ml was poured into a siliconized optically flat dish (diameter 6.8 cm) which just prior to the experiment had been carefully cleaned to be free from dust particles. The resulting solution layer had a depth of 0.56 ± 0.03 mm at the centre and was kept at 23 ± 0.1°C in the dish inserted into the 2D spectrophotometer [7].

2.2. *Space-resolved experiments*

The apparatus for space-resolved spectrophotometry in two dimensions consists of an optical precision set-up mounted on an optical table, a sensitive video camera with a broad spectral response and a video frame buffer linked to a computer system. A homogeneous parallel light beam is transmitted through the solution layer at 490 nm, imaged on the target of the camera, transferred to the video frame buffer and stored in the memory of the computer. One image consists of 512 x 512 picture elements (pixel) each having theoretically one out of 256 possible grey levels. With this raster and intensity resolution the information content of one full frame amounts to a quarter of a MByte. Image acquisition and storage is feasible at a rate of 30 frames per minute.

The spatial resolution depends on the photolens system and can be increased up to about 0.5 μm per pixel if an inverted microscope is used. For our purposes the maximum spatial resolution was chosen to be 15.6 μm/pixel. When recording a movie on a Umatic video recorder the frames are stored on tape at a rate of 25 frames per second, but images have significantly less spatial and intensity resolution and the lag time of the camera has to be taken into account [5,6]. Frequently, for the purpose of continuous documentation, the recording of single images with optimum intensity resolution and of video movies were performed simultaneously.

3. Archimedian and non-Archimedian geometry

Different types of spiral shapes are shown in Fig. 1. By decreasing the proton concentration, which also means that the excitability of the solution is decreased, one observes a transition from regular spirals (Fig. 1 A and B) to distorted spirals (Fig. 1 C and D).

The properties of the spirals shown in Fig. 1 A and B have been previously characterized by detailed spectrophotometric measurements. They are summarized as follows: the geometric shape is practically identical with that of an Archimedian spiral [8]. Isoconcentration lines of the pattern can be described by only 4 parameters: the two coordinates of the spiral centre, the pitch, and the momentary angular position. Involutes of a circle which exclude a circular region around the centre, fit this geometry as accurately as the Archimedian spiral. In view of the properties of the core of the spiral the involute appears to be more appropriate for comparison with realistic models.

The structure of the core region in these regular spirals was determined by an overlay technique as described in detail in [8]. To summarize, its main properties are (here exemplified for the case of Fig. 1B): at the centre of spiral rotation there is a singular site (diameter < 30 μm) that remains

Figure 1. Digital images of spiral patterns propagating in BZ-solution layers with decreasing excitability. Initial concentrations 0.12 M $CH_2(COOH)_2$, 0.33 M $NaBrO_3$, 0.06 M NaBr, 0.003 M ferroin for all systems; and 0.7 M (A), 0.37 M (B), 0.19 M (C), 0.15 M H_2SO_4 (D). Image size: 15x15 mm^2.

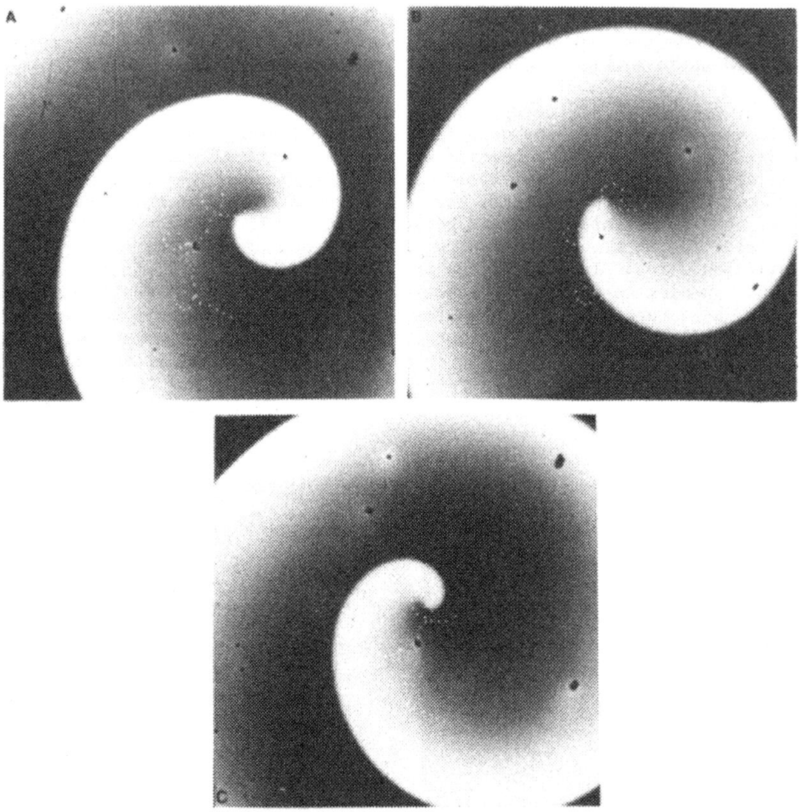

Figure 2. Three snapshots of the spiral tip in a BZ solution as in Fig. 1C with initially 0.19 M H_2SO_4. Time interval between successive images: 30 s, image size: 7x7 mm^2. The trace of three loops of the spiral tip motion for a time interval of 270 s is superposed (white dots).

quasi-stationary at a concentration slightly above the level of maximum reduction in the spiral pattern. For a large number of revolutions its position does not change in space. The spiral core is a circular region (diameter about 0.7 mm) in which a smooth and gradual transition takes place, namely from the stationary site towards the outer area of the pattern, whereby the amplitude of the emanating wave grows to maximum height. That is, only outside the spiral core each volume element becomes excited to a maximum degree of oxidation once per spiral revolution.

Spirals obtained for proton concentrations lower than ≈ 0.26 M differ from these properties in many respects (see Fig. 1 C and D). Their geometry is distorted, which means that the spiral pitch varies as a function of the arc length [12]. This implies a complex variation of the local curvature along the front. In fact, during rotation the overall shape of the spiral changes in that a displacement of the regions of smallest pitch takes place. In particular, the curvature of the front in the area closest to the tip is subjected to quite pronounced variations. Furthermore, the tip describes a more complex path with varying velocity. Some details of this behaviour are described in the next paragraph for a particular system.

4. Tip dynamics in non-Archimedian spirals

For the case of the system with an initial sulphuric acid concentration of 0.19 M (corresponding to the system shown in Fig. 1C), the motion of the tip of a rotating spiral was analyzed by tracing the turning point of a specifically enhanced isoconcentration level of a digital image series taken at 6 s intervals. Three images from the analyzed image series are shown in Fig. 2. The tip positions obtained for this series are superposed as small dots. During the corresponding time interval the tip performs three loops separated by a well defined distance, that is the trace of the tip alternates between an only slightly curved path, where the tip curvature is relatively small (Fig. 2 A,B), and a closed loop, where its curvature is much larger (Fig. 2C). On the whole, in this experiment the trace of the tip consists of eight loops, before it closes approximately on its starting point. It was shown that its trajectory resembles the geometry of a prolate epicycloid [12].

Figure 3. Digital overlay of 7 (A), 9 (B), 11 (C), 13 (D) spiral images as shown in Fig. 2 with initially 0.19 M H_2SO_4, taken successively at 6 s intervals. The image sequence corresponds roughly to the first loop of the spiral tip motion (indicated in Fig. 2 and in image A).

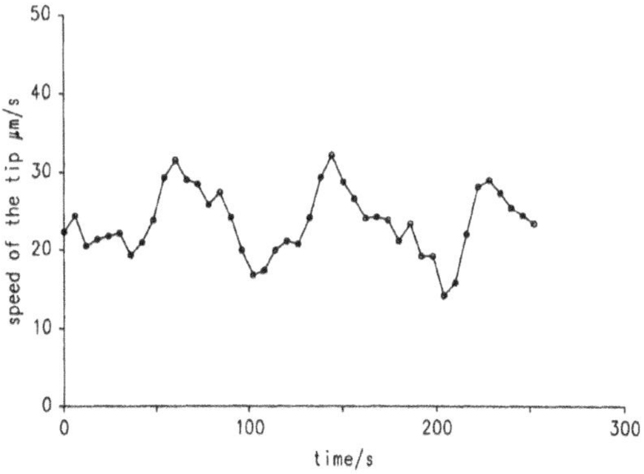

Figure 4. Speed of the spiral tip as a function of time, for the time interval covered by the three dotted loops in Fig. 2.

From the time series considered above, we selected a sequence of consecutive images showing the tip motion around the first loops of Fig. 2 for a digital overlay. Figure 3 presents 4 composite overlay images constructed from an increasing number of single snapshots, starting with 7 (Fig. 3A) and ending with 13 (Fig. 3B). Only the highest grey level, that is the highest degree of oxidation of the catalyst, was retained for each picture element during the time interval of the image sequence. This figure shows again how the path of the spiral tip changes from almost straight to highly curved. While the loop is building up, a dark spot appears around which it curls. Inside this spot there is a small site of minimum intensity which remains stationary for a limited period of time, during which the loop is completed. The properties of this temporary rotation centre are comparable to those found for the Archimedian case (compare [8]). At the centre of the dark spot the ferroin concentration remains unchanged, until after the tip has completed one turn and moves onto an adjacent location where it will perform the subsequent loop. There then follows a moment when a wave front arrives at this special site, leading its chemistry again to a newly excited state.

The complex path of the tip and its periodically changing curvature also results in changes of its speed. According to the results of Fig. 4, obtained for the 3 loops of Fig. 2, this speed varies periodically by almost a factor of two. It is at its maximum during the second half of the loop and minimum just before it enters a loop. A pronounced acceleration takes place during the first half of the loop.

5. Irregularities in the concentration distribution

The non-Archimedian spirals are characterized not only by their distorted geometry and the complexity of the tip dynamics, but also by irregular features with respect to the concentration distribution along the spiral whorls. In Fig. 5 a preliminary result is shown, which illustrates the shape of an arbitrarily chosen isoconcentration level in a spiral evolving from a solution containing initially 0.19 M sulphuric acid. While in Archimedian spirals any chosen concentration level obeys the same geometric law, here we are dealing with more complicated shapes which vary as a function of concentration level and time. The example of Fig. 5 shows where the area of

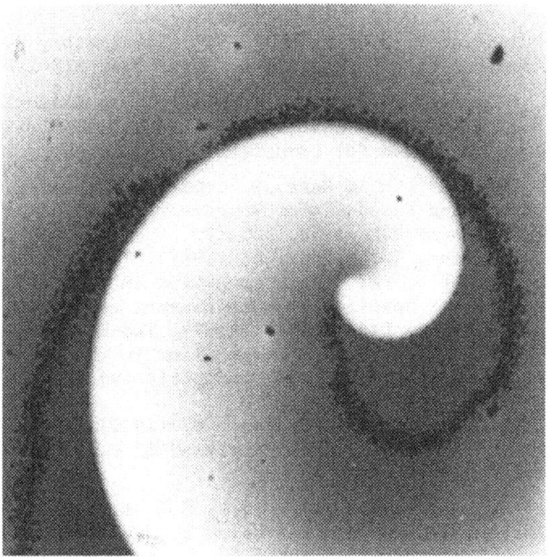

Figure 5. Isoconcentration level enhanced in black for one image selected from the spiral sequence used in Figs. 2 and 3.

maximum reduction is located. How this location is shifted in time is the issue of current analysis. Detailed knowledge about the discussed dynamics is certainly important for a better understanding of the spiral tip motion. This motion is governed by the local changes in chemical composition, as expressed in the theoretical concepts of curvature and dispersion of wave velocity.

Acknowledgements

The authors thank Prof. B. Hess for valuable discussions, Mr. U. Heidecke and Ms. A. Rietig for skillful help in the laboratory, Ms. G. Schultz for photography. This work was supported by the Stiftung Volkswagenwerk, Hannover, FRG.

References

[1] Agladze, K.I. (1983). *Proc. of the Biological Center of the Academy of Science, Puschino, USSR.*

[2] Agladze, K.I., Panfilov, A.V. & Rudenko, A.N. (1988). Nonstationary Rotation of Spiral Waves: Three-Dimensional Effect. *Physica* **D29**, 409-415.

[3] Field, R.J. & Burger, M. (eds.) (1985). *Oscillations and Travelling Waves in Chemical Systems.* Wiley: New York.

[4] Jahnke, W., Skaggs, W.F. & Winfree, A.T. (1989). Chemical Vortex Dynamics in the Belousov-Zhabotinskii Reaction and in the Two-Variable Oregonator Model. *J. Phys. Chem.* **93**, 740-749.

[5] Miike, H., Müller, S.C. & Hess, B. (1988). Oscillatory Hydrodynamic Flow Induced by Chemical Waves. *Chem. Phys. Lett.* **144**, 515-520.

[6] Müller, S.C., Plesser, Th. & Hess, B. (1985). Two-Dimensional Spectrophotometry with High Spatial and Temporal Resolution by Digital Video Techniques and Powerful Computers. *Anal. Biochem.* **146**, 125-133.

[7] Müller, S.C., Plesser, Th. & Hess, B. (1986). Two-Dimensional Spectrophotometry and Pseudo-Color Representation of Chemical Reaction Patterns. *Naturwissenschaften* **73**, 165-179.

[8] Müller, S.C., Plesser, Th. & Hess, B. (1987). Two- Dimensional Spectrophotometry of Spiral Wave Propagation in the Belousov-Zhabotinskii Reaction. I. Experiments and Digital Data Representation. *Physica* **D24**, 71-86; (1987). Two-Dimensional Spectrophotometry of Spiral Wave Propagation in the Belousov-Zhabotinskii Reaction. II. Geometric and Kinematic Parameters. *Physica* **D24**, 87-96.

[9] Nagy-Ungvarai, Zs., Tyson, J.J. & Hess, B. (1989). Experimental Study of the Chemical Waves in the Cerium- Catalyzed BZ-Reaction. 1. Velocity of Trigger Waves. *J. Phys. Chem.* **93**, 707-713.

[10] Nagy-Ungvarai, Zs., Tyson, J.J., Müller, S.C. & Hess, B. (1989). Experimental Study of the Chemical Waves in the Ce-Catalyzed BZ-Reaction. 2. Concentration Profiles. *J. Phys. Chem.* **93**, 2760-2764.

[11] Pagola, A. & Vidal, C. (1987). Wave Profile and Speed near the Core of a Target Pattern in the Belousov-Zhabotinskii Reaction. *J. Phys. Chem.* **91**, 501-503.

[12] Plesser, Th., Müller, S.C. & Hess, B. Spiral Wave Dynamics as a Function of Proton Concentration in the Ferroin- Catalyzed Belousov-Zhabotinskii Reaction. *J. Phys. Chem*, in press.

[13] Ross, J., Müller, S.C. & Vidal, Ch. (1988). Chemical Waves. *Science* **240**, 460-465.

[14] Winfree, A.T. (1973). Scroll-Shaped Waves of Chemical Activity in Three Dimensions. *Science* **181**, 937-939.

[15] Wood, P. & Ross, J. (1985). A Quantitative Study of Chemical Waves in the Belousov-Zhabotinskii Reaction. *J. Chem. Phys.* **82**, 1924-1936.

4. SYNERGETIC ANALYSIS OF SPATIO-TEMPORAL EEG PATTERNS

R. Friedrich, A. Fuchs and H. Haken

Institut für Theoretische Physik und Synergetik

Universität Stuttgart, Pfaffenwaldring 57 / IV
7000 Stuttgart 80, FRG

We present an analysis of the spatio-temporal patterns connected with the electrical activity of the brain which is measurable at the human scalp in the form of electroencephalograms (EEGs). In the case of the so-called Alpha activity the spatio-temporal EEG patterns show large scale spatial coherence which, however, undergoes a complex temporal evolution. We discuss the change of the spatio-temporal patterns during an epileptic seizure. Based on the synergetic theory of pattern formation in nonequilibrium systems we present a model of the Alpha EEG rhythm which reproduces the observed main spatial as well as temporal features.

1. Introduction

The efforts in understanding complex systems [1],[2],[3] have revealed that systems far from equilibrium exhibit the spontaneous emergence of spatial, temporal, as well as spatio-temporal patterns. It has turned out that systems of quite different nature, e.g. physical, chemical or biological systems, can exhibit quite similar behaviour. Recently, considerable interest has been addressed to the study of systems which form coherent spatial structures undergoing a complex and chaotic temporal evolution. Such behaviour is extensively studied, e.g. in the field of hydrodynamic instabilities or chemical reactions. It has long been recognized that similar phenomena can also be observed in biological systems. Prominent examples are the spatio-temporal patterns connected with the electrical activity of the brain measured in the form of electroencephalograms (EEG). So far the interest has mainly been devoted to the properties of the time signal obtained by measuring the potential difference between two locations on the scalp. It has been demonstrated by various groups [4,5] that in certain sleep stages or in the case of the so-called Alpha rhythm the underlying dynamics is essentially chaotic, involving only a few dynamical degrees of freedom. On the other hand, the fact that low dimensional temporal behaviour in a complex system like the brain is likely to exhibit coherent spatial patterns has not yet been sufficiently recognized in the literature dealing with dynamical properties of the EEG. However, it has been well-known in the neurophysiological literature that the EEG patterns are spatially coherent [6,7].

The present paper emphasizes the study of both the temporal and spatial properties of the EEG patterns which are intimately related with each other. The paper is divided into three parts. The first part reviews results of an analysis of spatio-temporal Alpha EEG patterns presented in a former paper [8]. We extend these results by applying the same methods to the analysis of EEG patterns derived from a human individual during an epileptic seizure. The main result may be summarized in the observation that the spatio-temporal behaviour of the EEG exhibits coherent spatial patterns which undergo a complex temporal evolution (Alpha EEG) or a more regular, time periodic evolution (epileptic seizure). These observations suggest a "macroscopic treatment" of these patterns in close analogy to pattern formation in nonequilibrium systems. The second part briefly discusses this "macroscopic approach" based on the synergetic treatment of pattern formation (for a detailed treatment cf. [1,2,3]). In the third part we shall derive a model of the spatio-temporal behaviour of Alpha EEG patterns on the "macroscopic level" along the lines suggested by the ideas formulated in part two. We shall present a dynamical system which describes the main characteristic features of Alpha EEG activity discussed in the first part in a phenomenological fashion, i.e. in a way which reproduces the observed spatio-temporal patterns as a collective property of the brain without reference to the behaviour of the single subunits of the system.

2. Analysis of spatio-temporal EEG patterns

The analyzed data sets have been derived by D. Lehmann. EEG signals were recorded in 16 channels located at different places on the scalp. The array scheme of the electrodes is shown in Fig. 1. More details are reported in references [6,8]. Each single electrode measures at its location $\underline{x}_i$ the value $V(\underline{x}_i, t)$ of the electric potential (with respect to a common reference).

It is evident that the spatial scales which can be resolved are of the same order of magnitude as the distance between two neighbouring electrodes. Thus, the detection of spatial structures is restricted to the large scales of the electric potential. However, a comparison of patterns (for the case of Alpha waves) derived with 48 electrodes [6] does not indicate that the patterns are strongly changed if the spatial resolution is increased. Thus, the patterns derived with 16 electrodes seem to give a reasonably good representation of the patterns at least with respect to the large scales.

For the following it has turned out to be convenient to map the surface of the scalp onto a circular disk, on which the electric potential defines a continuous spatio-temporal field $V(\underline{x}, t)$ (see Fig. 1). The complete field is reconstructed from the measured field at the 16 grid points by interpolation.

2.1. *Alpha-waves*

Alpha EEG activity is detected if the individual is resting and has his eyes

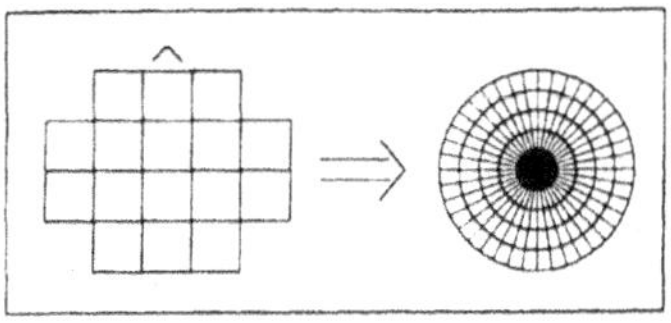

Figure 1. Array scheme of the electrodes.

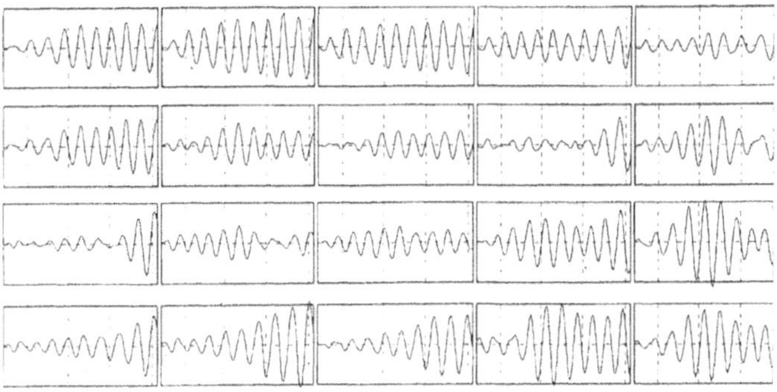

Figure 2. Time signal of the potential difference between two electrodes in the case of Alpha rhythm.

closed. The time signal of the potential between two different electrodes shows pronounced, more or less sinusoidal and regular oscillations with a dominant frequency of about 8-13 Hz, in the so-called alpha band (adult person) (see Fig. 2). The oscillations are modulated on a larger time scale of about one second. This modulation is known as "waxing and waning" and is apparently random. If the person starts some activity, e.g. opening eyes, a "blocking of the alpha-rhythm" is observed: the electrical signal becomes much weaker and the alpha-rhythm is more or less destroyed. It has been presumed [9] that "idle" nerve cells have the tendency to synchronize their activity, while the synchronization is destroyed under specific excitation of cell assemblies. The result of the synchronization in the case of the Alpha-rhythm is the emergence of coherent spatial patterns of the electrical potential [6,7,8].

Although the modulation of the alpha-rhythm seems to be rather complex, the corresponding spatial patterns as well as their temporal evolution turn out to be comparatively simple, as e.g. a movie reconstructed from the derived EEG data indicates [10]. The patterns at each time step consist mainly of two regions of different polarity. Less frequently two minima or maxima are observed. The patterns undergo, roughly speaking, two different kinds of motion. Usually the pattern rotates clockwise or anticlockwise in azimuthal direction. Thereby, the pattern is only slightly deformed (see Fig. 3). Besides this deformation it resembles a rotating wave motion. The frequency of the rotation is approximatley 10 Hz, i.e. the frequency of the alpha-rhythm. The duration of the time intervals, in between which the patterns rotate clockwise, or anticlockwise, seems to be random (see the discussion below). Less frequently one observes shorter periods of qualitatively different behaviour, where the rotation of the pattern stops and the polarity reverses periodically with a frequency of about 10 Hz (see Fig. 4): regions of positive and negative polarity are exchanged and the motion resembles a standing wave pattern.

Figure 3. Temporal evolution of Alpha EEG patterns: "Rotating wave motion" (duration of the sequence ≈ .1 sec).

Figure 4. Temporal evolution of Alpha EEG patterns: "Standing wave motion" (duration of the sequence ≈ .1 sec).

The following quantitative results have been obtained in the analysis of the patterns (see reference [8] for more details). By an expansion of the (interpolated) field into a complete orthonormal set of modes of the form

$$V(\underline{x},t) = \Sigma_{jm}\ \xi_{m,j}(t)\ \Psi_{jm}(r)\ \exp\,[im\phi] \tag{1}$$

and by a suitable superposition of the functions $\Psi_{jm}(r)$ it becomes possible to represent, at each time step t, the field $V(\underline{x},t)$ in the form

$$V(\underline{x},t) = \Sigma_{m=0,1,2}\ \xi_m(t)\ \psi_m(r)\ \exp[im\phi] + c.c. + R(\underline{x},t) \tag{2}$$

(c.c. means the complex conjugate of the first sum).
The remainder $R(\underline{x},t)$ thereby turns out to be small in the sense that

$$\|R\| \,/\, \|V\| < .05, \quad \|R\| = [\textstyle\int d^2\underline{r}\ R^2(\underline{x},t)]^{1/2} \tag{3}$$

Thus, it becomes possible to reconstruct the pattern at each time step by three (complex) collective modes up to an accuracy of $||R||/||V||<.05$.

A straightforward way to determine the collective modes of eq. (2) consists of applying the method of Karhunen & Loeve for feature selection in pattern recognition (see [3]). In a first step the correlation matrix $C(\underline{x},\underline{x}')$

$$\begin{aligned} C(\underline{x},\underline{x}') &= \langle V(\underline{x},t)\ V(\underline{x}',t)\rangle \\ &= 1/T \int_{\tau=0}^{\tau=T} d\tau\ V(\underline{x},\tau)\ V(\underline{x}',\tau) \end{aligned} \tag{4}$$

is calculated. (We use the notations for continuous fields $V(\underline{x},t)$. The corresponding expressions for the quantities defined on the finite dimensional grid are obvious.) The collective modes $\phi_j(\underline{x})$ are defined as the eigenvectors of the linear eigenvalue problem

$$\int d^2\underline{x}'\ C(\underline{x},\underline{x}')\ \phi_j(\underline{x}') = \lambda_j\ \phi_j(\underline{x}). \tag{5}$$

The modes corresponding to the largest eigenvalues contain the gross features of the patterns. The numerical treatment (see [8]) gives the results presented in Fig. 5. If we consider the patterns to be defined on a circular disk we see that the patterns are readily classified by their azimuthal spatial dependence:

$$\phi_j(\underline{x}) = \psi_{1m}(r)\ \exp\,[im\phi] \tag{6}$$

The most dominant group consists of modes which have an azimuthal dependence

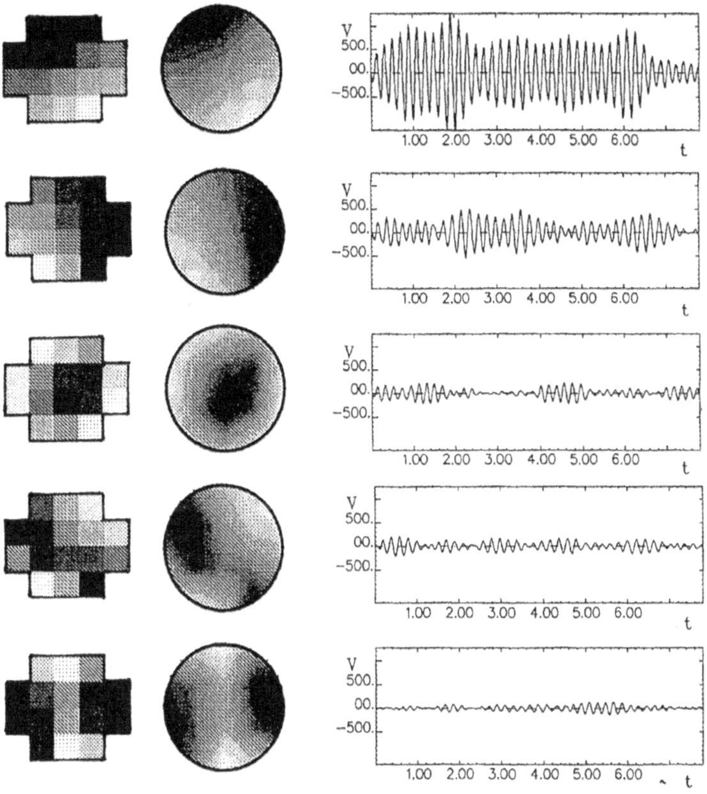

Figure 5. The most dominant modes for Alpha EEG activity as obtained by the Karhunen-Loeve-expansion.
Left: Patterns as obtained from the Karhunen-Loeve-expansion.
Right: Interpolated patterns. The azimuthal spatial dependence $\exp[im\phi]$ is significant.

like $\exp[\pm i\phi]$. The second mode is the one with nearly no azimuthal variation (m=0) and the third group is the one with an azimuthal dependence described by $\exp[\pm 2i\phi]$. A different way to determine the modes yielding the same results has been described in [8].

Since the contribution of the mode with $m=\pm 1$ turns out to dominate in the decomposition of eq. (1) the pattern usually consists of a region with positive and a region with negative polarity. If we further decompose the amplitude $\xi_1(t)$ of this mode into modulus $r(t)$ and phase $\phi(t)$ and plot phase $\phi(t)$ and modulus $r(t)$ as a function of time we obtain the graphs represented in Fig. 6. If the phase $\phi(t)$ increases with time the pattern rotates in an anticlockwise direction, whereas it rotates in a clockwise direction if it decreases. As a main result we see that the time segments of clockwise/ anticlockwise rotation is a stochastic quantity. In some sense the phase undergoes a kind of random walk. We mention that there are well-known physical systems where similar kinds of behaviour of a phase variable occur due to a

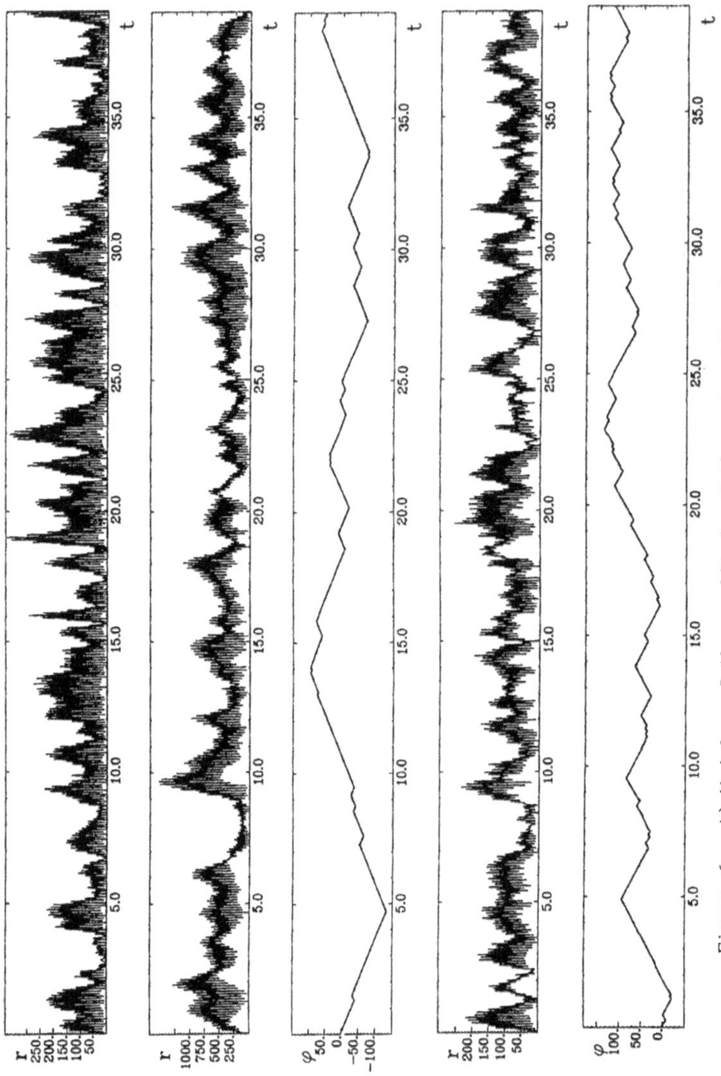

Figure 6. A) Modulus of the amplitude of the mode with m=0.
B) Modulus and phase of the amplitude of the mode with m=1 as a function of time.
C) Modulus and phase of the amplitude of the mode with m=2 as a function of time.

low dimensional dynamical system with chaotic dynamics.

This remark brings us to the next important observation in the analysis of EEG time series. It has been stressed by Babloyantz and her coworkers [4], Mayer-Kress et al. [5], and various other groups that the underlying dynamics of the EEG signals for various sleep stages as well as in the case of Alpha rhythm is chaotic ("Strange attractor in the human cortex"). In the case of Alpha rhythm Babloyantz et al. [4] estimated the correlation dimension to be

$$D_\alpha = 6.1 \pm .5. \qquad (7)$$

The signal analysis of the EEG is a rather subtle task which can never be expected to yield "rigorous" results. This fact has led to some controversial discussions in the literature [11]. Nevertheless, it seems to be well established that the underlying dynamics of the EEG in the Alpha rhythm is essentially deterministic and low dimensional.

2.2. *Spatio-temporal EEG patterns during an epileptic seizure*

It is a well-known fact that the EEG time signal during an epileptic seizure is dramatically changed. This has eminent importance for clinical applications. Although there are different kinds of epilepsia a common feature of the time signal in all cases seems to be the fact that it becomes "more" regular and nearly time periodic. This can also be seen in the data set provided by Lehmann (see Fig. 7), which we have analyzed in the same way as indicated in section 2.1 in the case of the Alpha rhythm. The corresponding evolution of the spatial pattern is shown in Fig. 8. In contrast to the Alpha waves, no rotation of the pattern is observed.

The modes obtained by the Karhunen-Loeve-expansion are exhibited in Fig. 9. The main result of this analysis is the fact that the temporal evolution of the patterns can be described by the following expression

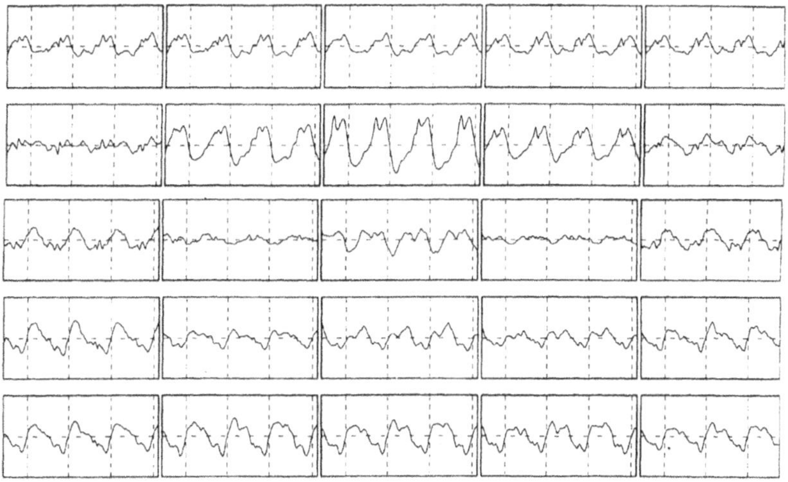

Figure 7. Time signal of the potential difference between two electrodes in the case of epileptic seizure.

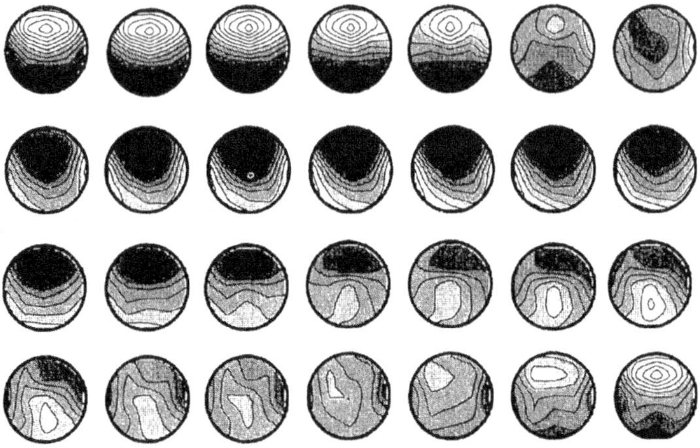

Figure 8. Time evolution of spatial pattern in the case of epileptic seizure (approximately 1 Period, ≈ .3 sec).

$$V(\underline{x},t) = \eta_1(t)\ \Psi_1(r)\ \exp[i\phi] + \eta_2(t)\ [a\ \Psi_2(r)\ \exp[2i\phi] + b\ \Psi_0(r)] + c.c. + R(\underline{x},t) \tag{8}$$

where $||R|| \,/\, ||V|| < .05$. In comparing the expressions eq. (8) for the epileptic case and eq. (2) for the Alpha rhythm, we see that the main change consists of the fact that the modes with m=0 and m=±2 couple in a rigid manner. Simultaneously, the dynamics turns from a chaotic one to a periodic one.

3. The macroscopic approach to complex systems

Over the past twenty years considerable insight has been gained into the basic mechanisms leading to the formation of spatio- temporal patterns in physical, chemical, and biological systems [1,2,3]. A systematic theory [1,2] has been developed to deal with the emergence of patterns close to nonequilibrium phase transitions. The main theory can be briefly summarized in the following way. Close to a nonequilibrium phase transition the behaviour of a system, which is described by a state vector $\underline{q}(\underline{x},t)$, is determined by

$$\underline{q}(\underline{r},t) = \Sigma_u\ \xi_u(t)\ \phi_u(\underline{x}) + \Sigma_s\ \xi_s[\xi_u(t)]\phi_s(\underline{x}) \tag{9}$$

The state vector is a superposition of suitably defined unstable modes $\phi_u(\underline{x})$ and stable modes $\phi_s(\underline{x})$. The essential point of this decomposition of the state vector is that the amplitudes ξ_s depend on the amplitude of the unstable modes which obey a closed set of evolution equations, the so-called order parameter equations (we neglect the presence of fluctuations, see [1,2]):

$$d_t\ \xi_u(t) = N_u[\xi_u] \tag{10}$$

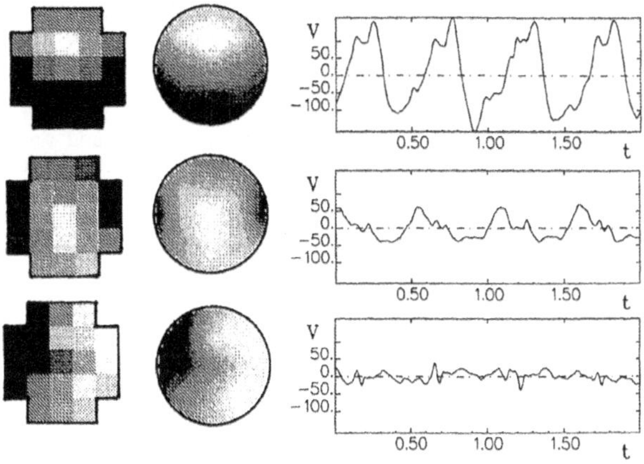

Figure 9. The modes obtained from the Karhunen-Loeve method in the case of epileptic seizure.
Left: Patterns as obtained from the Karhunen-Loeve-expansion.
Right: Interpolated patterns. The second mode from above is a superposition of the mode with m=0 and m=2.

The dynamical behaviour of the system is therefore entirely determined by the amplitudes of the (few) unstable modes, which are therefore called order parameters. We mention that the order parameters can undergo a chaotic dynamics so that the occurrence of coherent spatial patterns, essentially dominated by the unstable modes $\phi_u(\underline{x})$, connected with a complex dynamical chaotic evolution can be understood.

The representation of a state vector by unstable modes, stable modes, and order parameters allows a description of complex systems on a macroscopic scale. It is evident that a system showing the emergence of spatio-temporal patterns can be described by suitably defined order parameters and their dynamics without a specific reference to the behaviour of the single subsystems. Such an approach has proven to be successful in understanding, in a paradigmatic way, phase transitions in human hand movements [12]. We shall proceed in a similar fashion.

Let us briefly discuss this approach in the case of EEG patterns. The spatio-temporal EEG patterns are generated by the activity of a huge amount of single neurons interacting in a complex and, to a great extent, unknown fashion. A description of the resulting electrical activity in terms of single neurons and their mutual interactions is, due to the restricted knowledge of the characteristic properties, impossible and may, to some extent, be irrelevant with respect to the spatio-temporal EEG patterns. In contrast, we can try to understand these macroscopic patterns in a purely phenomenological way: we may try to identify the order parameters as well as the unstable modes, which essentially determine the spatial structures of the patterns. A determination of a dynamical system for the temporal evolution of the order parameters completes the "macroscopic description" of the patterns.

4. Spatio-temporal model of Alpha EEG waves

In the following we shall outline the modelling of Alpha EEG waves along the lines suggested by the synergetic treatment of pattern formation in nonequilibrium systems. Based on the observations gained through the analysis described in section 2 and by making additional assumptions to be specified below, we shall end up with a model which describes the essential characteristic features of the spatio-temporal behaviour of the Alpha rhythm. Once more we stress that the model is not based on a microscopic treatment of neuronal activity. Rather we take the macroscopic patterns and model their behaviour under the assumption that they are generated by the same basic mechanisms as patterns in other nonequilibrium systems. Thus, we have the task of identifying the order parameters and setting up a dynamical system, the order parameter equations, which governs their dynamics.

4.1 *The phenomenological derivation of the model*

The fact that the patterns of the Alpha waves can, at each time step, be nearly completely represented by the collective modes with azimuthal dependence m=0, m=±1, m=±2, suggests the identification of these modes as order parameters. We mention that the determination of these modes has been achieved by the Karhunen-Loeve method, which is a simple method for feature extraction in pattern recognition theory. We have applied this method here to a system where its usefulness has to be critically re-examined since we are dealing with nonlinear phenomena. It is well-known that, for instance in the case of a soft single mode nonequilibrium phase transition, the method to extract the relevant features, i.e. the order parameters, is in fact a nonlinear extension of the Karhunen-Loeve expansion [3]. Thus, an extension of the theoretical treatment to a situation including several order parameters undergoing chaotic time evolution should be performed. This treatment is rather involved and has not yet been successfully achieved. Therefore, we proceed by making the following assumption:

Hypothesis I:
The amplitudes of the collective modes $\Psi_m(r)\exp[im\phi]$ (m=0, m=±1, m=±2) are the order parameters.

This hypothesis implies that the field of the electric potential can be represented in the form

$$V(\underline{x},t) = \Sigma_{m=0,1,2}\, \eta_m(t)\, \Psi_m(r)\exp[im\phi] + c.c. \\ + U[\underline{x},\eta_m(t),\eta_m(t)^*,t] \tag{11}$$

and the dynamics of the amplitudes $\eta_m(t)$ is governed by a closed set of stochastic evolution equations:

$$d_t\, \eta_m(t) = N_m[\eta_m(t),\eta_m(t)^*,F] \tag{12}$$

Since we shall aim at deriving a purely deterministic model of the Alpha rhythm, which is motivated by the fact that the determination of the fractal dimension of the time signals indicates the existence of a low dimensional attractor, we shall neglect the dependence of N_m on the stochastic forces F.

In order to specify the dynamics of the order parameters η_m, i.e. to specify the vector field N_m, we make the assumption that the modes are essentially oscillating with the Alpha frequency. A possibility to incorporate this dynamics is to assume that the order parameters are excited by Hopf-bifurcations:

Hypothesis II:
The order parameters are excited by Hopf bifurcations.

This hypothesis specifies the linear part of the vector field N_m:

$$N_m = (\lambda_m + i\,\omega_m)\,\eta_m + O(\eta^2) \tag{13}$$

Here, λ_m denotes the linear growth rate and ω_m the bare frequencies of the modes, which are assumed to lie within the Alpha band (≃ 10 Hz).

The nonlinear part of the vector field is more difficult to establish. In order to constrain the possibilities we make an assumption on the symmetry properties of the system. The patterns of the Alpha waves emerge in a region which we have approximated by a circular disk. Therefore it is quite natural to assume that the medium which generates and supports the waves exhibits the same symmetry. Thus we shall assume that the emerging patterns break O(2)-symmetry.

Hypothesis III:

The emerging spatial patterns break O(2)-symmetry.

This assumption seems to be rather strange if one considers the architecture of the brain, but since the Alpha wave patterns are large scale patterns it seems to be justified to consider the phenomena which would occur due to deviations from the circular geometry and the O(2)-symmetry as secondary effects. Thus we use the assumption of O(2)-symmetry to isolate the dominant dynamics.

Two more technical assumptions are made in order to restrict the functional form of the order parameter equations. First, we assume that the nonlinearity N_m can be approximated by polynomials in $\eta_m(t)$, $\eta^*_m(t)$ up to cubic order. Second, we assume that quadratic terms as well as cubic terms except for terms $\eta\cdot\eta\cdot\eta$ do not need to be considered since they vary on time scales proportional to $\exp[\pm 2i\omega]$, $\exp[\pm 3i\omega]$ (rotating wave approximation). This assumption means that the fast Alpha oscillations decouple from the dynamics of the slow modulation.

The above assumptions readily lead to the following conclusion: the spatial patterns of the electrical potential are described by the expression

$$V(\underline{x},t) = \Sigma_{m=1,2}\,\{\xi^+_m(t)\,\psi^+_m(r)\,\exp[im\phi] + \xi^-_m(t)\,\psi^-_m(r)\,\exp[-im\phi]\}$$

$$+\ \text{c.c.} + \xi_0(t)\,\psi_0(r) + R(\underline{x},t) \tag{14}$$

(We have slightly rearranged the expressions compared to eq. (11).) The temporal evolution of the patterns is governed by the order parameter equations, which take the form

$$
\begin{aligned}
d_t\, \xi_0 &= \xi_0 \{\varepsilon_0 + i\omega_0 + \alpha_0[|\xi_0|^2, |\xi_1^+|^2, |\xi_1^-|^2, |\xi_2^-|^2, |\zeta_2^+|^2] \\
&\quad + C_1\, [\xi_2^+\, \xi_1^-\, \xi_1^{+*} + \xi_2^-\, \xi_1^+\, \xi_1^{-*}] + C_2\, [\xi_2^{-*}\, \xi_1^- \xi_1^- + \xi_2^{+*}\, \xi_1^+\, \xi_1^+] \\
&\qquad + C_3\, \xi_0^*\, \xi_1^+\, \xi_1^- + C_4\, \xi_0^*\, \xi_2^+\, \xi_2^- \\
d_t\, \xi_1^+ &= \xi_1^+ \{\varepsilon_1 + i\omega_1 + \alpha_1^+[|\xi_0|^2, |\xi_1^+|^2, |\xi_1^-|^2, |\xi_2^-|^2, |\xi_2^+|^2]\} \\
&\quad + \xi_1^{+*}\, [A_1\, \xi_2^+\, \xi_0] + \xi_1^-\, [A_2\, \xi_2^+\, \xi_0^* + A_3\, \xi_0\, \xi_2^{-*}] \\
&\quad + \xi_1^{-*}\, [A_4\, \xi_0\, \xi_0 + A_5\, \xi_2^+\, \xi_2^-] \\
d_t\, \xi_2^+ &= \xi_2^+ \{\varepsilon_2 + i\omega_2 + \alpha_2^+[|\xi_0|^2, |\xi_1^+|^2, |\xi_1^-|^2, |\xi_2^-|^2, |\xi_2^+|^2]\} \\
&\quad + B_1\, \xi_0\, \xi_1^+\, \xi_1^{-*} + B_2\, \xi_0^*\, \xi_1^+\, \xi_1^+ + B_3\, \xi_0\, \xi_0\, \xi_2^{-*} \\
&\quad + B_4\, \xi_2^{-*}\, \xi_1^+\, \xi_1^-
\end{aligned}
\tag{15}
$$

The equations for the amplitudes ξ_m^- and obtained from the set of equations by replacing ξ_m^+ by ξ_m^- and vice versa.

4.2. *Some results*

Instead of trying to classify the solutions and the behaviour of the order parameter equations independent of the coefficients A, B, C ... we shall proceed in a more direct way by keeping dominant terms of the normal form in such a way that the main features of the experimentally observed characteristics of the Alpha waves are modelled. The following model equations are sufficient to reproduce the essential features:

$$
\begin{aligned}
d_t\, \xi_0 &= \xi_0\, [\varepsilon_0 + i\omega_0 - |\xi_0|^2] + C\, [\xi_2^{-*}\, \xi_1^-\, \xi_1^- + \xi_2^{+*}\, \xi_1^+\, \xi_1^+] \\
d_t\, \xi_2^+ &= \xi_2^+\, [\varepsilon_2 + i\omega_2 - |\xi_2^+|^2 - \tau_2\, |\xi_2^-|^2] + B\, \xi_0\, \xi_1^+\, \xi_1^{-*} \\
d_t\, \xi_2^- &= \xi_2^-\, [\varepsilon_2 + i\omega_2 - |\xi_2^-|^2 - \tau_2\, |\xi_2^+|^2] + B\, \xi_0\, \xi_1^-\, \xi_1^{+*} \\
d_t\, \xi_1^+ &= \xi_1^+\, [\varepsilon_1 + i\omega_1 - |\xi_1^+|^2 - \tau_1\, |\xi_1^-|^2] \\
&\quad + A\, \xi_1^+\, [\xi_0^*\, \xi_2^+ - \xi_0\, \xi_2^{-*}] \\
d_t\, \xi_1^- &= \xi_1^-\, [\varepsilon_1 + i\omega_1 - |\xi_1^-|^2 - \tau_1\, |\xi_1^+|^2] \\
&\quad + A\, \xi_1^+\, [\xi_0^*\, \xi_2^- - \xi_0\, \xi_2^{+*}]
\end{aligned}
$$

If we put the coupling constants A, B, C equal to zero then the patterns with $m=0,\pm1,\pm2$ behave independently of each other. The patterns with $m=\pm1,\pm2$ thereby perform a rotating wave motion ($\tau_m > 1$) or a standing wave motion ($\tau_m < 1$). Thus, the observation that the EEG patterns evolve, during certain

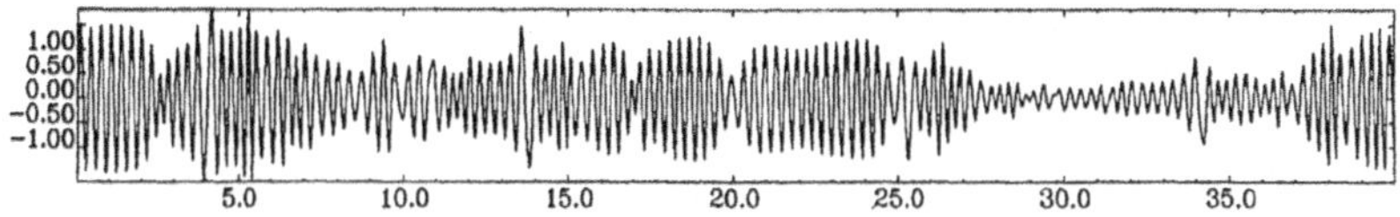

Figure 10. EEG time signal as obtained from our model.
The parameters are:
$\omega_0 = 0,\ \omega_1 = 10.1,\ \omega_2 = 10.2.$
$\varepsilon_0 = 2,\ \varepsilon_1 = 1.2,\ \varepsilon_2 = .1.$
$A = -3.2,\ B = .8,\ C = 4\ ;\ \ \tau_2 = 1.1,\ \tau_1 = 1.1.$

periods, like a rotating wave or a standing wave are the "pure" modes of the system if one neglects the nonlinear coupling of the modes. However, as the numerical integrations of the equations for nonvanishing coupling constants show, the nonlinear interaction among these simple behaviours in fact produces a complex and chaotic dynamics. Figure 10 shows the time signal $V(\underline{x},t)$ at a certain value $\underline{x}$ as a function of time. This signal is quite similar to the "real" EEG signals. The striking similarity becomes more evident if we look at the behaviour of the phase Φ of the modes with m=1, m=2 (see Fig. 11). The model reproduces the same characteristic features as the ones observed in the

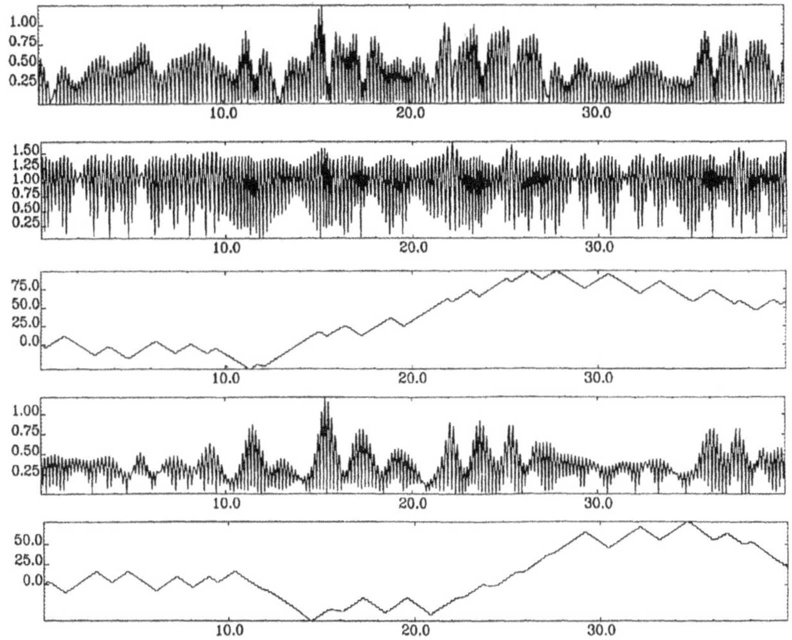

Figure 11. Time evolution as obtained from our model.
A) Modulus of the amplitude of the mode with m=0.
B) Modulus and phase of the amplitude of the mode with m=1 as a function of time.
C) Modulus and phase of the amplitude of the mode with m=2 as a function of time.

Figure 12. Spatio-temporal patterns of Alpha EEG activity as obtained from our model.

EEG analysis. Thus one of the most important features is reproducible by the above model, which is essentially a deterministic model to explain the complex spatio-temporal Alpha wave patterns. It is expected that further investigations of the properties of the underlying strange attractor as well as the dependence of the geometry of the attractor on the various parameters will stimulate a more detailed examination of the experimentally derived data sets as well as a refinement of the presented model. In fact a critical re-examination of the identification of the order parameters should become possible.

Let us comment on one additional fact, which, in our opinion, underlines the significance of the model equations. As has been outlined above, during epileptic seizure the modes with m=0 and m=±2 become rigidly coupled. As a result the mode with m=1 no longer undergoes rotations, and especially, no transitions between clockwise and counterclockwise rotation are observed. It is quite interesting to note that the terms in the model equations, which are essentially important for the reversals of direction of rotation, depend on the amplitudes of the modes with m=0 and m=±2. In fact, the occurrence of the reversals can be influenced and be prohibited by the amplitudes m=0, m=±2. This is a strong hint that our hypothesis to identify the order parameters, from which the derivation of the model of Alpha EEG activity starts and proceeds in a natural way, is justified. We hope that more complete knowledge will arise from a deeper investigation of the transition to epilepsia.

References

[1] Haken, H. (1983). *Synergetics. An Introduction.* 3rd edition. Springer Verlag: Berlin-Heidelberg-New York.

[2] Haken, H. (1987). *Advanced Synergetics.* 2nd ed. Springer- Verlag: Berlin-Heidelberg-New York-Tokyo.

[3] Haken, H. (1988). *Information and Selforganization.* Springer-Verlag: Berlin-Heidelberg-New York-Tokyo.

[4] Babloyantz, A. (1985). Strange Attractors in the Dynamics of Brain Activity. In *Complex Systems - Operational Approaches,* Haken, H. (ed.). Springer-Verlag: Berlin-Heidelberg-New York-Tokyo.
Babloyantz, A., Nicolis, C. & Salazar, M. (1985). Evidence of chaotic dynamics of brain activity during the sleep cycle. *Phys. Lett.* **A111**, 152.
Babloyantz, A. & Destexhe, A. (1986). Strange Attractors in the Human Cortex. In *Temporal Disorder in Human Oscillatory Systems,* Rensing, L. an der Heiden, U. & Mackey, M.C. (eds.). Springer-Verlag: Berlin-Heidelberg-New York-Tokyo.

[5] Mayer-Kress, G. & Holzfuss, J. (1986). *Analysis of the Human Electroencephalogram with Methods from Nonlinear Dynamics*, Rensin, L., an der Heiden, U. & Mackey, M.C. (eds.). Springer-Verlag: Berlin-Heidelberg-New York-Tokyo.

[6] Lehmann, D. (1971). Multichannel Topography of Human Alpha EEG Fields. *Electroenceph. clin. Neurophysiol.* **31**, 439.
Lehmann, D. (1972). Human Scalp EEG Fields: Evoked, Alpha, Sleep, and Spike-Wave Patterns. In *Synchronization of EEG activity in Epilepsia,* Petsche, H. & Brazier, M.A.B. (eds.). Springer-Verlag: Berlin-Heidelberg-New York-Tokyo.

[7] Harada, K., Kai, S. & Hirakawa, K. (1988). Spatio-temporal Behaviour of Brain Alpha Rhythms. *Sci. Form.* **3**, 113.

[8] Fuchs, A., Friedrich, R., Haken, H. & Lehmann, D. (1987). Spatio-Temporal Analysis of Multichannel 'α'-EEG Map Series. In *Computational Systems - Natural and Artificial,* Haken, H. (ed.). Springer-Verlag: Berlin-Heidelberg-New York-Tokyo.

[9] Anderson, P. & Anderson, S.A. (1968). *Physiological Basis of the Alpha Rhythm.* Appleton: New York.

[10] Movie: *Spatio-Temporal Alpha EEG Patterns,* produced by Fuchs, A., Friedrich, R. & Haken, H.

[11] Layne, S.P., Mayer-Kress, G. & Holzfuss, J. (1986). Problems Associated with Dimensional Analysis of Electroencephalogram Data. In *Dimensions and Entropies in Chaotic Systems,* Mayer-Kress, G. (ed.). Springer-Verlag: Berlin-Heidelberg- New York-Tokyo.

[12] Haken, H., Kelso, J.A.S. & Bunz, H. (1985). *Biol. Cybern.* **51**, 347.

5. ANALYSIS OF EEG SIGNALS AND THEIR SPATIAL CORRELATION OVER THE SCALP SURFACE

N. Flytzanis*, E. Yiachnakis** and J. Micheloyannis**

*Physics Department, University of Crete, Iraklion, Greece
** School of Medicine, University of Crete, Iraklion, Greece

Spectral electroencephalogram (EEG) analysis shows spatial coherence of α-waves over areas of the scalp. We use the correlation dimension D_2 to characterize the time evolution of the cortex activity. The dimension analysis of 8 channels suggests a uniform attractor over the surface with small but consistent differences in the correlation dimension. The analysis is greatly improved by eliminating the high degree of regularity for neighbouring points in the embedding space due to oversampling.

1. Introduction

Even though the cerebral cortex with its subcortical connections is a complicated biological system of structured neuronal networks, its electrochemical activity as registered by the EEG signal on the scalp surface presents a number of basic frequencies (δ,ϑ,α,β waves), as is evidenced from spectral analysis of EEG data. On the other hand it is quite complex despite the apparent regularity as observed in certain pathological cases as petit-mal epilepsy, Creutzfeld-Jakob disease etc. [1-3]. Nonlinear dynamics can give new statistically important parameters (dimension, entropy, etc.) to quantify this complexity [4,5] and differentiate between chaotic attractors that describe the state of the system. Evolution of brain activity (normal or abnormal) can be thought of as transitions between different strange attractors, caused by changes in the important "pacemaking" variables. The correlation dimension of an attractor [6] is of great interest because it is an estimate of the number of important variables (modes, pacemakers) and the complexity of the dynamics.

Considerable effort has been devoted recently to understanding the time evolution of potential mappings on the scalp using multichannel spontaneous EEG signals. The sequence of momentary scalp field maps, shows regular and abrupt transitions between stable configurations (with periodically reversed polarity) of varying duration [7,8]. These stationary segments could be the basic building blocks of brain information processing or could be simply transition states due to the self-organization of the system that passes through several instabilities that form a sequence of spatial patterns. Even though only the α-band power in the EEG is considered, there is a large variety of possible characteristic segments [7]. Of course the analysis of the whole spectrum gets complicated, but band filtering can drastically change the geometrical structure of the phase subspace.

The nonlinear analysis, which includes all spectral components, considers a global and uniform attractor to describe the cortex activity. Then data from a single channel with the time-delay [6] method can reconstruct the attractor. In this paper we will test the spatial uniformity of the dynamical system as far as the dimensionality is concerned. This will be done by analysing simultaneously obtained EEG signals along characteristic curves on the scalp surface, using techniques from both spectral analysis and nonlinear dynamics. In section 2 we describe the correlation dimension method and discuss some of its limitations. In section 3 we present the experimental results and their analysis, and in the final section we summarize the results.

2. Correlation dimension evaluation

The reconstruction of an attractor can even be achieved from a single time series x(t) using the method of time-delay coordinates. In this case one constructs a "state" vector X(t) embedded in an m-dimensional phase space with components

$$\vec{X}(t_i) = \{x(t_i),\ x(t_i+\tau), \ldots,\ x(t_i+(m-1)\tau)\} \tag{1}$$

The time interval τ between components is a multiple of the sampling time Δt in the data. For infinite data length the results should be independent of τ (as well as the embedding dimension m) but in practice this is so only over a range of τ-values, which varies with the analysis parameters and must be determined empirically. Several initial estimates for a best value for τ have been proposed for signals with strong periodic content, such as the zero of the autocorrelation function, its minimum or the minimum of the mutual information. Numerical work has shown that what is actually important for low-dimensional signals is the total period spanned by the vector $\vec{X}(t_i)$, i.e. $q=(m-1)\tau$ if $m \geq 2D_2+1$ where D_2 is the correlation dimension [9].

The correlation dimension is calculated from

$$D_2 = \lim_{r \to 0} \frac{\log C(r)}{\log r} \tag{2}$$

where C(r) is the correlation integral

$$C(r) = \frac{1}{N_c} \sum_{j=1}^{N_c} \frac{1}{N} \sum_{i=1}^{N} \theta(r - |\vec{X}_i - \vec{X}_j|) \tag{3}$$

with $\theta(z)$ the Heavyside function (equal to unity for positive and zero for negative argument). N is the total number of points in the time series and N_c is the number of reference points. In (3) the sum over i counts the number of points in the m-dimensional space that are within a distance r of the reference vector $\vec{X}_j$ and with the sum over j we average over the reference points to avoid any inhomogeneities in the attractor. To cut down on computer time N_c is much less than N. It has been shown that the choice $N_c/N \approx 1/3$ minimizes fluctuations in the correlation integral, but they remain almost constant for a much smaller ratio [10].

A plot of C(r) versus r in a log-log graph should be a straight line whose slope is the correlation dimension. The linear region is limited by saturation when $C(r)\to 1$ for large r-values, while on the low r region by noise. If the noise has an average amplitude then one expects a break in the curve for this r-value, so that the slope for smaller r increases with increasing embedding dimension [11]. The extent of the useful region increases with N and for a proper choice of the analysis parameters, but decreases with the correlation dimension.

3. Measurements and analysis

We analysed EEG measurements for a normal right-handed individual aged 22 years, under comfortable resting conditions with eyes closed. EEG signals were recorded using surface-electrodes from the points FP2, F4, C4, P4, O2, F8, T4, T6 (according to the international 10-20 system) each referred to interconnected (A1+A2) earlobe electrodes, denoted with 1-8 correspondingly in Fig. 1. The EEG can be registered as a digital signal (at 200 Hz) or an analog VIDEO signal. The latter gives the possibility of digitizing the signal at any desired rate using a 12-bit AD converter.

The 8 channels of artefact free data were filtered with a notch filter at 50 Hz, a high (50 Hz cutoff, down 3 dB at 70 Hz) and a low (cutoff at 0.5 Hz, down 3 dB at 0.16 Hz) frequency filter. The total time duration (250 sec) is sufficient to check for stationarity, and the segments analysed were about 20-30 sec epochs. On visual inspection the EEG was normal with predominant alpha rhythm and few visible artefacts in the temporal region. As shown in Fig. 2 (5 sec epoch) channel 6 has the largest and channel 5 the smallest amplitude. This is in accordance with the conclusion of relative power contribution analysis that the occipital area has a strong endogenous contribution to the alpha activity, which appears as exogenous in other areas of the scalp [12]. In Table 1 we present the results of stochastic signal spectral analysis [13] for the contribution in μV^2 of the main bands (alpha, delta and theta). The Fourier analysis involves the averaging of the spectra from 14 successive sections of 20 sec each. The first column is a background (not necessarily noise) in the averaging procedure due to phase variability in a deterministic chaotic system. There is an increased coherent α-activity in the frontal and posterior regions in accordance with the hypothesis of localized pacemakers. In general however coherence analysis shows no global coherence over all frequency bands.

Table 1. Contribution to the power in μV^2 from background, δ-band (0-2 Hz), ϑ-band (4-8 Hz), α-band (8-12 Hz).

Channel	Background	δ	ϑ	α
1	17.6	27.0	0.6	46.0
2	30.5	7.3	3.8	45.4
3	16.7	46.1	0.3	29.0
4	22.5	10.3	3.5	41.1
5	35.2	28.0	3.2	17.4
6	17.8	3.3	0.4	82.0
7	25.1	13.1	0.8	54.4
8	18.0	8.5	0.7	69.3

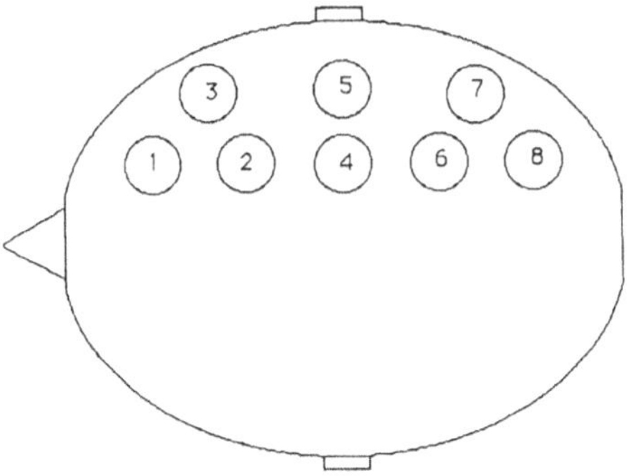

Figure 1. Diagram of channel positions on scalp.

In Table 2 we give the correlation dimension for all eight channels, with τ=20 msec. To verify data stationarity we analysed four different 20 sec epochs for each channel, and the comparison is given in Fig. 3 for channel 6. The errors are obtained from this sample. By including all the runs for various analysis parameters (delay time, data duration, etc.) the error estimate changes by less than 50%. There are variations in D_2, i.e. the lower values obtained for channels 1 and 6 where the possible pacemakers beam the α-rhythm. The saturation of the slope in channel 5 is imperfect and its value of D_2=8.4 is uncertain, since the validity of the Procaccia-Grassberger algorithm when applied to finite data is questionable at high correlation dimensions. The high D_2 is consistent with the large background in Table 1 and is probably due to muscular artefacts of small amplitude (even at resting conditions) and due to the large distance of channel 5 from the cortext surface. Due to attenuation in the intervening space the noise becomes an important part of the signal amplitude. The saturation of D_2 with m indicates that the noise in the system is not white, but it is dominated by low frequencies [14]. For longer epochs (up to 50 sec) and increasing τ (τ>10 msec) there is little variation in D_2, but the best plateau is for τ in the range 10-25 msec.

All the previous results used a sampling rate ν_s=200 Hz, but are valid at higher values. In Fig. 3 we have ν_s=500 Hz. Using the procedure in (3) gives a shoulder in the logC(r) plot at high embedding dimensions (Fig. 4a). It is due

Table 2. Correlation dimension for channels 1-8.

Channel :	1	2	3	4	5	6
D_2 :	6.2±.3	7.2±.3	7.2±.4	7.3±.4	8.6±.2	6.5±.2
Channel :	7	8				
D_2 :	7.5±.6	7.0±.3				

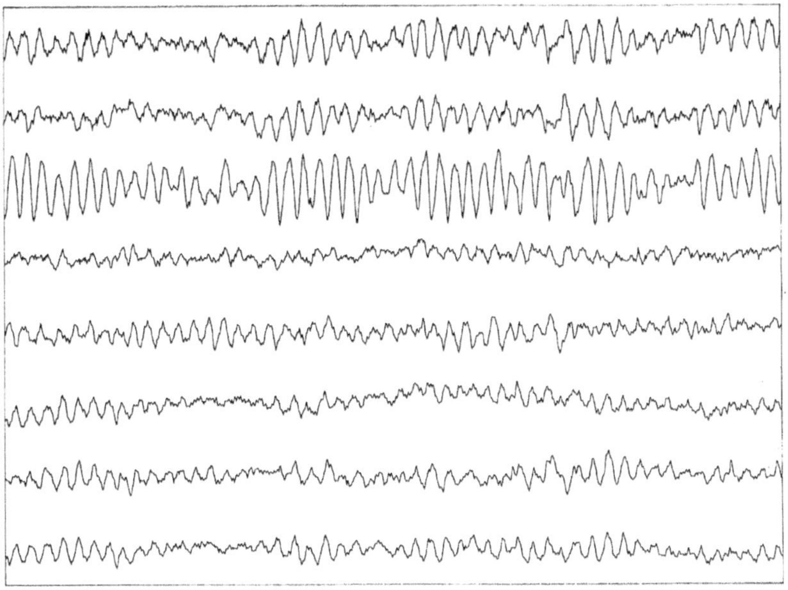

Figure 2. EEG signals (5 sec epoch) for all channels (1-8) from top to bottom.

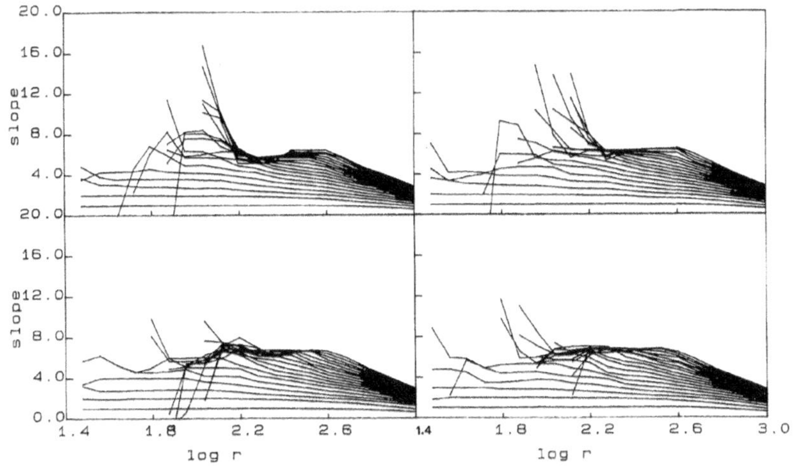

Figure 3. Slope of log C(r) vs. log for 4 different 20 sec epochs (τ=20 msec, m=1-20).

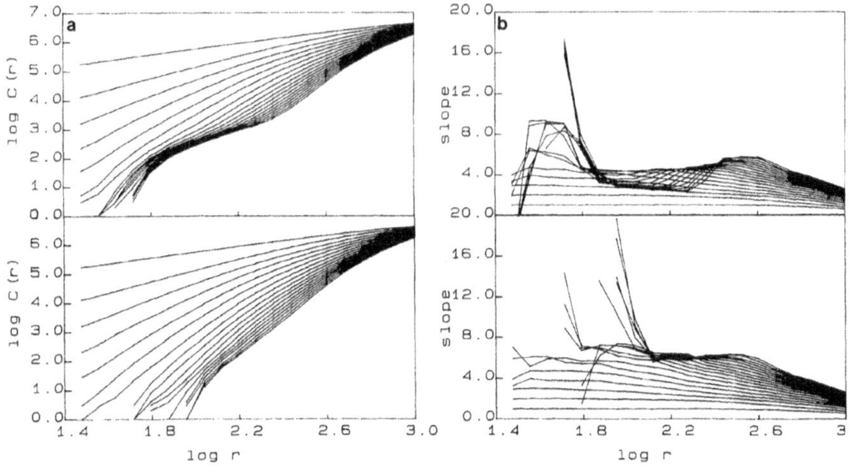

Figure 4. (a) log C(r) and slope vs. log r for channel 6 with νs=500 Hz, m-1-20, τ=20 msec and w=0. (b) Same with w=1.

to the oversampling since the large number of closely spaced points presents an artificial regularity (for finite data) which lowers D_2. When the first neighbours to the reference points in the sum over i in (3) are omitted (w=1), the shoulder is almost eliminated (Fig. 4b) increasing the effective linear region [15].

4. Conclusion and summary

The dimension analysis is a useful tool for the study of the cerebral activity, and the quantification of its complexity. The spatial variation of D_2 gives strong indications for the uniformity of brain activity as measured using electrical information from the scalp. There are, however, differences between the 8 channels which are outside the range of the rough error estimate. Part of the difference could be due to the relative increase of noise for smaller amplitude signals. Also due to anatomical and physiological reasons channels 1, 3, 5 and 7 are more susceptible to artefacts. The numerical values for D_2 are reasonably close with reported work on the α-rhythm [1-5]. This agreement, however, has a meaning only for a well controlled and selected sample of subjects under the same analysis parameters and then only for a comparative study. So while there is a general agreement on the existence of low dimensional chaos there is some spread of numerical values. Despite the fact that good estimates are obtained for short time series one should be careful in drawing conclusions before a bigger sample is examined.

The analysis involves a balance between accurate determination of the correlation dimension which necessitates long time series and stationarity which in general can only be guaranteed for finite time segments due to the possibility of artefacts or experimental limitations. The computation time increases like N^2 but can be decreased by (i) choosing as short a time series as necessary, (ii) using the maximum norm instead of the Euclidean in (3) and (iii) putting a maximum $r=r_0$ where the log $C(r)\approx 90\%$ (or $C(r)\approx 10\%$) of the saturation value. The choice of the parameters also depends on the experimental conditions (which must be explicitly stated) and the sources of noise. This can be inherent in the dynamical system, muscular artefacts, experimental or computational error arising from rounding off and the analog to digital conversion.

Having the proper sampling rate is very important to unravel the local foldings of the attractor in phase space, and it depends both on the main spectral components and the signal amplitude. The sampling of the attractor at infinitesimal distance gives the shoulders (small slope) of Fig. 4a. The elimination of the neighbourhood of the reference points removes the artificial regularity due to short time series.

In conclusion more careful analysis is required to verify the spatial variations of D_2. It is rewarding, however, that the points of low D_2 correspond to points of increased coherent α-activity. One can also investigate the correlation between the value of D_2 and the spectral content. Finally nonlinear dynamics along with spectral analysis of scalp electrical voltage can characterize spontaneous cortex activity.

Acknowledgement

The authors are indebted to Prof. J.G. Caputo for many useful discussions.

References

[1] Babloyantz, A. & Destexhe, A. (1987). In *Temporal Disorder in Human Oscillatory Systems,* Rensing, L., van der Heiden & Mackey, M.C. (eds.). Springer Series in Synergetics **B6**, 48.
[2] Babloyantz, Z. & Destexhe, A. (1987). In *From Chemical to Biological Organisation,* Markus, M., Müller, S. & Nicolis, G. (eds.), p. 307. Springer-Verlag.
[3] Babloyantz, A. & Destexhe, A. (1986). *Proc. Natl. Acad. Sci. USA* **83**, 3513.
[4] Watt, R.C. & Hameroff, S.R. (1988). *Int. Jour. Clin. Monit. Comp.* **5**, 3.
[5] Holzfuss, J. & Mayer-Kress, G., ibid Ref. [1], p. 114.
[6] Grassberger, P. & Procaccia, I. (1983). *Physica* **9D**, 189.
[7] Lehmann, D., Ozaki, H. & Pal, I. (1987). *Electroeceph. & Clin. Neurophys.* **67**, 271.
[8] Friedrich, R., Fuchs, A. & Haken, H. In this volume.
[9] Martinerie, J.M., Albano, A.M., Mees, A.I. & Rapp, P.E. Mutual information, strange attractors and optimal estimation of dimension, preprint.
[10] Theiler, J. Statistical precision of dimension estimators, preprint.
[11] Ben-Mizrachi, A., Procaccia, I. & Grassberger, P. (1984). *Phys. Rev.* **A29**, 957.
[12] Inouye, T., Shinosaki, K., Yagasaki, A. & Shimizu, A. (1986). *Electroenceph. clin. Neurophysiol.* **63**, 253.
[13] Dumermuth, & Molinari, L. (1987). In *Methods of Analysis of Brain Electrical and Magnetical Signals. EEG Handbook,* Gervin, A.S. & Rémond, A. (eds.). Elsevier Science.
[14] Osborne, A.R. & Provenzale, A. (1989). *Physica* **D35**, 357
[15] Theiler, J. (1987). *Phys. Rev.* **A36**, 4456.

6. SUSTAINED REACTION-DIFFUSION WAVES AND STATIONARY STRUCTURES IN AN OPEN REACTOR

J. Boissonade, Q. Ouyang, A. Arneodo, J. Elezgaray,
J.C. Roux and P. De Kepper

Centre de Recherche Paul Pascal/CNRS

Université de Bordeaux I, Avenue Schweitzer
33600 Pessac, France

A severe hindrance to the development of studies of chemical spatial structures is the lack of reliable open reactors to provide a permanent feed of fresh reactants and simultaneously preserve the diffusion process. The Bordeaux group has developed an original reactor which satisfies these prerequisites, namely the "Continuous Couette Flow Reactor", made of a circular Couette flow fed at both ends by CSTRs. The tunable effective diffusion rate is much larger than common molecular diffusion rates allowing for large size reaction-diffusion structures. We report sequences of bifurcations between indefinitely sustained spatio-temporal structures exhibiting waves travelling back and forth as well as non-trivial stationary states obtained with a variant of the chlorite-iodide reaction. A simple dynamical model, analogous to those used in the description of excitable media, provides a good basis for a theoretical interpretation.

1. Introduction

The possible emergence of sustained spatio-temporal concentration patterns resulting from the sole coupling between reaction and diffusion processes in single phase isothermal chemical systems has been predicted as early as 1952 [1]. Ever since, these dissipative structures have been thoroughly studied both in their own right, in consideration of imaginable applications in the field of biology [2-5], and as a valuable prototype of non-equilibrium spatial structures in the field of non-linear science [6]. The evolution of the concentrations c_i of the different species i is ruled by a system of Reaction-Diffusion equations:

$$\frac{\partial c_i}{\partial t} = f_i(\ldots,c_j,\ldots) + D_i \Delta c_i \tag{1}$$

where the non-linear functions f_i of the sole concentrations represent the contribution of reactive processes, the D_i are the diffusion coefficients and Δ is the Laplacian operator. The relative simplicity of these systems comes from the absence of **advection terms** associated with convective motions of the fluid. A number of chemical patterns, such as the so-called **mosaic structures** [7], have actually been shown to originate in parasitic convective motions [8] and will be discarded in the following. From a theoretical point of view, two classes of reaction- diffusion structures can be discriminated [9]:

- Structures resulting from destabilization of the homogeneous state by a non-homogeneous spatial mode [1,3,10-12]. They normally need differences between the diffusion coefficients of at least two different species, a generally unrealistic condition with small molecules for which the diffusion coefficient values are all about 10^{-5} cm^2/s.
- Structures associated with **fronts**, i.e. steep spatial changes of concentration which in turn correspond to a transition between two chemical states - in general an oxidized and a reduced state - with fast kinetics. Travelling waves in excitable media are the best known examples of such structures. This topic is extensively discussed in this book and in recent reviews [13-16]. Although they were generally observed as transient phenomena, indefinitely sustained wave patterns in chemical excitable media have recently been obtained experimentally [17-19]. A simple way to create a front structure is to impose a concentration gradient of initial reactants: if, for instance, the system is kept in the oxidized state at one position and in the reduced state at another position one expects a transition front to form between them. Such a gradient-imposed structure is somewhat trivial but we shall see that the front can be destabilized and start to move, and that additive stationary or moving fronts can form, leading to complex stationary or non-stationary patterns. The few theoretical studies of structures where the input species concentrations are fixed only on boundaries actually concern Turing type structures which are **controlled** by the gradient but there is almost no theoretical work on front structures intrinsically **induced** by this gradient [20-22].

Severe hindrances have delayed experimental research on sustained spatial chemical dissipative structures. Only recently the main practical problems have been solved and a few efficient reactors were designed to produce these structures [17,18,23,24]. We shall discuss results obtained with one of these pieces of apparatus, the **Couette Flow Reactor**. Another type of such reactor is presented in this conference by Dulos et al. [19]. The difficulties in devising a reactor to obtain a genuine reaction- diffusion process come both from the feeding procedure ensuring the non-equilibrium conditions and from the control of the transport processes (diffusion). We shall consider these questions successively.

- Difficulties raised by the feeding procedure: in most theoretical works on non-equilibrium chemical patterns the concentration of the input species is kept uniformly constant. This situation can only be approximated at the beginning of the reaction and cannot be sustained indefinitely without feeding the system with fresh reactants. Most practical feeding processes are associated with input streams and will most likely induce convective motions. An alternative, much more appropriate from an experimental point of view, is to feed the system **only at boundaries** where concentrations can easily be kept constant and let the reactants diffuse far into the core of the system where the influence of side effects induced by the feed is minimized. Since this easily achievable procedure implies the formation of concentration gradients, it is more likely suited to the study of the gradient-induced structures on which we shall focus in the following.
- Difficulties raised by the control of diffusion:
  - Any parasitic convection phenomena should be avoided. This can be

done in a convenient way by using a **permeable inert gel** as a support for the reactive medium [18,19] but this does not solve the problems coming now.
- Since we now limit ourselves to front structures, the diffusion coefficients D_i can be all equal (or of the same order of magnitude) to a single coefficient D but in order to study the role of spatial coupling in the emergence of patterns and eventually to follow the sequences of bifurcation as a function of diffusion rate, it is essential that this diffusion coefficient be **continuously tunable** over a large range. Such a condition is *a priori* unrealistic since the molecular diffusion cannot be controlled experimentally, except for small changes in the temperature which would anyway be accompanied by much larger changes in the rate constants.
- Moreover, due to the small typical value $D \sim 10^{-5}$ cm^2/s of molecular diffusion coefficients, the corresponding time and length scales τ and λ (related by $\lambda \sim \sqrt{D\tau}$) are somewhat experimentally inconvenient. For $\tau \sim 10^2$ s, a typical value for the period of an oscillating reaction which defines a macroscopic characteristic time scale, the expected structure "wavelength" λ is less than 1 mm and to a reactor size as small as λ = 1 cm corresponds a relaxation time τ of the sole diffusion profile longer than 30 hours! (see for example experiments of [19]).

If one notes that the diffusion in equation (1) is not necessarily "molecular diffusion" but could be any other process which obeys the Fick law, the solution to the problems above is clearly to search for another diffusive process with much higher controllable diffusion coefficients. Such a process is provided by the **Couette Flow Reactor** (CFR).

2. The Couette Flow Reactor

2.1. *Transport in the CFR*

The core of the reactor is a circular Couette flow made of two coaxial cylinders of length L, outer radius r and separated by a small annular gap of width a, filled with a reactive fluid of kinematic viscosity ν. In the following, we shall assume that $a \ll r \ll L$, a condition met in our reactor where a = 0.25 cm, r = 1.25 cm, L = 34 cm. The outer cylinder is fixed whereas the inner cylinder is rotated at rate Ω. The hydrodynamic regime in the gap is determined by the value of the Reynolds number $R_e = \frac{ar\Omega}{\nu}$. When $R_e > R_c \simeq 80$ the flow self-organizes into pairs of toroidal vortices inside which takes place a convective motion as sketched on Fig. 1 [25]. For larger rotation rates, when $R_e \gg R_c$, this motion becomes turbulent so that the radial and azimuthal mixing

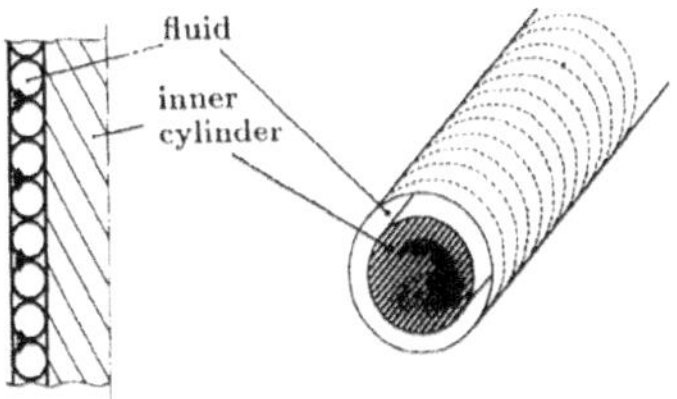

Figure 1. Scheme of a circular Couette flow.

achieved inside a toroidal cell is extremely fast and each of these cells can be considered as homogeneous. The axial mass transport from one cell to the neighbouring ones takes place on much larger time scales and the concentrations can change from one cell to the next. Provided that one does not sustain any net flow between the two ends of the cylinders, the total amounts of fluid exchanged between two adjacent cells are constant in time and equal in the two opposite directions. Thus the rate of transport of a given species is proportional to the difference of concentration between these two cells, the discretized form of the Fick law. At scales much larger than the gap, one thus expects that the system will behave as a quasicontinuum in the axial direction and follow a one-dimensional diffusion law; this has actually been corroborated by the experimental results of Tam & Swinney [26]. We must emphasize that, although the **local** transport is performed by convective turbulent processes of the solvent, the **effective** axial transport is simply diffusive on a scale $\ell \gg a$ and does not generate any advection term which would correspond to a net flow along the cylinders. As all the species are transported by the same solvent, all the effective diffusion coefficients D_i are equal to a single value D which increases with the rotation rate Ω. The range of possible values is limited at low rates by the requirement of keeping a diffusive regime and at high rate by the maximum accessible rotation rate. In our reactor, the coefficient D, determined by tracer experiments [27], ranges from 10^{-2} cm^2/s to 0.6 cm^2/s (Fig. 2). These values are 3 to 5 orders of magnitude larger than molecular diffusion coefficients.

2.2. *Feeding the CFR with chemicals: boundary conditions*

In order to feed the reactor with fresh chemicals at boundaries, the two ends of the cylinders open into two identical continuous stirred tank reactors (CSTR), kept homogeneous by the vigorous stirring. A schematic representation of the reactor is given in Fig. 3; for more technical details, see [27]. The input and output flows of each CSTR are carefully balanced in order to avoid any net flow along the Couette reactor and ensure the absence of advection terms in the evolution equations. In principle the CFR dynamics should be influenced by the dynamics inside the CSTRs. Nevertheless, provided that the volume of these CSTRs is large enough or, more exactly, that their residence time τ_R is small in regard to the diffusion exchange times through the Couette flow, the concentrations in the CSTRs are not in general significantly modified by the dynamics taking place in this Couette flow. Thus a stationary state in a CSTR constitutes a good approximation of a Dirichlet boundary condition. In all the experiments reported in this paper these conditions were met at both ends.

2.3. *Capabilities of the CFR*

Let us summarize all the characteristics of the CFR. We shall refer to the central part as **"the reactor"**, assumed to be operated in the turbulent regime, the CSTRs being the **"boundary conditions"**.

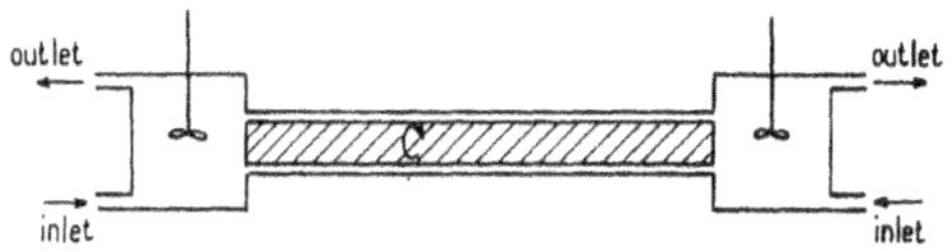

Figure 2. Scheme of the Couette flow reactor.

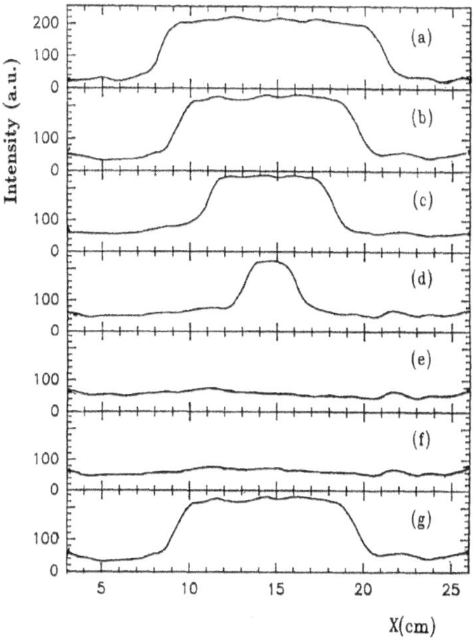

Figure 3. Spatial bistability and hysteresis as a function of diffusion (Light intensity profiles along the axis of the cylinders for increasing and decreasing rotation rates). Fixed control parameters: $[I^-]_0^{I,II} = 3.5\times10^{-3}$ M, $[H_2SO_4]_0^{I,II} = 4.5\times10^{-3}$ M, $[ClO_2^-]_0^{I,II} = 0.91\times10^{-3}$ M, T = 26°C, in and out flow rate of each CSTR. $F_0 = 30\ cm^3min^{-1}$.

- The system provides good azimuthal and radial mixing (homogeneous vortices). Joined with the condition $a \ll r \ll L$ this defines a **quasi 1-D system** where concentrations change only in the axial direction.
- There is **no axial advection term**.
- The axial transport is **diffusive**.
- The diffusion coefficients are all equal to a single value D, large with regard to molecular diffusion coefficient. This leads to space and time scales convenient for experimentation.
- This coefficient is continuously **tunable** by a simple change of the rotation rate Ω.
- The system is continuously fed with fresh reactants by diffusion from the ends where the concentrations can be kept approximately constant. The feed concentrations provide a second set of adjustable control parameters.

Finally the CFR is the practical realization of a system where the dynamics are given by a **1-D reaction-diffusion equation** with Dirichlet boundary conditions and where both reaction and transport parameters can be controlled.

2.4. *Operating conditions*

We have used several Red-Ox reactions involving fast switch between an oxidized state Ox and a reduced state Red. Here we shall limit ourselves to the Chlorite-Iodide reaction in acidic medium which exhibits many interesting dynamical properties [28]. Other results obtained with the B.Z. reaction are presented in [24], [27] and [29]. The changes of state are followed with a colour indicator, here a starch-like indicator enabling a change from yellow to blue with the I_3^- concentration. The Couette cylinders are made of optical quality glass so that the colour profile can be monitored all along with a video camera and reprocessed subsequently with an image analyser (NUMELEC PERICOLOR 2000) allowing for quantitative analysis and improvement of contrast and brilliance.

3. **Spatial structures in the CFR**

Various temporal and spatio-temporal structures have been obtained in the CFR. We only report two series of experiments which are specially representative. In each series, a selected control parameter, kept fixed during the experiment, is gradually changed from one experiment to the next.

3.1. *Spatial bistability*

The diffusion coefficient is used as the bifurcation parameter and the two CSTRs are fed with identical reactant compositions and both maintained in the reduced state. When the transport rate D is low the reactants diffuse slowly into the Couette flow. It takes a long while for them to reach the central region so that the region remains in the oxidized state, the final state when all the available species have reacted (chemical equilibrium). A steep transition from Red to Ox occurs at an intermediate point where a stationary front forms. Due to the symmetry, there are two such fronts leading to an inverted U-shaped colour profile. At high diffusion rates, the matter exchanges with the central region are too large for this region to be in a state different from their common CSTR state so that the colour profile is almost flat. At intermediate diffusion rates, the final profile depends on the initial state of the central region. The transition from Red to Ox occurs when some catalytic products of the reaction have accumulated beyond a critical threshold concentration. If the initial state is Red the transport rate is larger than the accumulation rate and the central region remains in state Red (flat profile); if the initial state is Ox, the catalytic products being already present in a large amount, the reaction rate is enhanced so that it can balance the losses, and the central region remains in state Ox (U-shaped profile). This is a **spatial bistability** where two stable steady state patterns - flat and U-shaped - exist for the same set of control parameters. Thus according to whether D is increasing or decreasing the transition between a flat profile and a U-shaped profile occurs for different values of D as shown in Fig. 3. Numerical simulations performed with a six variable model of the reaction also exhibit such an hysteresis and support the above analysis [27,29,30].

3.2. *Travelling waves and stationary structures*

In this series of experiments the control parameter is the chlorite concentration $[ClO_2^-]_0$ in one of the CSTRs (on the right in the figure). This $[ClO_2^-]_0$ is always high enough for this reactor to remain in state Ox whereas the other CSTR is kept in state Red (no input of chlorite). The diffusion rate was fixed to D = 0.31 cm^2/s (all other parameters as in Fig. 4). When the chlorite concentration increases the system goes through the sequence of bifurcations presented in Fig. 4. Since most profiles are nonstationary they are represented as a function of time. In order to provide better evidence for the fronts, only the two dramatic colour changes of the solution (black for blue colour, Red state; white for yellow colour, Ox state) are displayed; shades and full colour prints can be found in [23] and [27]. At low $[ClO_2^-]_0$ values the front between Ox and Red remains at a fixed position (Fig. 4a) but increasing these values makes the system go through a first bifurcation, equivalent to a Hopf bifurcation in homogeneous systems: the front location oscillates periodically along the reactor, the oscillation amplitude increasing with the control parameter (Fig. 4b). More complex sustained patterns of travelling waves form at larger $[ClO_2^-]_0$ values (Fig. 4c and 4d): a pair of additional travelling fronts appears and disappears periodically with different mechanisms well illustrated in the figure. At still larger values of the control parameter, the pattern is composed of three permanent oscillating fronts, the position of which oscillates with identical periods (Fig. 4e). Eventually the system still goes through another bifurcation leading to a three-front stationary pattern (Fig. 4f). This multipeaked pattern where Ox and Red alternate is highly atypical: it cannot be explained by the sole

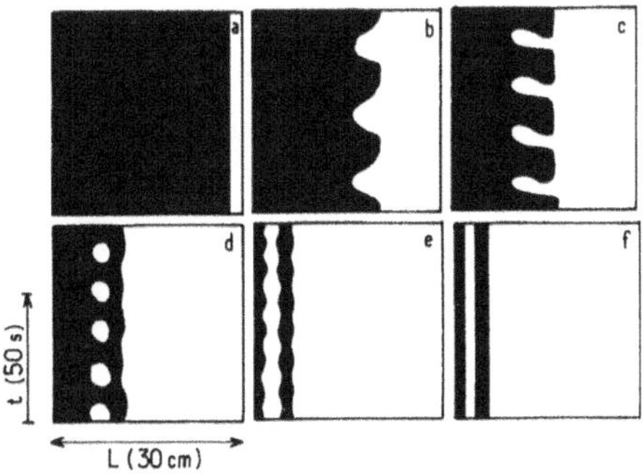

Figure 4. Space-time representation of the sustained spatial and spatio-temporal structures observed with the chlorite-iodide-malonic acid reaction. Constant control parameters: Ω=840 rpm (D ~ 0.31 $cm^2 s^{-1}$), T=26°C, input and output flow rate of each CSTR $F_0^{I,II}$ = 10 $cm^3 min^{-1}$, composition $[I^-]_0^{I,II}$ = 3.5×10^{-3} M, $[H_2SO_4]_0^{I,II}$ = 4.5×10^{-3} M, $[MA]_0^{I,II}$ = 2×10^{-3}, thiodene 2 g/l with $[ClO_2^-]_0^{I}$ = 0. Every pattern corresponds to a different Chlorite concentration in CSTR-II (right side). (a) Single stationary front: $[ClO_2^-]_0^{II}$ = 8×10^{-3} M; (b) simple oscillating front: $[ClO_2^-]_0^{II}$ = 9×10^{-3} M; (c-e) multiple oscillating fronts: c) $[ClO_2^-]_0^{II}$ = 11×10^{-3} M, d) $[ClO_2^-]_0^{II}$ = 12×10^{-3} M, e) $[ClO_2^-]_0^{II}$ = 14×10^{-3} M; f) Non-trivial stationary structure: $[ClO_2^-]_0^{II}$ = 15×10^{-3} M.

gradient of input species and results from the coupling of the nonlinear kinetics with the linear transport process. In the next paragraph we shall introduce some elements for the theoretical interpretation of such structures and bifurcation sequences.

Other sequences of patterns, including prechaotic [9,27,29] or fully chaotic [24] patterns, have been obtained in a CFR (note that in [24] the CSTRs are replaced by a direct controlled injection at the end).

4. A theoretical approach

It is not presently possible and it would most likely not be very informative to interpret these bifurcation sequences in terms of a realistic kinetic model of the chlorite-iodide-malonic acid reaction. Nevertheless some major features of the experimental results have been reproduced with a simple dynamical model [31-34] in much the same way as the FitzHugh-Nagumo model [35] provides an excellent support for the study of travelling waves in excitable media in spite of the over-simplifications used in the nerve impulse description. Our structures share many common features with these travelling waves: they both involve propagating fronts which, in the latter case, are known to result from the existence in phase space of a pleated slow manifold which captures all the trajectories in a time short with regard to all the other time scales. Thus, our model is a two-variable Van der Pol like system similar to those which are commonly used in the study of the travelling waves:

$$\frac{du}{dt} = \varepsilon^{-1}(v - f(u))$$
$$\frac{dv}{dt} = -u + \alpha \qquad (2)$$

where ε is a small positive parameter and α a free parameter. These equations actually ensure the existence of an S-shaped slow manifold $v = f(u)$ on which all trajectories are attracted in a time ~ $O(\varepsilon)$. The only steady state of the reaction term ($u_s = \alpha$, $v_s = f(\alpha)$) is located on the slow manifold; this steady state is stable for $\alpha < \alpha_L$ (lower branch), or $\alpha > \alpha_U$ (upper branch) while it is unstable for $\alpha \in [\alpha_L, \alpha_U]$ (Fig. 5). The critical values α_L and α_U correspond to a Hopf bifurcation leading to oscillatory behaviour. According to the specific shape f(u) of the slow manifold, this bifurcation can be either supercritical or subcritical [36]. To a classical cubiç function we shall prefer the following form of the slow manifold

$$f(u) = u^2 - u^3 + u^5 \qquad (3)$$

since this choice conditions the Hopf bifurcation in the reaction term to be subcritical, as is the case in most experimental situations. When adding a flux term to this Van der Pol like equation, bistability can also be recovered. Despite the fact that this model does not have all the properties required for a chemical scheme, u and v play the role of the concentration variables and we shall refer to the upper and lower branches of the slow manifold as analogous to the oxidized and reduced state branches of the Chlorite-Iodide reaction. In accordance with the characteristics of the Couette Flow Reactor, i.e. a single space variable z and all diffusion coefficients equal to D, the reaction-diffusion equations (1) become:

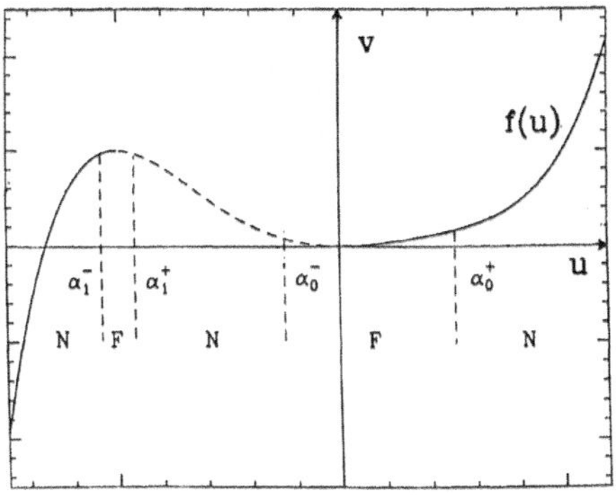

Figure 5. Sketch of the slow manifold $v = f(u) = u^2 - u^3 + u^5$. The single steady state $(u = \alpha,\ v = f(\alpha))$ is a focus (F) for $\alpha_1^- < \alpha_1^+$ or $\alpha_0^- < \alpha < \alpha_0^+$, and a node (N) everywhere else. Continuous line: stable steady state; dashed line: unstable steady state.

$$\begin{aligned} \frac{\partial u}{\partial t} &= D\frac{\partial^2 u}{\partial z^2} + \varepsilon^{-1}(v - f(u)) \\ & \qquad\qquad\qquad\qquad z \in [0,1] \\ \frac{\partial v}{\partial t} &= D\frac{\partial^2 v}{\partial z^2} - u + \alpha \end{aligned} \tag{4}$$

where the length of the CFR has been rescaled to unity for convenience. These equations have been ingegrated numerically by the method of lines (reduction to an ODE system by space discretization) and integrated with a fast ODE solver. In agreement with the experiments in section 3 we impose Dirichlet boundary conditions. The left-end CSTR $(z = 0)$ is set in a (reduced) lower-branch state while the right-end CSTR $(z = 1)$ is maintained in an (oxidized) upper-branch state:

$$\begin{aligned} v(z=0) &= f(u_0) \quad \text{with} \quad u_0 = u(z=0) < \alpha_L \\ v(z=1) &= f(u_1) \quad \text{with} \quad u_1 = u(z=1) > \alpha_U \end{aligned} \tag{5}$$

For the sake of simplicity, the value of α in system (2) is set independent of z, $\alpha > \alpha_U$, so that when switching off the diffusion process, all the intermediate cell points evolve asymptotically to the same stable steady state on the upper branch of the slow manifold. A more realistic model should take into account a spatial dependence of α where $\alpha(z = 0) = u_0$ and $\alpha(z = 1) = u_1$ but the practical implementation of this spatial constraint has not much meaning in this formal approach since it should strongly depend on the specific kinetics of the reaction. The parameters u_0 and ε are fixed to the values $u_0 = -1.5$, $\varepsilon = 0.01$ whereas u_1, D and α are taken as control parameters. In Fig. 6, we use the same space-time representation as in the experiments, except that the grey density is proportional to u in place of the all or none changes of Fig. 4.

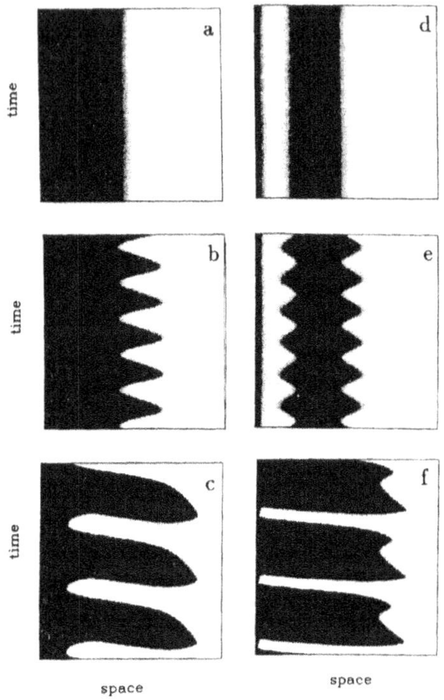

Figure 6. Space time representation of the u(z,t) profile coded on 32 intensity levels. $u_0 = -1.5$, $\varepsilon = 0.01$. (a) Stationary single front pattern ($u_1 = 1.1$, $D = 0.1$, $\alpha = 0.01$); (b) periodically oscillating single-front pattern ($u_1 = 1.1$, $D = 0.045$, $\alpha = 0.01$); (c) periodic alternation of a single front and a three-front pattern ($u_1 = 1.1$, $D = 0.01$, $\alpha = 0.01$); (d) stationary multifront pattern ($u_1 = 0.5$, $d = 0.08$, $\alpha = 0.2$); (e) periodically oscillating multifront pattern ($u_1 = 0.5$, $D = 0.06$, $\alpha = 0.2$); (f) periodic alternation of a single-front and a three-front pattern ($u_1 = 0.5$, $D = 0.02$, $\alpha = 0.2$).

For values of $u_1 \gg \alpha$, one observes only steady single front patterns which correspond to a switch between the two attracting branches of the slow manifold (Fig. 6a). When D is decreased, the transition front becomes sharper and sharper, until this stationary pattern loses its stability and starts to oscillate as shown in Fig. 6b. To gain some understanding of this instability, one may consider a spatially discretized version of our original continuous reaction-diffusion system (2). One can show [34,36] that the coupled elementary reactor cells, located at the front zone (where a steep gradient of concentration exists), can be driven by the diffusion coupling from a steady state to a limit cycle behaviour via a Hopf bifurcation by a physical mechanism which has been clearly identified in the direct diffusive coupling of two unsymmetrically constrained CSTRs [37] (the simplest two- point reaction-diffusion system). The soundness of an extrapolation to the continuous limit is supported by analytical studies on a piecewise linear approximation of the slow manifold [34]. When D is further decreased, the amplitude of the oscillations increases until a qualitative change occurs in the spatio-temporal evolution of the system (Fig. 6c). A three front profile alternates periodically with a single front proceeding from the coalescence of two neighbouring fronts. A similar pattern where the period is twice the period of the previous one is shown in Fig. 3f; although these two patterns correspond to completely different sets of control parameters, this observation provides a very promising indication for the existence of period-doubling bifurcations of these spatio-temporal patterns.

For $u_1 \sim \alpha$, we were able to freeze a pattern involving three spatial switches between the two branches of the slow manifold (Fig. 6d). A decrease of the diffusion coefficient induces a transition to a periodic structure (Fig. 6c).

The patterns collected in Fig. 6 do not constitute a bifurcation sequence since they are not obtained by a continuous variation of a single control parameter. However, they present some striking similarities with the experimental patterns of Fig. 4. In particular the initial and final stages of the experimental sequence are identical to the theoretical sequences 6a $\longrightarrow$ 6b and 6c $\longrightarrow$ 6d, both obtained by a continous change of the parameter D. Nevertheless, in the numerical simulations, the stationary multipeaked structure coexists with the oscillating single front patterns, in contrast with the experimental situation where these two patterns apparently take place in two different regions of the constraint space [27]. We have strong indications that these multipeaked structures arise from saddle-node bifurcations, i.e. at some distance from the "natural" steady state of the system, namely the stationary single front pattern [34,36]. Note that in the experiments we cannot exclude the presence of multiple stable states since the number of workable perturbations is limited to temporary changes in the boundary conditions or in the diffusion rate. The pattern 6c also exhibits similarities with the experimental pattern 4c. Moreover, although there is no pattern analogous to 6f in Fig. 3, spatial period doubling phenomena have been observed in experiments on the B.Z. reaction [9,27].

We have only reported here a small part of the results obtained with this model. In particular, we have found conditions where the oscillating front patterns undergo secondary instabilities leading to more complicated spatio-temporal behaviour including chaotically oscillating front structures. More details can be found in [33], [34] and [37].

5. Conclusion

The Couette Flow Reactor has proved to be a flexible tool for producing sustained reaction-diffusion structures where concentration fronts are induced and controlled by gradients of input reactants. In particular we have evidenced the first sustained multipeaked stationary structure in an isothermal single phase system and a sequence of bifurcations of spatio-temporal states leading to this unconventional structure. The diffusion coefficient which defines a characteristic length scale can be changed over a large range. Thus one can expect to study continuously the transition from small size to extended systems, i.e. the influence of the number of degrees of freedom on the development of complex strucures. The nature of the patterns presently observed in the CFR can be basically understood with an elementary two variable model which retains essentially the same dynamical ingredients as those involved in the excitable media theory. This clearly demonstrates that complex spatio-temporal patterns do not necessarily result from coupling of a continuum of oscillators or from nonlinear transport; the sole coupling through a linear process, namely the diffusion, of systems otherwise at steady state, is able to produce such patterns. In the future, this should confer to the CFR a privileged status in the experimental approach to problems such as transition to chemical turbulence.

Acknowledgement

This work has been supported by the B.P. Venture Research Unit.

References

[1] Turing, A.M. (1952). *Phil. Trans. R. Soc. Lond.* **B327**, 37.
[2] Nicolis, G. & Prigogine, I. (1977). *Self-organization in Nonequilibrium Chemical Systems.* Wiley: New York.
[3] Meinhardt, H. (1982). In *Models of Biological Pattern Formation*, Ch. 3. Academic Press: New York.
[4] Babloyantz, A. (1986). *Molecules, Dynamics and Life.* Wiley: New York.
[5] Markus, M., Müller, S.C. & Nicolis, G. (eds.) (1988). *From Chemical to Biological Organization.* Springer-Verlag.
[6] Haken, H. (1977). *Synergetics, an Introduction.* Springer- Verlag.
[7] Micheau, J.C., Gimenez, M., Borckmans, P. & Dewel, G. (1983). *Nature* **305**, 43.
[8] Avnir, D. & Kagan, M. (1984). *Nature* **307**, 717.
[9] Boissonade, J. In *Dynamic and Stochastic Processes. Theory and Applications,* Lima, R., Streit, L. & Vilela-Mendes, R. (eds.). Lecture Notes in Physics, Springer, to appear.
[10] Nicolis, G., Erneux, T. & Herschkowitz-Kaufman, M. (1978). *Adv. Chem. Phys.* **88**, 263.
[11] Pearson, J.E. & Horsthemke, W. (1989). *J. Chem. Phys.* **90**, 1588.
[12] Rovinsky, A.B. (1987). *J. Phys. Chem.* **91**, 4606; Rovinsky, A.B., in this volume.
[13] Keener, J.P. (1980), *SIAM J. Appl. Math.* **39**, 528.
[14] Field, D. & Burger, M. (eds.) (1985). *Oscillations and Travelling Waves in Chemical Systems.* Wiley: New York.
[15] Tyson, J.J. & Keener, J.P. (1988). *Physica* **32D**, 327.
[16] Zykov, V.S. (1989). *Simulation of wave processes in excitable media.* Manchester University Press: Manchester.
[17] Tam, W.Y., Horsthemke, W., Noszticzius, Z. & Swinney, H.L. (1987). *J. Chem. Phys.* **88**, 3395.
[18] Noszticzius, Z., Horsthemke, W., McCormick, W.D., Swinney, H.L. & Tam, W.Y. (1987). *Nature* **329**, 619.
[19] Dulos, E., Boissonade, J. & De Kepper, P., in this volume.
[20] Herschkowitz-Kaufman, M. & Nicolis, G. (1972). *J. Chem. Phys.* **56**, 1890.
[21] Dewel, G., Walgraef, D. & Borkmans, P. (1987). *J. Chim. Physique (Paris)* **84**, 1335.
[22] Boissonade, J. (1988). *J. Physique (France)* **49**, 541.
[23] Ouyang, Q., Boissonade, J., Roux, J.C. & De Kepper, P. (1989). *Phys. Lett.* **A134**, 282.
[24] Tam, W.Y., Vastano, J.A., Swinney, H.L. & Horsthemke, W. (1988). *Phys. Rev. Lett.* **61**, 2163.

[25] DiPrima, C. & Swinney, H.L. (1985). In *Hydrodynamic Instabilities and Transition to Turbulence,* Swinney, H.L. & Gollub, J.P. (eds), Ch. 6. Springer.
[26] Tam, W.Y. & Swinney, H.L. (1987). *Phys. Rev.* **A36**, 1374.
[27] Ouyang, Q. (1989). *Thesis,* Bordeaux.
[28] De Kepper, P., Epstein, I., Kustin, K. & Orbán, M. (1982). *J. Phys. Chem.* **86**, 170; Dateo, C.E., Orbán, M., De Kepper, P. & Epstein, I.R. (1982). *J. Am. Chem. Soc.* **104**, 504.
[29] De Kepper, P., Ouyang, Q., Boissonade, J. & Roux, J.C. In *Dynamics of exotic phenomena in chemistry,* Beck, M. & Kórös, E. (eds.), to appear in *Reac. Kinet. Catal. Lett.* (Budapest).
[30] Castets, V. *Personal communication.*
[31] Arneodo, A. & Elezgaray, J. (1989). In *Spatial Inhomogeneities and Transient Behaviour in Chemical Kinetics,* Gray, P., Nicolis, G., Baras, F., Borkmans, P. & Scott, S.K. (eds.). Manchester University Press, in press.
[32] Elezgaray, J. & Arneodo, A. In *New Trends in Nonlinear Dynamics and Pattern Forming Phenomena: The Geometry of Nonequilibrium,* Coullet, P. & Huerre, P. (eds.). Plenum: New York, in press.
[33] Arneodo, A. & Elezgaray, J. (1990). *Phys. Lett.* **A143**, 25.
[34] Elezgaray, J. (1989). *Thesis,* Bordeaux.
[35] FitzHugh, R. (1961). *Biophys. J.* **1**, 445; Nagumo, J., Arimata, S. & Yoshizawa, S. (1962). *Proc. IRE* **25**, 2061.
[36] Eckhaus, W. (1983). *Lecture Notes in Mathematics* **985**, 449.
[37] Elezgaray, J. & Arneodo, A., to be published.
[38] Boukalouch, M., Elezgaray, J., Arneodo, A., Boissonade, J. & De Kepper, P. (1987). *J. Phys. Chem.* **91**, 5843.

7. SPATIAL PATTERN FORMATION IN A CATALYTIC SURFACE REACTION: THE FACETING OF Pt(110) IN CO + O_2

R. Imbihl

Fritz-Haber Institut der Max-Planck-Gesellschaft
Faradayweg 4-6, D-1000 Berlin 33, West Germany

1. Introduction

The term "dissipative structures" which was introduced by Prigogine [17] describes a broad class of non-equilibrium systems where a constant flow of energy and/or matter leads to structures ordered in space or time, e.g. causes kinetic oscillations or spatial pattern formation. Temporal oscillations and spatial pattern formation are closely related since the same kinetic instabilities which lead to a periodic variation of the system variables in time may also induce a periodic variation in space. Consequently one often observes spatio-temporal structures, but one may also consider the case of a non-equilibrium structure which is only periodic in space. Such structures which rarely have been observed in chemical reaction systems were first discussed by Turing [20] and have been termed accordingly as "Turing structures".

In this report the formation of periodic facet structures during catalytic CO oxidation on a Pt(110) single crystal surface is presented as an example for the observation of a Turing structure in heterogeneous catalysis A number of observations suggest that the reaction-induced faceting of Pt(110) is a dissipative structure, e.g. a structure which is only stable under non-equilibrium conditions. This interpretation is in particular supported by the fact that the same system also exhibits kinetic oscillations under conditions very close to that of the faceting process. Moreover, the mechanism of the kinetic oscillations is based on periodic structural transformations of the surface which modulate the oxygen sticking coefficient (the probability that a molecule which hits the surface sticks) and hence the catalytic activity [5,13]. Periodic structural changes can occur on Pt(110) via the reversible CO-induced 1x1 $\rightleftarrows$ 1x2 surface phase transition which is also associated with the mass transport of Pt atoms. Since faceting requires a massive mass transport of Pt atoms, one can assume that the coupling between kinetic instabilities of the reaction and the surface phase transition might lead to the formation of a periodic facet structure.

Based on these ideas a computer simulation of catalytic CO oxidation on Pt(110) was carried out. The results of the simulation reproduce the formation of periodic facet structures and almost all of the essential experimental features thus confirming the interpretation given above. Since the individual

Nonlinear Wave Processes in Excitable Media
Edited by A. V. Holden *et al.*, Plenum Press, New York

steps of the simulation describe the surface processes on an atomic scale, the formation of a non-equilibrium structure can be followed by a detailed microscopic mechanism which has been derived from experimental evidence. The experiments which led to the formation of the model as well as the results of the computer simulation are presented in this report.

2. Experimental methods and reaction conditions

All experiments were carried out in an all-metal standard ultrahigh vacuum (UHV) system equipped with facilities for low energy electron diffraction (LEED), Auger Electron Spectroscopy (AES) for checking the sample cleanliness and a differentially pumped mass spectrometer for measurements of the reaction rate. Under reaction conditions, which typically were in the 10^{-4} Torr range, gases were introduced via leak valves while the system was continuously pumped by a turbomolecular pump. The whole UHV chamber was thus operated as a gradient free flow reactor. The Pt(110) single crystal which had the shape of a small disc of ~1 mm thickness and about 0.5 cm^2 surface area was cleaned in UHV in numerous oxidation cycles followed by sputtering with Ar^+ ions and annealing up to 1000°C. The effect of contaminants and oxide formation could thus be completely suppressed [11].

Ordered structures on a single-crystal surface can be detected by LEED, which due to the strong interaction of electrons with matter, is mainly sensitive to the topmost layer. The formation of ordered structures ("superstructures") on the surface can be detected by the appearance of additional (fractional-order) beams besides the 1x1 pattern for the substrate lattice.

3. Structural changes in the system Pt(110)/CO + O_2

3.1. *The basic LH mechanism*

It has been well established that the catalytic CO oxidation follows the Langmuir-Hinshelwood (LH) mechanism which proceeds along the following steps [6]:

$$(1) \quad CO + * \rightleftarrows CO_{ad}$$
$$(2) \quad O_2 + 2* \longrightarrow 2\,O_{ad}$$
$$(3) \quad O_{ad} + CO_{ad} \longrightarrow CO_2 + 2*$$

* denotes a free adsorption site.

Both reactants first have to adsorb before they can react to form CO_2 which is immediately released into the gas phase. The kinetics which are derived from the LH mechanism exhibit two different branches associated with a high and a low reaction rate as depicted in a plot of the reaction rate r_{CO_2} vs. p_{CO} in Fig. 1. Multiple steady states exist in the transition region between the high and the low reaction rate branch giving rise to a clockwise hysteresis in the reaction rate. In addition to the LH mechanism the kinetics of CO_2 production on Pt surfaces have to include the experimental findings that a CO adlayer inhibits the adsorption of oxygen, while CO can still adsorb and react on an oxygen covered surface. Therefore, the reaction rate first increases linearly

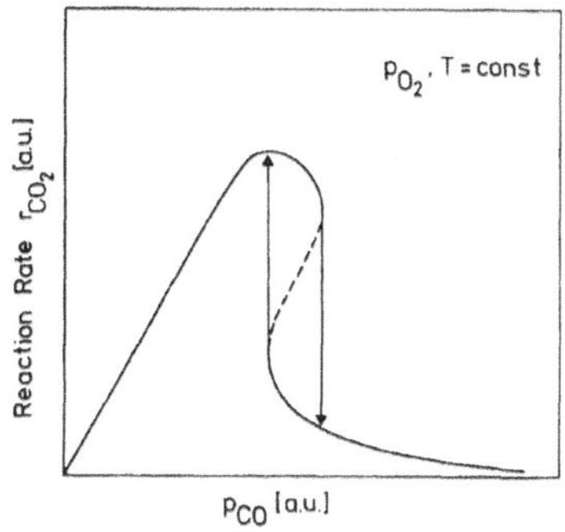

Figure 1. Schematic plot of the reaction rate r_{CO} vs. p_{CO} showing monostable and bistable regions in the kinetics of the Langmuir-Hinshelwood mechanism. Full lines indicate a stable steady state, while the dashed line represents an unstable steady state.

with p_{CO}, since the high reaction-rate branch is associated with an oxygen covered surface. As soon as more CO molecules are adsorbing from the gas phase than are consumed by the reaction, a CO adlayer forms, and the reaction rate drops to a low level.

Although the non-linearities which exist in the LH kinetics can predict multiple steady states and hence a hysteresis in the reaction rate, these are not strong enough to produce kinetic oscillations. These have to be created by an additional mechanism which causes a periodic switching of the reaction between the high and the low reaction-rate branch. On Pt surfaces under low pressure conditions this mechanism is provided by an adsorbate- induced surface phase transition as demonstrated in the next section with kinetic oscillations on Pt(110).

3.2. *Periodic structural changes*

The clean Pt(110) surface is not stable in its bulk-terminated 1x1 structure, but the surface reconstructs into a 1x2 structure where every second of the close-packed [1$\bar{1}$0] rows is missing [8,18]. Both surface structures are depicted in Fig. 2. One notices that the 1x2 "missing row" structure exhibits close-packed (111) microfacets on both sides of the [1$\bar{1}$0] ridges whose formation is presumably the driving force for the reconstruction. The 1x2

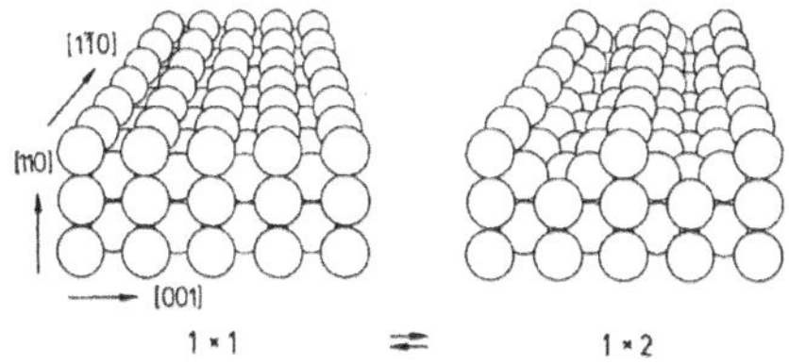

Figure 2. Structural model of the reconstructed and the non-reconstructed Pt(110) surface.

reconstruction is lifted upon adsorption of CO, and the Pt atoms revert into their original 1x1 positions. Since this transition is reversible, an adsorbate-induced 1x1 $\rightleftarrows$ 1x2 surface phase transition exists which is controlled by a critical CO coverage of ~0.2 [9,10,12]. The driving force for the lifting of the reconstruction by CO is simply the gain in adsorption energy, since E^{CO}_{ad} is substantially higher on the 1x1 than on the reconstructed surface. This has been demonstrated for the system Pt(100)/CO where a similar adsorbate-induced phase transition exists as on Pt(110) [1].

The change in the surface structure is also associated with a change in the adsorption properties. Since the oxygen-sticking coefficient s_{O_2} depends very strongly on the surface structure for Pt catalysts, this quantity is very strongly affected by the surface phase transition. In the transition region between the two branches of the LH kinetics (see Fig. 1) the rate of oxygen adsorption becomes rate-limiting for the catalytic CO oxidation, and the phase transition can therefore cause a switching between the two branches. On Pt(100) and Pt(110), s_{O_2} is higher on the 1x1 surface than on the reconstructed surface. The mechanism of kinetic oscillations based on a surface phase transition (PT) can be sketched as follows:

Starting with a CO-covered 1x1 surface, s_{O_2} and hence the reaction rate will be high. Due to the strong consumption of adsorbed CO by the surface reaction, Θ_{CO} will decrease. As Θ_{CO} decreases below the critical value for the PT, the surface will relax into its less active reconstructed configuration. With s_{O_2} being low, the rate of CO adsorption will exceed the reaction rate, and Θ_{CO} will rise again. Above the critical value for the PT the active 1x1 surface is established again, and a new cycle may start. An example which shows the oscillations in the reaction rate and the accompanying periodic structural changes on Pt(110) is displayed in Fig. 3.

Spatially resolving measurements demonstrated that Pt(100) and Pt(110) surfaces do not react homogeneously during kinetic oscillations, but one observes propagating reaction fronts of macroscopic dimensions [2,11,19]. These lead to a wave-like propagation of the structural changes during kinetic oscillations. The propagating reaction fronts arise in the reactive removal of the dense CO adlayer by oxygen as the result of the coupling between an autocatalytic step and the diffusion of adsorbed CO. The kinetic oscillations on Pt(110) and Pt(100) therefore represent spatio-temporal structures due to the formation of chemical waves in the reaction system. The absence of stationary spatial structures can be traced back to the very limited extent of structural changes during kinetic oscillations. In the simplified picture of kinetic oscillations used above, all structural transformations are periodic and they are restricted to the first layer of the substrate where the phase transition takes place. In the following it shall be shown that conditions also exist where the structural changes involve several layers of the substrate and lead to the formation of a stationary spatial pattern on the surface in a faceting process.

3.3. *The faceting of Pt(110) in CO + O_2*

3.3.1. *Conditions for faceting.* If one exposes a Pt(110) surface to a constant

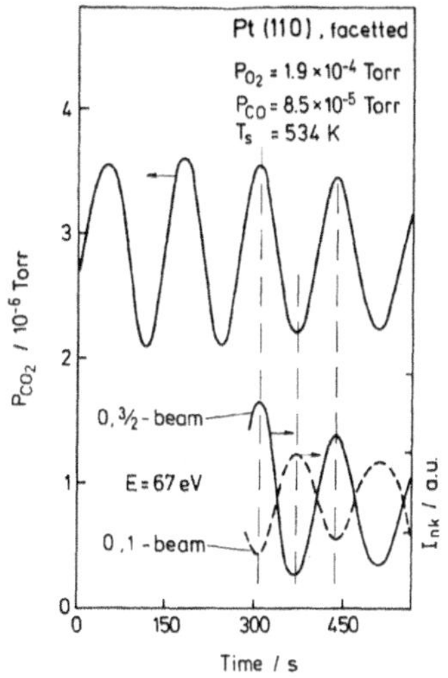

Figure 3. Periodic structural changes during kinetic oscillations on Pt(110) as monitored through the variation of the 0,3/2-beam of the 1x2 surface. (After ref. [16].)

flow of CO and O_2 under conditions close to kinetic oscillations, one observes in LEED a continuous structural transformation of the surface as demonstrated by a series of beam profiles shown in Fig. 4 [15]. The integral order beams

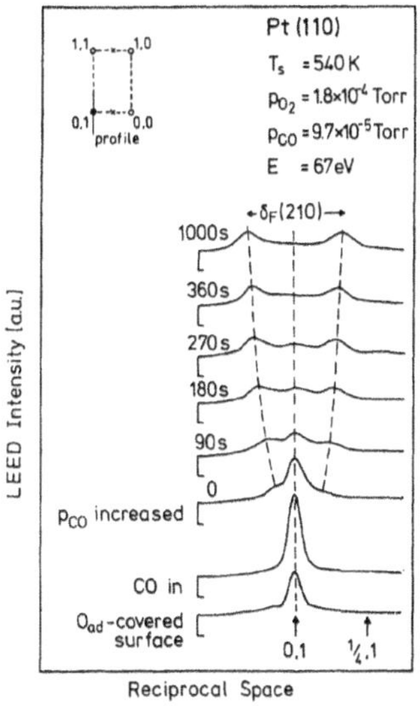

Figure 4. Development of faceting during the catalytic CO- oxidation on Pt(110). LEED beam profiles served to monitor the progress of faceting. (After ref. [15].)

Figure 5. Ball model showing a (210) facet in the rear section and a 1x2 "missing row" structure of Pt(110) in the front section.

first broaden and then split along the [1$\bar{1}$0] direction, and after a period of ~20 min a stationary state is reached. The analysis of the diffraction pattern reveals that facets with (210) orientation have been formed. A ball model of the (210) surface which is displayed in the rear section of Fig. 5 shows that the facet can be built up with two structural units by alternating (100) steps with unreconstructed (110) terrace units. By varying the number of (100) step units per (110) terrace units the inclination angle of the facet with respect to the (110) plane can be increased (decreased) which then corresponds to a larger (smaller) splitting of the integral order beams in LEED. The surface orientations which can be built up with this simple variational principle all belong to the [001] zone. Which one of the possible orientations is formed in the faceting process, depends on the reaction conditions, e.g. on p_{O_2}, p_{CO}, T, but in general, less steep facets than Pt(210) are obtained.

The faceting of a Pt single crystal surface is not an unusual phenomenon and has often been observed - usually with thermodynamically less stable high-index planes under the influence of a strong adsorbate such as oxygen. The driving force can be explained by equilibrium thermodynamics, as the faceting there either reduces the surface free energy or strengthens the bonding of the adsorbate to the substrate. However, a faceting of Pt(110) could not be observed with one of the gases alone, but only under reaction conditions when both gases are present. The lack of a thermodynamical driving force is also demonstrated by the thermal instability of the faceted surface. Heating up a faceted Pt(110) surface above T ~ 500 K to 600 K results in the removal of the facets and in the restoration of the flat surface. The thermal reordering process is also present under conditions where the faceting of Pt(110) takes place, since stopping the flow of the reactants also leads to a slow restoration of the flat surface. Accordingly the faceting of Pt(110) should be described as a dynamical equilibrium as depicted in Fig. 6, where the competition between reaction-induced growth and thermal annealing of facets determines the degree of faceting for a given set of p_{O_2}, p_{CO}, T-parameters. Since the growth rate for faceting is limited by the reaction rate and hence p_{O_2}, p_{CO}, while thermal annealing can increase very strongly

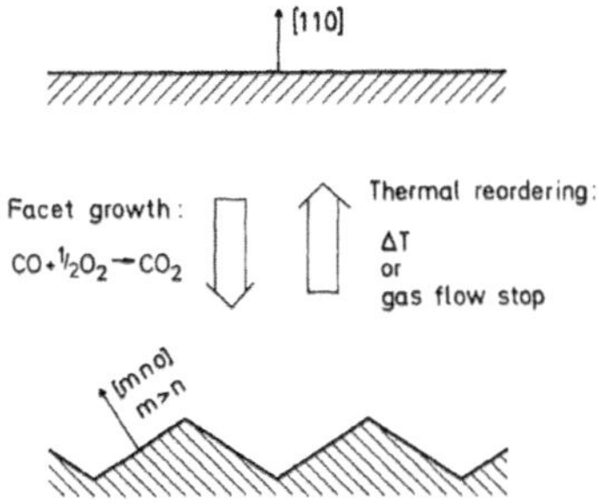

Figure 6. Scheme illustrating the dynamical equilibrium between reaction-induced facet growth and thermal restoration of the flat surface during the catalytic CO-oxidation on Pt(110).

with increasing temperature, a high temperature limit exists for the faceting of Pt(110). Beyond this temperature limit which is at 530 K for $p_{O_2} = 1.5 \times 10^{-4}$ Torr, the thermal reordering process becomes dominant and keeps the surface essentially flat [15].

The faceting of Pt(110) is restricted to a certain parameter range in the kinetics of the catalytic CO oxidation which are depicted in Fig. 7 in an r_{CO_2} vs. p_{CO} plot similar to Fig. 1. Faceting is observed only in the p_{CO} range beyond the rate maximum, where the transition from the high reaction-rate branch to the low reaction-rate branch takes place. Since on Pt(110) this transition is associated with the change from an oxygen-covered 1x2 surface to a CO-covered 1x1 surface, the 1x1 $\rightleftarrows$ 1x1 PT is always involved in the faceting process and evidently plays an essential role there.

The change in the surface structure by faceting is associated with an increase in catalytic activity which in the r_{CO_2} vs. p_{CO} plot of Fig. 7 shows up as a shift of the rate maximum towards a higher p_{CO} value. The faceting of Pt(110) can therefore be followed in situ simply by measuring the increase of the reaction rate at constant p_{CO}, as is indicated by the arrow in Fig. 7. Adsorption experiments on the faceted surface demonstrated that the increase in catalytic activity results from a higher oxygen- sticking coefficient due to the presence of steps on the faceted surface. One notices that the position of the new rate maximum in Fig. 7 belonging to the faceted surface is just right at the end of the arrow which marks the reaction rate after completion of the faceting process. Evidently the extent of the structural changes (for a given p_{O_2}, p_{CO}, T) is determined by an extremum principle such that the surface facets only to that extent that is necessary to make the reaction switch to the high rate branch.

3.3.2. *Formation of periodic facet structures.* For a realistic description of the faceted surface, quantitative data are needed about the dimensions of the

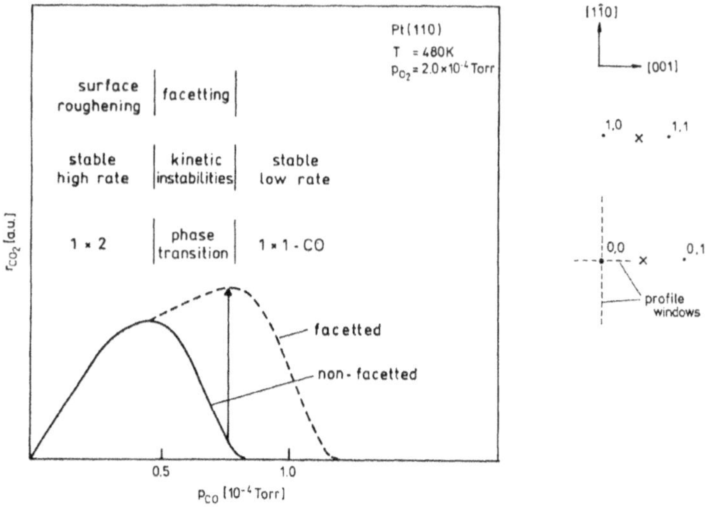

Figure 7. Relation between the conditions for faceting and the kinetics of the catalytic CO-oxidation on Pt(110). The full line indicates the rate curve for the non-faceted surface, while the dashed curve indicates the increase in catalytic activity after strong faceting of the surface. The different regions indicated on top of the rate curve all refer to the non-faceted Pt(110) surface. The insert shows a schematic of the LEED pattern with the 1x2 reconstruction. (After ref. [7].)

facets, the uniformity of size and orientation of the facets and the roughness of the surface. This kind of information is in principle available from an evaluation of the LEED beam profiles, since these are sensitive to the degree of order on the surface. Large ordered regions give rise to sharp beam profiles, while disorder causes a broadening of the spots. In the case of a conventional LEED instrument this information is, however, obscured by instrumental limitations, as the finite width of the primary electron beam eliminates details of the profiles. For this reason a LEED beam profile analysis of the reaction- induced substrate changes on Pt(110) was carried out with a high resolution instrument (SPALEED) by Falta et al. [7].

A series of profiles of the 0,0 beam in Fig. 8 which were recorded from a strongly faceted Pt(110) surface, displays very strong changes, as the primary electron energy E is varied. By changing E and thus the wavelength of the electron one moves from in-phase scattering conditions (S = 2.0, S being the scattering vector), where the different terrace levels exhibit constructive interference, to out-of-phase scattering conditions (S = 2.5), where the interference is destructive. The sharp profile at S = 2.0 is not sensitive to the surface roughness, since all three Laue conditions are fulfilled at this particular energy. The observation of 1st and 2nd order satellites in the fivefold split 0,0 beam at S = 2.10 indicates the formation of a periodic structure with considerable long-range order in the [1$\bar{1}$0] direction. From an evaluation of the profiles, which yields a (430) orientation of the facets and a regular spacing of 70 lattice units along the [1$\bar{1}$0] direction, the

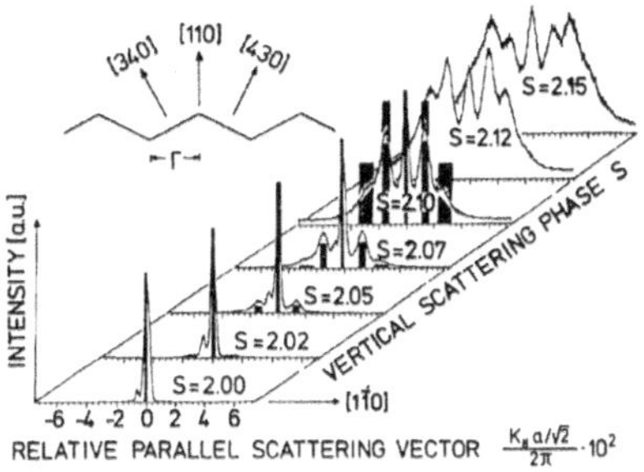

Figure 8. Spot profiles of the 0,0-beam after strong faceting of the surface for different scattering vectors $S(k_{\perp} \cdot d = 2\pi S;)$. The bars represent the intensity distribution calculated from the geometrical model displayed in the insert. (After ref. [7].)

sawtooth-like structure, depicted in the insert of Fig. 8, has been derived as a model for the faceted surface. Facets of uniform size and orientation form a regular structure whose spacing 2 Γ is equal to 70 lattice units in the [1$\bar{1}$0] direction. The consistency of this model can be tested by comparing the rms roughness of 2.8, which can be derived from the geometry of the model with the rms roughness obtained from an evaluation of the beam profiles. The latter yields an rms roughness of 3.0 which is in excellent agreement with the proposed model.

Under reaction conditions of the high-rate branch in Fig. 7, no faceting develops, but the detailed profile analysis gives an rms roughness of 2.6 similar to the rms roughness of the faceted surface. So the surface reaction alone already causes a roughening of the surface, but specific conditions are required to produce long-range ordering such that faceting occurs.

4. Computer simulation

4.1. *The surface model*

The conditions under which faceting occurs demonstrate that the 1x1 $\rightleftarrows$ 1x2 PT evidently is an essential step in the faceting mechanism. The role of the PT can be made plausible with a simple ball model of Pt(110), as is depicted in Fig. 5. If one wants to create a 1x2 structure, then starting with a 1x1 surface every second of the [1$\bar{1}$0] rows has to be removed. The extra atoms have to be put on top of the original layer thus opening up a new layer which starts with a (100) step. Repeating the 1x1 $\rightleftarrows$ 1x2 sequence several times produces a number of (100) steps which suffices to build up the small (210) facet shown in the rear section of Fig. 5. The PT can thus provide the mass transport of Pt atoms which is necessary for faceting via the different densities of the 1x1 and 1x2 structure. This mechanism alone would only cause a roughening of the surface, and an additional mechanism is required which directs the structural changes such that regular structures result.

The idea which is underlying the model presented here is that the coupling between kinetic instabilities and the adsorbate-driven PT renders the flat Pt(110) surface unstable against structural fluctuations. Faceting occurs accordingly as the result of a Turing instability in the system

Pt(110)/CO + O_2. Instabilities can arise very easily in the system Pt(110)/CO + O_2, since the coupling between the surface reaction and the substrate changes leads to a complex feedback behaviour of the kinetics. This is provided essentially by the properties of the CO-induced 1x1 $\rightleftarrows$ 1x2 PT which is controlled by critical CO coverages and thus depends on the reaction conditions. The phase transition, on the other hand, influences the surface reaction, since the mass transport of 50% of the surface atoms that is associated with the formation of the 1x2 "missing row" structure, necessarily creates steps. These exhibit a higher sticking coefficient s_{O_2} for oxygen adsorption than flat terraces and hence increase the reaction rate.

It is evident that the complexity of the real system which involves many degrees of freedom for the substrate and adsorbate changes had to be reduced enormously in order to keep the problem tractable. The Monte Carlo method was used for the simulation of the surface changes which were divided into an adsorbate part and a substrate part, and an adsorbate and substrate lattice, respectively. Changes in the adsorbate and in the substrate were carried out subsequently. Each adsorption/reaction cycle was followed by a substrate cycle in which the surface structure was changed according to the distribution of CO_{ad} on the surface and the rules of the PT. So the simulation tries to mimic the surface diffusion of Pt atoms under the influence of the catalytic CO oxidation similar to rearranging the surface structure in a ball model.

Modelling of the surface reaction which proceeds along the LH mechanism was carried out on a two-dimensional array of 200 x 5 sites with each site being either empty, CO-covered or oxygen- covered. The individual steps of the LH mechanism included adsorption and desorption of CO, dissociative adsorption of O_2, surface diffusion of adsorbed CO via site hopping and the reaction between adjacent CO_{ad} and O_{ad} to form CO_2 which desorbs immediately. Since the faceting as well as the self-diffusion of Pt takes place preferentially along the $[1\bar{1}0]$ direction, a one- dimensional array of sites was used for the simulation of the substrate changes which represents a sideview of the adsorbate lattice as shown in Fig. 9a. Each site in the substrate lattice therefore corresponds to a row of 5 Pt atoms whose status can be either α for a 1x1 row or β for a 1x2 row. The mapping of the real structure onto the substrate lattice is demonstrated with a ball model displayed in Fig. 9a. Note that the direction of the 1x2 reconstruction is perpendicular to the direction in which the faceting occurs. The status of each row is determined from the CO coverage Θ_{CO}^{j} averaged over the 5 sites of the corresponding row j and the two critical CO coverages for the PT in both directions.

The choice of only a small number of sites over which the CO coverage was averaged was based on the results of a recent investigation by Scanning Tunnelling Microscopy (STM) which clearly demonstrated the local nature of the 1x1 $\rightleftarrows$ 1x2 PT [9]. An STM picture of the Pt(110)-1x2 surface after partial lifting of the 1x2 reconstruction by adsorbing some CO is reproduced in Fig. 10. This picture clearly shows that the removal of the 1x2 reconstruction proceeds in a very localized process just by moving Pt atoms over small distances of only a few lattice units.

The way the PT was treated in the one-dimensional model used here is demonstrated in Fig. 9b. If an α row was to be converted into a β row ($\Theta_{CO}^{j} < \Theta_{crit1}$), an additional β row had to be created in order to compensate for the different density of Pt atoms which was then placed adjacent to the original

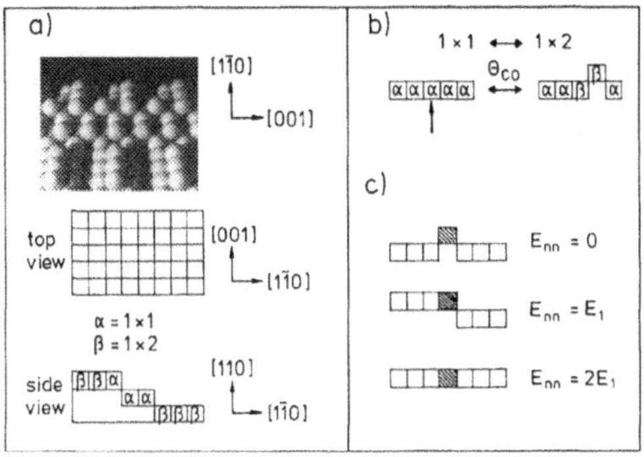

Figure 9. Surface model used for the simulation of reaction-induced substrate changes on Pt(110). (After ref. [14].)
a) Ball model of a faceted (110) surface and the corresponding two model surfaces used for the simulation of changes in the adsorbate and substrate.
b) Modelling of the 1x1 $\rightleftarrows$ 1x2 phase transition using the one-dimensional surface model shown in a).
c) Interaction energy of a row for three possible row configurations in the one-dimensional substrate model.

row as indicated in Fig. 9b. This way, steps could be either created or annihilated by the PT according to the rule of mass conservation. Double steps were only permitted as structural intermediates to be dissolved in the subsequent cycle. Besides the PT, changes in the substrate were also allowed to take place via thermally activated site hopping of rows. This mechanism describes the thermal annealing process which is apparently governed by equilibrium thermodynamics, as it restores the flat (110) surface if left unbalanced. Therefore, an attractive interaction between the rows was introduced which was described by a single nearest-neighbour interaction energy E_1 as shown in Fig. 9c.

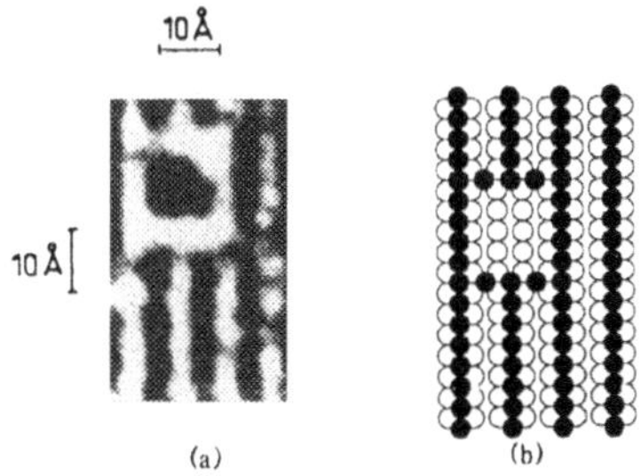

Figure 10. STM image of the Pt(110)-1x2 surface after partial lifting of the reconstruction by adsorbed CO.
a) STM image.
b) Ball model corresponding to a) (after ref. [9]).

The jump probability W_j was then calculated from the change in the interaction energy ΔE_{nn} according to the algorithm of Metropolis:

$$W_j = 1, \quad \text{if } \Delta E_{nn} \leq 0$$

$$W_j = \exp(-\Delta E_{nn}/kT), \quad \text{if } \Delta E_{nn} > 0.$$

The process just described will lead to a thermal reordering of a rough surface towards a flat surface as long as no other forces are present and as long as $E_1 > kT$.

A flowchart of the simulation program is displayed in Fig. 11. Each adsorption/reaction cycle is followed by NTHERM substrate cycles where the PT and thermal reordering takes place. The value for NTHERM takes care of the different time scales of the adsorption/reaction process and the PT which for $p_{O_2} = 1 \times 10^{-4}$ Torr and T = 480 K were estimated to be 0.1 s and 0.01 s, respectively. The time scale for the adsorption/reaction process is determined by the adsorption rate which is ~ 10 monolayers/s at 10^{-4} Torr, while the time scale of the PT has been estimated on the basis of experimental data [9]. Before the next adsorption/ reaction cycle was started, the oxygen-sticking coefficients $s_{O_2}{}^j$ of each row j were updated in order to account for the structural changes of the substrate. A higher s_{O_2} was assigned to the bottom row of each step ($s_{O_2,step} = 1$), while all other rows were treated as terrace rows with a lower s_{O_2} ($s_{O_2,terrace} = 0.25$). Periodic boundary conditions (PBC) were applied for the adsorbate as well as the substrate lattice.

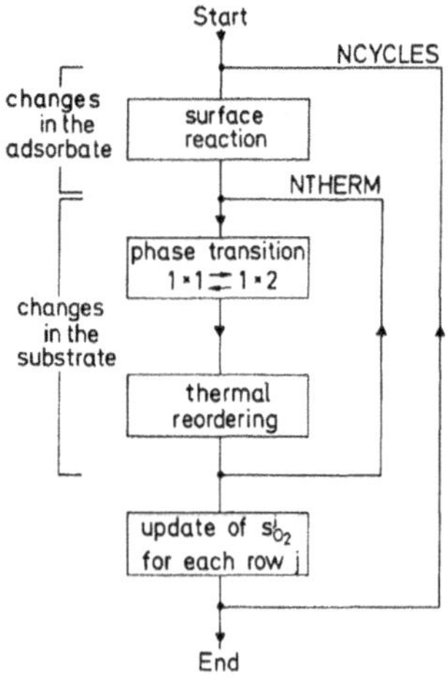

Figure 11. Flowchart for the simulation program.

4.2. *Results*

If the surface is exposed to a constant flow of CO and O_2, the initially flat surface becomes roughened as local variations in the CO coverage caused by the surface reaction initiate local structural changes and hence the mass transport of Pt atoms. If one selects conditions where Θ_{CO} fluctuates around the critical coverages for the PT, one observes the formation of regular structures whose development is depicted in Fig. 12. A stationary state is reached after ~4000 cycles, beyond which the structure changes only marginally (tested up to 8000 cycles). Equally spaced facets of roughly uniform size form a periodic structure with a similar sawtooth-like shape as observed in the experiment (see Fig. 8). The lateral periodicity λ_f of ~40 lattice units allows for an integer number of modes similar to the eigenmodes of a string. Simulation runs carried out with varying lattice size demonstrated that λ_f is almost independent of the number of rows n_r, as long as $n_r >> \lambda_f$ is fulfilled. No significant dependence of the final structure on the initial conditions was detected, even if an arbitrarily faceted surface was taken as the starting point.

The adsorbate coverages and the 1x2 reconstruction averaged over the whole surface both exhibit strong fluctuations during the development of faceting as demonstrated in Fig. 13. The rms roughness which can serve as a measure for the degree of faceting increases continuously up to a value of 2.8 which is reached after 4000 cycles. At this point, the gas flow was stopped in the computer simulation in order to demonstrate that the surface reaction itself is the only driving force for the faceting of the surface. The thermal

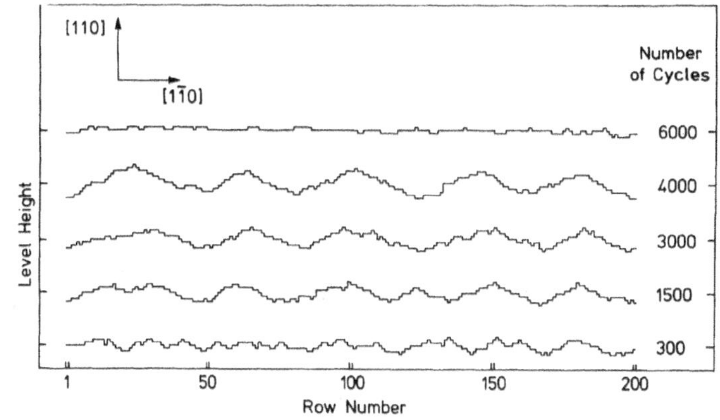

Figure 12. Development of a regular facet structure during the catalytic CO oxidation followed by a restoration of the flat surface after stopping the gas flow at 4000 cycles. The following probabilities P were used: $P_{ads,terr}(O_2) = 0.25$, $P_{ads,step}(O_2) = 1.0$, $P_{ads}(CO) = 0.4$, $P_{des}(CO) = 0.04$, $P_{diff}(CO) = 1.0$, $P_{react} = 1.0$, $E_1 = 3$ kcal/mol, T = 500 K, $\Theta^{CO}_{crit1} = 0.3$ (1x1 → 1x2), $\Theta^{CO}_{crit2} = 0.5$ (1x2 → 1x1), NTHERM = 20. (After ref. [14].)

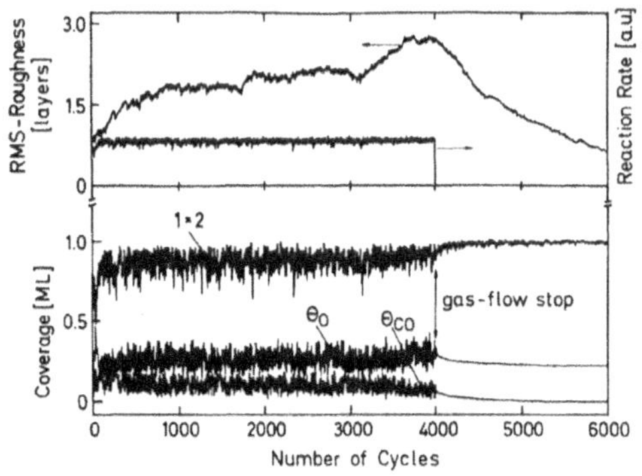

Figure 13. Variation of the adsorbate coverages, 1x2 reconstruction, reaction rate and rms roughness during the development of the faceted surface shown in Fig. 12. (After ref. [14].)

reordering process being dominant then alone leads to a slow restoration of the flat Pt(110) surface as evidenced by the decrease of the rms roughness in Fig. 13 and the flattening of the substrate profile in Fig. 12. This result shows that the flat Pt(110) surface is in fact the thermodynamically stable state in the absence of dissipative forces.

The role of the surface reaction in the faceting process can best be rationalized in connection with the kinetics of CO_2 formation which are displayed in a r_{CO_2} vs. p_{CO} plot in Fig. 14b. The reaction rate exhibits the same behaviour that has also been observed in the experiment (see Fig. 7) and which can be traced back to the LH mechanism over which the catalytic CO oxidation proceeds. Kinetic instabilities exist in the transition region between the high and the low reaction rate branch, where also the 1x1 $\rightleftarrows$ 1x2 PT proceeds. Under appropriate conditions hysteresis effect and/or kinetic oscillations can be observed in this region. The same region is also important for the faceting process, as one can see if one plots the rms roughness which is obtained after 2000 cycles together with the reaction rate vs. p_{CO} in Fig. 14a. The rms roughness is strongly peaked around the region of kinetic instability, whereas the surface remains essentially flat in the stable regions of the kinetics. Faceting is therefore limited to the region beyond the rate maximum, where also the 1x1 $\rightleftarrows$ 1x2 PT takes place, as has been observed in the experiment. The plateau of constant rms roughness on both sides of the faceting region has been identified as thermal disorder which is independent of the reaction conditions. One notices that the rate maximum of the faceted surface in Fig. 14b is shifted towards a higher p_{CO} value as compared to the flat surface which reflects the increase in catalytic activity due to the presence of steps.

The above results demonstrate that the model reproduces all essential features of the experiment, e.g. the growth conditions for facets, the increase in catalytic activity due to faceting and the thermal instability of the facets. The most interesting result, however, is clearly the reproduction of the periodic facet structures which before had been observed in the experiment. The comparison of the experimental values of 70 lattice units for

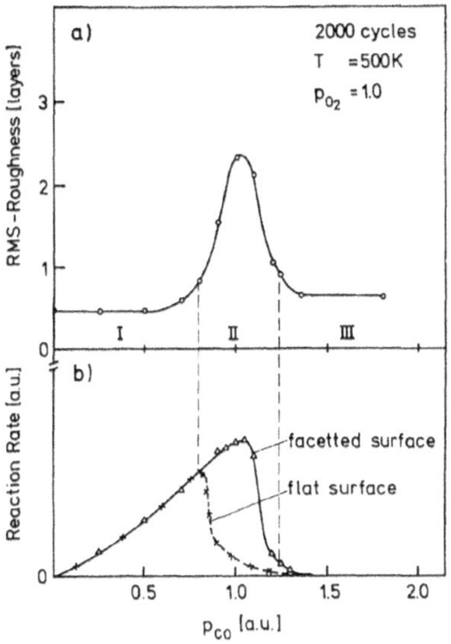

Figure 14. Dependence of the faceting on the reaction conditions. The circles in a) denote the rms roughness after exposing the surface to reaction conditions for 2000 cycles, while the triangles in b) show the reaction rate which is obtained after 2000 cycles. The adsorption probabilities P_{ads} are connected to the partial pressure p via $P_{ads}(CO,O_2) = p_{CO,O_2} \cdot s(O_2,CO)$. (After ref. [14].)

λ_f and 3.0 for the rms roughness with the results of the simulation which yields 40 lattice units for λ_f and 2.8 for the rms roughness demonstrates almost quantitative agreement. Even the time scale for the faceting of Pt(110) agrees very well with the experiment, since 4000 cycles correspond to ~400 s in real time - about the same time which is needed in the experiment for the development of a faceted surface. So although it is evident that the real microscopic mechanism will be very much different from the one implemented here, the basic assumptions of the model appear to be correct.

The mechanism for faceting is clearly based on the large CO coverage fluctuations which occur in the instability region of the reaction and which cause the mass transport of Pt atoms each time the surface is forced to undergo a phase transition. This effect alone, however, would just cause a roughening of the surface similar to thermal roughening without introducing long-range order. The formation of large facets can be understood by considering the flow of Pt atoms on a partially faceted surface as depicted in Fig. 15. If the facet size is to increase the outward flow J_{out} has to be larger than the inward flow J_{in} which is accomplished in the following way. Since steps exhibit a higher s_{O_2}, the reaction kinetics on the facet will correspond to the stable region I in Fig. 14, while the flat part of the

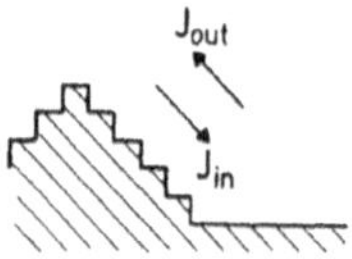

Figure 15. Flow of Pt atoms on a partially faceted surface.

surface is still in the region II of kinetic instability. Consequently CO coverage fluctuations will destabilize the flat (110) terrace and cause J_{out} to be higher than J_{in} thus increasing the size of the facet.

The consistency of the proposed mechanism can be tested by using the same s_{O_2} for step and terrace sites. The surface reaction then causes a mere roughening of the surface without the development of facets in agreement with the proposed mechanism. Increasing the temperature beyond ~500 K also inhibited the formation of facets and led to a roughening of the surface. At this temperature the thermal mobility of the rows of Pt atoms becomes high enough to dominate over facet growth such that the surface remains essentially flat. The model reproduces quite well the high temperature limit of 530 K which has been determined experimentally for faceting at $p_{O_2} = 1.5 \times 10^{-4}$ Torr.

Due to the complexity of the system, the influence of only a few parameters on the development of periodic facet structures has been investigated systematically. Since a mass transport of Pt atoms is required for faceting, the lateral periodicity λ_f depends on the parameters which control the mobility of the rows of Pt atoms. Increasing NTHERM or increasing the temperature were both found to lead to a larger lateral periodicity in agreement with the general properties of a reaction-diffusion system. The structures which are obtained are mostly stationary spatial patterns with some amount of structural fluctuations. The results of the simulation therefore confirm the interpretation of the faceted Pt(110) surface as a non-equilibrium structure of the Turing type.

5. Conclusions

An example has been presented where a heterogeneously catalyzed reaction causes the formation of a stationary spatial pattern on the surface of the catalyst. This has been observed in the system Pt(110)/CO + O_2, where the catalytic CO oxidation induced a regular facet pattern characterized by a lateral periodicity of the order of ~100 Å. Spatial pattern formation arises here via a mechanism which is very similar to that of kinetic oscillations in the same system, e.g. a coupling between structural changes and changes in the catalytic activity. Under conditions where the system exhibits oscillatory behaviour the structural changes are necessarily periodic and limited to the topmost layer of the substrate, but a change in the conditions may also provide a situation where the flat surface is no longer stable. The restructuring of the surface can then go several layers deep and faceting occurs as the result of a Turing instability. Computer simulations confirmed this interpretation and moreover they allow one to follow the formation of a dissipative structure on an atomic level. Thus the faceting of Pt(110) in CO + O_2 represents an example where spatial pattern formation can be understood by a detailed microscopic picture.

Acknowledgements

The author would like to thank S. Wasle for the preparation of the drawings as well as M. Reimers and I. Reinhardt for typing the manuscript.

References

[1] Behm, R.J., Thiel, P.A., Norton, P.R. & Ertl, G. (1983). *J. Chem. Phys.* **78**, I7437; II7448.
[2] Cox, M.P., Ertl, G. & Imbihl, R. (1985). *Phys. Rev. Lett.* **54**, 1725.
[3] Eiswirth, M. & Ertl, G. (1986). *Surf. Sci.* **177**, 90.
[4] Eiswirth, M., Möller, P., Wetzl, K., Imbihl, R. & Ertl, G. (1989). *J. Chem. Phys.* **90**, 510.
[5] Eiswirth, M., Krischer, K. & Ertl, G., submitted to *Appl. Phys. A.*
[6] Engel, T. & Ertl, G. (1979). *Adv. Catal.* **28**, 1.
[7] Falta, J., Imbihl, R. & Henzler, M., submitted to *Phys. Rev. Lett.*
[8] Fery, P., Moritz, W. & Wolf, D. (1988). *Phys. Rev.* **B38**, 7275.
[9] Gritsch, T., Coulman, D., Behm, R.J. & Ertl, G. (1989). *Phys. Rev. Lett.* **63**, 1086.
[10] Hofmann, P., Bare, S.R. & King, D.A. (1982). *Surf. Sci.* **117**, 245.
[11] Imbihl, R., Cox, M.P. & Ertl, G. (1986). *J. Chem. Phys.* **84**, 3519.
[12] Imbihl, R., Ladas, S. & Ertl, G. (2988). *Surf. Sci.* **206**, L903.
[13] Imbihl, R. (1989). In *Optimal Structures in Heterogeneous Reaction Systems*, Plath, P.J. (ed.), Springer Series in Synergetics. Springer: Berlin.
[14] Imbihl, R., Reynolds, A.E. & Kaletta, D., in preparation.
[15] Ladas, S., Imbihl, R. & Ertl, G. (1988). *Surf. Sci.* **197**, 153.
[16] Ladas, S., Imbihl, R. & Ertl, G. (1988). *Surf. Sci.* **198**, 42.
[17] Nicolis, G. & Prigogine, I. (1977). *Self-Organization in Nonequilibrium Systems.* Wiley: New York.
[18] Niehus, H. (1984). *Surf. Sci.* **145**, 407.
[19] a) Rotermund, H.H., Jakubith, S., von Oertzen, A. & Ertl, G. (1989). *J. Chem. Phys.* **91**, 4942.
b) Rotermund, H.H., Engel, W., Kordesch, M. & Ertl, G. (1990). *Nature* **343**, 355.
[20] Turing, A.M. (1952). *Trans. Roy. Soc. Lond.* **B237**, 37.

8. EXCITABLE CHEMICAL REACTION SYSTEMS

M. Marek, M. Dolnik, I. Schreiber and J. Finkeova

Department of Chemical Engineering
Prague Institute of Chemical Technology
Suchbatarova 5, 16628 Prague 6
Czechoslovakia

1. Introduction

Well defined and thoroughly studied chemical reaction systems such as the Belousov-Zhabotinskii (BZ) reaction [4] often enable a comparison of experimental results with the results of modelling both in oscillatory and excitable systems [2,5,8]. In this paper we make an attempt to demonstrate that properties of distributed forced excitable BZ systems can be understood on the basis of the response of the corresponding lumped parameter system (CSTR) to a single or periodic pulse forcing. We introduce an experimental technique for the construction of the phase excitation curve (PEC), based on double pulse experiments and demonstrate its use for the modelling and interpretation of the dynamical response of excitable systems to external forcing.

2. Experiments

2.1. *Experiments on a Petri dish*

A shallow layer of liquid reaction medium placed in a Petri dish forms a simple model of a spatially quasi-two-dimensional system. Various wave patterns can be supported by an excitable medium; the circular wave is the most common one. Controlled initiation of such waves by an immersed silver wire in a Petri dish was studied by Showalter et al. [11] and by Marek et al. [8].

A modified Belousov-Zhabotinskii reaction mixture (composition 0.286 M $HBrO_3$, 0.007 M KBr, 0.05 M $CH_2(COOH)_2$ and 0.00375 M ferroin) was used in experiments performed in a thermostated (15°C) shallow layer of liquid. The height of the layer was 1 mm. Circular waves were initiated by a periodically switched voltage applied between the silver and platinum electrodes immersed in the reaction mixture. When the silver electrode was negatively biased (-1.7 V) no generation of waves was observed while a positive bias (1.9 V) could cause the generation of travelling concentric circular waves. The positive voltage difference elicits the dissolution of silver ions into the reaction mixture, silver ions react with bromide ions and the resulting decrease of bromide concentration below a critical level causes the wave generation. The overall forcing period T consisted of two unequal time periods, $T = LP^- + LP^+$, the time of the positive bias LP^+ was equal to 90 ms.

The forcing period T was chosen as a variable parameter. Successive time instants T_k, k = 0, 1, 2, ..., at which waves passed through a fixed location from the end of the silver electrode and time intervals $\delta T_k = T_k - T_{k-1}$, k = 1, 2, ..., between passages of two successive waves were recorded. The number of generated waves r and the number of applied impulses s in the course of an experimental run were counted. The firing number ν defined as an average number of waves generated per one forcing period was indicated by the periodicity of the sequence $\{\delta T_k\}$. In this case, ν is equal to p/q, where p is the period of $\{\delta T_k\}$ and q is the number of pulse perturbations necessary to create p waves.

The dependence of the firing number on the forcing period is shown in Fig. 1. The dependence has a typical stepwise character where the steps correspond to individual resonances. Both periodic and aperiodic patterns of wave trains were observed.

2.2. *Experiments in a tubular reactor*

A tubular reactor with periodic stimulation at one end forms an essentially one-dimensional system [2]. A thermostated (20°C) tubular reactor was used in experiments with modified BZ reaction mixture (0.250 M $HBrO_3$, 0.05 M $CH_2(COOH)_2$, 0.007 M KBr, 0.00375 M ferroin). The methodology of experiments is quite analogous to that of the experiment in the Petri dish. The waves were initiated repetitively at the left end of the reactor by periodically switching between a negative and a positive voltage applied to the silver/platinum pair of electrodes immersed in the reaction mixture. No generation of waves occurred when the silver electrode was negatively biased (-1 V), while a travelling pulse wave could be found when the silver electrode was positively biased (+2 V). The time of the positive bias LP^+ was set to 3 s. The waves were observed as thin blue strips moving to the right from the silver electrode.

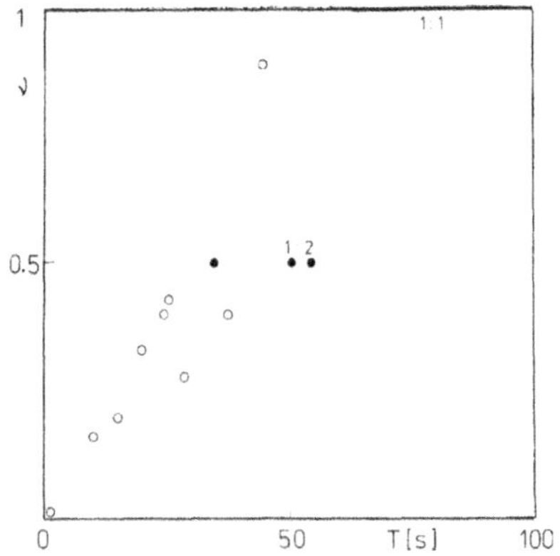

Figure 1. The experimental dependence of the firing number ν on the forcing period T measured on a Petri dish. The full circles denote periodic regimes and the empty ones irregular regimes or regimes with a higher order resonance (q > 2).

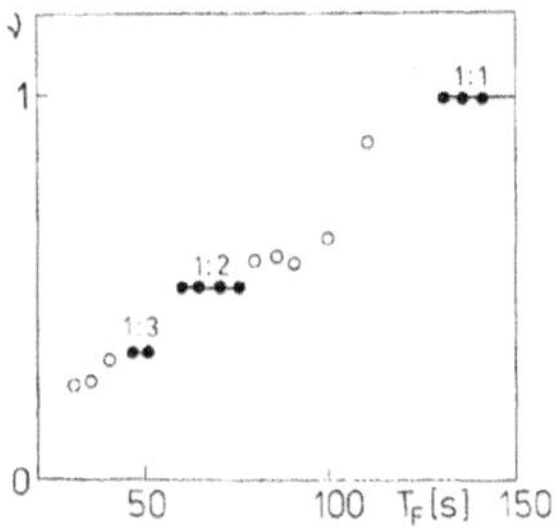

Figure 2. The experimental dependence of the firing number ν on the forcing period T measured in a tubular reactor. The full circles denote periodic regimes and the empty ones irregular regimes or regimes with a higher order resonance ($q > 3$).

The firing number was evaluated in the same way as in section 2.1. The dependence of the firing number ν on the forcing period T is shown in Fig. 2. We can again observe a typical stepwise dependence similar to the devil's staircase function [1] with evident plateaus of 1/3, 1/2 and 1/1 resonances.

2.3. *Experiments in CSTR*

2.3.a. *Periodic forcing.* The well stirred flow-through continuous reactor (CSTR) is a lumped parameter model system most often used in studies of nonlinear chemical dynamics of multiple steady states, periodic and aperiodic oscillations, and studies of forced oscillatory systems [1,3,9] but it was not so far used for the investigation of excitable systems. Generally, excitable systems ("excitators") have no apparent natural rhythm similar to the autonomously oscillating systems ("oscillators"); however, when the excitator is periodically forced it is possible to find dynamical regimes similar to those which are observed in forced oscillators. Similarly as in the case of autonomous oscillators we are then interested in the classification and characterization of dynamical regimes resulting from variations of the forcing amplitude and period.

A slightly modified type of the apparatus described in a different paper [5] was used in the present experiments. The reactor was made from organic glass and its reaction volume was 45 ml. A specially designed stirrer of the Rushton type was used, the stirring intensity was 700 rpm. The residence time was 39 minutes and the temperature was kept constant at 20°C. Experiments were performed with the same type of BZ reaction mixture as the experiments in the Petri dish and the tubular reactor [2,8]. The following input concentrations of reactants were used in CSTR experiments: 0.2 M $HBrO_3$, 0.05 M $CH_2(COOH)_2$, 0.007 M KBr and 0.00375 M ferroin. The reactants were pumped into the reactor separately by four input channels of calibrated pumps.

The concentration of Br^- and the redox potential (proportional to the Fe^{3+}/Fe^{2+}) were measured by a bromide selective electrode and a platinum wire electrode with a standard calomel electrode with a salt bridge as a reference one, respectively. The perturbation was realized by a fast addition of controlled constant volume of $AgNO_3$ solution to the reacting medium. The volume of pulse additions was held constant at 0.5 ml and the concentration of added $AgNO_3$ was varied from 0.01 to 0.0225 M; hence the amount of $AgNO_3$ added changed from 5×10^{-6} to 1.125×10^{-5} mol.

An excitable steady state characterized by a constant time course of the Br^- concentration and of the redox potential was established in the CSTR after a transient period. Then the $AgNO_3$ solution was periodically added with the period T. The observed dynamical regimes were qualified by the firing number similarly as in the experiments with distributed systems. Hence r, s, p, q have the same meaning as before but instead of travelling waves we take a rapid increase of redox potential as an excitation event. Most often periodic regimes of 1/q-type were observed.

A typical dynamical behaviour showing periodic regime with the firing number $\nu = 1/5$ is depicted in Fig. 3. The time dependence of both measured electrode potentials is given in Figs. 3a,b and the corresponding phase portrait is shown in Fig. 3c. The comparison of Figs. 3a and 3b illustrates well the all or none character of the response of the excitable system. The signal corresponding to the redox potential (Fig. 3a) fires only at each fifth perturbation. This can also be clearly seen in the corresponding phase portrait (Fig. 3c).

The dependence of the firing number on the forcing period T ("excitation diagram") is shown in Fig. 4. The devil's staircase-like structure of the plot

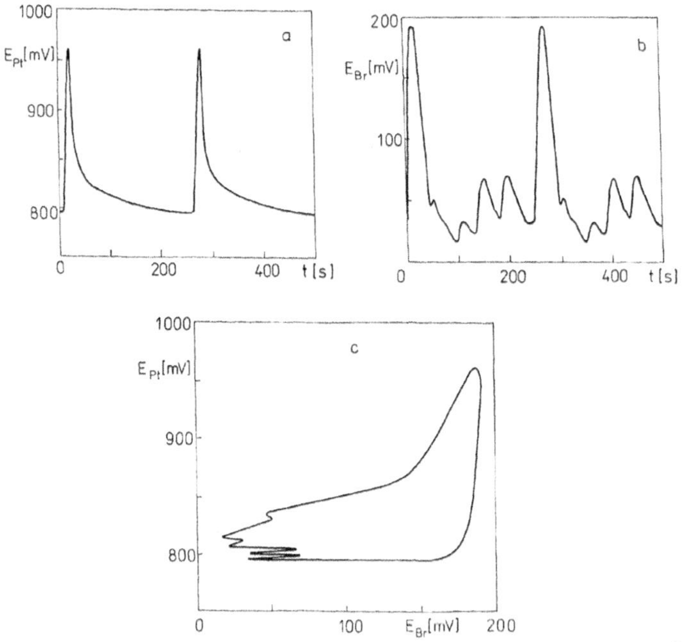

Figure 3. Dynamic regime 1/5 measured in a CSTR, $A = 1.0 \times 10^{-5}$ mol $AgNO_3$, T = 60 s.

a) Time evolution of the redox potential.
b) Time evolution of the bromide selective electrode potential.
c) Planar phase portrait.

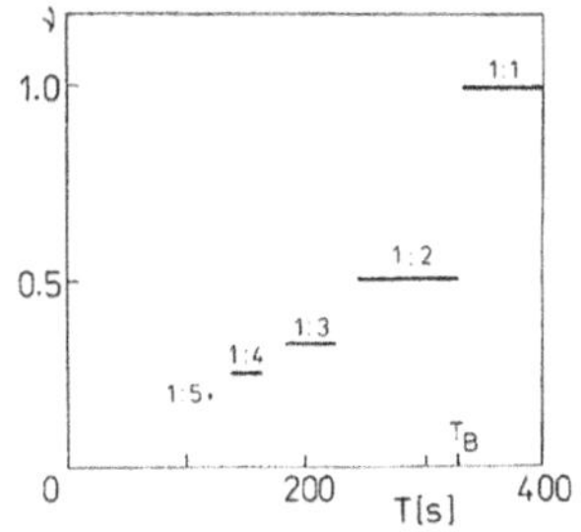

Figure 4. "Excitation diagram" obtained from CSTR experiments. The dependence of the firing number ν on the forcing period T for the constant amplitude of pulse A = 5.0×10^{-6} mol $AgNO_3$.

is evident. Only periodic regimes were observed for the amplitude of perturbation 5×10^{-6} mol. Both periodic and aperiodic regimes occur for the perturbation amplitude 1.125×10^{-5} mol [5].

2.3.b. *Phase excitation curves.* The periodically oscillating BZ reaction in a CSTR exhibits positive or negative phase shifts of oscillations when a single pulse perturbation by a bromide or a silver ion is used [3]. By plotting old (φ) and new (φ') phases we obtain a "phase transition curve" (PTC). The shape of the PTC varies with the amplitude. When the single pulse affects only actual "period" (which is often the case in strongly relaxational systems) the results of periodic forcing can be modelled by a repeated iteration of the corresponding PTC [3]. In principle it is also possible to determine the "phase excitation curve" (PEC) by applying the pulse experiments and use it for the prediction of the behaviour of the excitable medium when exposed to continuous forcing.

The experimental evaluation of the PEC was in this case based on the "method of two subsequent pulses", i.e. on the method based on two additions of $AgNO_3$ of a chosen amplitude and with varying time delay between the pulses T_o. The excitation diagrams firing number-forcing period showing the location of the 1/1 firing number were used to determine conditions of the perturbation. The first pulse elicits an excitation cycle, the second pulse applied with a varying time delay (at different values of the phase of the excitation cycle) then causes a phase shift of the excitation cycle (i.e. a lengthening or a shortening of the actual refractory period, i.e. of the time when the system regains excitability). We have chosen as a basic refractory period the lowest period T_B for which an excitation with the firing number 1/1 is still possible (determined on the basis of the corresponding excitation diagram).

The PEC is then defined by means of the relation between the old phase $\varphi = T_o/T_B$ and the new phase $\varphi' = 1 + \varphi - T_R/T_B$; here T_R denotes the actual refractory period after the pulse addition (i.e. the shortest time elapsed from the first pulse when the system becomes again excitable). The phase shift depends both on T_o and on the amplitude of the pulse. The method of two subsequent pulses can be used in situations where the length of the refractory period can be determined directly, e.g. when the excitable system reaches after the perturbation the state defined by a certain value of the bromide ion concentration, c_{Br}, or of the redox potential. Otherwise the value of T_R has to be determined indirectly, by testing the recovery of the excitability by means of an application of a third, testing pulse [5]. The experimentally determined PEC is depicted in Fig. 5. We can observe that the perturbation by $AgNO_3$ causes an increase of the refractory period for low values of φ and a decrease of the refractory period for high values of φ.

3. **Modelling**

3.1. *Spatially one-dimensional model*

The main qualitative features which combined together lead to wave patterns observed in experiments on the Petri dish and in the capillary are excitability and diffusion. Comparing the results with experiments in the CSTR it seems that excitability itself is the more determining active dynamical property. This is reflected in such a quantity as the firing number while the diffusion determines the actual shape of the wave trains. In order to support this hypothesis we have performed numerical experiments based on the model of a spatially one-dimensional reaction- diffusion system and its spatial discretizations corresponding to a chain of coupled CSTR's. We are interested in a qualitative interpretation of experimental observations and therefore we do not use a kinetic model of the BZ reaction but instead we take another excitable system - the SH model [10]. This choice enables us to draw qualitative conclusions without suffering numerical difficulties with the stiffness of the BZ reaction.

Let us consider the following two-component one-dimensional reaction diffusion system,

$$\begin{aligned} \frac{\partial X}{\partial t} &= D_X \frac{\partial^2 X}{\partial z^2} + \frac{\nu_o + X^\gamma}{1 + X^\gamma} - X\,(1 + Y), \\ \frac{\partial Y}{\partial t} &= D_Y \frac{\partial^2 Y}{\partial z^2} + X\,(\beta + Y) - \delta Y\,, \quad 0 < z < L \end{aligned} \tag{1}$$

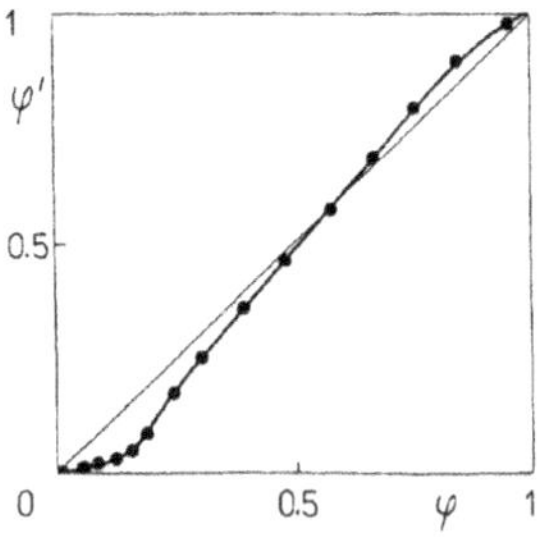

Figure 5. Phase excitation curve, $A = 5.0 \times 10^{-6}$ mol $AgNO_3$.

where z is the spatial coordinate and α, δ, $\nu_o > 0$, $\gamma, \beta > 1$ are kinetic parameters. We set $\alpha = 12$, $\beta = 1.5$, $\gamma = 3$, $\delta = 1$, $\nu_o = 0.01$, $D_X = 0.008$, $D_Y = 0.004$ and $L = 2.5$ for numerical computations. We suppose zero flux boundary conditions for the component Y at z=0, z=L and for X at z=L,

$$\frac{\partial Y(0,t)}{\partial z} = \frac{\partial Y(L,t)}{\partial z} = \frac{\partial X(L,t)}{\partial z} = 0. \tag{2}$$

The periodic forcing is introduced via mixed boundary conditions for X at z=0,

$$(1 - p(t)) \frac{\partial X(0,t)}{\partial z} + p(t)\ (A - X(0,t)) = 0. \tag{3}$$

The parameter A can be interpreted as the forcing amplitude and p(t) is a T-periodic piecewise constant function,

$$p(t) = \begin{cases} 1 \text{ for } 0 < t(\text{mod } T) < LP \\ 0 \text{ for } LP < t(\text{mod } T) < T. \end{cases} \tag{4}$$

Thus the forcing is realized by a nonzero flux of X at z=0 for the time period of the length LP which can be interpreted as the length of the pulse while the system is relaxing for the rest of the forcing period T. We set LP = 0.6 for computations.

Choosing some fixed A and T, and starting from a homogeneous concentration profile of X and Y, the model (1)-(4) yields wave patterns for which the firing number ν can be computed in a way similar to that used in the evaluation of the experimentally observed wave patterns. The plot of ν versus T produces a devil's staircase-like structure. An example is shown in Fig. 6.

3.2. *Chain of coupled CSTR's*

To study the effects of excitability in more detail we can discretize the system (1). The resulting system of N coupled CSTR's is described by 2N equations,

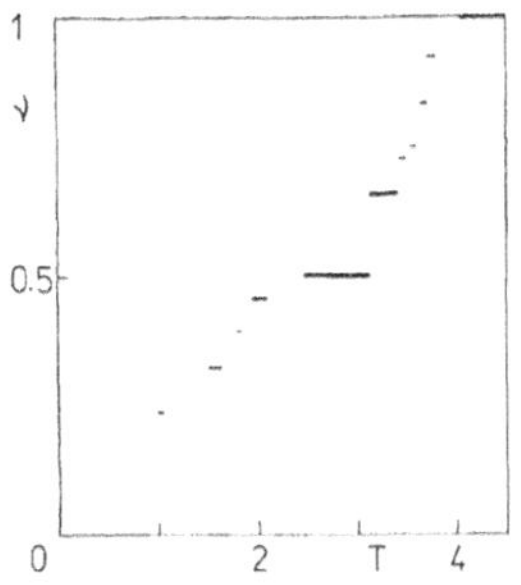

Figure 6. The dependence of the firing number ν on the forcing period t for the spatially one-dimensional model (1)-(4), A = 0.8.

$$\frac{dX_n}{dt} = D_1(X_{n+1} + X_{n-1} - 2X_n) + \frac{\nu_o + X_n^\gamma}{1 + X_n^\gamma} - X_n(1 + Y_n),$$

$$\frac{dY_n}{dt} = D_2(Y_{n+1} + Y_{n-1} - 2Y_n) + X_n(\beta + Y_n) - \delta Y_n, \tag{5}$$

$$n = 1,\ldots,N$$

with boundary conditions

$$Y_0=Y_1,\ Y_N=Y_{N+1},\ X_N=X_{N+1},$$

$$(1-p(t))\ (X_1 - X_0) + p(t)\ (A- X_0) = 0. \tag{6}$$

Let us mention that the transport coefficients D_1, D_2 have no direct relation to D_X, D_Y unless the number of cells is large enough. The unforced single cell (i.e. N=1, LP=0) system of Eqs. (5),(6) possesses for the chosen parameter values three steady states, an attractor, a repellor and a saddle. This type of excitability is associated with an invariant closed curve consisting of the attractor, saddle and its unstable separatrices. This corresponds to one of the three simplest phase portraits yielding excitability as described in [6].

The forced single CSTR can be interpreted as a system with infinitely fast diffusion coupled to an external reservoir by finite rate diffusion. Thus the wave patterns are lost, but the forcing term retains the same character as in the spatially distributed system (1)-(4).

We fixed the so far unspecified parameters at the following values: LP = 0.6, D_1 = 32 (D_2 can be arbitrary if N=1), and computed firing number ν by choosing a sequence of values of T and counting the number of excitation events over a large number of forcing periods as it was realized in CSTR experiments. The stable rest state of the unforced system was chosen as the initial condition. Remarkably, this system reveals qualitatively the same dynamical behaviour as the distributed one. Figure 7 shows ν versus T plot possessing again the devil's staircase-like structure.

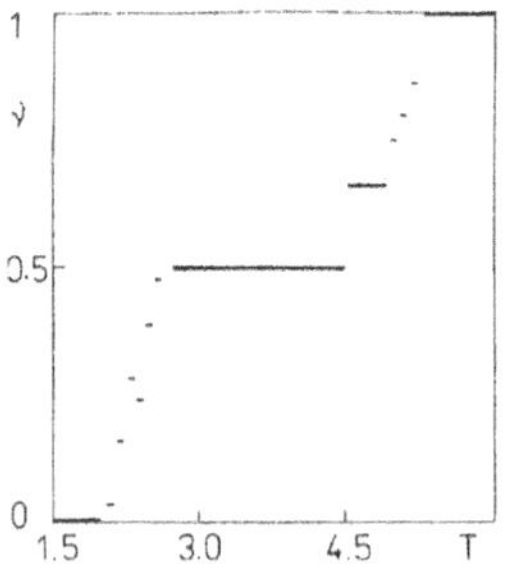

Figure 7. The dependence of the firing number ν on the forcing period T for the lumped parameter model (4)-(6), N = 1, a = 0.4.

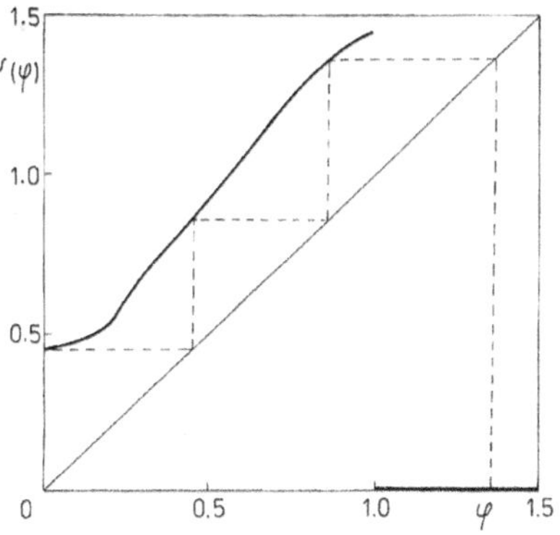

Figure 8. One-dimensional mapping constructed with the use of PEC, A = 5.0×10^{-6} mol $AgNO_3$, T = 150 s. 1/3-periodic regime is denoted by dashed line.

3.3. *An iterative map based on the PEC*

The old phase-new phase relationship extracted from the double pulse experiments with the CSTR (see section 2.3) in the form of the phase excitation curve - PEC - can be used to simulate the periodically forced system. Let the PEC be represented by the mapping $\vartheta(\varphi)$. We need to find a mapping which relates two phases φ_k, φ_{k+1} corresponding to two subsequent external pulse perturbations. This is accomplished by adding $\vartheta(\varphi_k)$ to the value of the dimensionless forcing period T/T_B. If the resulting phase φ_{k+1} at the next pulse perturbation exceeds 1 then the pulse elicits the excitation event and in the first approximation, the value of $\vartheta(\varphi_{k+1})$ can be set to zero. Therefore the mapping which relates φ_k and φ_{k+1} can be written as

$$\varphi_{k+1} = \psi(\varphi_k) = \begin{cases} T/T_B + \vartheta(\varphi_k) & \text{for } 0 < \varphi < 1, \\ T/T_B & \text{for } \varphi \geq 1. \end{cases} \qquad (7)$$

This mapping is discontinuous at $\varphi = 1$ and, in addition, independent of φ for $\varphi \geq 1$, see Fig. 8. Thus the asymptotic dynamical behaviour produced by successive iterations of ψ is always periodic because of the presence of the constant part of the map. Two successive iterates φ_k, φ_{k+1} such that $\varphi_k < 1 < \varphi_{k+1}$ correspond to the occurrence of the excitation event, hence the firing number can be easily computed. The nondecreasing character of $\psi(\varphi)$ in the interval $0 \leq \varphi \leq 1$ implies that every q-periodic sequence $\{\varphi_k\}$ has the firing number $\nu = 1/q$.

We have used cubic splines to interpolate the experimental data in Fig. 5, and applied Eq. (7) to obtain the ν versus T plot, see Fig. 9. Obviously, this devil's staircase is discontinuous as it possesses only values of the form $\nu = 1/q$ but it seems that all q's are present in this structure. One encounters

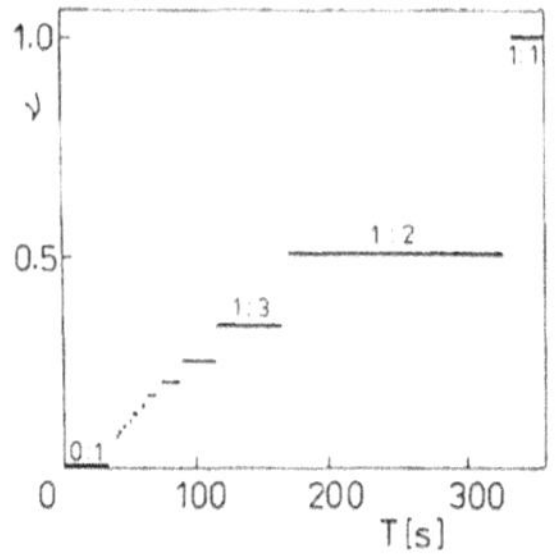

Figure 9. The dependence of the firing number ν on the forcing period T for the iterated map (7), $A = 5.0 \times 10^{-6}$ mol $AgNO_3$.

here a kind of intermittent but globally periodic behaviour.

We can expect that the previously described discontinuous return map is only an approximation to a continuous return map with a very steep descending part. For example, such a map was obtained from a model of the BZ reaction in the oscillatory mode [12] and we expect the same to hold for the forced excitable BZ reaction. To demonstrate this at a qualitative level we can use the single cell model (5),(6) which produces a one-dimensional stroboscopic Poincare map if we let $D_1 \to \infty$ [7,8]. An example of such a map relating the Y values immediately after two successive pulse perturbations is shown in Fig. 10. To simplify the analysis of the dynamical behaviour of this map the steeply descending part can be deleted and we get qualitatively the same situation as in Fig. 8.

4. Conclusions

The firing number-forcing period dependence reveals the devil's staircase-like structure observed experimentally both in a lumped parameter and distributed

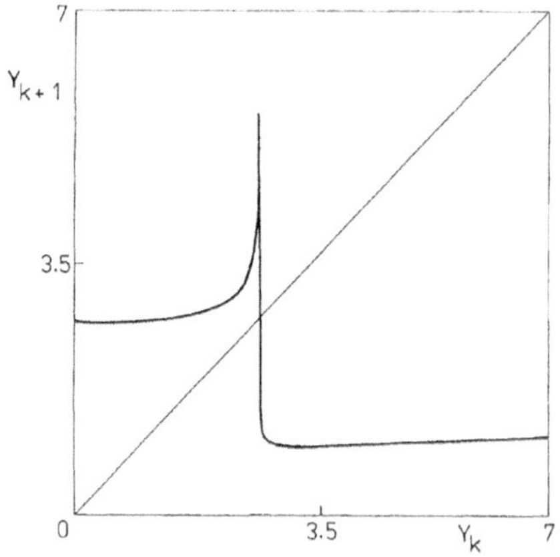

Figure 10. One-dimensional map obtained from the model (4)-(6), N = 1, A = 0.6, T = 3.1.

systems (both quasi-one-dimensional and two-dimensional) with the BZ reaction medium. This structure can be interpreted by means of relatively simple models. The dynamics observed both in experiments and models reveal alternating periodic and aperiodic regimes. The staircase structures and chaotic dynamics of excitations were also observed in forced cardiac cells, cf. Jaliffe et al. [this volume]. The interpretation of observed dynamics by means of one-dimensional return maps based on the use of phase excitation curves introduced in this paper seems promising and appears to be equally valid for both forced oscillatory and excitatory systems.

References

[1] Bak, P., Bohr, T. & Jensen, M.H. (1988). Circle Maps, Mode-locking and Chaos. In *Directions in Chaos 2*, Hao Bai-Lin (ed.), pp. 16-45. World Scientific: Singapore.

[2] Dolnik, M., Finkeova, J., Schreiber, I. & Marek, M. (1989). Dynamics of Forced Excitable and Oscillatory Chemical Reaction Systems. *J. Phys. Chem.* **93**, 2765.

[3] Dolnik, M., Schreiber, I. & Marek, M. (1986). Dynamic Regimes in a Periodically Forced Reaction Cell with Oscillatory Chemical Reaction. *Physica* **21d**, 78.

[4] Field, R.J. & Burger, M. (eds.) (1985). *Oscillations and Traveling Waves in Chemical Systems.* Wiley: New York.

[5] Finkeova, J., Dolnik, M., Hrudka, B. & Marek, M. Excitable Chemical Reaction Systems in a CSTR. Submitted to *J. Phys. Chem.*

[6] Marek, M., Finkeova, J. & Schreiber, I. (1989). Devil's Staircase in Chemical Excitable Systems. *Workshop "Irreversible Processes and Selforganization", Rostock, Feb. 1989.* To appear in Teubner Verlag, Leipzig, ed. W. Ebeling.

[7] Marek, M. & Schreiber, I. (1987). Formation of Periodic and Aperiodic Waves in Reaction-Diffusion Systems. In *Bifurcation: Analysis, Algorithms, Applications*, Kupper, T., Seydel, R. & Troger, H. (eds.), pp. 201-212. Birkhauser: Basel.

[8] Marek, M., Schreiber, I. & Vroblova, L. (1988). Complex and Chaotic Waves in Reaction-Diffusion Systems and on the Effects of Electric Field on Them. In *Structure, Coherence and Chaos in Dynamical Systems*, Christiansen, P.L. & Parmentier, R.D. (eds.), pp. 233-243. Manchester University Press: Manchester.

[9] Schneider, F.V. (1985). Periodic Perturbations of Chemical Oscillations. *Ann. Rev. Phys. Chem.* **36**, 347.

[10] Selkov, E.E. (1970). Two Alternative Autooscillating Stationary States in Thiole Metabolism - Two Alternative Types of Cell Reproduction: Normal and Malignant Ones. *Biofizika* **XV**, 1065-1073 (in Russian).

[11] Showalter, K., Noyes, R.M. & Turner, H. (1979). Detailed Studies of Trigger Wave Initiation and Detection. *J. Am. Chem. Soc.* **101**, 7463.

[12] Schwartz, I.B. (1984). Random Mixed Modes Due to Fluctuations in the Belousov-Zhabotinskii Reaction. *Phys. Lett.* **102A**, 25.

9. PULSE STIMULATION OF COUPLED CHEMICAL OSCILLATORS

Miloš Dolnik and Miloš Marek

Prague Institute of Chemical Technology
Department of Chemical Engineering
Suchbatarova 5, 166 28 Prague 6
Czechoslovakia

1. Introduction

Stimulated single and coupled reaction cells serve as well defined models of behaviour of dynamical regimes of biological cells and tissues. Results of experimental and theoretical studies of responses to single and periodic pulse stimulations of an oscillating open (flow-through) reaction cell were discussed in earlier papers [1,2]. It was demonstrated that single pulse stimulation can cause both positive and negative phase shift of concentration oscillations. The phase transition curves (PTC), i.e. the dependence of the new (shifted) phase of the oscillation on the phase when the stimulation has been applied, were constructed from experimental data. For strong stimulations PTC's were of the topological type "0" and for weak stimulations of the type "1" [2].

A one-dimensional deterministic model based on experimentally determined [2] and numerically modelled [5] phase transition curves described results of periodically forced experiments rather well. The existence of periodic and aperiodic regimes of concentration oscillations in a single reaction cell with periodic pulse forcing were thus fully confirmed.

A system of two coupled well stirred reaction cells serves as a discrete model of a continuous reaction diffusion system. Periodic and aperiodic regimes were also experimentally observed in two coupled cells with mutual mass exchange [3]. However, completely new phenomena - the transient and permanent extinction of oscillations [4] and the out of phase synchronization of oscillations [3] - appear when the pulse forcing is applied to coupled cells.

Here we shall discuss the use of a discontinous map for the description and interpretation of experimental results on forced coupled cells.

2. Single pulse stimulation

The Belousov-Zhabotinskii (BZ) reaction was used in experiments. The system of two coupled cells is schematically shown on Fig. 1. The coupling by mass exchange is denoted by D, flow into and out of the reactors by F_i and F_o,

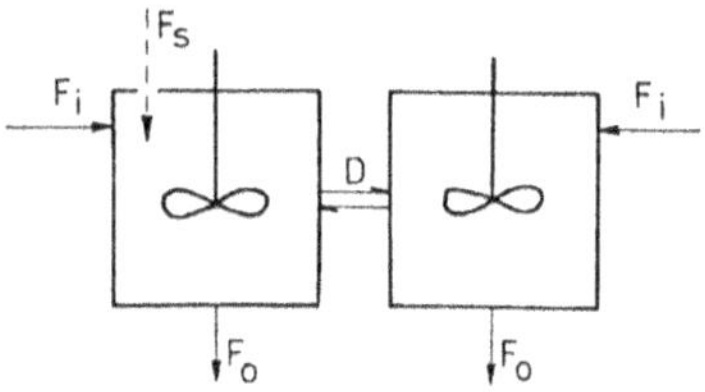

Figure 1. System of two coupled flow-through reaction cells with pulse stimulation - schematically. D represents mass exchange between the cells. F_i - input flow, F_o - output flow of reaction species. F_S - pulse stimulation.

respectively. Only one of the reactor cells was stimulated (F_s) by bromide ions.

The effects of pulse stimulation on oscillatory BZ reaction were studied in dependence on the intensity of the interaction between the cells, the amplitude and the phase of the stimulation.

For a single pulse stimulation which did not cause a cessation of oscillations the old phase φ_i and the new phase φ'_i were determined from the relations:

$$\varphi_1 = \varphi_2 = \frac{T_o}{T_B}$$
$$\varphi'_i = \varphi_i - \frac{(T^i_p - T_B)}{T_B} = \vartheta_i(\varphi_i) \qquad i = 1,2 \tag{1}$$

where T_B is the average period of unaffected oscillations, T_o is the time elapsed from the reference event to the pulse addition, T^i_p is the time interval in the reactor i between two successive reference events when the pulse perturbation was applied.

Examples of three phase transition curves for different values of the intensity of the interaction D are shown in Fig. 2. At the high values of the coupling intensity (Fig. 2a) the phase transition curves in both reactors are identical. Only the synchronization in phase with positive and/or negative phase shifts was observed. For intermediate value of the coupling (D = 6.1×10^{-3} s^{-1}) also the extinction of oscillations occurred. In this case the phase transition curves cannot be defined according to the schema shown in Fig. 2b. When the intensity of the coupling between the reactors is further decreased the phase transition curves for individual reactors start to differ and also become discontinuous. A single pulse stimulation then elicits an out of phase synchronization in this case (Fig. 2c).

3. Modelling of single pulse stimulation

The eight variable model of two coupled flow-through reactors, based on the expanded Oregonator kinetic model of the BZ reaction [6], was used to model

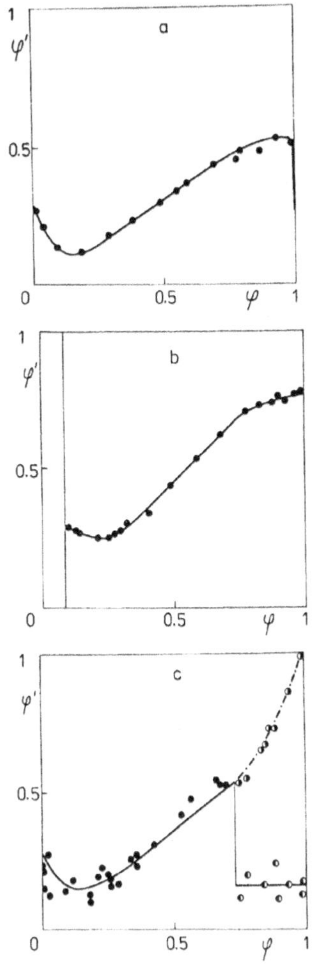

Figure 2. Phase transition curves - experimental measurement. Amplitude of stimulation a) A = 5 x 10^{-3} M, D = 0.09 s^{-1}, PTC's in individual reactors are the same. b) A = 5 x 10^{-4} M, D = 6.1 x 10^{-3} s^{-1}, extinction of oscillations occurred - region where the PTC's are not defined. c) A = 5 x 10^{-3}, D = 3.0 x 10^{-3} s^{-1}. The PTC's for individual reaction cells are different and discontinuous. Points - experimentally obtained data ◐ - PTC for the "forced" cell, ◑ - PTC for the "unforced" cell. Regression curves - functions $\vartheta_i(\varphi_i)$ were used in model (3).

the results of pulse stimulations. The model can be written in the form:

$$d\vec{X}_1/dt = R(\vec{X}_1) + k_o(\vec{X}_1^o-\vec{X}_1) + D(\vec{X}_2-\vec{X}_1) + \mathcal{A}(X^2,t)$$
$$d\vec{X}_2/dt = R(\vec{X}_2) + k_o(\vec{X}_2^{\,0}-\vec{X}_2) + D(\vec{X}_1-\vec{X}_2) \qquad (2)$$

Here $\vec{X} = \{X^1,X^2,X^3,X^4\}$ and concentrations X^i are defined as X^1 = $[HBrO_2]$, X^2 = $[Br^-]$, X^3 = $[Ce^{4+}]$ and X^4 = [HOBr]. The subscript 1 denotes the concentrations in the "forced" reaction cell, the subscript 2 in the "unforced" reaction cell and the superscript o denotes the input concentrations. $R(\vec{X})$ stands for the reaction kinetics term, k_0 is a reciprocal value of the residence time in both reactors, D represents the intensity of the interaction and $\mathcal{A}(X^2,t)$ describes a defined jump in the phase space (δ - pulse stimulation in the direction of the Br^- concentration axis). The phase transition curves were evaluated from simulations of the model (2).

Examples of three numerically obtained PTC's are depicted on Fig. 3. For high values of the intensity of interaction the PTC's are identical in both

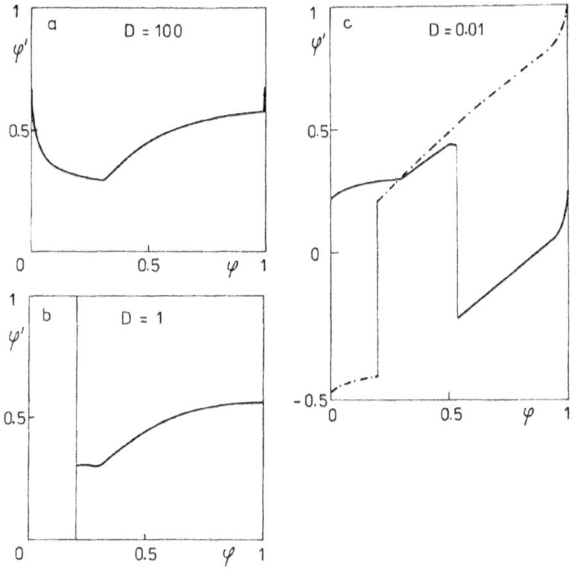

Figure 3. Phase transition curves - results of numerical simulation of model (2). Amplitude of stimulation A = 5.0×10^{-5}. a) D = 100 s^{-1}, high intensity of the interaction, the same PTC's for both reaction cells. b) D = 1 s^{-1}, intermediate intensity of interaction; region where PTC's are not defined and the extinction of oscillations occurred. c) D = 0.01 s^{-1}, low intensity of interaction - discontinous and different PTC's for individual reaction cells. Out of phase synchronization occurred for large phase delay between oscillations in individual cell.

reactors (Fig. 3a), for lower values of the coupling there are regions of extinction of oscillations (Fig. 3b) and for still lower value of the coupling intensity the PTC's are discontinuous and out of phase synchronization occurred (Fig. 3c). Thus complete qualitative agreement between the simulated and experimental results has been reached.

4. Periodically stimulated coupled cells

Both periodic and aperiodic regimes in a periodically stimulated single chemical cell were successfully modelled by a one-dimensional discrete map [2]. The description of the dynamic behaviour of the system of two coupled oscillators with periodically repeated pulse stimulation requires the use of at least two-dimensional discrete map. We have used the model based on the phase transition curves and on the phase delay between the reactors after the single pulse stimulation. The model is in the form:

$$\varphi_{1,k+1} = \frac{T + \vartheta_1(\varphi_{1,k}) + \dfrac{\Delta^1 - \Delta^{n1}}{2} \pmod 1}{1 - \dfrac{\Delta^{n1} - \Delta^{n1+1}}{2}} = f(\varphi_{1,k},\varphi_{2,k})$$

$$\varphi_{2,k+2} = \frac{T + \vartheta_2(\varphi_{2,k}) + \dfrac{\Delta^{n2} - \Delta^1}{2} \pmod 1}{1 - \dfrac{\Delta^{n2+1} - \Delta^{n2}}{2}} = h(\varphi_{1,k},\varphi_{2,k}) \qquad (3)$$

$$T = T_F/T_B \quad , \quad n1 = [T + \vartheta_1(\varphi_{1,k})] \quad , \quad n2 = [T + \vartheta_2(\varphi_{2,k})]$$

The values of n1 and n2 must be shifted by ±1 so that $\varphi_{k+1} \in \langle 0,1\rangle$. Here T_F is the period of stimulation, ϑ the new phase, φ the old phase and [ ] denotes the integer part. The symbol Δ denotes the phase delay between the reactors. The initial value of the phase delay Δ^1 is determined from the difference

$$\Delta^1 = \vartheta_2(\varphi_{2,k}) - \vartheta_1(\varphi_{1,k}) \qquad (4a)$$

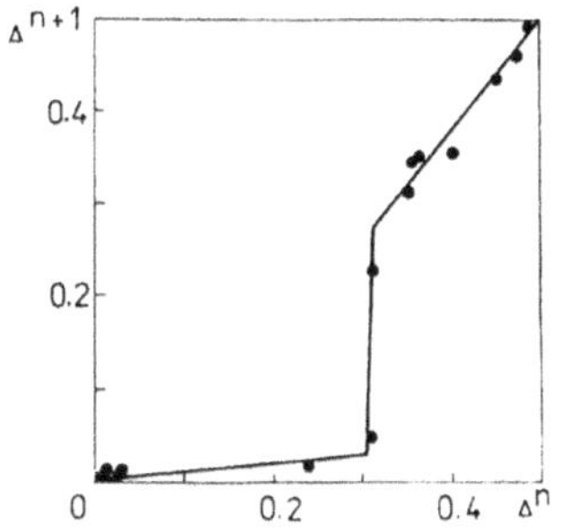

Figure 4. The dependence of the phase delay between oscillations in individual reaction cells on preceding ones - experimental measurements. Intensity of interaction D = 0.003 s^{-1}. Lines correspond to the function $g(\Delta)$ in model (3).

Successive values of phase delays depend for the constant intensity of the interaction only on preceding ones.

$$\Delta^{n+1} = g(\Delta^{n}) \tag{4b}$$

The function g was constructed on the basis of experimental data and for $D = 3.0\times10^{-3}\ s^{-1}$ is shown on Fig. 4. Dynamical regimes predicted by model (3) were characterized by rotation numbers in individual reactors evaluated from the computed sequences $\{\varphi_{1,k}\}$ and $\{\varphi_{2,k}\}$. The results are summarized in Fig. 5a,b. In the dependence of rotation numbers on the period of stimulation were found regions where the rotation numbers in both reactors are the same (regions with $\rho_1/\rho_2 = 1$ in Fig. 5b). Such regions correspond to synchronized periodic regimes in coupled reactors. Different rotation numbers in individual cells ($\rho_1/\rho_2 \neq 1$ in Fig. 5b) demonstrate the presence of desynchronized aperiodic and/or periodic regimes.

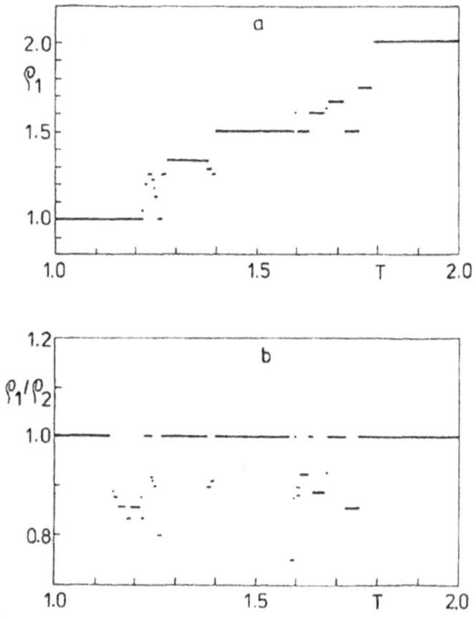

Figure 5. Example of results obtained from two-dimensional model (3). The sequence of phases $\{\varphi_{1,k}\}$ and $\{\varphi_{2,k}\}$ were computed on the basis of experimentally obtained PTC's for amplitude of stimulation $A = 5 \times 10^{-3}$ M and intensity of interaction $D = 0.003\ s^{-1}$. a) Dependence of rotation numbers ρ_1 on the relative period of stimulation T. b) Dependence of the ratios of rotation numbers in individual reaction cells on the period of stimulation T.

References

[1] Dolnik, M., Schreiber, I. & Marek, M. (1984). Experimental observations of periodic and chaotic regimes in a forced chemical oscillator. *Phys. Lett.* **100A**, 316-319.

[2] Dolnik, M., Schreiber, I. & Marek, M. (1986). Dynamic regimes in a periodically forced reaction cell with oscillatory chemical reaction. *Physica* **21D**, 78-92.

[3] Dolnik, M., Padušakova, E. & Marek, M. (1987). Periodic and aperiodic regimes in coupled reaction cells with pulse forcing. *J. Phys. Chem.* **91**, 4407-4410.

[4] Dolnik, M. & Marek, M. (1988). Extinction of oscillations in forced and coupled reaction cells. *J. Phys. Chem.* **92**, 2452-2455.

[5] Dolnik, M., Finkeova, J., Schreiber, I. & Marek, M. (1989). Dynamics of forced excitable and oscillatory chemical reaction systems. *J. Phys. Chem.* **93**, 2764-2774.

[6] Ruoff, P. & Noyes, R.M. (1986). An amplified Oregonator model simulating alternative excitabilities, transitions in types of oscillations and temporary bistability in a closed system. *J. Chem. Phys.* **84**, 1413-1423.

10. PERIODIC FORCING OF A SPATIALLY ONE-DIMENSIONAL EXCITABLE REACTION-DIFFUSION SYSTEM

Lenka Ketnerova, Hana Sevcikova* and Miloš Marek

Department of Chemical Engineering
Prague Institute of Chemical Technology
Suchbatarova 5, 166 28 Prague 6
Czechoslovakia
*Hostalkova 73, 169 00 Prague 6
Czechoslovakia

1. Introduction

Spatially distributed excitable reaction-diffusion systems are used to model excitable media of different physical nature [2,4,9], as, for example, neurophysiological systems, which are sometimes difficult to study directly. The most often studied models are either experimental [1,4,6] (based on the Belousov- Zhabotinskii reaction medium arranged as pseudo-one, -two or -three dimensional systems) or theoretical [7,8,9] (based on the description of a system where nonlinear reaction and Fickian diffusion take place). The spatially distributed excitable media are known to support travelling waves of excitation, which appear as responses of the system to local stimuli. The present paper refers to results of theoretical studies of the relationship between the character of the elicited wave patterns and both the character of the stimulus and the type of the excitability of the medium.

2. Mathematical model

The studied mathematical model of the reaction-diffusion one-dimensional system uses the SH reaction scheme [5] and has the form [8]:

$$\frac{\partial X}{\partial t} = \frac{D_X}{L^2}\frac{\partial^2 X}{\partial z^2} + \alpha\,\frac{(\nu_o + X^\gamma)}{(1 + X^\gamma)} - X(1 + Y)$$

$$\frac{\partial Y}{\partial t} = \frac{D_Y}{L^2}\frac{\partial^2 Y}{\partial z^2} + X(\beta + Y) - \delta Y \qquad (1)$$

where the local stimulus is modelled by the time dependent boundary conditions at z=0 representing a local periodic perturbation of X variable:

$$X(0,t) = A_f \quad \text{for } t \in \langle nT_f;\ nT_f + \Delta\tau)$$
$$\partial X(0,t)/\partial z = 0 \quad \text{for } t \in \langle nT_f + \Delta\tau;\ (n+1)T_f) \tag{2}$$
$$\partial Y(0,t)/\partial z = 0 \quad \text{for } t \geq 0 \quad n = 0,1,2,\ldots,$$

Here X and Y represent the concentrations of two reaction species, t is time, $z \in \langle 0,1 \rangle$ dimensionless spatial coordinate, L denotes a characteristic length of the system and D_X, D_Y are diffusion coefficients of the corresponding reaction species. α, β, γ, δ and ν_o are positive kinetic constants, $\Delta\tau$ is the time length of the perturbation and A_f and T_f stand for the amplitude and the period of the perturbation, respectively. In this study the fixed values of parameters were: L=1.0, D_X=0.008, D_Y=0.004, $\Delta\tau$=0.6, α=12.0, β=1.5, γ=3.0 and ν_o=0.01. The suitable choice of kinetic parameters for which the reaction medium exhibits the excitability resulted from the former stationary analysis of the system without diffusion terms [8] (D_X=D_Y=0.0).

Two different values of the parameter δ were used to model two different types of the excitable kinetics.

Type I excitability (δ = 1.0) is represented on a schematic phase portrait of a system without diffusion terms in Fig. 1a. The lower fixed point $S_L(X_L,Y_L)$ represents stable, but excitable steady state where the separatrix of a saddle point determines the excitability threshold. Superthreshold perturbation (point B) leads to an excitation cycle in a system without diffusion terms which, in connection with a diffusion in a spatially distributed system (Eqs. 1) can evoke a travelling wave of excitation.

The phase portrait of the second type of excitable medium (δ = 1.6, type II excitability) contains an additional upper stable fixed point $S_U(X_U,Y_U)$ (surrounded by an unstable limit cycle) which is not excitable (cf. Fig. 1b). Except for the all or none response to the perturbation of the lower stable state (points B or A, respectively) the transition from the lower stable state

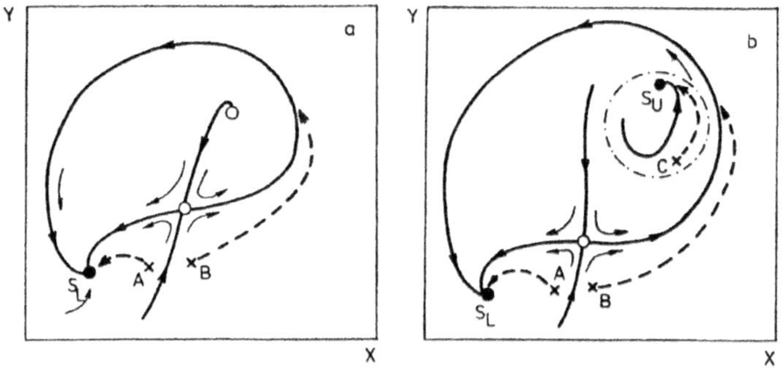

Figure 1. Schematic phase portraits of a system without diffusion terms (Eqs. 3), D_X=D_Y=0).
• (o,resp.) - stable (resp. unstable) stationary states;–·–·– - unstable limit cycle; – – – – phase trajectories after perturbations A,B or C.
a) Type I excitability (δ = 1.0).
b) Type II excitability (δ = 1.6).

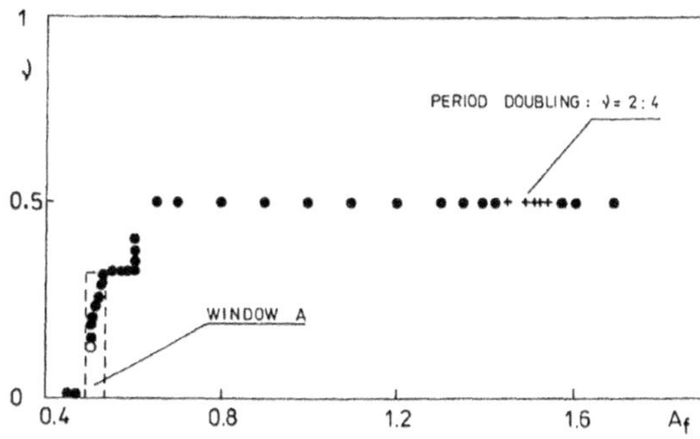

Figure 2. The dependence of the firing number on the forcing amplitude for the excitable medium of type I for the fixed value of T_f = 2.75. Points denote the calculated solutions.

to the upper one can be evoked when a suitable perturbation (point C in Fig. 1b) is applied.

To study the generation and propagation of wave trains the Eqs. (1) with boundary conditions (2) and the Neumann boundary conditions at z=1 were solved starting from the initial steady concentration profile $X(z,0)=X_L$ and $Y(z,0)-Y_L$ for $z \in <0,1>$. The numerical method with implicit three point finite difference scheme was applied using 128 spatial points and the time step equal to 0.025.

Generated wave trains were characterized by the firing number ν computed as [3]:

$$\nu = \limsup_{q \to \infty} p/q \tag{3}$$

where p is a number of travelling waves generated by the q applied perturbations.

3. Results

Figures 2 and 3 show the firing number as a function of the forcing amplitude

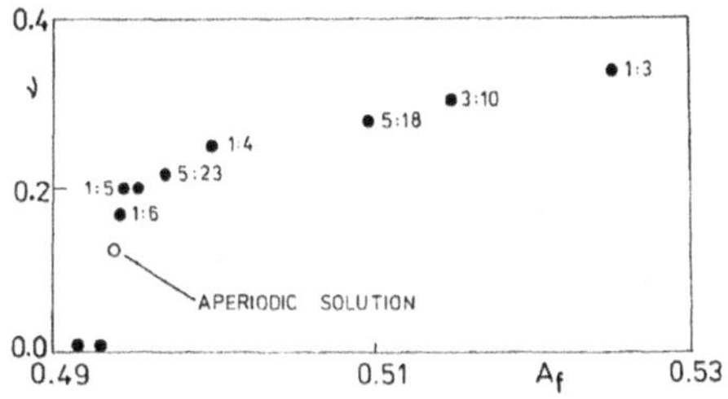

Figure 3. Window A.

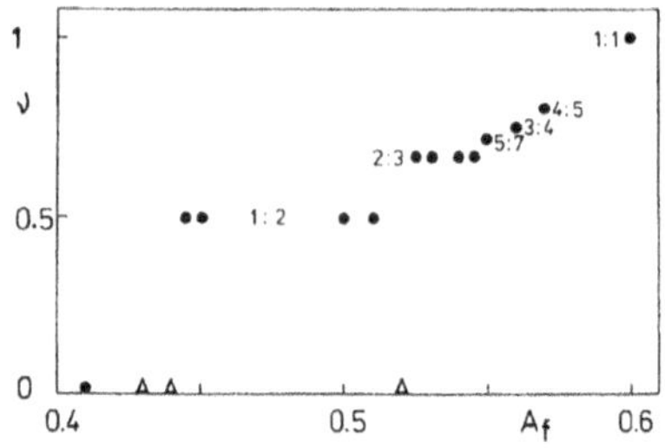

Figure 4. The dependence of the firing number on the forcing amplitude for the excitable medium of type II for the fixed value of T_f = 5.0. Points denote the calculated solutions. Δ - periodic nonhomogeneous solutions oscillating near the S_U state (Fig. 7).

A_f obtained for the medium of type I excitability for the fixed period of forcing T_f= 2.75. Similar dependence for the medium with type II excitability is represented in Fig. 4. Both dependences have the devil's staircase-like structure and the similar structure also appears on the dependence of the firing number on the forcing period for the fixed value of the amplitude. Several main types of solutions can be recognized:

(i) Periodic solutions without travelling waves: These solutions arise in the excitable medium of type I when either the amplitude and/or the period of forcing are lower than certain critical values. The firing numbers of such solutions are equal to zero as no travelling waves are generated. The final solutions have the form of spatially nonhomogeneous profiles of X and Y oscillating periodically in the close vicinity of the steady state S_L with the period $T=T_f$. The example of the periodically oscillating spatially nonhomogeneous X profile (obtained for T_f=2.75 and A_f=0.49) is shown in Fig. 5. One can see that the amplitude of oscillations decreases with the distance from the point of perturbation and, finally, approaches zero.

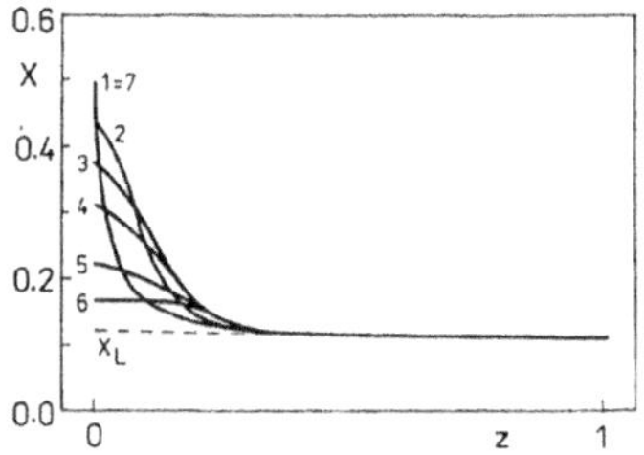

Figure 5. Time sequence of the spatial profiles of the variable X representing the periodic solution without travelling waves (ν = 0) obtained when the subthreshold perturbation is applied (type I excitability, A_f = 0.49, T_f = 2.75). Curves numbered 1 to 7 correspond to time values: 63.375, 64.0, 64.5, 65.0, 65.5, 66.0 and 66.125 respectively.

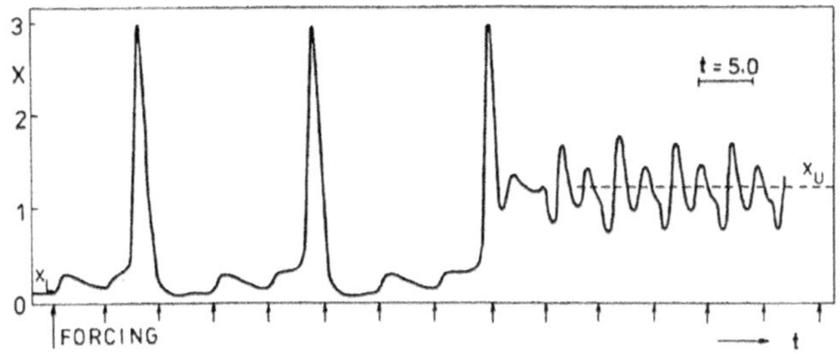

Figure 6. Time dependence of the X variable at z = 0.07 representing a transition from the initial steady state $S_L(X_L, Y_L)$ to the periodic solution without travelling waves (ν = 0) established in the neighbourhood of the upper steady state $S_U(X_U, Y_U)$. (Type II excitability, A_f = 0.43, T_f = 5.0).

For the excitability of type II these solutions exist only when the low amplitude perturbations are applied (cf. Fig. 4). The higher perturbation amplitudes evoke a transition from the initial steady state S_L to the spatially nonhomogeneous state periodically oscillating in the vicinity of the upper steady state S_U (see Fig. 6). In the course of the transition a certain number of waves (depending on the values of forcing parameters) is generated before the final periodic regime is established in the system, see Fig. 7. These solutions appear as traps on the generally growing dependence of the firing number on forcing parameter (see Fig. 4).

(ii) Periodic solutions with travelling waves: These solutions exhibit the firing number $\nu \in \langle 0,1 \rangle$. The periods T of such solutions were found to be equal to an m multiple of the forcing period T_f, in the course of which the n waves are generated (n and m being integers). Solutions with

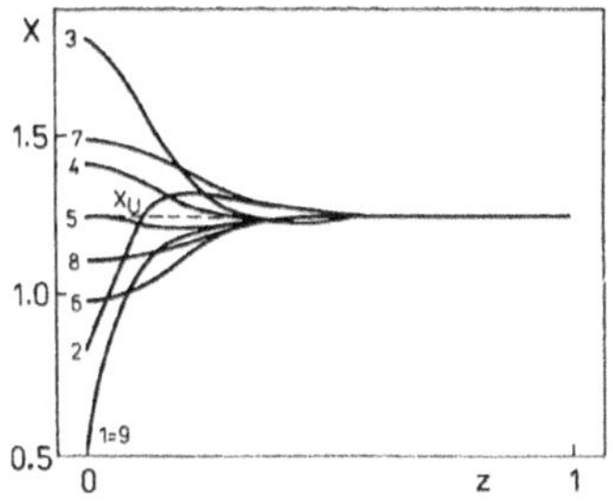

Figure 7. Time sequence of the spatial profiles of the variable X representing the periodic solution without travelling waves (ν = 0) obtained for the excitable medium of type II (A_f = 0.43, T_f = 5.0). Curves numbered 1 to 9 correspond to the time values: 265.2, 266.0, 266.6, 267.0, 267.2, 267.8, 268.8, 270.0 and 270.2, respectively.

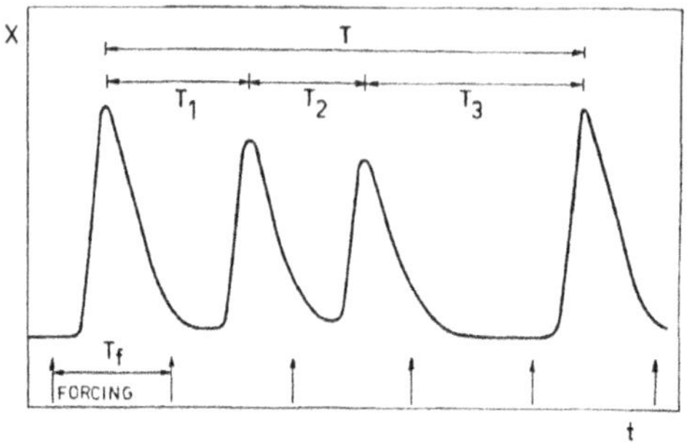

Figure 8. The schematic picture of an unequally spaced wave train.

n = 1 are equally spaced wave solutions as the mutual periods T_i between the successive waves are always the same and equal to an m multiple of the forcing period, $T_i = mT_f = T$. If n > 1 the nonequally spaced wave solutions are obtained composed from n travelling waves differing in their mutual periods T_i. The example of the solution with $\nu = 3:4$ is illustrated in Fig. 8. The relation:

$$T = \sum_{i=1}^{n} T_i = mT_f \tag{4}$$

was found to hold for the period of these solutions.

The organization of periodic solutions in dependence on the forcing amplitude follows the Farey ordering as it can be seen in Fig. 3 and Fig. 4. The period doubling for the periodic solutions with $\nu = 0.5$ (bifurcation from the equally spaced solution with $\nu = 1:2$ to the solution with $\nu = 2:4$) was found when the parameter A_f was varied.

(iii) Aperiodic solution with travelling waves: Such a solution was found for type I excitability at the transition between periodic solutions with

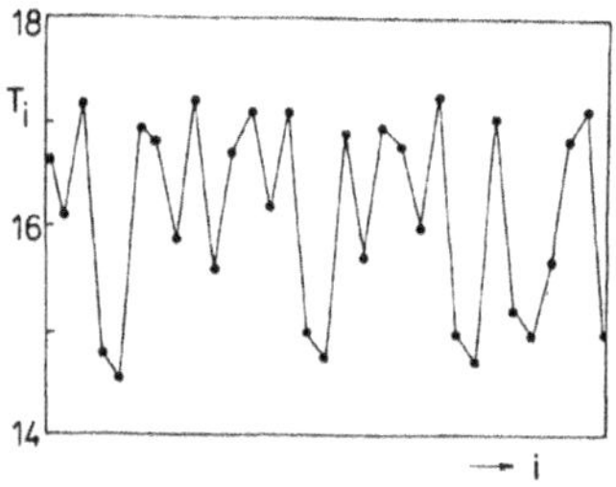

Figure 9. The aperiodic sequence of mutual periods T_i between waves travelling in the excitable medium of type I ($A_f = 0.494$, $T_f = 2.75$, $\nu = 0.125$).

the firing numbers ν = 0 and ν = 1:6. The aperiodicity is illustrated in Fig. 9 where the sequence of mutual periods T_i is plotted.

4. Conclusions

The type I excitable medium containing single excitable stable steady state represents the most commonly studied type of excitability. The type II excitable medium with two stable states where a single one is excitable had not been studied until now. The observed disappearance of travelling waves due to the attraction of the nonexcitable state is a new phenomen observed in this type of excitability. Both equally and nonequally spaced periodic wave trains as well as the aperiodic ones have been found to propagate in both types of excitable systems and to possess the devil's staircase-like structure in dependence of the firing number on the forcing amplitude.

References

[1] Field, R.J. & Burger, M. (1985). *Oscillations and Travelling Waves in Chemical Systems*. Wiley: New York.
[2] Krinski, V.I., (ed.) (1984). *Self-Organization. Autowaves and Structures far from Equilibrium*. Springer-Verlag: Berlin.
[3] Marek, M., Schreiber, I. & Vroblọva, L. (1987). Complex and Chaotic Waves in Reaction-Diffusion Systems and the Effects of Electric Field on Them. In *Structure, Coherence and Chaos*, Christiansen, P.L. & Parmentier, R.D. (eds.). Manchester Univesity Press: Manchester.
[4] Marek, M. & Sevcikova, H. (1988). Dispersion Curves and Pulse Wave Propagation in Excitable Systems. In *From Chemical to Biological Organization*, Markus, M., Müller, S.C. & Nicolis, G. (eds.), pp. 103-112. Springer-Verlag: Berlin.
[5] Selkov, E.E. (1970). Two Alternative Autooscillating Stationary States in Thiole Metabolism - Two Alternative Types of Cell Reproduction: Normal and Malignant Ones. *Biofizika* **XV**, 1065-1073 (in Russian).
[6] Sevcikova, H. & Marek, M. (1983). Chemical Waves in Electric Fields. *Physica* **9D**, 140-156.
[7] Sevcikova, H., Kubicek, M. & Marek, M. (1984). Concentration Waves - Effects of an Electric Field. In *Mathematical Modelling in Sciences and Technology*, Avula, X.J.R., Kalman, R.E., Ljapis, A.I. & Rodin, E.Y. (eds.), pp. 477-482. Pergamon Press: New York.
[8] Sevcikova, H. & Marek, M. (1986). Chemical Waves in Electric Field-Modelling. *Physica* **21D**, 61-77.
[9] Zykov, V.S. (1984). *Modelling of Wave Processes in Excitable Media*. Nauka: Moscow.

11. TRAVELLING WAVE SOLUTIONS OF A SIMPLE NERVE CONDUCTION EQUATION FOR INHOMOGENEOUS AXONS

A.K. Schierwagen

Karl-Marx-Universität Leipzig
Sektion Informatik, FG Neuroinformatik
Karl-Marx-Platz 10/11, Leipzig, DDR-7010

1. Introduction

Helmholtz's measurement of impulse conduction velocity in frog sciatic nerve which he accomplished in 1850 marked the beginning of the exact, quantitative description of excitation phenomena in nerves (Fig. 1). Currently, nerve impulse conduction represents one of the simplest, and because of this, most fully studied nonlinear wave phenomena in excitable media.

Usually, theoretical investigations of action potential (AP) propagation assume uniform electrical and geometric properties along the axon. From these analyses, much insight into propagation mechanisms has been gained, suggesting constant shape and velocity of the AP in homogeneous axons. A linear or square root relationship between velocity and axonal diameter for myelinated or unmyelinated nerve fibres, respectively, was deduced [6,9]. However, experimenters have noted several effects which could not be explained with this theory. Examples are blocking of impulse conduction, frequency modulation and changes of AP shape in regions of nonuniform axon geometries (for review, see [19]). Motivated by these observations, several groups have performed numerical computations in order to explore the effects of changing axonal geometry upon AP propagation (e.g. [3,7,12,14]). They could largely reproduce the experimental findings, and in addition new predictions were obtained.

A major drawback of the computer simulation approach, however, is the impossibility of exploring analytically how the various physical parameters describing the inhomogeneous axon affect the solution.

Goldstein & Rall [7] instead used results from dimensional analysis [5] to compare theoretical axons having different values of physical parameter but identical nonlinear membrane properties. The following analysis is inspired by this approach. In some sense it corresponds to the attempts in the general theory of excitable media to characterize reaction-diffusion equations possessing stable solitary wave solutions solely in terms of the attached reaction equations [20].

The present study is based on a simple version of the FitzHugh- Nagumo equations [4,13] for impulse propagation in excitable media, as used by Scott

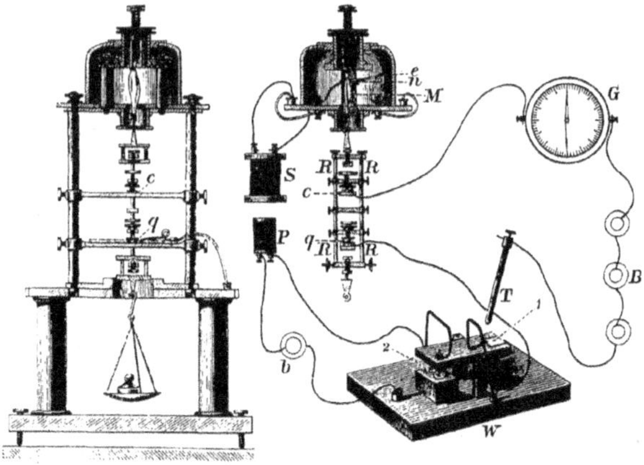

Figure. 1. Experimental arrangement used by Helmholtz in 1850 to measure the impulse velocity on a nerve fibre. A frog nerve-muscle preparation is set up so that, when the primary circuit (P) breaks, simultaneously a nerve impulse is initiated and time measurement on the ballistic galvanometer (G) starts. When the muscle (M) contracts, a contact (c) is broken and the measurement terminates. The impulse velocity is obtained if the difference of times measured for distal (e) and proximal (n) inputs is divided by the corresponding distance along the nerve fibre. (Reproduced from [2].)

[17] and others [15,16]. Applying certain variable transforms, we demonstrate the existence of a class of inhomogeneous axon cables which can be reduced to the homogeneous case, and for which analytical solutions can be derived. A transcendental equation for the wavefront velocity is obtained and approximately solved.

2. The nerve conduction model

A model of an axon can be set up by combining the nonlinear ordinary differential equations for an excitable membrane with the parabolic partial differential equations for a core conductor. Here the case of a continuous axon having its whole surface membrane excitable will be considered. Figure 2 schematically displays the core conductor, together with its equivalent electrical circuit from which the following equations are obtained:

$$\begin{aligned} &\frac{\partial}{\partial x}\left(\frac{1}{r_a(x)}\frac{\partial V}{\partial x}\right) - c(x)\frac{\partial V}{\partial t} = j_i(V, w_1, \ldots, w_n) \\ &\frac{\partial w_k}{\partial t} = f(V, w_1, \ldots, w_n) \quad \text{for } k = 1, \ldots, n \end{aligned} \tag{1}$$

where j_i is the membrane ion current per unit length, V is the transmembrane potential and $w_1, \ldots, w_n$ are auxiliary variables describing the transport of ions such as K^+, Na^+ and Cl^- across the membrane. j_i and $f = (f_1, \ldots, f_n)$ are

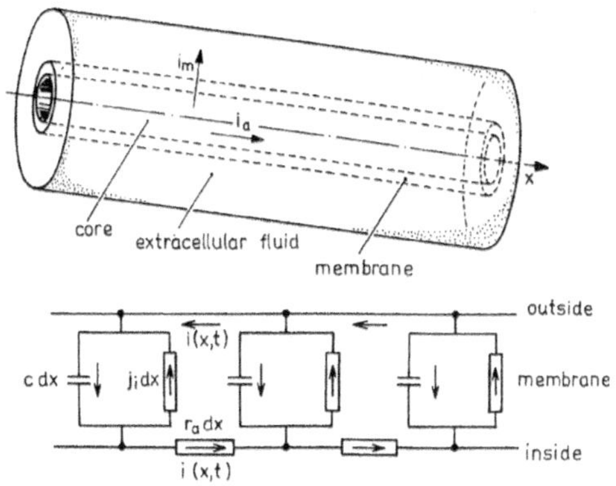

Figure 2. Scheme of a core conductor and its equivalent electrical circuit. Inside the membrane is the conducting core consists of axoplasm, outside the extracellular fluid. In the equivalent circuit, the extracellular resistivity is neglected. For explanation of symbols, see text.

continuous functions in V and $w = (w_1, \ldots, w_n)$; $r_a(x)$ and c(x) are the axial resistance per unit length and the membrane capacitance per unit length, respectively.

For suitable chosen j_i and f systems (1) have been shown to simulate essential features of the AP conduction.

(1) the bistable (wavefront) equation [15,17] (n=0)
(2) the Fitzhugh-Nagumo equations [4,13] (n=1)
(3) the Goldstein-Rall equations [7] (n=2)
(4) the Hodgkin-Huxley equations [8] (n=3).

Common to all these special cases is the property that in the uniform case (r_a and c constant) travelling wave solutions exist, i.e. nontrivial solutions $V(x,t)=V_u(y)$ of (1) depending only on the moving spatial coordinate $y = x - u \cdot t$, $u > 0$ such that $(V,w) \longrightarrow (0,0)$ for $y \longrightarrow \infty$.

Here, the assumption of axon homogeneity is dropped because we focus on AP conduction in axons of variable geometry.

In that case, changes in shape and velocity of the AP are to be expected [7] suspending travelling wave solutions of (2). In the following, we consider an axon cable of variable, circular cross-section where the specific electrical parameters R_i (intracellular resistivity, kΩ cm), C_m (membrane capacitance, $\mu F/cm^2$) and G_m (membrane conductance, mS/cm^2) are fixed along the axon. The corresponding parameters per unit length of cable then vary as

$$r_a(x) = 4R_i/[\pi \cdot d(x)^2]$$

$$c(x) = C_m \cdot \pi \cdot d(x) \qquad (2)$$

$$g(x) = G_m \cdot \pi \cdot d(x)$$

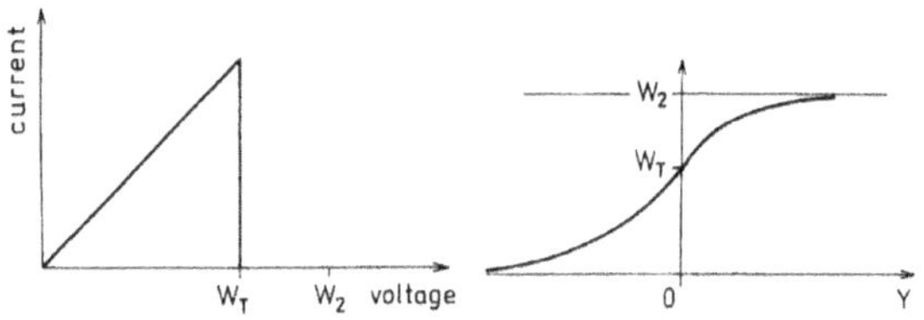

Figure 3. Piecewise linear approximation for the membrane ionic current (left) and travelling wave solution which moves to the left and changes the voltage level (right).

where d(x) is the axon diameter.

The membrane ionic current j_i in (1) is represented by a piecewise linear function (Fig. 3)

$$j_i\,(V) = g\,(x)\cdot h\,(V),$$
$$h\,(V) = V\cdot\{1 - H\,(V - V_T)\} \tag{3}$$

where H is the Heaviside step function and $V_T = V_T(x)$ is the space-dependent threshold value to be determined below (cf. paragraph 3). With these assumptions, equation (1) is linear both below and above V_T. The nonlinearity of the problem then appears only at the threshold $V_T(x)$ where the membrane switches from one functional state into the other one, as described by (3).

In concentrating on the barest essentials of the conduction process, we will consider the case n=0 in (1), i.e. the bistable equation in which AP recovery is neglected. Thus the nerve conduction equation to be studied is

$$\frac{\partial}{\partial x}\left(\frac{1}{r_a(x)}\frac{\partial V}{\partial x}\right) - c(x)\,\frac{\partial V}{\partial t} = g(x)\,h(V). \tag{4}$$

3. Transformation into normal form

Applying the variable transforms (see [18] for motivation)

$$T \doteq t/\tau,\quad \tau = c(x)/g(x)$$
$$Z = \int_0^x \frac{ds}{\lambda(s)},\quad \lambda(x) = [r_a(x)g(x)]^{-\frac{1}{2}}, \tag{5}$$

equation (4) can be written in dimensionless form:

$$\frac{\partial^2 V}{\partial Z^2} + q\,\frac{\partial V}{\partial Z} - \frac{\partial V}{\partial T} = h(V) \tag{6}$$

where

$$q = \frac{3}{2}\,\{\ln d(Z)\}' \tag{7}$$

and the prime denotes differentiation with respect to Z. Equation (6) can be simplified by changing the dependent variable V according to

$$V(Z,T) = W(Z,T) \cdot F(Z) \tag{8}$$

where

$$F(Z) = \exp\left(-\frac{1}{2}\int q\,dz\right) = d(Z)^{-\frac{3}{4}}. \tag{9}$$

The new variable W is the solution of the simpler equation

$$\frac{\partial^2 W}{\partial Z^2} - \frac{\partial W}{\partial T} - p \cdot W = h(W) \tag{10}$$

or equivalently,

$$\frac{\partial^2 W}{\partial Z^2} - \frac{\partial W}{\partial T} - I(Z) \cdot W = 0. \tag{11}$$

In analogy to the theory of second-order linear differential equations [10], we call (11) the **normal form** of (4) and I(Z) the **invariant**. Equations with the same normal form are said to be **equivalent**. The invariant is related to q(z) in (7) by the **special Riccati equation**

$$I(Z) = p + m, \quad p = \frac{q^2}{4} + \frac{q'}{2} \tag{12}$$

where m is given by

$$m = \begin{cases} 0 \text{ for } W > W_T \\ 1 \text{ for } W < W_T \end{cases} \tag{13}$$

and W_T is the (constant) threshold value related to the corresponding value V_T of the original variable V through (5) and (8). Given that the invariant possesses a continuous derivative, there is a unique solution of (11) which assumes and continuously depends on prescribed continuous initial and boundary values [1]. Hence, there is a unique solution of (11) in each region $W < W_T$ and $W > W_T$.

4. Analytically treatable axon geometries

Various classes of nonuniform axon geometries are obtained by choosing the function I(Z) in (11) and solving for the axon diameter d(Z) via (7). The simplest class to consider are the axons for which I(Z) = p + m, with p a constant (cf. [11,18] for related problems in passive electrical networks).

This class can be determined by solving the Riccati equation

$$2q' = 4\,p - q^2 \tag{14}$$

obtained from (12). Equation (14) is integrable by separation of variables, and its solutions can be expressed by elementary functions. In dependence on the constant p, six different solutions exist for (14). These solutions and the diameter variations which they represent are given in Table 1.

Table 1. Axon geometries as defined by the solutions of the special Riccati equation $2q^1 = 4p - q^2$, $q = \frac{3}{2}\{\ln d\,(Z)\}^1$.

Type of geometry	p	q(Z)	Variation of diameter d(Z)
uniform	$p=0$ $q=0$	0	d_0
power	$p=0$ $q\neq0$	$2/(Z-C)$	$d_0\cdot(1-Z/C)^{4/3}$
exponential	$p>0$ $q_1^2=4p$	q_1	$d_0\cdot\exp(\frac{2}{3}\,q_1Z)$
hypberbolic sine	$p>0$ $\lvert q\rvert>q_1$	$q_1\coth[q_1(Z-C)/2]$	$d_0\cdot\{\sinh[q_1(Z-C)/2]\}^{4/3}$
hyperbolic cosine	$p>0$ $\lvert q\rvert<q_1$	$q_1\tanh[q_1(Z-C)/2]$	$d_0\cdot\{\cosh[q_1(Z-C)/2]\}^{4/}$
trigonometric cosine	$p<0$	$-[q_1]\tan[\lvert q_1\rvert(Z-C)/2]$	$d_0\cdot\{\cos[\lvert q_1\rvert(Z-C)/2\}^{4/3}$ $\lvert q_1\rvert(Z-C)\neq\pm(2n+1)\pi$, $n\in\mathbb{N}$

5. Propagation of the "leading edge"

The axon diameter variations as given in Table 1 all lead to nerve conduction equations (4) which are equivalent under the transforms (5), i.e. they possess the same normal form (11). As it is well known, (11) admits travelling wave solutions. Since recovery is neglected in the present model, the wave is a transition front corresponding approximately to the "leading edge" of an AP [17]. Using the moving spatial coordinate $Y = Z - \Theta \cdot T$, (11) becomes the ordinary differential equation

$$\frac{\partial^2 W}{\partial Y^2} - \Theta\,\frac{dW}{dY} - (p+m)\cdot W = 0 \tag{15}$$

where Θ is the unknown propagation velocity.

For simplicity, we choose $Y = 0$ to be the point where $W = W_T$. Following

Scott [17], a leading edge which makes a transition between the two W-levels, $W=0$ and $W=W_2$ (see Fig. 3), and satisfies (15) is easily constructed.

Let $p > 0$ (the case $p \leq 0$ can be discussed in an analogous way). Then the solution to (15) must be of the form

$$W(Y) = A_1 \exp(\alpha_1 Y) + A_2 \exp(\alpha_2 Y) + A_3 \tag{16}$$

where

$$\alpha_{1/2} = -\frac{\Theta}{2} \pm \sqrt{\frac{\Theta^2}{4} + (p+m)} \tag{17}$$

The functions W(Y) in the two regions can be specified from (16) and (17):

$$W(Y) = W_T \exp(\beta_1 Y) \quad \text{for } W < W_T \tag{18}$$

and

$$W(Y) = W_2 - (W_2 - W_T) \exp(\beta_2 Y) \quad \text{for } W > W_T \tag{19}$$

where

$$\beta_1 = -\frac{\Theta}{2} + \sqrt{\frac{\Theta^2}{4} + p+1} \tag{20}$$

and

$$\beta_2 = -\frac{\Theta}{2} - \sqrt{\frac{\Theta^2}{4} + p} \;. \tag{21}$$

One way to compute the unknown propagation velocity Θ consists of demanding continuity of dW/dY at $Y = 0$ which implies

$$\beta_1 W_T = -\beta_2 (W_2 - W_T). \tag{22}$$

(22) is a transcendental equation in Θ which can be solved only approximately. For p large, we find

$$\Theta \approx \pm \sqrt{\frac{p(2W_T - W_2)^2}{W_T W_2 - W_T^2}} \tag{23}$$

where the different signs in (23) correspond to wavefronts moving in either direction.

Acknowledgements

I would like to thank the organizer of this meeting for generous financial support including travelling costs.

References

[1] Babitsch, W.M. et al. (1972). *Lineare Differential- gleichungen der mathematischen Physik.* Akademie- Verlag: Berlin.

[2] Bernstein, J.L. (1900). *Lehrbuch der Physiologie.* Verlag F. Enke: Stuttgart.

[3] Dodge, F.A. & Cooley, J.W. (1973). Action potential of the motoneuron. *IBM J. Res. Develop.* **17**, 219-229.

[4] FitzHugh, R. (1969). Mathematical models of excitation and propagation in nerve. In *Biological Engineering,* pp. 1-85, Schwan, H.P. (ed.). McGraw-Hill: New York.

[5] FitzHugh, R. (1973). Dimensional analysis of nerve models. *J. theor. Biol.* **40**, 517-541.

[6] Goldman, L. & Albus, J.S. (1968). Computation of impulse conduction in myelinated fibres: theoretical basis of the velocity-diameter relation. *Biophys. J.* **8**, 596-607.

[7] Goldstein, S.S. & Rall, W. (1974). Changes of action potential shape and velocity for changing core conductor geometry. *Biophys. J.* **14**, 731-757.

[8] Hodgkin, A.L. & Huxley, A.F. (1952). A quantitative description of membrane current and its application to conduction and excitation in nerve. *J. Physiol. (Lond.)* **117**, 500-544.

[9] Hunter, P.J., McNaughton, P.A. & Noble, D. (1975). Analytic models of propagation in excitable cells. *Prog. Biophys. Molec. Biol.* **30**, 99-144.

[10] Kamke, E. (1967). *Differentialgleichungen Bd. I.* Geest & Portig: Leipzig.

[11] Kelly, J.J. & Ghausi, M.S. (1965). Tapered distribution RC networks with similar immittances. *IEEE Trans. Circuit Theory* **CT-12**, 554-558.

[12] Khodorov, B.I. & Timin, E.N. (1975). Nerve impulse propagation along non-uniform fibres (investigations using mathematical models). *Prog. Biophys. Molec. Biol.* **30**, 145-184.

[13] Nagumo, J., Arimoto, S. & Yoshizawa, S. (1962). An active pulse transmission line simulating nerve axon. *Proc. IRE* **50**, 2061-2070.

[14] Parnas, I., Hochstein, S. & Parnas, H. (1976). Theoretical analysis of parameters leading to frequency modulation along an inhomogeneous axon. *J. Neurophysiol.* **39**, 909-922.

[15] Pawelussen, J.P. (1980). Heteroclinic waves of the Fitzhugh-Nagumo equations. Amsterdam, preprint.

[16] Rinzel, J. & Keller, J.B. (1973). Traveling wave solutions of a nerve conduction equation. *Biophys. J.* **13**, 1313-1337.

[17] Scott, A.C. (1975). The electrophysics of a nerve fiber. *Rev. Mod. Phys.* **47**, 487-533.

[18] Schierwagen, A.K. A non-uniform equivalent cable model of membrane voltage changes in a passive dendritic tree. *J. theor. Biol.,* in press.

[19] Waxman, S.G. (ed.). *Physiology and Pathobiology of Axons.* Raven Press: New York.

[20] Zykov, V.S. (1987). *Simulation of Wave Processes in Excitable Media.* Manchester University Press: Manchester.

12. THE OPTIMAL CONTROL OF AN EXCITABLE NEURAL FIBRE

J.E. Rubio and A.V. Holden

Departments of Mathematics and Physiology
University of Leeds
Leeds, U.K.

We consider the optimal control of a neural fibre, described by a nonlinear diffusion equation with a polynomial nonlinearity. An iterative scheme is established to compute a minimum-energy control, in which at each step of the iteration a linear problem is solved by means of measure-theoretical methods and linear programming. The method converges for moderate values of the nonlinear terms. Some numerical results are given.

1. **Introduction**

We develop in this paper a procedure for the design of an optimal control for a nonlinear diffusion equation which describes the behaviour of a neural fibre. This equation has been much studied [1,2,6], but no serious mathematical attempt seems to have been made to develop a useful method for the design of optimal controls for it.

The nonlinearity will be considered to have a polynomial form; if y is the potential of the membrane, the nonlinear term will be taken as $f(y)=\gamma[y(a-y)+z]$, where γ, a and z are numerical parameters much discussed in the literature, especially by Casten et al. [1]. This function has been used by some authors to model other phenomena such as combustion or the spread of epidemics [6].

Our approach is a direct extension of some of our work on the control of the linear diffusion equation [4,5]. After choosing a performance criterion, an iterative scheme is established so that at each iteration a linear problem is solved, by means of an approach involving measure theory and linear programming. The scheme converges for moderate values of the constant γ which multiplies the nonlinearity f, and a good approximation to the optimal control for the nonlinear equation is thus obtained.

2. **The problem**

Consider a nonlinear diffusion equation

$$y_{xx}(x,t)=y_t(x,t) + \gamma\, f(y(x,t)),\quad (x,t)\in(0,1)\times(0,T) \tag{1}$$

where y is the variable representing potential difference, γ is a numerical parameter, T is the final value of the time, and $f:\mathbb{R}\to\mathbb{R}$ is a nonlinear polynomial function of the form

$$f(y) = y(a-y)(1-y)+z \tag{2}$$

where a and z are numerical parameters to be chosen below.

The boundary conditions are:

$$y_x(0,t)=0, \quad t\in[0,T],$$
$$y_x(1,t)=u(t), \quad t\in[0,T], \tag{3}$$
$$y(x,0)=0, \quad x\in[0,1];$$

here $u(t)$, $t \in [0,T]$, is the control, which can be interpreted as the current being injected into the fibre at the end defined by x=1; the other end is insulated. The control u will be termed **admissible** if it is measurable, $u(t)\in[-1,1]$ ae and $y(x,T)=g(x)$, a given function which is the desired final state, on [0,1]. That is, we put a (normalized) amplitude constraint on the input current, and we wish to reach a given state g at the final time T. For instance, it could be that we wish to stop a possible wild behaviour of the fibre by applying a control that will bring this to a stop at the final time T. The set of all admissible controls, to be denoted by U, is assumed to be nonempty.

Let $f^0\in\text{Lips}(\Omega)$, the space of Lipschitz-continuous functions on $\Omega\equiv[0,T]\times]-1,1]$. The control problem consists in finding a control $u\in U$ so as to minimize the functional

$$J(u) = \int_0^T f^0(t,u(t))dt; \tag{4}$$

that is, we are interested in reaching the final state g by means of a control with values in [-1,1], while at the same time minimizing the integral criterion (4), which may itself represent the **total energy** used in the process - if the integrand is u^2 - or the **total charge** - if the integrand is $|u|$.

3. Linear problems and measures

We shall develop an algorithm to solve this problem for values of the parameter $\phi\in[0,\gamma_0]$, for some $\gamma_0\in\mathbb{R}$; this algorithm will consist of an iterative procedure, in which a sequence of related **linear problems** are solved. Firstly - before introducing iteration - we shall consider the linear problem associated with that above. Let $F\in C((0,1)\times(0,T))$, and let

$$y_{xx}(x,t)=y_t(x,t) + F(x,t), \quad (x,t)\in(0,1)\times(0,T), \tag{5}$$

with the same boundary conditions as in (2). This is a linear partial differential equation, with an explicit solution given by:

$$y(x,T) = \sum_{n=0}^{\infty} a_n\cos(n\pi x), \tag{6}$$

with

$$a_0=\int_0^T [u(t)+F_0(t)]dt, \quad a_n = \int_0^T [(-1)^n\ 2u(t)+F_n(t)]\Psi_n(t)dt \tag{7}$$

$$F_0(t)=\int_0^1 F(x,t)dx \quad F_n\ (t)=2\int_0^1 F(x,t)\cos(n\pi x)dx \quad \Psi_n(t)=\exp(-n^2\pi^2(T-t))$$
$$n=1,2,\ldots$$

If we wish to attain the state $g\in L_2(0,1)$ at the final time T, therefore, the control u must satisfy the following equalities which determine its **moments**; we assume that g has the half-range Fourier series $g(x)=\Sigma_0^\infty a_n\cos(n\pi x)$:

$$\int_0^T u(t)dt = a_0 - \int_0^T F_0(t)dt \equiv c_0$$
$$\int_0^T (-1)^n 2u(t)\ \Psi_n(t)dt = a_n - \int_0^T F_n(t)\Psi_n(t)dt \equiv c_n,\ n=1,2,\ldots \tag{8}$$

Thus, our linear control problem can be said to be reduced to finding an admissible control satisfying (8) while minimizing the performance criterion (4). In general, a solution to this problem may not exist; as in [4] we replace this problem by one in which the minimum of a linear functional is sought over a set of Radon measures on Ω. We notice that, for a fixed control function u, the mapping $f \rightarrow \int_0^T f(t,u(t))dt$ defines a positive linear functional on $C(\Omega)$. Thus, by the Riesz representation theorem, there exists a unique Radon measure μ on Ω such that $\int_0^T f(t,u(t))dt = \int_\Omega f\ d\mu \equiv \mu(f)$ for all $f\in C(\Omega)$. We now replace the original minimization problem by one in which we seek the minimum of $\mu(f^0)$ over a set Q of positive Radon measures on Ω. This set is defined by the following equalities:

$$\mu(\varphi_n)=c_n\ ,\ n=0,1,2,\ldots$$

$$\mu(h)=a_h,\ h\in C'(\Omega). \tag{9}$$

Here we have put $\varphi_0(t,u)=u$, $\varphi_n(t,u)=(-1)^n 2u\Psi_n(t)$, $n=1,2,\ldots$ The space $C'(\Omega)$ is that subset of $C(\Omega)$ composed of functions which depend only on the variable t, and a_h is the integral of h over $[0,T]$; thus, the second set of equalities in (9) indicate that the measures in Q should project on $[0,T]$ as the Lebesque measure. We shall need this property below (see also [5]).

It is not difficult to show that an optimal measure $\mu^*\in Q$ exists which minimizes the function $\mu\rightarrow\mu(f^0)$. Also, by means of a theorem of Ghouila-Houri [3] we can show that, given a number $\varepsilon>0$ and an integer $k>0$, we can find a piecewise-constant control u, defined of course on $[0,T]$, so that

$$\left|\int_0^T f^0(t,u(t))dt - \mu^*(f^0)\right| \le \varepsilon$$
$$\left|\int_0^T \varphi_n(t,u(t))dt - c_n\right| \le \varepsilon \ , \ n=0,1,\dots,k. \tag{10}$$

The last set of inequalities implies that the $L_2[0,t]$ norm of the difference between the actual final state and the desired state g can be made as small as necessary, choosing ε sufficiently small and k sufficiently large. Finally, we address ourselves to the problem of estimating numerically an approximate control - really, a sub-optimal control - such as u in (10). We have published elsewhere [4] a detailed account of this method - applied to a simpler problem, but trivially extendable to the present case - so we will just describe briefly the main ideas, leaving the reader interested in the proofs to consult our references. Firstly, we should recognize that the problem of minimizing $\mu\to\mu(f^0)$ over the set Q defined by (9) is one of **infinite-dimensional linear programming**; it is possible to approximate its solution μ^* by the solution of a simple finite linear programming problem, that is, one for which there is a finite number of equations and unknowns. In turn, this solution enables us to construct a suboptimal control such as the one above. We shall take T=1.

Indeed, let N_1, N_2 be positive integers, and consider a set of equations just like (9), but with N_1+1 equations of the first kind (those involving the functions φ_n) and N_2 equations of the second, involving the functions h; these will not be continuous functions, but rather pulse-like, lower-semicontinuous functions so that - changing the notation in an obvious way - h_i =1 in the subinterval $[(i-1)/N_2, i/N_2) \subset [0,1]$, $i=1,\dots,N_2$, and zero elsewhere. The measures to be considered here are to be defined not on the whole of Ω but on a finite subset, to be denoted by ω. Let N_3 be another positive integer, and take N_3-1 equidistant points on the time interval [0,1], and $2N_3-1$ points on the control set [-1,1]; the finite subset $\omega\subset\Omega$ is composed of $N\equiv(N_3-1)(2N_3-1)$ points z in Ω with coordinates (t,u) defined by the points on the axis. Of course, as $N_3\to\infty$ the points in this set get more and more numerous and closer to each other, while no region in Ω is left uncovered. The measures to be considered are those of the form

$$\mu = \sum_{j=1}^{N} \alpha_j \delta(z_j), \tag{11}$$

where $z_j\equiv(t_j,u_j)$, $j=1,\dots,N$, are the points in the set ω, $\delta(z)$ is a unitary atomic measure with support $z\varepsilon\omega$, and the coefficients α_j, $j=1,\dots,N$, are nonnegative real numbers; they define the measure and will be the unknowns which will define our optimal measure below. Indeed, we consider the minimization of

$$\sum_{j=1}^{N} \alpha_j f^0(z_j) \tag{12}$$

over the set of measures of the form (11); that is, over the set of n-tuples of nonnegative real numbers $[\alpha_1,\alpha_2,\ldots,\alpha_N]$ satisfying the equalities

$$\sum_{j=1}^{N} \alpha_j \varphi_n(z_j) = c_n, \quad n=0,1,\ldots,N_1$$
$$\sum_{j=1}^{N} \alpha_j h_i(z_j) = 1/N_2, \quad i=1,\ldots,N_2. \tag{13}$$

We have, therefore, a linear programming problem with $M\equiv N_1+N_2$ equations and N unknowns. It is well-known that the optimal measure is defined by a set of at most M nonzero unknowns; it happens that this problem is regular, and there are **exactly** M nonzero α_j's in the optimal measure; their indices define, of course, the corresponding values $z_j\equiv(t_j,u_j)$. We can now construct a piecewise-constant control u, as in (10) above; because of the second set of equalities in (13) there is at least one value of t_j in each subinterval of length $1/N_2$ defined above in connection with the definition of the functions a_i; the sum of all the α_j's associated with each of these intervals adds up to $1/N_2$, so we partition each of these subintervals further, by the corresponding α_j's in any order; the control is defined as having the value u_j in the subinterval defined by the corresponding α_j. It can be proved that for sufficiently large N_1,N_2 and N_3 the left-hand side of the inequalities (10) can be made as small as desired; the infimum and the desired final state can be approximated in this manner, so that we can refer to the control function thus constructed as a suboptimal control.

4. Iteration

In order to estimate an optimal control for our original nonliner problem (1)-(4), we shall solve a sequence of **linear** problems such as (5); at each step of the iteration, we shall estimate the solution, to be denoted by y^k, and then, in the next step, the one whose solution would give us y^{k+1} as well as the corresponding optimal control u_{k+1}; we put $F(x,t)\equiv\gamma f(y_k(x,t))$, $(x,t)\in(0,1)\times(0,1)$. Indeed, consider the scheme:

$$\begin{aligned}
&y^0(x,t)=0, \quad (x,t)\equiv(0,1)\times(0,1)\\
&y^{k+1}_{xx}(x,t)=y^{k+1}_t(x,t) + \gamma\, f(y^k(x,t)), \quad (x,t)\in(0,1)\times(0,1)\\
&y_x(0,t)=0, \quad t\in[0,1]\\
&y_x(1,t)=u^*_{k+1}(t), \quad t\in[0,1]\\
&y(x,0)=0, \quad x\in[0,1],\\
&y(x,1)=g(x), \quad x\in(0,1).
\end{aligned} \tag{14}$$

where u^*_{k+1} is the control which minimizes the performance criterion (4)

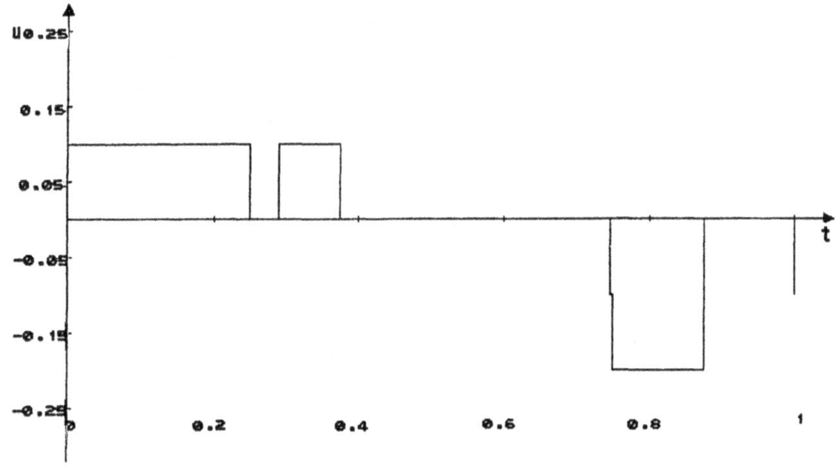

Figure 1. Graph of the nearly optimal control function u. It is piecewise-constant but not bang-bang.

satisfying the equations and boundary conditions in (14); this is a problem just like the one treated in Section 3. Note that if the scheme does converge, then the limit solution and the corresponding control represent a solution to our original problem.

Since at each step of the iteration we have a linear problem of the kind solved in Section 3, we can at each step estimate a suboptimal control $\hat{u}_{k+1}$, rather than a strictly optimal one; if the sequence of operations converges, the controls towards the end of the sequence will be suboptimal for the original nonlinear problem. At each step, the modified simplex system solves a liner programming problem such as (12)-(13), a control is estimated as described at the end of Section 3, the new trajectory y^{k+1} is estimated by means of equations such as (6)-(7); the moments of the nonlinear term $\gamma f(y^{k+1}(x,t))$ are then computed, and the moments c_n, $n=0,1,\ldots,N$, recomputed.

We have tried a numerical implementation of this process, with the following values for the parameters and functions:

$$g(x)=0.01 + 0.01 \cos\pi x$$

$$\gamma=0.2,\ N_1=3,\ N_2=8,\ N_3=10,\ a=0.5,\ z=0$$

$$f^0(u)=u^2$$

After 8 iterations, the change in the moments c_n changed by less than about 2%, so we took the corresponding control as a suboptimal one; its graph can be seen in Figure 1. The main computational difficulties were, as in other applications of this method [4] that the finite-dimensional linear programming problems may be unfeasible, even if, as in this case, the infinite-dimensional ones are feasible. This happened seldom here, so it was perfectly acceptable to use the corresponding results - which were nearly feasible anyhow; the corresponding basis was nearly complete - as starting point for the simplex method in the next iteration.

We can see that the optimal control is of coure piecewise constant, but not bang-bang - perhaps a surprising result. Of course, current has to be put into the membrane as well as taking it out so as to reach the given final state. The control u in Figure 1 achieves this using - nearly - a minimum amount of energy to do so.

References

[1] Casten, R., Cohen, H. & Langerstrom, P. (1975). Perturbation analysis of an approximation to Hodgkin-Huxley theory. *Quart. Appl. Math.* **32**, 365-378.

[2] Fitzhugh, R. (1969). Mathematical models of excitation and propagation in nerve. In *Biological Engineering,* Schwan, P. (ed.). McGraw-Hill: New York.

[3] Ghouila-Houri, A. (1967). Sur la géneralization de la notion de commande d'un système guidable. *Revue Française d'Automatique, Informatique, et Recherche Operationelle* **4**, 7-32.

[4] Rubio, J.E. (1986). *Control and Optimization.* Manchester University Press.

[5] Wilson, D.A. & Rubio, J.E. (1977). Existence of optimal controls for the diffusion equation. *Journal of Optimization Theory and its Applications* **22**, 91-100.

[6] Zykov, V.S. (1988). *Simulation of Wave Processes in Excitable Media.* Manchester University Press.

13. A NUMERICAL MODEL FOR REENTRY IN WEAKLY COUPLED PARALLEL EXCITABLE FIBRES

J.Brindley[1], A.V.Holden[2], and A.Palmer[1]

[1]Department of Applied Mathematical Studies
[2]Department of Physiology
The University, Leeds, LS2 9JT

The phenomenon of reentry in excitable tissues, where a single wavefront excites the same region of tissue repeatedly, has been suggested as a mechanism for several cardiac disorders.

Numerical simulations of biophysically realistic excitation systems in two or more dimensions are computationally demanding, and so we examine a caricature that preserves the essential features of reentry in excitable tissues. We simulate reentry in two weakly coupled but otherwise separate parallel fibres described by the FitzHugh-Nagumo equations to model each fibre (FitzHugh, 1961). This model was first used by Hramov *et al* (1984) to study rotor drift in parallel fibres with non uniform conductivity.

$$\partial V_1/\partial t = \partial^2 V_1/\partial x^2 - V_1(V_1-1)(V_1-a) - w_1 + c(V_2-V_1)$$

$$\partial w_1/\partial t = \varepsilon\ (V_1-bw_1)$$

$$\partial V_2/\partial t = \partial^2 V_2/\partial x^2 - V_2(V_2-1)(V_2-a) - w_2 + c(V_1-V_2)$$

$$\partial w_2/\partial t = \varepsilon\ (V_2-bw_2),$$

where a acts as a threshold, c an ohmic conductance, V as a voltage and w a recovery variable (see fig.1). To establish reentry we start the computation with two uncoupled fibres. A single action potential is initiated at the end of one fibre while the other fibre remains at rest. When the impulse is about

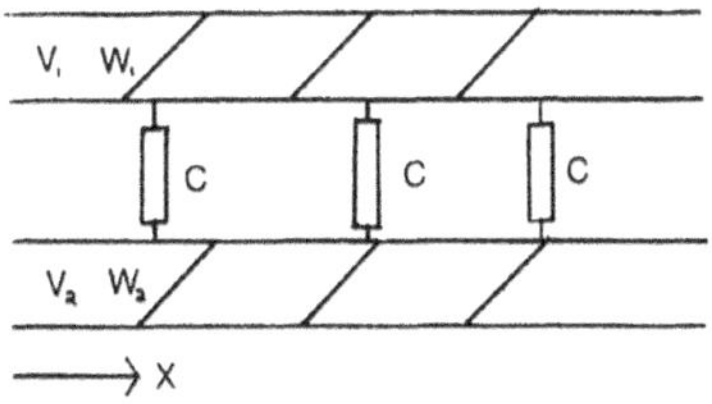

Figure 1. Two parallel fibres with linear coupling

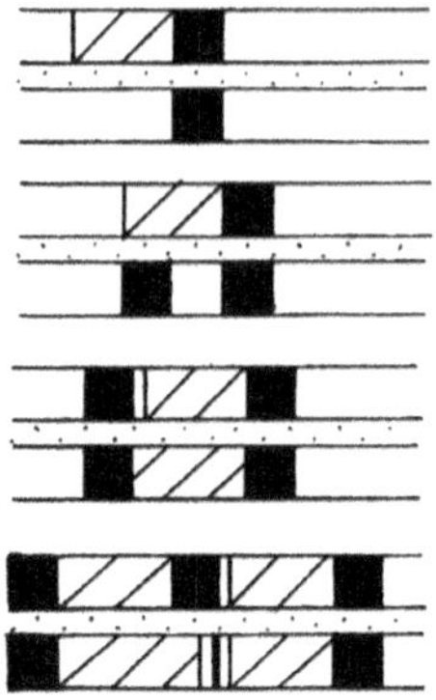

Figure 2. One cycle of reentry in two parallel fibres

half way along the active fibre the coupling is switched on. For certain ranges of the parameters a and c, a rotor of excitation is established in the fibres (see fig.2). A rotor occurs if two impulses traveling in opposite directions are formed at the time that the coupling is switched on. One impulse travels in the opposite direction to the original impulse and may restimulate the first fibre when it has passed the refractory tail of the original impulse.

This process can either continue for many cycles, or can terminate spontaneously. In figure 3 we have reentry continuing for at least six rotor cycles. Figure 4 shows the failure of reentry after two cycles. We have mapped how the number of impulses generated depends on the parameters corresponding to activation threshold, a, fibre coupling, c, and recovery rate, ε (figures 5 and 6). For low coupling (high resistance between fibres) and low threshold (high membrane excitability), a rotor is formed which generates many pairs of

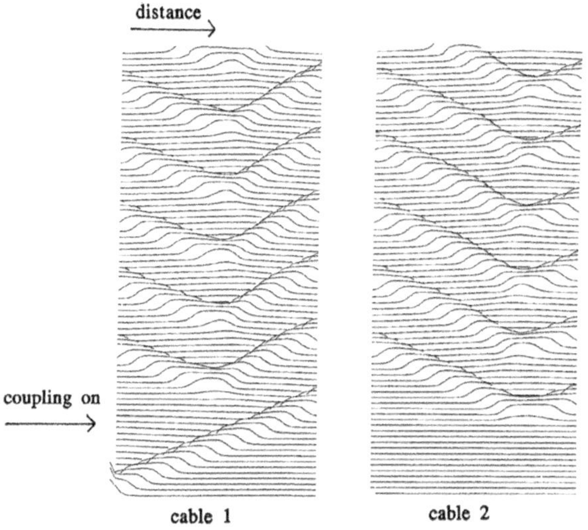

Figure 3. Six cycles of reentry. a=0.04, b=2.5, c=0.03, e=0.008

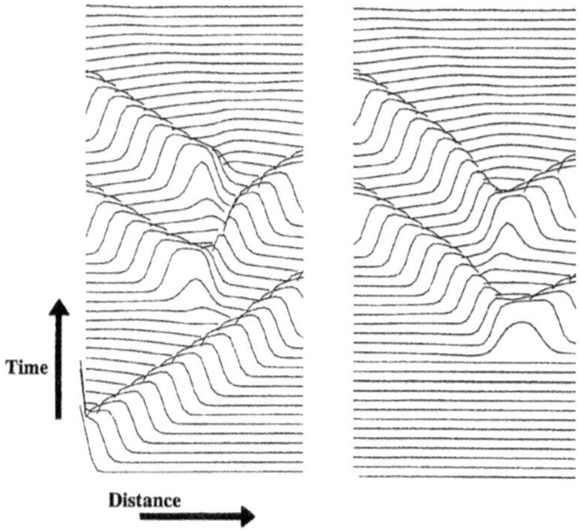

Figure 4. Failure of reentry after two cycles.
a=0.02, b=2.5, c=0.05, e=0.008

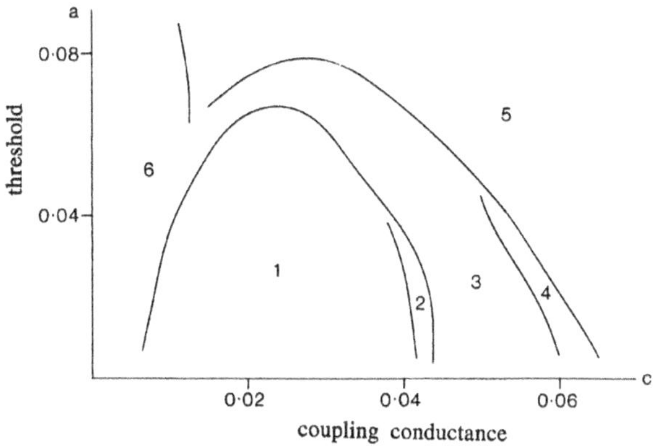

Fig 5. Dependence of behaviour on coupling and threshold, b fixed at 2.5, e fixed at 0.008
1. More than two cycles of reentry; 2. Two cycles of reentry
3. One cycle of reentry; 4. One and a half cycles of reentry
5. One coupled pair of impulses formed; 6. No impulse formed on second fibre

coupled pulses and as coupling and threshold are increased the rotor enters a region where only a few coupled impulses are generated. For high coupling or low threshold the coupling of the fibres does not induce a rotor.

Giving different properties to each fibre can lead to surprising results. In figure 7 the fibres have different conduction speeds due to different radii. This has allowed the formation of a rotor where fibre 1 excites fibre 2 twice for each time that fibre 2 excites fibre 1.

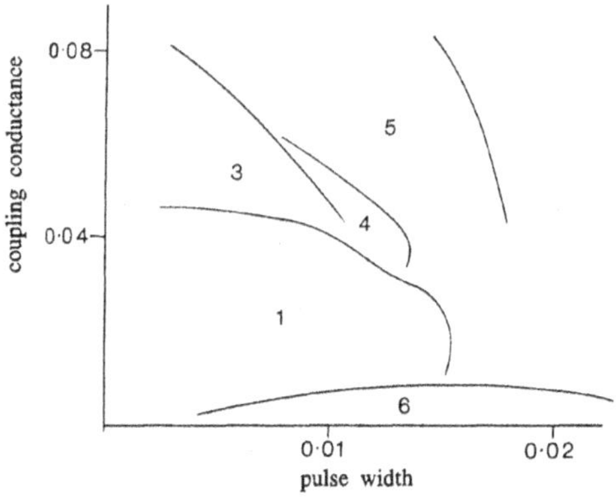

Figure 6. Dependence of behaviour on coupling and recovery rate b fixed at 0.008, a fixed at 0.02
Key as for fig 5

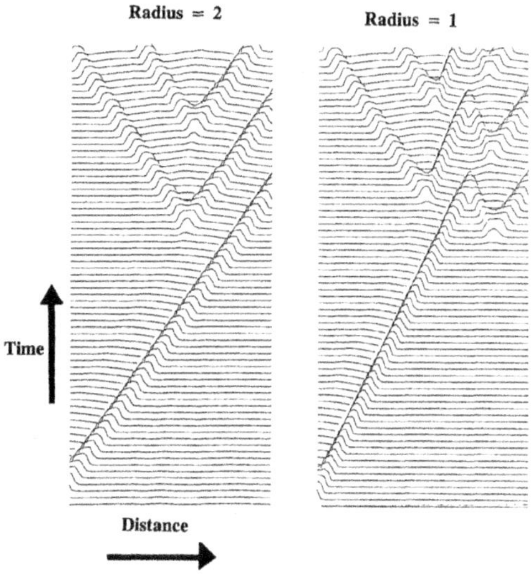

Figure 7. Reentry in coupled fibres of different radii

The numerical simulations were carried out using an implicit second order finite difference scheme (Crank Nicolson) with an explicit first order step to obtain initial predictions for the reaction terms. The step sizes used were 0.5 for the space step and 0.25 for the time step.

Acknowledgement: AP is supported by a SERC research studentship

References

FitzHugh,R. (1961), Biophys. J. **1** 445-66
Hramov, R.N., Rudenko, A.N., Panfilov,A.V., Krinskii, V.I. (1984), Studia Biophysica **102** 69-74

14. KINEMATICS OF WAVE PATTERNS IN EXCITABLE MEDIA

A.S. Mikhailov

Department of Physics
Moscow State University
117234 Moscow, USSR

The definitive property of an excitable medium is that it possesses a single steady state of rest, which is stable under small perturbations, and it supports undamped propagation of solitary pulses (in one dimension) or waves - such that the medium is found in the same state of rest before and after the wave has passed. Hence, in a two-dimensional medium (e.g. in a thin layer of the Belousov-Zhabotinskii solution) propagating waves might have a break.

What would be the subsequent evolution of a broken wave?

If we start with such an initial condition, further development would depend critically on the degree of excitability of our medium (which is correlated with the propagation velocity of pulses). Numeric simulations of Zykov [13,28] show that in a medium with poor excitability the broken wave contracts (Fig. 1a), while at sufficiently high excitability the wave sprouts from its free end (Fig. 1c,d) and this leads to creation of a spiral wave. In effect, this method of producing spiral waves by breaking of wavefronts is well-known and is commonly used in experiments with excitable media.

In a series of publications [1-14,21-28] Brazhnik, Davydov, Mikhailov, Morozova, and Zykov proposed a phenomenological kinematic theory of spiral waves and scroll rings in excitable media. This theory was used to describe steady circulation of spiral waves [4,13-14,22,,25] and to analyze their stability [6,8,9]. It explains drift of spiral waves in nonhomogeneous media [8,9] and predicts [8,9] a resonance effect for spiral waves in media with periodically modulated excitability (the latter effect was observed by Agladze, Davydov and Mikhailov [2] for the Belousov-Zhabotinskii reaction). When we additionally take into account the dependence of the propagation velocity on the time after passage of the previous wavefront, this theory predicts, within a certain interval of parameters, instability of steady circulation and periodic meandering of spiral waves [27-28]. The rotation frequency and other properties of spiral waves on spherical surfaces were found in [6]. The kinematic theory was also applied to describe complicated drift motion of spiral waves on landscapes with varying curvatures [12] and in anisotropic media [11]. In three dimensions this theory describes collapse and inflation of scroll rings [4-5,7], as well as existence of stable scroll rings [7,13]. It predicts resonance of scroll rings [1]. The kinematic theory was used to calculate the rotation frequency of twisted scrolls and to study propagation of twisting waves along a straight scroll [7,10].

Nonlinear Wave Processes in Excitable Media
Edited by A. V. Holden *et al.*, Plenum Press, New York

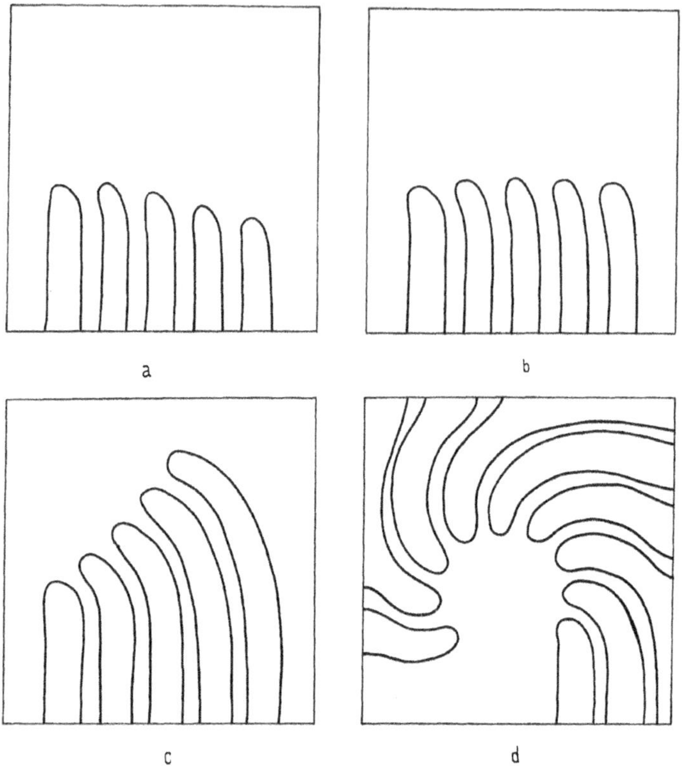

Figure 1. Evolution of a broken excitation wave for a two-component reaction-diffusion model (7.36) in the monograph by Zykov [25]. Lines of constant activator concentration E are shown at subsequent times for the four different excitabilities of the medium, controlled by the parameter ε. a - poor excitability ($\varepsilon = 0.4$), b - marginal excitability ($\varepsilon = 0.388$), c - supercritical excitability ($\varepsilon = 0.35$), d - high excitability ($\varepsilon = 0.3$). Computations by Zykov [13,28].

1. Properties of spiral waves

Kinematics of broken wavefronts in two-dimensional excitable media can be formulated as follows. Any wave is specified by indication of the oriented curve of its front. Most conveniently this can be done by using a so-called **natural equation** of a curve $k = K(l)$ that establishes a relationship between the length l of the curve arc, measured from the end point, and front curvature K at the corresponding point. Since a natural equation determines the curve only up to its position on a plane, the position of the end point and the tangent direction at this point should be provided additionally. If the form of a curve changes in time, $K = K(l,t)$.

a) Every segment of a front moves in its normal direction at the velocity

$$V = V_0 - DK. \tag{1}$$

b) The break of a wavefront contracts ($G < 0$) or smoothly sprouts in the tangent direction at the velocity

$$G = \gamma(K_{cr} - K_0), \tag{2}$$

where $K_0 = \lim_{l \to 0} K(l,t)$.

These two conditions are sufficient to find an equation for $K(l,t)$:

$$\frac{\partial K}{\partial t} + \left(\int_0^l KV\, d\xi + G\right) \frac{\partial K}{\partial l} = - K^2 V - \frac{\partial^2 V}{dl^2} \tag{3}$$

where the velocity G of sprouting is determined by the curvature $K_0(t)$ at the free end, and the velocity V of the normal front motion is determined by the local curvature K of the front, i.e. $V = V(K(l,t))$. Thus, we obtained a closed equation for $K(l,t)$ that governs evolution of a broken wavefront. Equation (3) was derived by Zykov [21] and later by Davydov & Mikhailov [4,14]. Recently it was used by Meron & Pelce' [18] who assumed that G is some constant parameter. In a slightly different form (namely, decomposed into two equations) for $G = 0$ it is given by Zykov in [25].

When G and V are given by (1) and (2), we find

$$\frac{\partial K}{\partial t} + \left[\int_0^l KV\, d\xi + \gamma(K_{cr} - K(0,t))\right] \frac{\partial K}{\partial l} = K^2 V_0 + D\, \frac{\partial^2 K}{\partial l^2}. \tag{4}$$

This equation is derived with an assumption that all curvatures are sufficiently small to satisfy the condition $DK/V_0 \ll 1$. Therefore, we have replaced V by V_0 everywhere in (4). However, we keep the last term in (4). Although it is proportional to a small parameter $\beta = DK/V_0$, this term involves a highest derivative and we should be particularly careful in dealing with it. Below it is shown that such a term does play an important role in a spiral wave solution.

Equation (4) has a trivial stationary solution $K(l,t) \equiv 0$ for $l \geq 0$, that corresponds to a flat broken front which propagates with velocity V_0. However, this solution is absolutely unstable with respect to small perturbations. Numerical simulations of (4) reveal (Fig. 2) that, independently of the form of the initial perturbation, a flat halfwave evolves to a steadily rotating spiral wave.

By using (4), we can find the basic properties of steadily rotating spiral waves.

For a spiral wave that rotates at a fixed angular velocity, the form of a front is constant; only its position on the plane changes in time. Therefore, K does not depend on t. Moreover, stable steady circulation is possible only if the front neither grows nor contracts at a break point. This implies that the curvature at the break point should be equal to the critical one, i.e. $K_0 = K_{cr}$. The circle around which the break point moves is the core of the spiral wave; note that that broken front is orthogonal to the core.

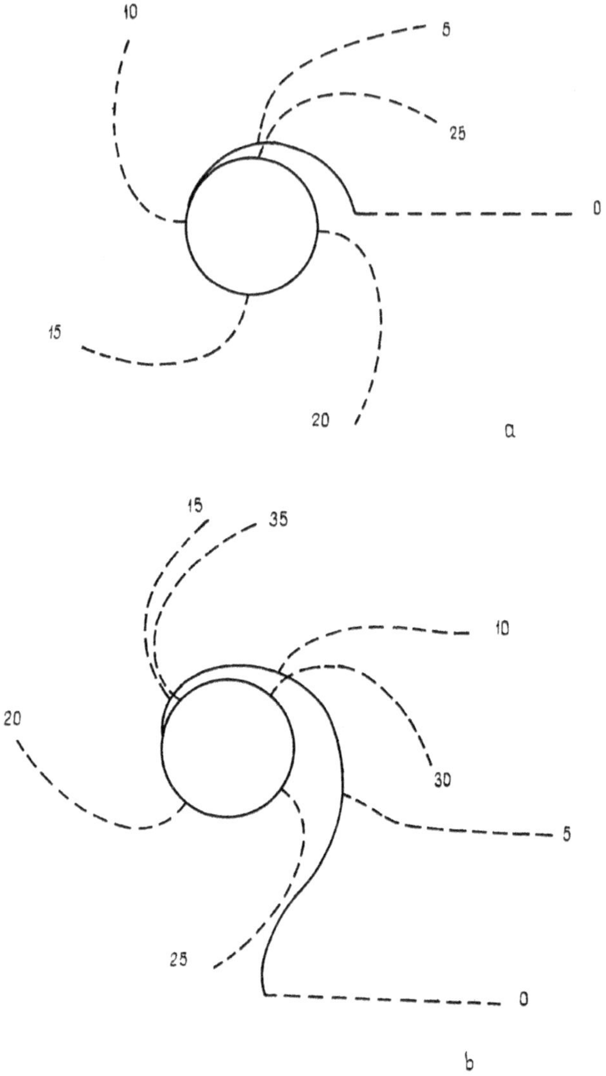

Figure 2. Evolution of an initially flat broken wave with a small positive (a) or negative (b) curvature perturbation near its end; subsequent front positions are shown. Computations by Zykov [13,28] for the kinematic equation (4).

Equation (4) describes, as well, circulation of a spiral wave around a hole in an excitable medium. If the hole boundary is not permeable to the diffusion flux, then the front is orthogonal to it. The form of a wavefront is then determined by a stationary limit of (4) in the absence of the term that describes sprouting of the free end (there is no free end in this case since the front touches a hole).

For a steady circulation (when $\partial K/\partial t = 0$) equation (4) can be integrated once which gives

$$K\int_0^l KV_0 \, d\xi - D\,(dK/dl) = \omega. \qquad (5)$$

The integration constant ω is simply a rotation frequency of the spiral wave. This can be easily recognized, if we note that the front is orthogonal to the core circle (or to the hole). Hence, we see that $dV/dl = -D(dK/dl)$ at $l = 0$ is the rotation frequency of a spiral wave front.

First we consider what is found in a model where V does not depend on the curvature, so that $D = 0$ and the second term in the left side of (5) vanishes. Then equation (5) can be easily solved analytically. Its solution is

$$K = (\omega/2lV_0)^{1/2}. \qquad (6)$$

Suppose that a hole of radius R is present in the medium. Then, the rotation frequency of a spiral wave that circulates around this hole is equal to $\omega = V_0/R$ and (6) gives

$$K = (2Rl)^{-1/2} \qquad (7)$$

A curve defined by this natural equation represents an involute of a circle of radius R. Far from a hole (at $l \gg R$) it approximates an Archimedian spiral with a constant step $h = 2\pi R$. Thus in a model with $V \equiv 0$ we reproduce the results of Wiener & Rosenblueth [29].

Note that at $l \to 0$ the solution (7) has $K \to \infty$. This implies that, if the coefficient D in the dependence $V = V_0 - DK$ is small but yet nonvanishing, the dependence of the propagation velocity on the curvature cannot be neglected at sufficiently small l's, close to the boundary of a core or a hole.

Now we turn to the problem of spiral waves in a homogeneous medium (in the absence of any holes) and take into account the dependence of the propagation velocity of a front on its local curvature. The rotation frequency $\omega = \omega_0$ of a spiral wave, that appears in (5) as an unknown parameter, is determined from the following arguments.

Far from the centre of a spiral, the front curvature should tend to zero, i.e. $K \to 0$ as $l \to \infty$. In effect, this condition already defines in a unique way a certain phase trajectory for the differential equation (5) at a given value of ω. Thereby, the value of the curvature $K(0)$ at $l = 0$, i.e. at the break of the wavefront, is uniquely determined. Demanding now that $K(0) = K_{cr}$, we obtain an equation which allows us to find ω_0.

For sufficiently large values of l, we can neglect the term with the derivative in the right side of (5). In this region the solution is given by

(6). On the other hand, it follows from (5) that at $l = 0$ we should have $dK/dl = -\omega/D$. Therefore, for sufficiently small values of l the natural equation of our curve is

$$K \approx K_{cr} - (\omega/D)l. \tag{8}$$

Equations (6) and (8) represent the outer and the inner approximations for the same phase trajectory, and therefore they should match at an intermediate point. Let us require that at some point $l = l_0$ the values of both functions, as well as their derivatives, coincide. These conditions yield

$$\omega_0 = (2/3)^{3/2}(DV_0)^{1/2}K_{cr}^{3/2}, \tag{9}$$

$$l_0 = (1/2)(3/2)^{1/2}(D/K_{cr}V_0)^{1/2}. \tag{10}$$

Matching of the outer and the inner solutions is a standard approach for the equation (5) that has a small parameter at the higher derivative. A more accurate estimate for the rotation frequency ω_0 of spiral waves can be obtained by scaling arguments which yield

$$\omega_0 = \nu(DV_0)^{1/2}K_{cr}^{3/2}, \quad \nu \approx 0.69\ldots \tag{11}$$

Up to the terms of order DK_{cr}/V_0, the core radius R_0 of a spiral wave can be estimated from the condition $\omega_0 R_0 = V_0$ to be

$$R_0 = \nu^{-1}\,(V_0/D)^{1/2}K_{cr}^{-3/2} \tag{12}$$

Now we discuss the results. We see that at $K_{cr} \to 0$ the rotation frequency ω_0 of a spiral wave diminishes and tends to zero; at the same time the core radius R_0 of the spiral wave increases indefinitely. Everywhere, except in a narrow boundary layer of width l_0 at the core boundary, the wavefront of the spiral represents an involute of a circle of radius R_0; its form is given by (7), with $r = R_0$. The width l_0 of the boundary layer is small compared with the core radius, i.e. $l_0/R_0 \sim (DK_{cr}/V_0) \ll 1$.

In a similar way we can find a solution for a spiral wave rotating around a hole. Note that the end of a spiral wave that moves along the impermeable boundary of the hole cannot be considered to be free. Since there is no diffusion flux in the direction along the front at its end, the conditions of front propagation are then the same as those for the solid, unbroken front. At the hole boundary the front curvature can therefore exceed the critical value K_{cr} for a free end of a wavefront, but it should remain less than the critical curvature K^* for the propagating unbroken front. This explains the hysteresis effect observed by Pertsov, Panfilov & Ermakova [19] in computer simulations.

In a certain interval of the hole radii ($R^* < R < R_0$), two circulation regimes with different rotation frequencies coexist. In the first regime, the end of a wavefront moves along the boundary of a hole. In the second regime, the wavefront has a free end that moves along a large core of radius R_0 and does not feel the presence of a hole located in the centre. If a large hole of radius $R > R_0$ is created in the medium at an initial time moment and later we gradually diminish its size, the spiral wave will remain stuck to the hole until the value $R = R^*$ is reached. After that the end of a wavefront will separate from the hole and the rotation frequency will abruptly decrease, reaching the value (11) of the rotation frequency of spiral waves in homogeneous media. The latter value of the rotation frequency will be maintained for $R < R^*$. If we move in the reverse direction, starting from a vanishingly small value for the hole radius R and then slowly increasing it, the regime with a free end of a spiral wave is maintained till $R = R_0$; the end of a spiral wave sticks to the hole border only when $R > R_0$.

In the framework of the kinematical model used it was assumed that the normal propagation velocity V depends only on the local curvature of a given segment of the wavefront, i.e. $V = V(K)$. Generally, this velocity V can depend, as well, on the distance from the end point (i.e. on the arc length l). This possibility was examined by Brazhnik, Davydov & Mikhailov [6].

Note that all results found in this Section are applicable only in the case when the core perimeter is much larger than the width of a single pulse and the front does not run into its own tail.

2. Resonance, drift, and meandering of spiral waves

The fundamental equation of wavefront kinematics (4) can be used to study various time-dependent wave phenomena in excitable media. We consider first the process of establishing steady circulation of spiral waves in homogeneous media. Suppose that a small perturbation is introduced into the form of a spiral wave, localized at a distance l from the free end of the wavefront. It follows from (4) that this perturbation will drift away from the centre to the periphery, simultaneously spreading and fading in a diffusional manner, with an effective "diffusion" constant D. According to (4), near the free end (where $K \not\approx K_{cr}$) the velocity of such drift is approximately estimated as $K_{cr}V_0l$. Hence, this perturbation moves away from the end point at a speed about $K_{cr}V_0l$ but at the same time its width increases as $(Dt)^{1/2}$. What is the distance at which such a perturbation should be localized initially if, despite the drift, it reaches the end point of the front? The latter happens if the width $L_s \approx (Dt)^{1/2}$ of a spreading perturbation exceeds (during some time interval) the distance $L_c \approx l + K_{cr}V_0lt$ from its centre to the end point of a wavefront. Comparing these two expressions, we find that L_s can exceed L_c only until the initial distance l is less than $(D/K_{cr}V_0)^{1/2}$. But, according to (10), this is precisely the length l_0 of a boundary layer near the core!

Consequently, the free end motion can be influenced only by those front perturbations that are created within a narrow boundary layer, at a distance not larger than l_0 from the end point. These perturbations of the front form are damped within the time $\tau_D \sim l_0^2/D \sim (K_{cr}V_0)^{-1}$. Note that the characteristic damping time τ_D for perturbations of the wavefront form that can influence motion of a free end is always smaller than the rotation period of a spiral wave. Indeed, it follows from (11) that $\omega_0\tau_D \sim (DK_{cr}/V_0)^{1/2} \ll 1$.

There is, also, another characteristic time τ_G related to the effects of front sprouting or contraction.

If the cofficient γ in the expression (2) for the speed of sprouting were equal to zero, K_{cr} would not enter at all into the fundamental equation of kinematics (4). In this case, many spiral wave solutions exist, differing by the value K(0) of the front curvature at a free end. Every such solution describes a certain spiral wave that steadily circulates with a frequency ω given by (11) with a replacement of K_{cr} by K(0). Then, by locally perturbing the form of a wavefront within the boundary layer, we can initiate transition to a new circulation regime with the value of the front curvature at a free end which differs from the previous one by a small correction term $\delta K(0)$.

When the coefficient γ is nonvanishing and positive, all deviations of the curvature K(0) from K_{cr} are damped. However, when γ is sufficiently small, the characteristic time-scale τ_G of such damping is large and a **quasistationary regime** with $\tau_G \gg \tau_D$ is realized. In this regime, the form of a wavefront near the core adjusts adiabatically to the curvature K(0,t) at the free end. Brazhnik et al. [8-9] showed that, under the conditions of a quasistationary regime, the curvature $K_0 \equiv K(0,t)$ at a free end obeys the equation

$$\frac{dK_0}{dt} = -\nu\gamma(V_0/D)^{1/2}K_0^{3/2}(K_0 - K_{cr}). \tag{13}$$

For small perturbations $\delta K_0 = K_0 - K_{cr}$, Eq. (13) can be linearized so that it takes the form

$$(d/dt)\delta K_0 = -\delta K_0/\tau_G, \tag{14}$$

where the characteristic relaxation time τ_G is $\tau_G = D/\gamma\omega_0$.

Note that $\omega_0\tau_G = D/\gamma$ and, therefore, the quasistationarity condition $\tau_D \ll \tau_G$ reads as

$$\gamma/D \ll (V_0/DK_{cr})^{1/2}. \tag{15}$$

Estimates for the known examples of excitable media show that usually γ varies from zero to about D. Therefore, since $(DK_{cr}/V_0)^{1/2} \ll 1$ is assumed, condition (15) is almost always satisfied and the quasistationary approximation is applicable.

As mentioned above, a natural equation K = K(l,t) defines the form of a curve, but not its position on a plane. To specify evolution of a wavefront, we should supplement this equation by equations which determine an orientation of this curve and indicate the position of its initial point.

Since the end point moves simultaneously in the normal and in the tangent direction (because of the front sprouting or contraction), its Cartesian coordinates X_0 and Y_0 change in time according to equations

$$\dot{X}_0 = - V(0) \sin \alpha_0 - g \cos \alpha_0,$$
$$\dot{Y}_0 = V(0) \cos \alpha_0 - G \sin \alpha_0. \tag{16}$$

Furthermore, evolution of an angle α_0 that specifies the direction of a front tangent at the end point obeys an equation

$$\frac{d\alpha_0}{dt} = - D \left.\frac{\partial K}{\partial l}\right|_{l = 0} + GK_0. \tag{17}$$

Together with the fundamental equation of kinematics (4), equations (16) and (17) completely define the motion of a wavefront with a free end.

The law of motion of an end point is especially simple in a quasistationary approximation, when we can distinguish the effects of sprouting from the effects of establishing the steady form of a wavefront near the core. In this case $\partial K/\partial l \approx - \omega/D$ at $l = 0$, where ω is given by (11) with the replacement of K_{cr} by K_0. Hence, (17) reads as

$$\dot{\alpha}_0 = \nu\, (V_0 D)^{1/2} K_0^{3/2} + \gamma K_0 (K_{cr} - K_0). \tag{18}$$

Therefore, when condition (15) holds, the trajectory of the end point of a spiral wave can be found by solving four ordinary differential equations (13), (16), and (18).

By varying the parameters of a medium, we can change its excitability and, therefore, the critical curvature K_{cr}. Suppose that this quantity changes periodically in time as

$$K_{cr}(t) = K_{cr} + K_1 \cos (\omega_1 t + \varphi), \tag{19}$$

where $K_1 \ll K_{cr}$ and the modulation frequency ω_1 is close to the rotation frequency ω_0 of a spiral wave. Then, as it was shown in [30], a **resonance effect** is observed.

In the presence of a periodic modulation of K_{cr}, the centre of rotation of a spiral wave is not fixed but performs a circular motion. This effect can be demonstrated (see Fig. 3) by a direct numeric integration of equations (13), (16), and (18) with $K_{cr}(t)$ given by (19). An approximate analytical theory of the resonance effect is also possible when $K_1 \ll K_{cr}$. In this case, variations of K_0 are small and the linearized version of these equations can be used. Straightforward, but somewhat tedious calculations performed by Brazhnik et al. [9] yield the following results.

The radius R_{res} of a circle along which the centre of a spiral wave migrates is

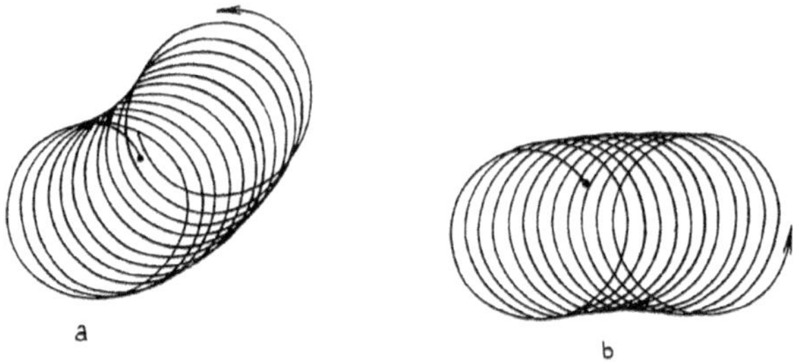

Figure 3. Resonance of a spiral wave under periodic modulation of the excitability of the medium. The trajectory of the free end point is shown. The spiral wave frequency is $\omega = 0.3186$, parameter K_{cr} varies periodically with frequency (a) $\omega_1 = 0.315$ and (b) $\omega_1 = 0.32$. Computations by Zykov [13,28].

$$R_{res} = \frac{(3/4)\ V_0(K_1/K_{cr})}{|\omega_1 - \omega_0|[1 + (D/\gamma)^2]^{1/2}} \qquad (20)$$

Note that R_{res} grows when ω_1 is approaching ω_0. The velocity of motion of a spiral wave centre is

$$V_{res} = \frac{(3/4)\ V_0\ (K_1/K_{cr})}{[1 + (D/\gamma)^2]^{1/2}}\ ; \qquad (21)$$

it is proportional to the modulation amplitude K_1.

When the two frequencies coincide (i.e. under the condition $\omega_1 = \omega_0$ of a complete resonance), the spiral wave centre moves at a constant velocity (21) along a straight line. The direction of the centre motion is then determined by the initial modulation phase φ and by the direction of rotation of this spiral wave (clockwise or counterclockwise).

Resonance of spiral waves was observed by Agladze, Davydov & Mikhailov [2] in the experiment using a photosensitive modification of the Belousov-Zhabotinskii reaction.

If the medium contains two spiral waves with the opposite directions of rotation, then by adjusting the initial phase of modulation, these two waves can be made to migrate one to another and, finally, to annihilate.

Next we consider spiral wave drift in inhomogeneous excitable media which was analyzed in the framework of kinematics by Brazhnik et al. [8,9]. We assume that the critical curvature K_{cr} varies along the x-axis, but so slowly that the change of K_{cr} on a distance of a core radius R_0 is small: $|\partial K_{cr}/\partial x| \ll K_{cr}/R_0$. When the free end of a spiral wave moves in such an inhomogeneous medium, it passes through regions with different values of K_{cr}. Therefore, the free end experiences periodic variation of a critical curvature and the problem of drift is effectively reduced to that of resonance of the spiral wave, with a modulation frequency coinciding with the frequency of the spiral wave. The spiral wave centre will slowly drift along a straight line at some angle θ to the x axis. The drift velocity is

$$V_d = \frac{(3/4)V_0}{[1 + (D/\gamma)^2]^{1/2}} \cdot \frac{R_0}{K_{cr}} \cdot \left|\frac{\partial K_{cr}}{\partial x}\right|. \tag{22}$$

The direction of the drift is determined by the angle θ, where $\tan\theta \approx -\gamma/D$.

Spiral wave migration in **anisotropic** media with an angular dependence of the propagation velocity V_0 was investigated recently in the framework of kinematics by Davydov & Zykov [11].

When a spiral wave rotates not on a plane, but on a **curved surface**, the fundamental equation of kinematics (3) should be replaced (see Brazhnik, Davydov & Mikhailov [6]) by

$$\frac{\partial K}{\partial t} + \left(\int_0^1 KV\, d\xi + G\right) \frac{\partial K}{\partial l} + K^2 V + \frac{\partial^2 V}{\partial l^2} = - WV, \tag{23}$$

where W is the local Gaussian curvature of the surface. The solutions of (2.11) in the form of steady spirals with immobile centres are possible only for the surfaces with constant W, e.g. for the sphere and the plane (W = 0). Brazhnik, Davydov & Mikhailov [6] showed that stable solutions on a sphere always consist of **two** spiral waves with the cores in the opposite poles, which rotate in the same direction. The rotation frequency depends on the radius of the sphere.

If the local curvature W of the supporting surface varies, spiral waves migrate in a complicated manner. These effects were analyzed by Davydov & Zykov [12] within the quasistationary approximation in the case of small curvatures W.

In the above mentioned effects of drift and of resonance of spiral waves, migrations of spiral waves are induced by some external factors, such as a gradient in the excitability or its periodic uniform modulation. However, in a slightly more general kinematic model spiral waves can become inherently unstable with respect to migrations of their centres. This instability explains **meandering** of spiral waves.

In a simplest kinematic model we assumed that the propagation velocity of an excitation wave depends only on the local curvature of its front. Therefore, the above results are valid only for sufficiently sparse wave patterns, with distances between any two consecutive waves much larger than the characteristic width of an excitation pulse, including its tail. This limitation can be lifted, to some extent, if we introduce into our model the dependences of the propagation velocity V_0 and the critical curvature K_{cr} on the time interval T which is measured after the last passage of an excitation wave through a given point of the medium (a similar approach used recently by Keener & Tyson [15-16,20] did not include the notion of a broken front).

The specific form of functions $V_0(T)$ and $K_{cr}(T)$ is determined by a particular reaction-diffusion model. As a possible approximation, one can use the expressions

$$V_0(T) = V_0^{(0)}(1 - \mu/T),$$
$$K_{cr}(T) = K_{cr}^{(0)}(1 - \Gamma/T). \tag{24}$$

Here $V_0^{(0)}$ and $K_{cr}^{(0)}$ are the corresponding quantities for a solitary propagating wave (i.e. in a limit $T \to \infty$); μ and Γ are some positive coefficients.

Functions $V_0(T)$ and $K_{cr}(T)$ can be used instead of the parameters V_0 and K_{cr} in the equations for front kinematics. These equations should be supplemented by an equation for T as a function of coordinates and time. To construct $T(x,y,t)$ we notice the moment $T^*(x,y,t)$ when an excitation wave last passed through a given point with the coordinates x and y. Then function $T(x,y,t)$ is given by $T = t - T^*(x,y,t)$.

For a steadily circulating spiral wave, the time interval T between any two consecutive waves is the same for all points of the medium, it is simply the rotation period of the spiral wave $T = 2\pi/\omega$. Hence, such spiral waves are described in precisely the same manner as in Section 1. The only difference is that now we should substitute $V_0(T)$ ad $K_{cr}(T)$ instead of V_0 and K_{cr}. To find the rotation frequency, we can use (11) which now becomes an equation for the rotation frequency ω of spiral waves.

Note that no excitation fronts propagate inside the central core region of a spiral wave. Therefore, $T = \infty$ in the core region. On the other hand, in the remaining part of the medium we have $T = T_0$. Hence, a steadily rotating spiral produces a special distribution of T's which is everywhere constant except for a circular central region of radius R_0 where T goes to infinity.

This effective nonuniformity of medium properties creates the conditions for **meandering** of spiral waves which was found in numerical simulations by Zykov [26] and later explained by him within the framework of wavefront kinematics [27-28]. Suppose we slightly perturb a spiral wave, so that its tip enters the core region. Then its motion is determined by the laws of kinematics with a higher propagation velocity V_0 and a higher critical curvature K_{cr} (since $T = \infty$ there). This implies that such a tip will move quickly along a trajectory of a small radius. Such motion will continue until the front does not run into its own tail, i.e. does not come into a region with a small value of T. When this happens, K_{cr} sharply decreases. As a result, further propagation of this portion of the front, which has high curvature, is impossible. The end point shifts by a jump along the front to the point with a curvature less than the new value of K_{cr} and propagation is allowed. After that the end point will start to move at a lower velocity along the trajectory of higher radius. It will result in a slow increase in the value of T.

The above considerations indicate that a steady circulation of spiral waves can be unstable, resulting in a complicated motion of the spiral wave tip. This effect is revealed by direct numerical integrations of the kinematic equation (Fig. 4).

3 Wave patterns in three dimensions

The kinematic approach developed for two-dimensional excitable media can easily be generalized to three dimensions (Brazhnik, Davydov, Mikhailov & Zykov [4-5]). We assume that a wave is completely defined by its oriented front surface. Any surface in a three-dimensional space is characterized, at its given point, by two principal curvature radii R_1 and R_2. For small curvatures, the normal propagation velocity of a small surface segment depends linearly on the sum of the two principal curvatures, i.e. on the doubled mean curvature $2H = 1/R_1 + 1/R_2$, so that $V = V_0 - 2DH$, where D is a certain proportionality factor.

A front surface can sprout or contract at its free edge (where it is cut). The velocity G of such tangent motion depends not only on the mean front curvature H at the edge but, also, on the geodetic curvature æ of the edge line. For small curvatures, we obtain [5] in the linear approximation $G = -\gamma_1(2h - K_{cr}) - \gamma_2\text{æ}$. Here γ_1 and γ_2 are some positive coefficients; the geodetic curvature of the edge line is taken to be positive if the edge line is convex in respect to the front surface.

In the framework of the kinematic theory, we can analyze the evolution of a simple (untwisted) scroll ring. We assume that the scroll ring radius R is much larger than the thickness of its filament, which is about the diameter of a spiral wave core. The detailed calculation by Brazhnik et al. [5] shows that, in the linear 1/R approximation, the scroll ring radius changes in time according to the equation

$$\dot{R} = -\frac{D}{R}\left(1 - \frac{3}{4\beta} \cdot \frac{q_1^2 + q_2}{q_1^2 + 1}\right), \tag{25}$$

where we introduced the notations $q_1 = \gamma_1/D$ and $q_2 = (\gamma_2 - D)/D$. At the same time such a ring drifts in the vertical direction (i.e. along its symmetry axis) with the velocity

$$\dot{Z}_0 = -\frac{D}{R}\left(\frac{q_2}{2\nu\beta} + \frac{3}{4\beta} \cdot \frac{q_1 q_2}{q_1^2 + 1}\right). \tag{26}$$

If a scroll rotates in the opposite direction, the direction of its drift along the symmetry axis is also reversed.

Therefore, a scroll ring is almost always unstable. It either shrinks (if $dR/dt < 0$) and disappears, or inflates (if $dR/dt > 0$). Collapse or inflation are accompanied by the drift of a ring along its symmetry axis.

There is, however, a narrow interval of the parameters q_1 and q_2 where dR/dt is very small and changes its sign -see (24). Within this interval we should take into account in the expression for dR/dt the terms of higher orders in 1/R. A careful examination reveals [17] that the terms of the order

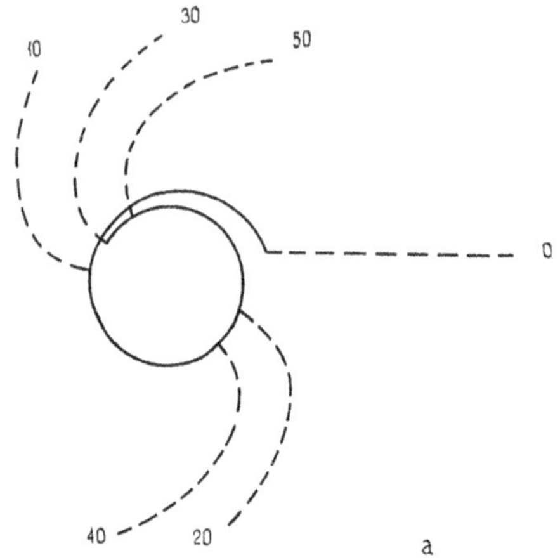

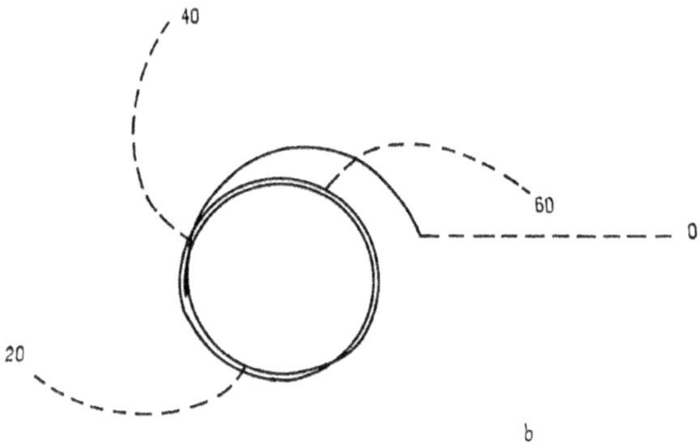

Figure 4 a and b (for legend see facing page).

$1/R^2$ vanish after averaging over the rotation period, whereas the nonvanishing cubic term $1/R^3$ enters into the expression for dR/dt with a positive coefficient. Therefore, if in the linear approximation a scroll ring is shrinking but the speed of shrinking is very small, this nonlinear positive term will prevent collapse and stabilize the scroll ring.

When $q_1^2 \gg q_2$, the radius of a stable scroll ring is given by

$$R = R_0 \frac{[(13/32)\nu\beta^{1/2}q_1]^{1/2}}{[q_1^2/\beta - 3/4]^{1/2}} , \qquad (26)$$

where $R_0 = V_0/\omega_0$ is the filament radius of the scroll ring which coincides, in this approximation, with the spiral wave core radius in the same excitable medium.

Equation (26) implies that stable scroll rings exist only for $q_1^2 >$

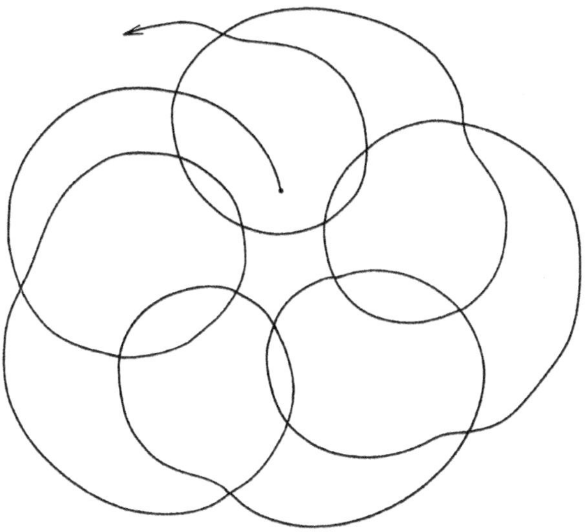

c

Figure 4. Evolution of a broken flat wave in media with different refractoriness. When refractoriness is low ($\mu = \Gamma = 1$) steady circulation is established (a). At a higher refractoriness ($\mu = \Gamma = 5$) small meandering of the tip around a core circle appears (b). At still larger refractoriness ($\mu = \Gamma = 20$), which is comparable to the rotation period $T_0 = 18$ of the spiral wave, the wave tip moves along a cycloid trajectory and the amplitude of meandering of a spiral wave is large (c). Computations by Morozova & Zykov [28].

$(3/4)\beta$. The radius of a stable scroll ring becomes infinite at $q_1^2 = (3/4)\beta$; it rapidly diminishes for larger values of q_1. Equation (27) remains valid while $R \gg R_0$. It is expected that the solution for a stable scroll ring is lost when R is about R_0. Note that, according to (26), stable scroll rings drift slowly with a velocity of about $\beta^{1/2}(D/R)$ along their symmetry axes.

Stable scroll rings were observed by Zykov in a numerical simulation [29] for a particular model of an excitable medium.

Although the above results were found for circular scroll rings, they can be used, as well, to describe in the simplest approximation the time evolution of any vortex with a slightly deformed filament. Indeed, if we consider a small section of such a vortex, it will look like a part of a scroll ring of the same radius.

Suppose that initially we had a straight scroll and then we have locally perturbed its filament. If the parameters of a given medium are such that scroll rings shrink there, this perturbation will diminish in time. Hence, we can say that a vortex filament possesses in this case a certain "elasticity", tending to shorten its length. In the opposite case, when a scroll ring inflates, any protrusion of a filament will increase in time, leading to the filament elongation. Consequently, a straight cylindrical scroll turns out to be unstable with respect to small local deformations of its filament. This instability could lead to formation of very complex (and, possibly, even chaotically organized) wave patterns in such media.

The resonance of scroll rings was studied by Abramychev, Davydov & Mikhailov [1]. By applying periodic modulation of the medium excitability and adjusting the amplitude and the phase of such modulation, it is possible to stabilize inflating scroll rings and control the velocity of their drift along a central axis.

To conclude this Section, we discuss properties of twisted vortices. For a cylindrical twisted vortex, the front edge curve represents a screw line that is wound onto a cylinder (the vortex filament). The local twistedness of such a vortex is characterized by a parameter $\mu = 2\pi/h$, where h is the step of a screw line. This parameter can be positive or negative, depending on whether the screw is right or left.

The kinematic theory predicts that the rotation frequency of a twisted scroll should be higher than that for a simple scroll. For small values of μ, the rotation frequency of a twisted scroll wave was found by Brazhnik et al. [8,10]

$$\omega = \omega_0 \; [1 + (3/2)(\gamma_2/\gamma_1)(|\mu|/K_{cr})]. \tag{28}$$

If we twist a vortex nonuniformly, so that $\mu = F(z)$, the "waves of twistedness" will start to propagate. This was studied by Brazhnik et al. [3,11]. In the first approximation, the pulses of twistedness propagate without changing their form at a speed

$$u = (3/2)(\gamma_2/\gamma_1)(\omega_0/K_{cr}). \tag{29}$$

The direction of propagation is determined by the sign of μ. When two propagating pulses collide, an analog of a shock structure is produced.

Note that the above results are valid, as well, for twisted scroll rings whose filaments are closed into circles. When the radii of such rings are large enough, they shrink (or inflate) and drift along a central axis in the same manner as the untwisted rings.

References

[1] Abramychev, A.Yu., Davydov, V.A. & Mikhailov, A.S. On the theory of resonance in excitable media. *Biofizika*, to be published.

[2] Agladze, K.I., Davydov, V.A. & Mikhailov, A.S. (1987). Observation of

the resonance of spiral waves in an excitable distributed medium. *Pisma Zh. Eksp. Teor. Fiz.* **45**, 601-603. (English transl.: *Sov. Phys. - JETP Letters.*)

[3] Brazhnik, P.K. (1988). *Geometric Methods in the Theory of Autowave Patterns.* Ph.D. Thesis (Dept. Physics, Moscow State Univ.).

[4] Brazhnik, P.K., Davydov, V.A.& Mikhailov, A.S. (1986). Spiral waves and vortex rings in combustion with subsequent recovering of the initial properties of the medium. In *Kinetics and Combustion, Proc. VIII Soviet Symp. on Combustion and Explosion, Tashkent, 1986,* pp. 39-43. (Inst. Chem. Phys., Chernogolovka.)

[5] Brazhnik, P.K., Davydov, V.A., Zykov, V.S. & Mikhailov, A.S. (1987). Vortex rings in excitable media. *Zh. Eksp. Teor. Fiz.* **93**, 1725-1736. (English transl.: *Sov. Phys. - JETP.*)

[6] Brazhnik, P.K., Davydov, V.A. & Mikhailov, A.S. (1987). Kinematic approach to description of autowave processes in active media. *Teor. Mat. Fiz.* **74**, 440-447. (English transl.: *Sov. Phys. - Theor. Math. Phys.*)

[7] Brazhnik, P.K., Davydov, V.A., Zykov, V.S. & Mikhailov, A.S. (1987). Dynamics of three-dimensional autowave patterns. In *Proc. II Soviet Conf. on Mathematical and Computational Methods in Biology, Pushchino,* pp. 118-119.

[8] Brazhnik, P.K., Davydov, V.A., Zykov, V.S. & Mikhailov, A.S. (1987). Evolution of spiral waves in nonhomogeneous and nonstationary excitable media. In *Proc. II Soviet Conf. on Mathematical and Computational Methods in Biology, Pushchino,* pp. 119-120.

[9] Brazhnik, P.K., Davydov, V.A., Zykov, V.S. & Mikhailov, A.S. (1988). Drift and resonance of spiral waves in excitable media. *Izv. VUZ. Radiofizika* **31**, 574-584. (English transl.: *Sov. Phys. - Radiophysics.*)

[10] Brazhnik, P.K., Davydov, V.A. & Mikhailov, A.S. (1989) Twisted vortex in excitable medium. *Izv. VUZ. Radiofizika* **32**, 289-293. (English transl.: *Sov. Phys. - Radiophysics.*)

[11] Davydov, V.A. & Zykov, V.S. (1989). Spiral waves in anisotropic excitable media. *Zh. Eksp. Teor. Fiz.* **95**, 139-148. (English transl.: *Sov. Phys. - JETP.*)

[12] Davydov, V.A. & Zykov, V.S. Drift of spiral wave on nonuniformly curved surface. *Zh. Eksp. Teor. Fiz.*, to be published.

[13] Davydov, V.A., Mikhailov, A.S. & Zykov, V.S. (1989). Kinematical theory of autowave patterns in excitable media. In *Nonlinear Waves in Active Media,* Engelbrecht, Yu. (ed.). Springer: Berlin.

[14] Davydov, V.A. & Mikhailov, A.S. (1987). Spiral waves in distributed active media. In *Nonlinear Waves. Patterns and Bifurcations,* pp. 261-279, Gaponov-Grekhov, A.V. & Rabinovich, M.I. (eds.). Nauka: Moscow.

[15] Keener, J.P. (1986). A geometrical theory for spiral waves in excitable media. *SIAM J. Appl. Math.* **46**, 1039-1056.

[16] Keener, J.P. & Tyson, J.J. (1986). Spiral waves in the Belousov-Zhabotinsky reaction. *Physica* **D21**, 300-324.

[17] Khrustova, N.A. (1989). Diploma Thesis (Dept. Physics, Moscow State University).

[18] Meron, E. & Pelce', P. (1988). Model for spiral wave formation in excitable media. *Phys. Rev. Lett.* **60**, 1880-1883.

[19] Pertsov, A.M., Panfilov, A.V. & Ermakova, E.A. (1984). Numerical simulation of spiral waves in active media. *Physica* **D14**, 117-125.

[20] Tyson, J.J. & Keener, J.P. (1987). Spiral waves in a model of myocardium. *Physica* **D29**, 215-222.

[21] Zykov, V.S. (1975). Investigations of some properties of self-sustained acitvity in excitable medium. In *Controlling Complex systems,* pp. 59-66, Tsypkin, Ya.A. (ed.). Nauka: Moscow.

[22] Zykov, V.A. (1980). Kinematics of steady circulation in excitable medium. *Biofizika* **25**, 319-322. (English transl.: *Sov. Phys.Biophysics.*)

[23] Zykov, V.A. (1980). An analytical estimate of dependence of propagation speed of excitation wave in two-dimensional excitable medium on the curvature of its front. *Biofizika* **25**, 888-892. (English transl.: *Sov. Phys. - Biophysics.*)

[24] Zykov, V.S. & Morozova, O.L. (1980). Circulation frequency of spiral wave and form of excitation pulse. *Biofizika* **25**, 1071-1076. (English transl.: *Sov. Phys. - Biophysics.*)

[25] Zykov, V.S. (1984). *Simulation of Wave Processes in Excitable Media.* Nauka: Moscow. (English translation: Manchester University Press, 1987).

[26] Zykov, V.S. (1986). Cycloidal circulation of spiral waves in excitable medium. *Biofizika* **31**, 862-865. (English transl.: *Sov. Phys. - Biophysics.*

[27] Zykov, V.S. (1987). Kinematics of nonsteady circulation of spiral waves in excitable medium. *Biofizika* **32**, 337-430 (English transl.: *Sov. Phys. - Biophysics.*)

[28] Zykov, V.S. & Morozova, O.L. (1988). Kinematic method of investigation of stability of spiral autowaves. Preprint, Inst. of Control Science, Moscow.

[29] Wiener, N. & Rosenblueth, A. (1946). The mathematical formulation of the problem of conduction of impulses in a network of connected excitable elements, specifically in cardiàc muscle. *Arch. Inst. Cardiol. Mex.* **16**, 205-265.

15. THE EFFECT OF WAVEFRONT INTERACTIONS ON PATTERN FORMATION IN EXCITABLE MEDIA

E. Meron

Department of Chemical Physics
Weizmann Institute of Science
Rehovot 76100, Israel

The nonlinear interactions of solitary wavefronts in excitable media are determined by the manner of recovery to the rest state. The distance between a pair of wavefronts tends to lock at one of countably many possible values in the case of oscillatory recovery, while it increases indefinitely when the recovery is monotonic. We derive these results from the basic reaction diffusion equations and study the implications on pattern formation in one and two space dimensions. In particular we demonstrate how spatiotemporal complexity may arise in one dimension, and discuss possible consequences of the interplay between wavefront-interactions and curvature in two dimensions.

1. Introduction

Excitable media provide good examples of nonequilibrium systems where spatially extended patterns can be understood in terms of simple building blocks: solitary waves or impulses in one space dimension, solitary wavefronts and vortices in two-dimensions and so on. The basic reason for that lies in the localized nature of these structures. Thus a given impulse in a wavetrain can be viewed as an independent, particle-like entity that propagates in the **perturbative** field of nearby impulses. Following further the analogy to particles we would like to know how to determine the degrees of freedom of localized structures, and how to obtain their time evolution in the presence of other structures in the neighbourhood. Once the dynamical laws of individual localized structures are known we can address the many body problem and study spatiotemporal behaviours of extended patterns.

The key to these questions rests in the continuous symmetries of the system. Assuming a homogeneous medium (or translational invariance), the degree of freedom of an impulse is the position of that impulse at a given time, x(t), or the time it passes through a given location, t(x). In the presence of a perturbing field this degree of freedom acquires a slow component, $\chi(t) = x(t) - c_0 t$, where c_0 is the propagation speed of the unperturbed impulse. An equation of motion of this component is derivable as a solvability condition which guarantees that the unperturbed form of the impulse that propagates at speed $c_0 + \dot{\chi}$, is an approximate solution of the perturbed problem.

In section 2 I will show how this approach can be used to study the interaction between a pair of impulses and to understand the emergence of complex spatial structures in one space dimension. In section 3 I will proceed to patterns in two dimensions and consider dynamical aspects of rotating spiral waves. More specifically, I will show how the interplay between wavefront interactions and curvature may lead to destabilization of steady rotation and to spiral waves whose cores expand in time. A brief discussion of these results in section 4 will conclude this presentation.

2. Wavetrains of impulses in one space dimension

Consider homogeneous excitable media that support solitary waves propagating at constant speed c_0. These media are described by reaction-diffusion equations (rde's) of the general form [1]

$$\partial_t \mathbf{U} = \mathbf{M}(\mathbf{U}) + \mathbf{D}\partial_x^2 \mathbf{U}, \tag{1}$$

where $\mathbf{U}$ represents a set of fields, $\mathbf{M}(\mathbf{U})$ is the reaction part and $\mathbf{D}$ is a matrix of transport coefficients (diffusion, conduction, etc.). It is assumed that $\mathbf{M}(\mathbf{0}) = \mathbf{0}$. The solution $\mathbf{U} = \mathbf{0}$ represents the quiescent state of the medium.

Let $\mathbf{U} = \mathbf{H}(x - c_0t)$ denote a solitary wave solution of (1). An alternative notation, $\mathbf{U} = \mathbf{S}(t - x/c_0)$, will be useful when viewing time as the dependent variable. Being localized, the solitary wave solution decays off exponentially as $\chi \equiv x - c_0t \longrightarrow \pm\infty$ or as $\tau \equiv t - x/c_0 \longrightarrow \mp\infty$. The tail of the solitary wave ($\chi \longrightarrow -\infty$ or $\tau \longrightarrow \infty$) represents the manner in which the medium recovers to the rest state after excitation. Two principal forms of recovery are possible: **monotonic**, in which case the tail assumes a pure exponential form, $\mathbf{H}(\chi) \propto \exp(\eta_L\chi)$, and (damped) **oscillatory**, $\mathbf{H}(\chi) \propto \exp(\eta_L\chi)\cos(\nu_L\chi + \phi_L)$. In the latter case the system undergoes a succession of super and subnormal periods until complete recovery is attained [2,3]. The head of the solitary wave is assumed to have in both cases a pure exponential form, $\mathbf{H}(\chi) \propto \exp(-\eta_R\chi)$ ($\chi \longrightarrow \infty$).

Imagine now two impulses propagating in the positive x direction. If the spacing, λ, between the two is considerably larger than their width, η_L^{-1}, each impulse can be viewed as propagating in the perturbative field of the other. In such a case we may try a solution in the form of a superposition of displaced solitary waves:

$$\mathbf{U}(x,t) = \mathbf{H}(\chi - \chi_1(t)) + \mathbf{H}(\chi - \chi_2(t)) + \mathbf{R}, \tag{2}$$

where $\mathbf{R}$ is a small correction term. Equations of motion for the displacements, χ_i, follow from solvabiilty conditions which remove singularities from $\mathbf{R}$. Expressed in terms of the impulse positions, $x_i = c_0t + \chi_i$, these equations read [4]

$$\dot{x}_1 = c_0 + a_R \exp[-\eta_R(x_1 - x_2)] \tag{3a}$$

$$\dot{x}_2 = c_0 - a_L \exp[-\eta_L(x_1 - x_2)]\cos[\nu_L(x_1 - x_2) + \psi_L], \tag{3b}$$

where the overdot denotes differentiation with respect to time and it is assumed that $x_1 > x_2$. Notice that the functional forms appearing in (3) are, essentially, the asymptotic forms of the solitary-wave solution. We have assumed here the more general case of an oscillatory tail. Monotonic recovery can be dealt with by setting $\nu_L = \psi_L = 0$ in (3b). The last term on the right hand side (rhs) of (3b) represents the effect exerted on the second impulse by the refractory wake of the first one. The last term on the rhs of (3a) represents the reverse effect: diffusion of a chemical (conduction of ions, etc.) from the location of the second impulse facilitates the propagation of the first one. The latter effect is normally negligible in excitable media for in most cases $\eta_L \ll \eta_R$. I will use this fact to neglect the last term on the rhs of (3a).

Using (3) we can easily derive an equation for the interpulse spacing $\lambda \equiv x_1 - x_2$:

$$\dot{\lambda} = a_L \exp(-\eta_L\lambda)\cos(\nu_L\lambda + \psi_L) \equiv F_L(\lambda). \tag{4}$$

We can study now the interaction between a pair of impulses. Consider first the case of monotonic recovery ($\nu_L = \psi_L = 0$). Evidently, no steady state (constant speed) solution with finite spacing is possible. The interpulse spacing either increases or decreases according to the sign of a_L. The theory which leads to (3) provides also an analytical expression for a_L. For our purposes, however, it is sufficient to use the asymptotic form (large λ) of the dispersion relation, $c = c_0 - F_L(\lambda)$ [5], to deduce the sign of a_L. Dispersion relations evaluated for the Belousov-Zhabotinsky reaction [6,7] and many other systems imply $a_L > 0$. Integrating (4) we find for this case that the interpulse spacing grows logarithmically with time. For $a_L < 0$ the interpulse spacing decreases but only down to values where absolute refractoriness is approached. The recovery, in this case, is non-oscillatory but also non-monotonic, exhibiting one supernormal period during which an excitability higher than that of the rest state is realized. This case calls for a nonlinear evaluation of the impulse tail and will not be considered here (see however ref. [5]).

When the recovery is oscillatory, infinitely many steady state solutions exist: $\lambda_n = (2\nu_L)^{-1}[(2n-1)\pi - 2\psi_L]$ where n is any integer number. The condition for stability, $F_L'(\lambda) < 0$, implies that any second solution, as n runs over the integers, is stable. The stable solutions represent "bound" states of impulse pairs. We shall see in the next section, when discussing two-dimensional patterns, that the tendency of impulses or wavefronts to lock at fixed distances from the preceding ones, has interesting consequences.

The analysis presented above can be extended to any number, N, of impulses. The resulting equations of motion are

$$\dot{x}_1 = c_0; \quad \dot{x}_i = c_0 - F_L(x_{i-1} - x_i) \quad i = 2,\ldots,N. \tag{5}$$

In the case of monotonic tails and for finite N, the only constant speed solution is the solitary impulse, i.e. the limit of infinite spacing. Oscillatory tails, on the other hand, allow a multitude of stable, constant speed solutions. The stability criteria are $F'_L(\lambda_i) < 0$ for any spacing $\lambda_i \equiv x_i - x_{i+1}$ in the pattern, and there are many ways of choosing sequences of spacings that obey these criteria and the constant speed requirement: $F_L(\lambda_i) = 0$ for any i. For large N the system develops extreme sensitivity to initial conditions and to external noise.

In physiological contexts patterns of impulses are generally formed by natural pacemakers or by local stimulations. To handle such formation events, it is useful to reformulate the theory in terms of the times, $t_i(x)$, at which impulses pass through a given point x [5]. A pacemaker located at x = 0 defines then a set of "initial conditions" for the variables t_i. To this end we propose a solution in the form

$$\mathbf{U}(x,t) = \sum_{i=1}^{N} \mathbf{S}(\tau - \tau_i(x)) + \mathbf{R}, \tag{6}$$

insert it in (1) and employ solvability conditions to retain the smallness of **R**. The outcome is a set of differential equations for the timings $t_i(x) = x/c_0 + \tau_i(x)$ which are similar in form to (5). Figure 1 presents numerical simulations on equations which describe a medium with oscillatory recovery. A pacemaker located at x = 0 generates impulses periodically. As demonstrated in the figure, the response of the medium away from the pacemaker is not necessarily periodic. The particular pattern that is realized is sensitive to the period of the pacemaker: a slight change (less than 2%) leads to a completely different pattern (compare Fig. 1a with 1b). There is no bifurcation associated with this tradition; the two patterns coexist and the pacemaker's frequency serves only as a means to select one pattern over the

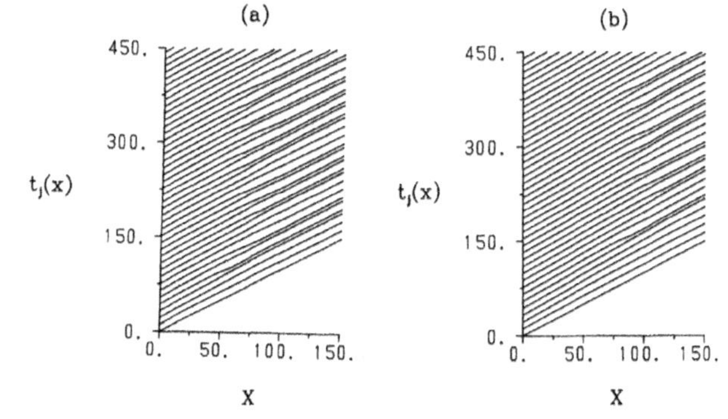

Figure 1. Impulse timings as functions of space in the case of oscillatory recovery. Periodic stimulations at x = 0 generate the patterns. Impulse interactions destroy periodicity during propagation. A slightly larger stimulating frequency in (b) leads to a completely different asymptotic pattern.

other. In fact there are **many** coexisting stable patterns (including uniformly spaced), that are realizable with other values of stimulating frequency or when external noise is applied.

3. Dynamical aspects of spiral waves

The results obtained for one-dimensional patterns carry over to two dimensions with impulses being replaced by solitary wavefronts. There are, however, two new considerations in two-dimensional space: curvature of wavefronts may become important [1], and phase singularities can form [8]. The latter are manifested as rotating spiral waves. In the following I will suggest two cross-effects of wavefront-interactions and curvature in spiral waves, that may lead to dynamical behaviours other than simple steady rotation.

The effect of curvature on the normal velocity of a wavefront in excitable media is well known: normal velocity decreases linearly as curvature increases [1]. Thus the normal velocity along a spiral wavefront decreases as the distance to the spiral core gets smaller. This velocity gradient plays an important role in sustaining the rotational motion of spiral waves. We also know from the analysis of the previous section that wavefronts that tail off in an oscillatory manner tend to lock at fixed distances from the preceding ones. Consider now a segment of the spiral arm that contains the tip. As the distance from the segment just ahead of it becomes constant, the velocity profile becomes ever flatter and the tendency to curl ever weaker. The outcome is an outward motion of the spiral tip, instead of simple rotation along a circle, and a spiral wave whose core expands in time.

Curvature normally acts to stabilize wavefronts against short wavelength perturbations. This is, however, no longer the case near the spiral tip: if we straighten a small segment that contains the tip, curvature is reduced, normal velocity is enhanced, and further straightening is favoured. Thus in the vicinity of the tip curvature becomes a destabilizing factor. Figure 2 illustrates these considerations. A similar argument holds for perturbations that tend to increase the curvature near the tip. The repulsive interaction with the wavefront that surrounds the tip (in the case of monotonic recovery) acts to quench such perturbations, and it is the competition between this

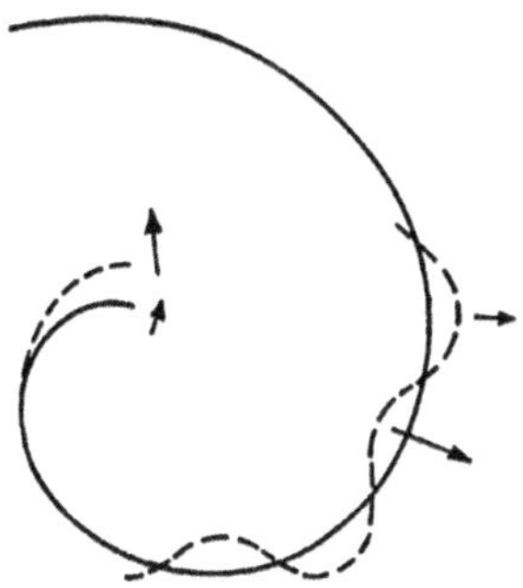

Figure 2. Schematic illustration of the effect of curvature. Normal velocity (indicated arrows) decreases as curvature increases. As a result localized or short wavelength perturbations (dashed curves) that occur sufficiently far from the tip always decay, while those involving the tip may grow.

interaction and curvature that determines the stability of simple steady rotation. Factors that make the effect of curvature stronger may render simple rotation unstable. This effect may possibly be the mechanism which leads to tip meandering [8,9].

In order to substantiate these suggestions let us consider equation (1) again, but with the second spatial derivative replaced by the two-dimensional Laplacian $\partial_x^2 + \partial_y^2$. It is advantageous to work in a coordinate system having a spiral structure and rotating with a fixed angular velocity. A convenient choice is that of an involute of a circle

$$X = \rho_0 \cos(\sigma - \omega t) + \rho_0\sigma \sin(\sigma - \omega t), \tag{7a}$$

$$Y = \rho_0 \sin(\sigma - \omega t) - \rho_0\sigma \cos(\sigma - \omega t), \tag{7b}$$

parametrized by $\sigma = \sqrt{(\rho^2/\rho_0^2 - 1)}$ where $\rho = |\mathbf{X}| = \sqrt{X^2 + Y^2})$, and rotating with an angular velocity ω. The involute coordinate system (σ, r) is defined by the relation

$$\mathbf{x} = \mathbf{X}(\sigma, t). + r\hat{\mathbf{r}}(\sigma, t), \tag{8}$$

where $\hat{\mathbf{r}}$ is a unit vector normal to the spiral. The range of the normal coordinate r is chosen to be $(-d/2, d/2)$ where $d = 2\pi\rho_0$ is the pitch of the involute spiral. The coordinate σ is chosen to be larger than some small value $\sigma_{min} > 0$ to avoid the singularity in the curvature of the involute at $\sigma = 0$. Figure 3 illustrates the notations.

The involute coordinate system allows us to view a spiral wave as an array of nearly parallel wavefronts. To see this, we first split the range of σ into intervals, $\sigma_{min} + 2\pi(\ell-1) \le \sigma_\ell < \sigma_{min} + 2\pi\ell$, representing parallel segments of the spiral arm (see Figure 3). Solitary wavefronts which are peaked on these segments can now be constructed: $\mathbf{H}_k(\sigma, r) \equiv \mathbf{H}(r + n(\sigma)d - kd)$ where $n(\sigma_\ell) \equiv \ell$. Using this array of wavefronts we write a spiral solution in the form

$$\mathbf{U}(\sigma, r, t) = \sum_k \mathbf{H}_k(\sigma, r - \zeta_k) + \mathbf{R}(\sigma, r, t), \tag{9}$$

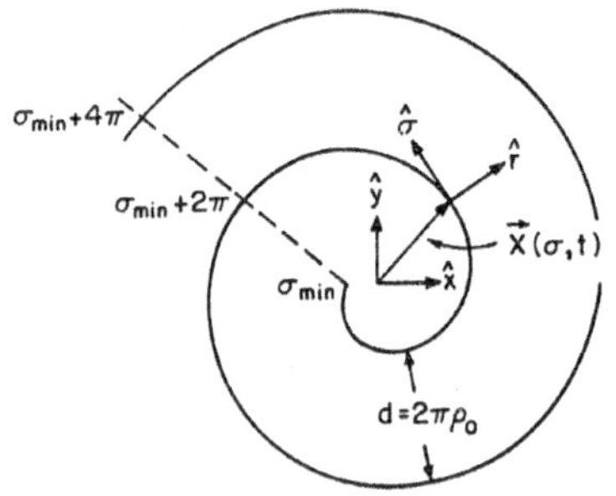

Figure 3. The involute coordinate system (σ, r). The position vector $\mathbf{X}$ is given by equations (2).

where the displacements $\zeta_k \equiv \zeta(\sigma_k, t)$ are evaluated at $\sigma_k = \sigma + 2\pi[k - n(\sigma)]$. To understand this form better, consider, for example, the first interval $\sigma = \sigma_1$. On this interval $n(\sigma) = 1$ and

$$U(\sigma,t) = H[r - \zeta(\sigma,t)] + H(r - d - \zeta(\sigma + 2\pi,t)] + \ldots + R. \tag{10}$$

Thus the main contribution comes from the wavefront that is peaked on $\sigma = \sigma_1$ (first term on the rhs of (10)), next in order of importance is the contribution from the wavefront ahead of it (second term) and so on.

The geometrical form of the spiral wave is given by $x = X(\sigma,t) + \zeta(\sigma,t)\hat{\mathbf{r}}(\sigma,t)$, and our next objective is to derive partial differential equations for the displacements ζ_ℓ. Inserting (9) into (1) and employing solvability conditions as before, we find [10]

$$\partial_t\zeta_\ell = -DK_\ell - \Omega_\ell\partial_{\sigma\ell}\zeta_\ell + DK_\ell^2\partial_{\sigma\ell}^2\zeta_\ell - F_L(\zeta_{\ell+1} - \zeta_\ell + d), \tag{11}$$

where

$$K_\ell = (\rho_0\sigma_\ell + \zeta_\ell)^{-1} \text{ and } \Omega_\ell = \omega - \rho_0 DK_\ell^3.$$

In deriving (11) we have assumed, for the sake of simplicity, that all diffusion constants are equal, thus D in (11) is a scalar. The generalization to nonequal diffusion coefficients is straightforward. We also assumed, as in section 2, that $\eta_L \ll \eta_R$. Equation (11) can be used to derive an expression for the normal velocity of the spiral wavefront, $v_r = \dot{\mathbf{x}}\cdot\hat{\mathbf{r}}$, where $\mathbf{x} = X(\sigma,t) + \zeta(\sigma,t)\hat{\mathbf{r}}(\sigma,t)$. Expressed in terms of the spiral-wave curvature, $K(\sigma,t)$, the normal velocity reads

$$v_r(\sigma_\ell,t) = c_0 - F_L(\zeta_{\ell+1} - \zeta_\ell + d) - DK(\sigma_\ell,t), \tag{12}$$

where $c_0 = \omega\rho_0$.

Numerical simulations, directed at demonstrating core expansion and destabilization of simple rotation, have been carried out for a system (11) consisting of two wavefronts. Figure 4a displays tip trajectories obtained with monotonic (solid curve) and oscillatory (dashed curve) recoveries. The

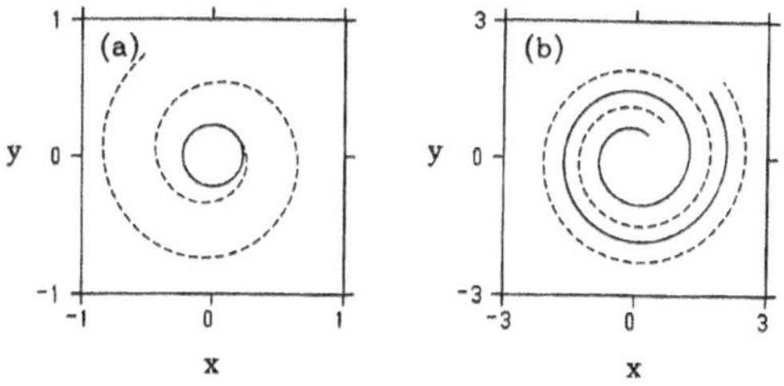

Figure 4. (a) Dynamics of spiral tips. Simple rotation obtained with a monotonic tail (solid curve) versus an outward spiral motion obtained with an oscillatory tail (dashed curve). (b)Demonstration of core expansion. Spiral wave at an early stage (solid curve) and at a later time with larger core (dashed curve).

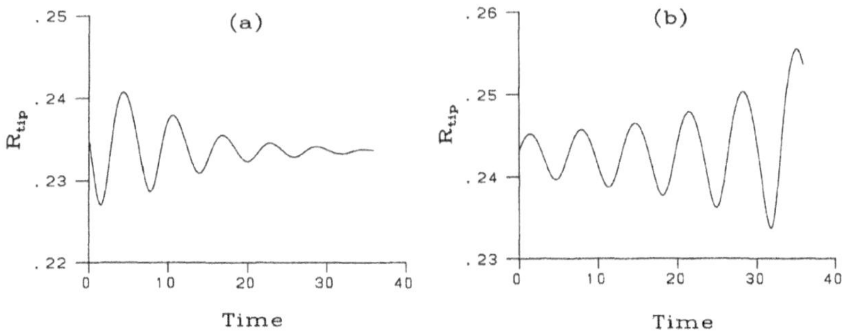

Figure 5. Destabilization of simple rotation upon increasing the diffusion coefficient D. (a) Stable simple rotation: the distance of the tip from the origin, R_{tip}, converges to a constant value. (b) Growing oscillations of R_{tip} at higher D values imply unstable rotation.

outward spiral motion of the tip supports our expectations to observe core expansion in media that give rise to oscillatory recoveries. Figure 4b further illustrates this phenomenon by displaying two forms of the spiral wave at different instances. Figure 5 shows the time evolution of $R_{tip} = |\mathbf{x}(\sigma_{min},t)|$, the distance of the tip from the origin, for two values of the diffusion coefficient D. For D small enough (Fig. 5a) R_{tip} converges to a constant value and simple rotation is attained. As D is increased growing oscillations develop (Fig. 5b), suggesting destabilization of simple rotation and the appearance of a second frequençy (along with D, ρ_0 has been increased as well since $c_0 = \omega\rho_0 \sim \sqrt{D}$). These results are in accord with our expectations that factors that make the effect of curvature stronger, may lead to destabilization of simple rotation. The instability of Fig. 5b develops at longer times to large amplitude, irregular dynamics, rather than to smooth periodic oscillations. That this is a result of constrained tip dynamics that is implicit in the present approach. I will return to this point in the next section.

4 Conclusion

I have tried in this presentation to advance an approach which emphasizes the importance of wavefront interactions in understanding pattern formation in excitable media. The approach exploits the localized nature of solitary wavefronts and the existence of a continuous translational symmetry, to construct extended patterns and to derive their dynamics. It has been shown that wavefront interactions may completely alter the structures of patterns during propagation and may lead to surprising spatiotemporal behaviours when coupled to curvature effects.

The analysis of spiral waves presented in section 2 is still incomplete. The spiral tip pertains to a phase singularity where the field U(x,y,t) vanishes. The tip is therefore a two- dimensional structure whose dynamics is governed by two translational degrees of freedom. In forcing a constant angular velocity ω, we imposed tip dynamics that involves only one degree of freedom (in the rotating frame), $\zeta(\sigma_{min}, t)$. This constraint should be relaxed in order to obtain regular tip dynamics that involves two frequencies. Work in that direction is in progress.

The nature of wavefront interactions depends on the manner of recovery after excitation. Oscillatory recovery has been observed in neural [11] and cardiac [12] systems, but so far, not in chemical media. It is expected to appear in a parametric range that precedes the bifurcation point to uniform oscillations and the experimental search for it is very much encouraged. A successful effort will allow us to test the emergence of complex wavetrains in one dimension and spiral-core expansion in two dimensions.

Acknowledgements

I wish to thank Christian Elphick, Ed Spiegel and John Rinzel for the fruitful collaboration we had in studying wave phenomena in one-dimensional systems.

References

[1] Tyson, J.J. & Keener, J.P. (1988). *Physica D (Amsterdam)* **32**, 327.
[2] Miller, R.N.' & Rinzel, J. (1981). *Biophys. J.* **34**, 227.
[3] Rinzel, J. & Maginu, K. (1984). In *Non-Equilibrium Dynamics in Chemical Systems*, Vidal, C. & Pacault, (eds.). Springer: Berlin.
[4] Elphick, C., Meron, E. & Spiegel, E.A. (1988). *Phys. Rev. Lett.* **61**, 496; Patterns of propagating pulses, *Siam J. Appl. Math.*, in press.
[5] Elphick, C., Meron, E., Rinzel, J. & Spiegel, A. Impulse patterning and relaxational propagation in excitable media, submitted to *J. Theor. Biol.*
[6] Dockery, J.D., Keener, J.P. & Tyson, J.J. (1988). *Physica D (Amsterdam)* **30**, 177.
[7] Pagola, A., Ross, J. & Vidal, C. (1980). *J. Chem. Phys.* **92**, 163.
[8] Winfree, A.T. (1987). *When Time Breaks Down.* Princeton University Press: New Jersey; Electrical Instability in Cardiac Muscle: phase singularities and rotors, *J. Theor. Biol.*, in press.
[9] Agladze, K.I., Panfilov, A.V. & Rudenko, A.N. (1988). *Physica D (Amsterdam)* **29**, 409.
[10] Meron, E. (1989). *Phys. Rev. Lett.* **63**, 684.
[11] Gardner-Medwin, A.R. (1972). *J. Physiol.* **222**, 357.
[12] Spear, J.F. & Moore, E.N. (1974). *Circ. Res.* **35**, 782.

16. MODEL FOR SPIRAL WAVE FORMATION IN EXCITABLE MEDIA

P. Pelcé

Laboratoire de Recherche en Combustion
Universite de Provence-St. Jerome
13397 Marseille, France

1. Introduction

Geometrical models are simple caricatures that are used to capture the essential features of a complex dynamics. For instance in dendritic growth the motion of a liquid-solid interface is determined by a free-boundary problem for a diffusion field (temperature, concentration of an impurity). This kind of problem is difficult to solve since it involves nonlinear integral equations.

Great progress was made in the understanding of the solutions of this problem by considering a geometrical model for the motion of the interface: the motion of a closed curve whose normal velocity is a function of its curvature. The study of this model, more tractable from a mathematical point of view, was crucial to the understanding of how surface tension effects determine growth velocities. Later it appeared that the same mechanism determined growth velocities in the full free-boundary problem. A similar approach was followed in the context of flame propagation.

We want to apply the same ideas for the problem of spiral formation in excitable media. The complete problem is difficult to solve since it involves two-dimensional time-dependent solutions of reaction-diffusion equations. Nevertheless due to the existence of different time scales between excitation and refractory processes, it is possible to describe the formation of a spiral in terms of the motion of an opened curve (the line where the excitation process occurs) moving with a normal velocity function of the local curvature and curvilinear coordinate. Furthermore, contrary to the problem of the growth of a crystal, where the interface is closed, tangential velocity for the end points must be specified. Then, by starting from an initial condition which is a straight segment, we study how the spiral forms, and determine scaling laws for the time of formation and the distance between centres.

Nonlinear Wave Processes in Excitable Media
Edited by A. V. Holden *et al.*, Plenum Press, New York

2. Metastable state, excitable state

2.1. *Metastable state*

A metastable state is an equilibrium state which disappears after a perturbation of finite amplitude.

When a seed crystal is introduced in a supercooled liquid, it develops, leading successively to various crystal shapes, from the more regular at low velocity (spherical, faceted) to the more complex at large velocity (dendrites). In the case of the growth in a pure melt, growth is limited by the diffusion of latent heat, so that the crystal shape (the liquid-solid interface) appears as a moving boundary for the temperature field.

When a reactive mixture is ignited, the combustion propagates towards the fresh mixture. Combustion occurs in a very thin zone which separates fresh and burned mixture. This zone can be considered as an interface which, at low propagation velocity is regular (spherical shape) and at larger velocity becomes more complex (cellular shapes, turbulent shapes). Here too, growth is limited by the diffusion of temperature, i.e. flame shape appears as a moving boundary for one diffusion field.

2.2. *Excitable state*

An excitable state is an equilibrium state which disappears during a finite time after a perturbation of finite amplitude.

Consider for instance a B.Z. reagent in the excitable state. If the concentration of a reactant is locally modified from the equilibrium value, the spatial inhomogeneity develops, leading to the propagation of a reactive interface. Contrary to the case of the metastable state, the medium left behind the front comes back to the initial equilibrium state. Thus, if for some reason, the interface breaks at some point, the two free extremities roll up behind the propagating front, leading to the formation of two contrarotating spiral waves. Here, at least two diffusion fields are needed to describe the dynamics: one trigger variable, which varies on a fast scale, and a recovery variable, usually varying very slowly, which drives the excited state to its initial equilibrium value.

3. The spirit of a geometrical model

In all cases very different time and spatial scales are observed.

In crystal growth, the thickness of the liquid-solid interface is $d_0 = 10\mathring{A}$ which is much smaller than the tip radius of the crystal $\rho = 1\ \mu m$. In the case of flame propagation, the reaction zone of the flame of size $d \approx .1$ mm is much smaller than the diffusive length of size $l \approx 1$ mm. In excitable media, the excitation zone is much smaller than the recovery zone.

It follows that the dynamics of growth can be determined by a solution of a free-boundary problem, i.e. an equation for the slow field with boundary conditions on the region where the fast field varies, assimilated to a sheet of negligible thickness.

In general, a free-boundary problem is difficult to solve. A simpler problem consists in studying motions of curves moving with some given normal velocity function of the geometrical quantities (arc length, curvature) and to forget the field which is outside. This is the spirit of the geometrical models.

The curve can be either closed, as in crystal growth or in the case of flame propagation, or opened as in the case of spirals in excitable media. The velocity of a point of the curve is decomposed as

$$\frac{d\vec{X}}{dt} = (\vec{v}\cdot\vec{n})\vec{n} + (\vec{v}\cdot\vec{\tau})\vec{\tau}$$

where n and τ are the vectors normal and tangent to the curve.

3.1. *The curve is closed*

Consider first the case where the curve is closed. To describe the motion of the curve, one needs to specify the velocity of its points. As an example, take a growing circle. In polar coordinates, (r,θ)

$$\vec{X} = r(\theta)\vec{u}$$

the velocity of a point of the circle is:

$$\frac{d\vec{X}^{\cdot}}{dt} = \frac{dr}{dt}\vec{u} + r\frac{d\theta}{dt} v$$

If the normal velocity of the interface satisfies

$$\frac{dr}{dt} = \frac{D}{r}$$

one sees that for any choice for the tangential velocity

$$r\,\frac{d\theta}{dt} = f(r,\theta,t)$$

one obtains the same growing circle, whose radius grows like $r = \sqrt{Dt}$.

Thus, there is a gauge invariance in the dynamics: to determine the dynamics of a closed curve, one needs only to specify the normal velocity of the points. The choice of the tangential velocity fixes the parametrisation.

3.2. *The curve has two free extremities*

In this case, the tangential velocity of the free extremities must be specified so that only the parametrisation which gives the correct tangential velocities at the extremities is allowed.

4. **Specific laws**

4.1. *Crystal growth*

Brower et al. [1] considered the following relation for the normal velocity of the closed interface:

$$\vec{v}\cdot\vec{n} = \kappa + d_0 \cdot \frac{\partial^2\kappa}{\partial^2 s}$$

The first term must be related to the Mullins-Sekerka instability [2]. The second mimics the stabilizing effects of surface tension. This model cannot be derived from the original free-boundary problem but the same mathematical results can be derived. For instance, when $d_0 = 0$ (no surface tension effects), there is a continuum of solutions for interfaces of stationary shape propagating with constant velocity [1]. When $d_0 \neq 0$ and depends on the angle of the normal to the interface with the direction of propagation, there is a breaking of this continuum, i.e. the velocity of the crystal is determined [3].

4.2. *Flame propagation*

Contrary to the case of crystal growth, the geometrical model can be derived rigorously from the reaction-diffusion equations when the curvature of the interface is weak [4]:

$$\vec{v}\cdot\vec{n} = 1 + (\alpha - 1)\kappa + (1 + \frac{1}{2}\alpha^2)\kappa^2$$

$$+ (2\alpha + 5\alpha^2 - \frac{1}{3}\alpha^3)\kappa^3 + \alpha^2(\alpha + 3)\frac{\partial^2\kappa}{\partial s^2}$$

where α is a function of the Lewis number of the reactive mixture and the reduced activation energy of the reaction.

4.3. *Excitable media*

As for the case of flame propagation, it can be shown from the initial reaction-diffusion equations that, if the transverse concentration flux is neglected, relative variations of the normal velocity are proportional to curvature [5]:

$$\vec{v}\cdot\vec{n} = c - \kappa$$

This is true far from the free extremities of the spiral. Close to the extremities we propose the following model:

- corrections to the normal velocity:
- effect of tangential flux: $\delta(v\cdot n) = \left(\exp -\frac{s}{d} + \exp\left(-\frac{(1-s)}{d}\right)\right)$
- effect of refractory tail: $\delta(v\cdot n) = \frac{\kappa}{1-w\kappa}$

tangential velocity: $(v\cdot\tau) =$ const.

where s is the arc length.

5. Time-dependent solutions

5.1. *The closed curve*

For simplicity, we choose the gauge which corresponds to the simplest relation for the velocity of the points of the curve, i.e. the tangential velocity vanishes everywhere [1]. Define the corresponding parametrisation as σ. Then, the velocity of a point of the curve is

$$\frac{d\vec{X}}{dt} = (\vec{v}\cdot\vec{n})\vec{n}$$

where n is the unit vector normal to the curve. Consider the motion of the curve in a plane and define θ the angle of the tangent vector τ with a fixed axis, $h = (\tau.\tau)^{1/2}$ the metric. Then,

$$\tau = h \exp i\theta, \qquad n = i \exp i\theta$$

By using the identity

$$\frac{\partial\tau}{\partial t} = \frac{\partial}{\partial\sigma}\left(\frac{d\vec{X}}{dt}\right)$$

one finally obtains two coupled equations for the metric and the curvature as:

$$\frac{\partial h}{\partial t} = -h\kappa(\vec{v}\cdot\vec{n})$$

$$\frac{\partial\kappa}{\partial t} = \left(\kappa^2 + \frac{1}{h^2}\frac{\partial^2}{\partial\sigma^2}\right)(\vec{v}\cdot\vec{n})$$

These equations can be integrated numerically and the curve reconstructed from the data of h and κ.

5.2. *Curve with free extremities*

As mentioned before, contrary to the case of the closed curve, the gauge of zero-tangential velocity is not compatible with the finite velocities of the free extremities. Thus we take the simplest gauge compatible with the non zero velocity of the extremities [6]:

$$\frac{d\vec{X}}{dt} = (\vec{v}\cdot\vec{n})\vec{n} + \sigma\vec{\tau}$$

Then, by using the same steps, one obtained coupled equations for the metric and the curvature as:

$$\frac{\partial h}{\partial t} = 1 - h\kappa(\vec{v}\cdot\vec{n})$$

$$\frac{\partial\kappa}{\partial t} = \left(\kappa^2 + \frac{1}{h^2}\frac{\partial^2}{\partial\sigma^2}\right)(\vec{v}\cdot\vec{n}) + \frac{\sigma}{h}\frac{\partial\kappa}{\partial\sigma} - \frac{1}{h^2}\frac{\partial(\vec{v}\cdot\vec{n})}{\partial\sigma}\frac{\partial h}{\partial\sigma}$$

We learned at this conference that similar derivation of kinematic equations was done by Zykov (see for instance [7]).

6. Results

The initial condition is a segment of length L. With the dimensions of the relevant quantities of the problem one can build two dimensionless numbers: the Peclet number $u = D\tau/L^2$ and the ratio $v = T/\tau$. Here, D is the coefficient of diffusion, τ the excitation rise time and T the refractory time. The model is relevant for determining quantities which do not involve interactions between successive fronts such as the time to form the core or the distance between the centres of the two contrarotating spirals, as a function of the parameters u and v [6].

References

[1] Brower, R.C., Kessler, D.A., Koplik, J. & Levine, H. (1983). *Phys. Rev. Lett.* **51**, 1111.
[2] Mullins, W.W. & Sekerka, R.F. (1964). *J. Appl. Phys.* **3**, 444.
[3] Langer, J.S. (1986). *Phys. Rev.* **A33**, 435.
[4] Frankel, M.L. & Sivashinsky, G.I. (1987). *J. Phys.* **48**, 25.
[5] Markstein, G.H. (1951). *J. Aero. Sci.* **18**, 199.
[6] Meron, E. & Pelcé, P. (1988). *Phys. Rev. Lett.* **60**, 1880.
[7] Zykov, V.S. (1988). *Modelling of wave processes in excitable media.* Manchester University Press: Manchester.

17. RANDOMIZED AUTOMATA FOR ISOTROPIC MODELLING OF TWO- AND THREE-DIMENSIONAL WAVES AND SPATIOTEMPORAL CHAOS IN EXCITABLE MEDIA

M. Markus, M. Kraf'czyk and B. Hess

Max-Planck Institut für Ernährungsphysiologie
Rheinlanddamm 201, D-4600 Dortmund 1, FRG

1. Introduction

In the present volume, a large variety of excitable media in chemistry, physics, biology and even astronomy is presented. This volume also contains a number of tools for the modelling of wave propagation in such media. Among these tools, cellular automata have proven for a long time to show a number of advantages: they are intuitively appealing, being easily described and understood, they can easily be programmed and they run relatively fast on a computer.

The simplest automata reported so far are based on a periodic grid, consisting of squares and hexagons in two dimensions or cubes in three dimensions [1-4]. Also, fractal grids have been proposed [5]. However, these automata suffer from the drawback that the shape of the cells propagates into the macroscopic scales, thus leading to waves of the same (or related) shape. In other words, the wave propagation is anisotropic.

As an example of cell shape propagation, Figs. 1a and 1b illustrate a two-dimensional cellular automaton with square cells. It is assumed that all eight neighbours of an excited cell can be excited ("Moore neighbourhood"). Figure 1a shows the propagation of a target pattern starting from a single excited (black) cell. It can clearly be seen in this figure how anisotropy arises due to the propagation of the square shape of the cells: the velocity in the direction of the cell diagonals is $\sqrt{2}$ times larger than in the direction of the cell edges. Figure 1b shows the anisotropic development of a spiral wave in the same automaton. As a further example, Figs. 1c and 1d show how hexagonal cells lead to hexagonal waves and thus also to anisotropy.

In an attempt to attain isotropy, we tested the effect of an aperiodic Penrose lattice [10] on wave propagation. We assumed that all neighbours sharing at least one corner with an excited cell also become excited. The result is shown in Fig. 2 for a target pattern. One sees that the wave is irregular for small times and tends to be decagonal for large times. Thus, the aperiodicity of this lattice is not sufficient for isotropic propagation, but leads to a tenfold symmetry, as is also observed in diffraction patterns

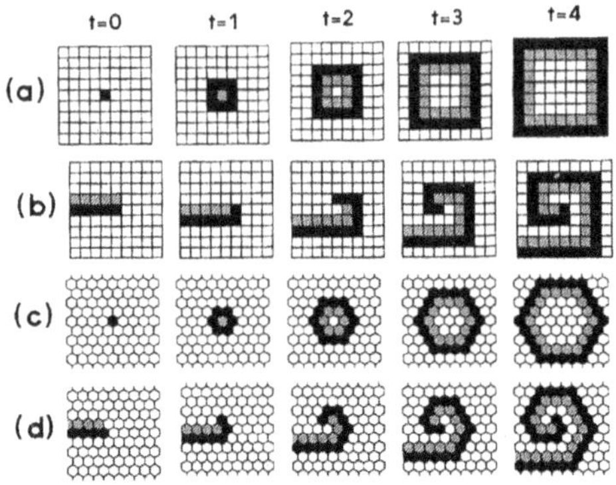

Figure 1. Cellular automata with periodic cell distributions demonstrating the propagation of the cell shapes into the wave shapes. Excited cells are shown in black, refractory cells are dashed and receptive cells are white. (a,b): Square cells. (c,d): Hexagonal cells. (a,c): Target patterns initiated by one excited cell. (b,d): Spiral waves initiated by a free wave end.

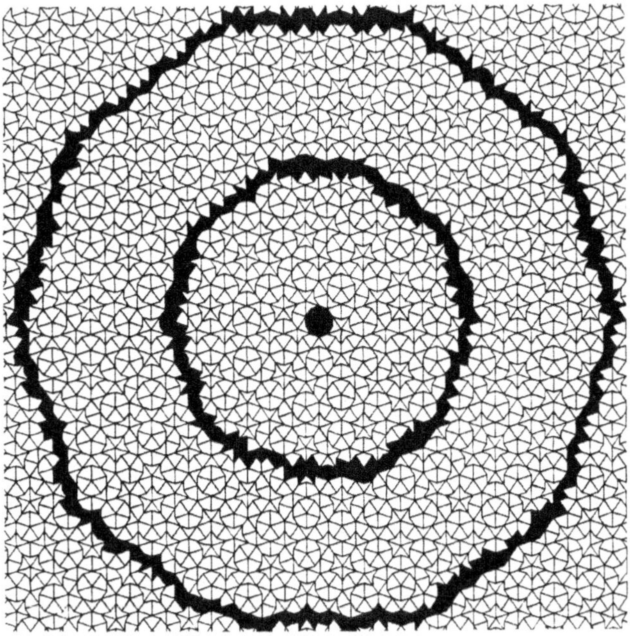

Figure 2. Development of a target pattern in an automaton with a Penrose lattice consisting of "kites" and "darts" [10]. The excited wavefront is shown at t=0, 10 and 20. In spite of the aperiodicity of the grid, decagons are obtained instead of circular waves.

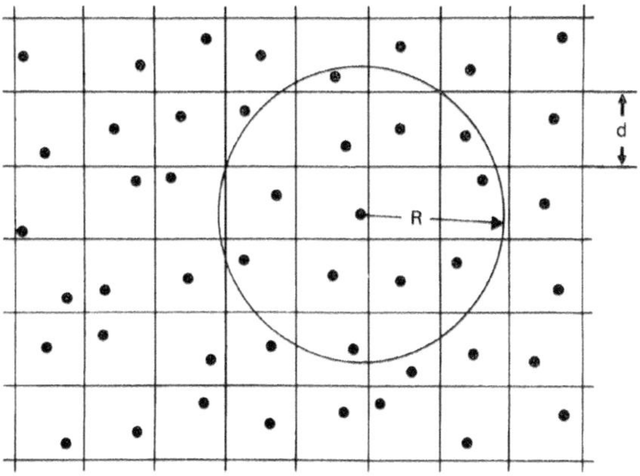

Figure 3. Geometry of the cellular automata used in the present work for two dimensions. Each point is randomly distributed within each cell. The point at the centre of the circle with radius R can only be excited by points within this circle.

measured with some aperiodic alloys [11].

The procedure of averaging over the cells within the neighbourhood, as proposed in [15,16] (also used in [17,18]) leads to a propagation closer to being isotropic since the wavefront corners appear rounded. However, the shapes of the waves still contain the information about the shape of the cells, and thus isotropy is not attained.

The problem of anisotropic cellular automata is also encountered in fluid dynamics. In fact, the Hardy-Pomeau-dePazzis model [12], which uses square cells, is highly anisotropic. The Frisch-Hasslacher-Pomeau model [13], which uses an hexagonal lattice, is isotropic but is confined to two dimensions. The so-called "face-centred hypercubic" model is isotropic in three dimensions, but it operates in a four-dimensional space, thus requiring large computational effort (see [14]).

In this volume, two other solutions to the anisotropy-problem are proposed. In one of them [6], isotropy is obtained at the cost of an almost-continuum of states (256 states), which is a partial renunciation of the original idea of discretization in cellular automata. In the other approach [7], wave propagation is made isotropic by a random occupation of cells in a square lattice. However, in our own simulations using this procedure we needed cell numbers orders of magnitude larger than in our model to avoid patchiness due to the arising inhomogeneities

2. Models and graphical display

In order to solve the anisotropy problem we have suggested [8,9] randomizing the geometry of the model as shown in Fig. 3. The two-dimensional (resp. three-dimensional) medium is divided into square (res. cubic) cells with side length d. One point is placed randomly in each cell. Although randomness is introduced primarily to attain isotropy, it is also justified by being more similar to natural cell distributions than spatial periodicity. The "quasi-homogeneity" constraint of one point per cell is imposed to reduce the number of cells, while ensuring isotropy.

Each point of the automaton (see Fig. 3) may assume n+2 states S: S=0 (receptive state), S=1,2,...,n (refractory states) and S=n+1 (excited state). The states S(t+1) are determined iteratively (t:time) at each point from the states S(t) as follows: first, a circle (or sphere, in the three-dimensional case) C with radius R is drawn around the point under consideration (in analogy with Huygen's principle) and the number ν of excited cells within C is counted. In an approximation in which excitability of the refractory states is neglected, the following rules are applied (model I):

a) If S(t)=0 and $\nu \geq m$, then $\sigma(t)=n+1$ (m is a given threshold);
b) If S(t)=0 and $\nu < m$, then $\sigma(t)=0$;
c) If S(t)=1,2,...,n,n+1, then $\sigma(t)=S(t)-1$;
d) If $\sigma(t)=1$ or n+1, then $S(t+1)=\sigma(t)$;
e) If $\sigma(t)=2,3,\ldots,n$, then $S(t+1)=[\langle\sigma(t)\rangle]$, where the brackets <> indicate the average over all points within C, and the Gaussian brackets [] indicate taking the next integer. σ is an intermediate state in the algorithm.

Altogether, model I contains only three control parameters: r=R/d, n and m. A larger number of parameters is needed to include excitability of the refractory states leading to dispersion (dependence of the velocity of planar waves on the period [19]). In that case, we make the ansatz $m = m_o + p\,S(t)$, where $S(t)=0,1,2,\ldots,S_{max}$. Here, it is considered that for increasing S the distance from the threshold in phase space also increases, and thus increasingly large perturbations are needed to trigger the system into the excited state. The states $S_{max}+1$, ... , n , n+1 are assumed to be non-excitable. The rules now have to be rewritten as follows (model II):

a) If $S(t)=0,1,2,\ldots,S_{max}$ and $\nu \geq m_o + p\,S(t)$, then $\sigma(t)=n+1$;
b) If S(t)=0 and $\nu < m_o$, then $\sigma(t)=0$;
c) If $S(t)=1,2,\ldots,S_{max}$ and $\nu < m_o + p\,S(t)$, then $\sigma(t)=S(t)-1$;
d) If $S(t)=S_{max}+1,\ldots,n,n+1$, then $\sigma(t)=S(t)-1$;
e) If $\sigma(t)=1$ or n+1, then $S(t+1)=\sigma(t)$;
f) If $\sigma(t)=2,3,\ldots,n$, then $S(t+1)=[\langle\sigma(t)\rangle]$.

Model II contains five control parameters: r, n, m_o, p and S_{max}.

For the graphical representation of the results in two dimensions, one pixel (picture element) is assigned to each square and is given a grey value according to the state of the point within that square. The darkness of the grey shading increases with decreasing S (S=n+1, n,...,1, 0). In three dimensions, only the state S=n+1 (excited state) is shown in the back half of the medium and the darkness of the grey shading increases as cells lay deeper in the direction perpendicular to the plane of the figure.

3. Results and discussion

3.1. *Two-dimensional circular waves and spirals*

Typical examples of simulations of waves in two dimensions using model I are shown in Fig. 4. The isotropy of the model can be visualized in this figure by the "round" shape of the waves (Fig. 4b), in contrast to the polygonal (or nearly polygonal) shapes shown in Figs. 1 and 2 or reported elsewhere (see references in the Introduction). A single spiral (upper left in Fig. 4) is initiated by a rectangular array of excited cells (width R) buffered on one

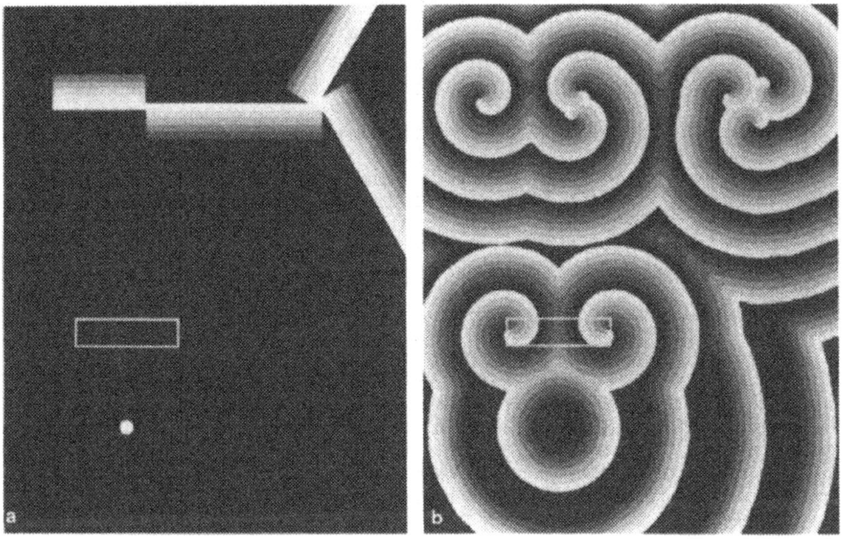

Figure 4. Simulations of two-dimensional waves (height: 750 cells, model I, r=10, m=1, n=4 except inside the rectangle with white borders, where n is set to 16 for 10 ≤ t ≤ 26. (a): Initial configuration (t=0). (b): t=68.

Figure 5. Simulation of a wave rotating around a round obstacle (indicated by crisscrossed lines), comparable to measurements in the rat cortex [24] (320x320 cells, model I, r=6, n=6, m=1, t=35, 45, 55, 65 and 75).

side by rectangular arrays of the same width and decreasing S. The single spiral develops at the free ends of the rectangular arrays, analogously as in experiments where spirals form at the free ends of disrupted chemical waves

[20]. A double spiral (upper part of Fig. 4, left of the centre) is initiated by two excited rectangular arrays meeting at a common cross section but buffered on opposite sides, and the three-armed spiral (upper right of Fig. 4) by three buffered rectangular arrays meeting at a common cross section.

Measurements of two- and three-armed spirals have been performed by Agladze & Krinsky [21] and are comparable to the simulations shown in Fig. 4. A particular result, both in the present simulations and in experiments, is the occurrence of oscillations of the distance between the arm-tips of two- and three-armed spirals. This result is also obtained in ref. [6]. Setting here $n=6$, $r=6$ and $m=1$ in model I, the distance between the tips of a two-armed spiral has a period equal to 9 iterations and an amplitude equal to 0.18 times the spiral wavelength far from the centre. The period and amplitude of these oscillations has not been determined in experiments as yet.

The target patterns emerging from the lower left part of Fig. 4 are initiated by a single, periodically excited pacemaker point, which operates analogously to the sinus node of the heart. The period of the pacemaker is set here to 10 iterations. In the rectangular domain (indicated by white borders), n is temporarily set to 16 for $33 \leq t \leq 46$, while at other times and elsewhere, $n=4$. This transient inhomogeneity of n simulates here an inhomogeneity of the myocard, which appears at some stages of infarction and has been described as the cause of the sudden increase of heart rate during paroxysmal tachycardia (see e.g. [22]). The mechanism of this tachycardia can be understood as follows. The inhomogeneity, which appears at $t=33$, is first passed by a circular wave at $t=35$ and has not recovered its excitability when the next circular wave passes through it ($t=46$). Thus, the second wave is broken. When the rectangle has recovered its excitability ($t > 46$), the free ends of the broken second wave curl into spirals. These spirals have a higher frequency (smaller wavelength and same velocity) than the target patterns. Therefore, the spirals annihilate the waves emerging from the pacemaker, and thus an overall pulsation at a higher frequency takes over. It is noteworthy that the spiral frequency is unique for a given medium, while the pulsation due to the pacemaker may (in general) have a broadband spectrum (not shown in Fig. 4). Thus, the spirals may account for the observed narrowing of the ECG spectrum obtained during ventricular fibrillation. It has been pointed out [23] that this narrowing contradicts the widely supposed notion that fibrillation is related to chaos: actually one observes a so-called "pathological periodicity" instead of a "chaotic disease".

3.2. *Wave rotating around an obstacle and its interaction with spiral waves*

Figure 5 shows a simulated wave rotating around a non-excitable round obstacle ("hole") at five consecutive time steps separated by $\Delta t=10$ iterations. The white arrow indicates the direction of rotation. This result is comparable to measurements of increased extracellular $[K^+]$ (with suppressed neural activity) rotating around an obstacle caused by thermocoagulation in the rat cortex [24].

In the BZ reagent, experiments have been performed on the interaction of a wave rotating around an obstacle and a pair of spiral waves [25]. It was shown that the latter annihilate the first. However, if after this annihilation the spirals are destroyed by a drop of KBr covering their centres, the wave rotating around the obstacle resuscitates. These phenomena are illustrated in Figs. 6 and 7. Figure 6 shows the annihilation of the wave rotating around a round obstacle (upper right of the picture) by a pair of spirals (centred at the bottom left of the picture). In Fig. 7 a drop of KBr covering the spiral centres is simulated by a rectangular region (lower left)

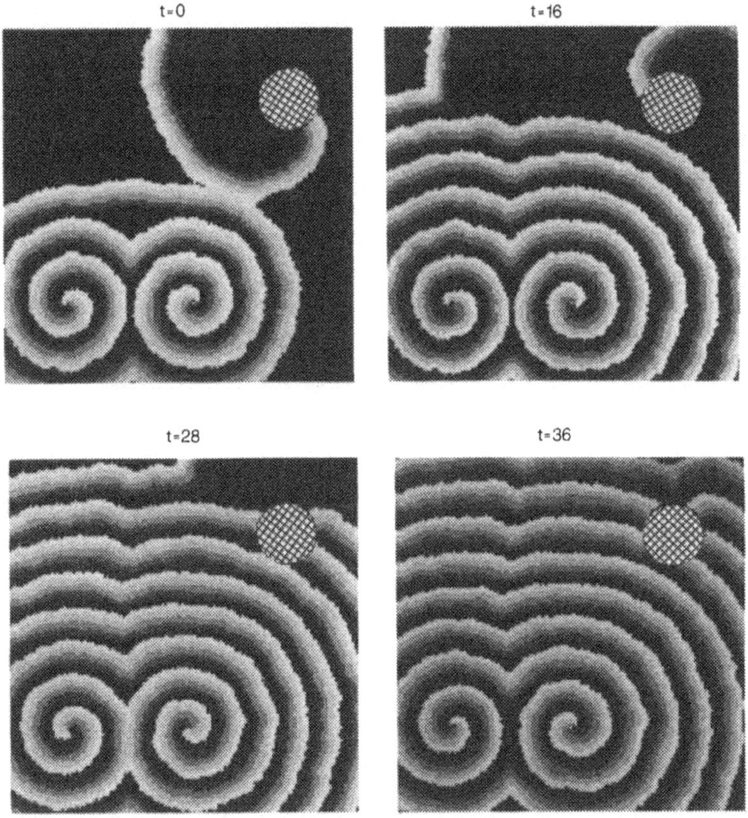

Figure 6. Annihilation of a wave rotating around an obstacle (indicated by crisscrossed lines on the upper right) by a pair of spirals (centred at the lower left). (170x170 cells, model I, r=3, n=3, m=1).

with larger refractoriness. After this intervention, the spiral waves are gradually destroyed and the wave rotating around the obstacle is restored.

3.3. *The spiral core*

Waves rotating around an obstacle, as discussed in the last section, were actually the first waves in an excitable medium simulated with a cellular automaton [1]. These historical simulations were intended to mimic rotating waves during heart arrhythmias. Later on, however, it was found that the presence of an obstacle is not a necessary condition for this type of wave. It was shown that waves may rotate around a region of low and constant excitation, which develops autonomously: the spiral core. Figure 8 shows the core (indicated by a white circle) calculated with the present automaton (model I). The visualization of the core is possible here due to the high resolution of the simulation. The tip of the spiral - defined by the point of the excited front where the curvature changes its sign - rotates around the

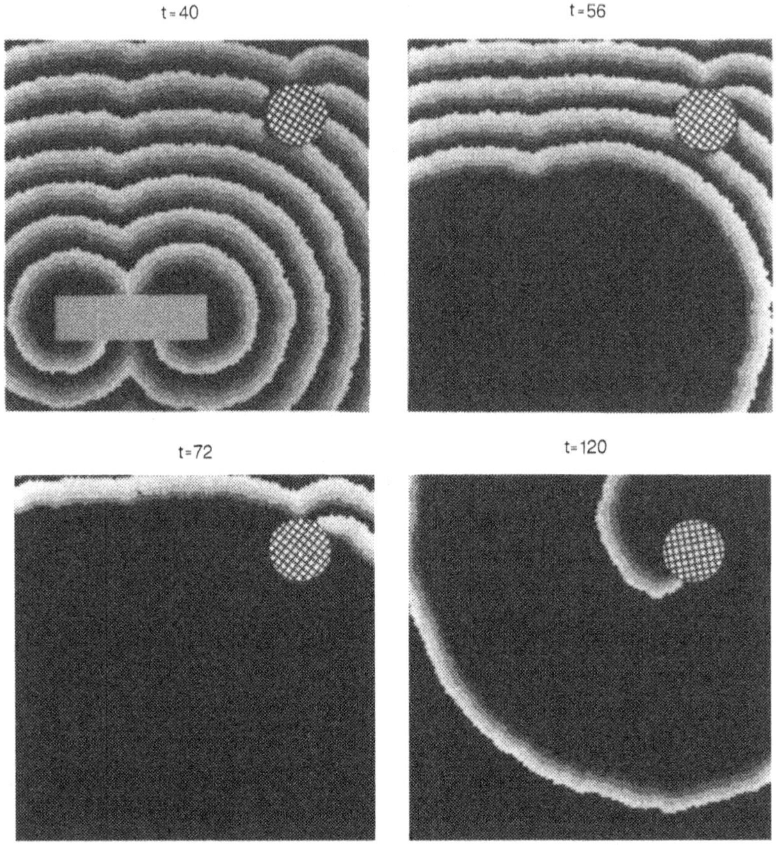

Figure 7. Self-restoring of a wave rotating around an obstacle (previously annihilated by a pair of spirals, as shown in Fig. 6). The spirals are destroyed by imposing high refractoriness on the rectangle at the lower left. this rectangle simulates a drop of KBr in experiments [25]. (Parameters as in Fig. 6.)

white circle. Within this circle, S is relatively low and almost constant, as shown in Fig. 9a. In this figure, S is displayed as a function of the cell abscissa index for all iterations during one rotation of the spiral. The ordinate-index passes through the centre of the core and is set constant. The core is determined by the region where never $S=n+1$ ($=13$ in this case). This region is given in Fig. 9a by the interval $184 \leq x \leq 203$.

The numerical determination of the spiral core is a remarkable outcome since alternative calculations [19] have required the introduction of the core radius as an empirical parameter, instead of rendering its value as a result of calculations.

The simulations shown in Fig. 9a are comparable to measurements with the BZ reagent given in Fig. 9b. The latter diagram shows profiles of the

Figure 8. Simulation of a spiral wave rotating around a core. The perimeter of the core is indicated by a white circle (height: 356 cells, model I, r=12, n=12, m=1).

intensity of light (given in grey levels) passing through a thin layer at equidistant times during one spiral rotation [20].

3.4. *Isotropy*

The isotropy of the present model, which is made qualitatively visible in the simulations illustrated here, can be quantified by the measure

$$\alpha = (V_{max} - V_{min})/V_{min},$$

where V_{max}, resp. V_{min}, are the maximum, resp. minimum, velocities over all angles of propagation [36]. In order to perform this quantification, target patterns were generated in two and three dimensions starting from a single excited cell. Examples of these target patterns are given in Fig. 10 for two different automaton resolutions. The left pattern (r=6) yields α=0.015 and the right one (r=12) α=0.007. The lower part of Fig. 10 shows the percentual deviation of the mean distance, measured from the centre of the initially excited cell to the edge of the wavefront, as a function of the angle. Taking into consideration that the automaton is symmetric with respect to the x- and y-axis, the averages over the four quadrants were plotted on the figure. The isotropy is visualized here by the fact that the plotted percentual deviations are independent of the abscissa, except for small fluctuations, which are due to the random distribution of the points within the cells.

In some cases it may be interesting to include anisotropy in the simulations, e.g. for cardiac tissue [37]. This deliberate inclusion of anisotropy has nothing to do with the anisotropy arising from the periodicity of the cells, which is a numerical artifact. To a model like ours, where the anisotropy problem has been solved, one may add specific rules to mimic the anisotropy observed in natural systems.

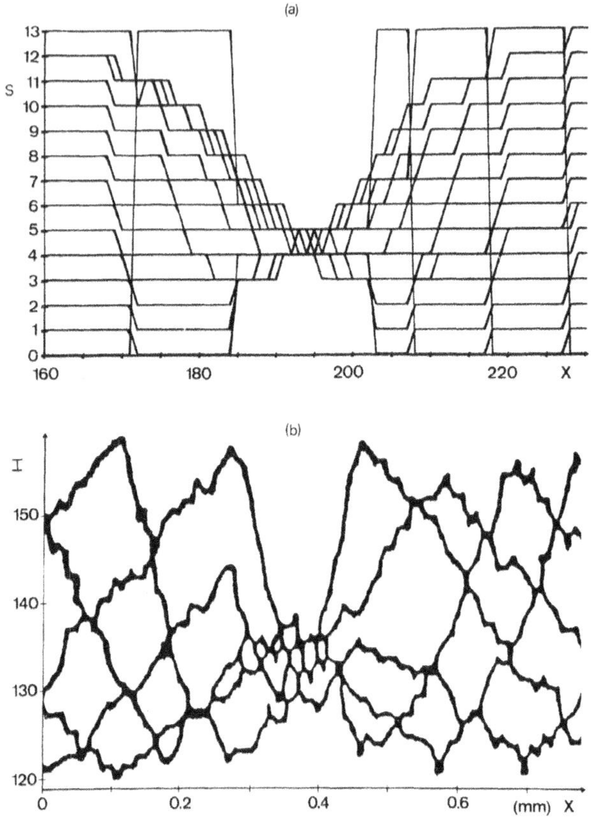

Figure 9. Simultaneous display of profiles of the state of the system along a cross section through the spiral core at equidistant times during one spiral rotation. (a): Simulations (parameters as in Fig. 8). (b): Experimental (BZ reagent; Zs. Nagy- Ungvarai, S.C. Müller and B. Hess, submitted for publication).

3.5. *The eikonal equation*

The "eikonal equation" in two dimensions is defined as the dependence of the normal velocity N on the curvature K of the wavefront [19]. Curvature effects are taken into account in our automaton by the parameter m. N and K were determined from the simulated waves as was done in experiments [26,27]. Figure 11, resp. Fig. 12, show N as functions of K for $K > 0$, resp. $K < 0$, for two values of m in model I. For m=154 our model clearly yields a linear relationship $N = c - DK$. Such a linear relationship has been obtained in a different theoretical approach [19] and in experiments with the BZ reagent [26,27]. c is the velocity of planar waves and D corresponds to the diffusion coefficient of the autocatalytic species of the system.

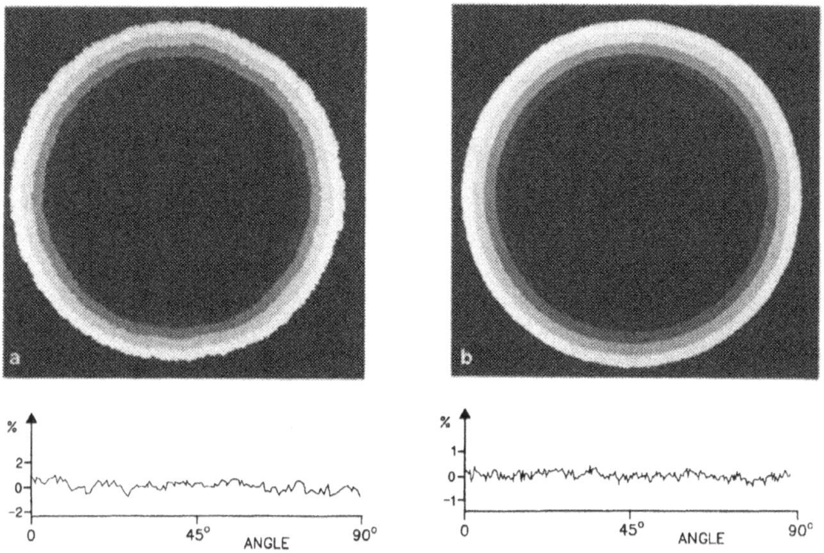

Figure 10. Demonstration of the isotropy of the model used in this work. Upper: simulated circular waves. Lower: percentual deviation from the mean distance travelled by the excited wavefront (averaged over the four quadrants) as a function of the angle. Model I, m=1, n=3, t=15. (a): 200x200 cells, r=6. (b): 400x400 cells, r=12.

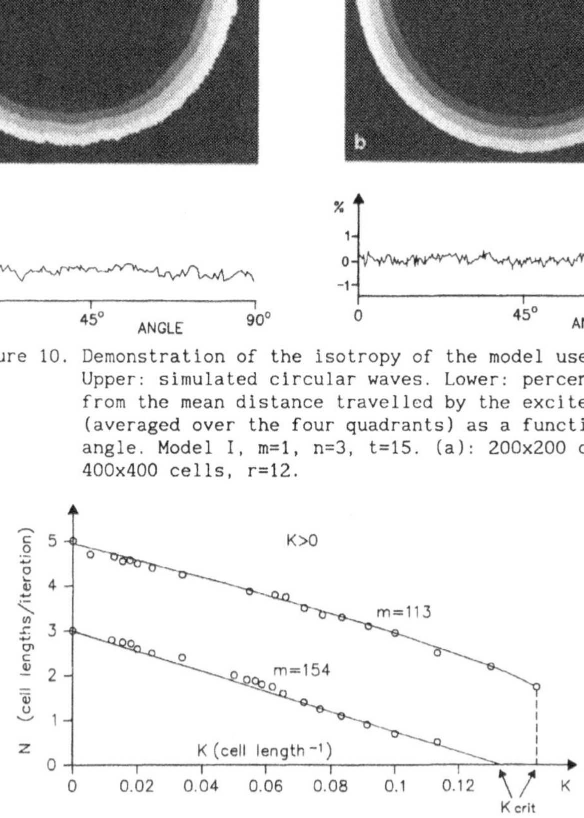

Figure 11. Simulated dependence of the normal velocity N on the curvature K of convex wavefronts (model I, r=12, n=12).

We would like to include here a note on nomenclature. In references [19], [26] and [27], $K > 0$ is used for convex, and $K < 0$ for concave wavefronts. In references [28-30], the signs of K are defined oppositely. We adopt here the former nomenclature.

From the simulations with m=154, we obtain $c \approx 3$ cell lengths/iteration and $D \approx 22$ (cell lengths)^2/iteration. For m=113 the relationship between N and K is nonlinear, and for $K > 0$ there is a discontinuous transition to N=0. Similar nonlinearities have been obtained in an analysis applicable e.g. to nerve

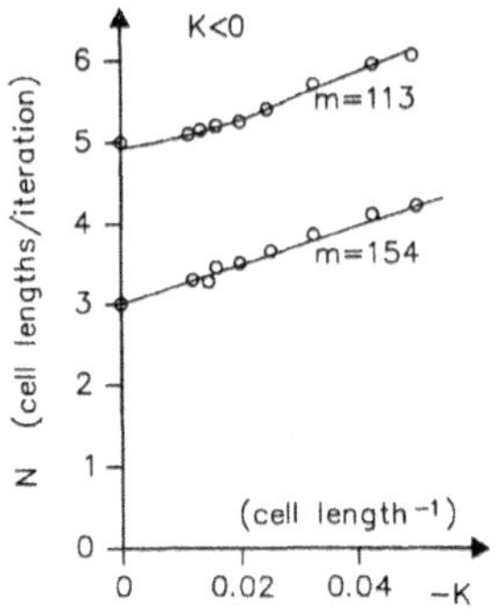

Figure 12. As Fig. 11, but for concave wavefronts.

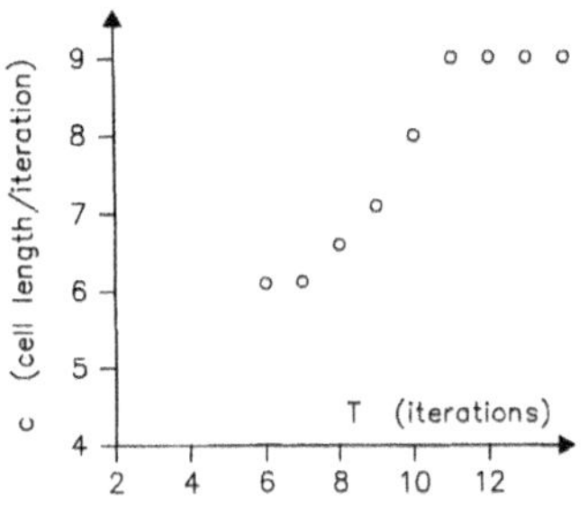

Figure 13. Simulated dispersion relation, i.e. dependence of the velocity c of planar waves on the period T (model II, r=9, n=9, m_o=1, p=5, S_{max}=6).

membranes [29,30]. In general, the curvature at which N=0 is called the critical curvature $K_{crit}=\rho_{crit}^{-1}$. ρ_{crit} is the radius below which no propagation is possible.

For the BZ reagent, measurements yielded $D\approx 2\times10^{-5}$ cm^2/sec, $\rho_{crit}\approx 23$ μm [26,27]. In heart tissue, the following values were estimated [28]: D≈0.6 cm^2/sec, $\rho_{crit}\approx 200$ μm. A circle with radius ρ_{crit} contains about 50 myocardial cells. Thus, ectopic foci in heart tissue must have at least this number of cells in order to trigger a spreading wave of contraction. For the spread of rabies among foxes the following values were estimated: D≈200 km^2/year and $\rho_{crit}\approx 4$ km, a circle of radius ρ_{crit} containing the home ranges of about 50 foxes (see [28]). For the aggregating slime mould *Dictyostelium discoideum* experiments rendered $D\approx 0.66\times10^{-5}$ cm^2/sec [31] and $\rho_{crit}\geq 200$ μm, corresponding to a circle containing at least 500 cells [32].

3.6. *Dispersion relation*

Using the extended model which includes excitability of the refractory states (model II) one obtains a dispersion relation analogous to that from other theoretical approaches [6,19] and from experiments (*D. discoideum* [33]; BZ reagent: unpublished results by A.T. Winfree displayed in [19]), namely a monotonously increasing, saturation dependence of the velocity c of the planar waves on the period T. In fact, simulating planar waves by periodically exciting the cells along an edge of the medium we obtain Fig. 13, which has the expected shape. For $T < 6$ iterations, no propagation takes place.

3.7. *Spatiotemporal chaos*

Transitions from spatiotemporal order to chaos, and back to order have been observed experimentally in the BZ reagent [34,35]. In these observations, chaos arises due to coupling of the chemical waves with convection, while suppression of convection leads to self-organization into spiral waves.

These experiments can be roughly simulated with our model by mimicking the convection rolls through a spatially periodic refractory time n. As indicated on the left of Fig. 14, we let n vary periodically (period δ) in the direction of the ordinate in a piecewise linear manner between two values n_1 and n_2. Figure 15 shows the result of the application of this periodic inhomogeneity. For $t \geq 73$, aperiodicities seem to have fully developed. In order to show that these aperiodicities actually correspond to deterministic chaos, the maximum Lyapuňov exponent λ_{max} was calculated using the method of Wolf et al. [38]. Phase spaces were constructed by considering 9 points p_i ($i=1,2,\ldots,9$) placed at a fixed distance from each other and from the edges of the two-dimensional medium, as indicated by the crosses on the right of Fig. 14. The phase variables v_i were defined as the average of S within circles of radius 9d and centred at the p_i. Figure 16a shows λ_{max} as determined in phase spaces with dimensions $D_E=1,2,\ldots,9$, each space being defined by the phase variables v_j, $j=1,\ldots,D_E$. The bars in the figure indicate the range within which λ_{max} oscillated after convergence in time. An example of this convergence is shown in Fig. 16b. Within the error given by the bar length, λ_{max} was independent of the parameter EVOLV (see [38]) for $13 \leq \text{EVOLV} \leq 19$ iterations. The fact that λ_{max} resulting from our calculations is approximately independent of D_E for $4 \leq D_E \leq 9$ and significantly greater than zero shows that these spatiotemporal patterns are indeed chaotic. Analogously

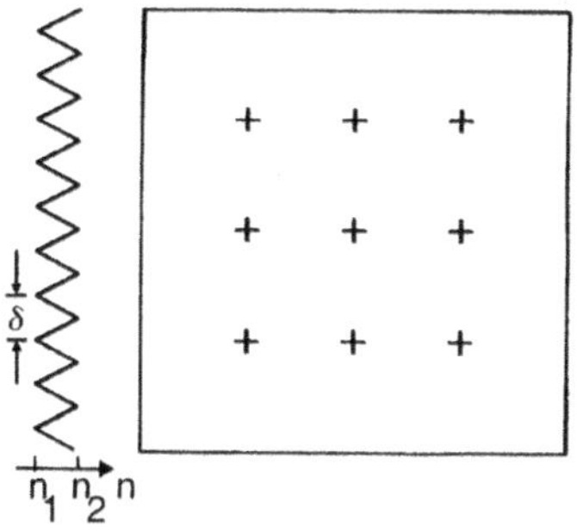

Figure 14. Left: space dependence of refractory time n. Right: border of the medium and points $p_1,\ldots,p_9$ (crosses). The phase variables are defined as the averages of the state S in circles with radius 9d and centred at the p_i.

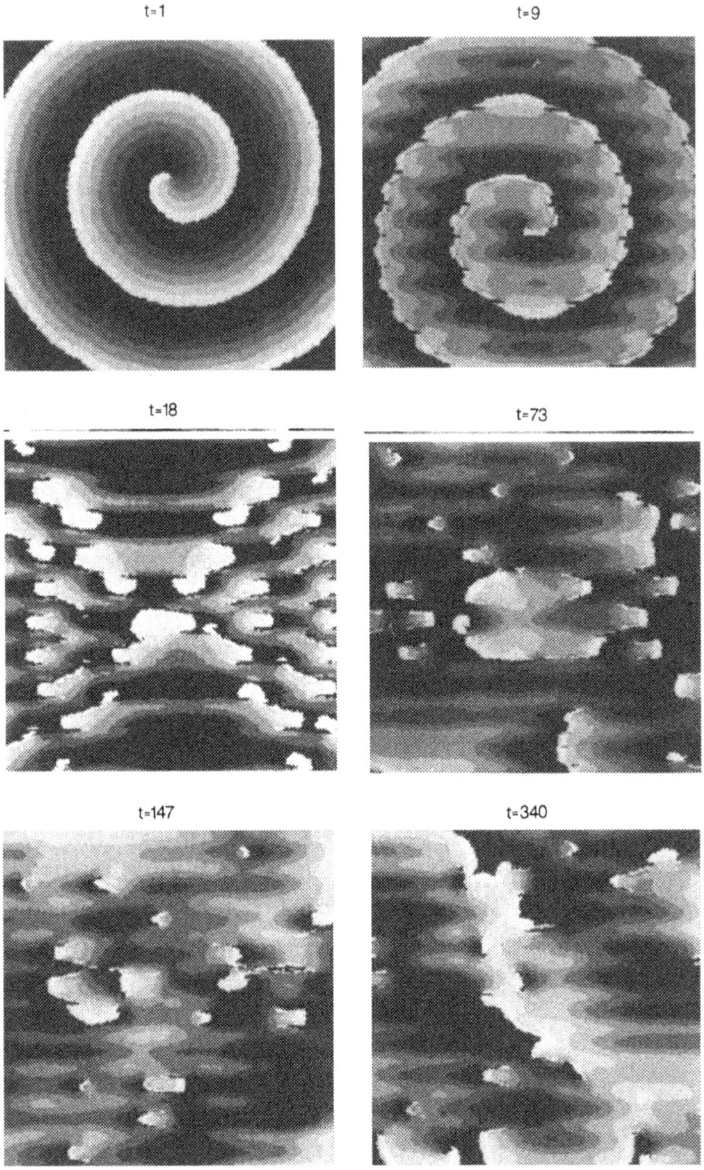

Figure 15. Simulation of the development of spatiotemporal chaos (200x200 cells, model I, r=6, m=1). n=6 for t= 1; n varies as shown in the left of Fig. 14 with δ=10d, n_1=4 and n_2=20 for $1 < t \leq 340$.

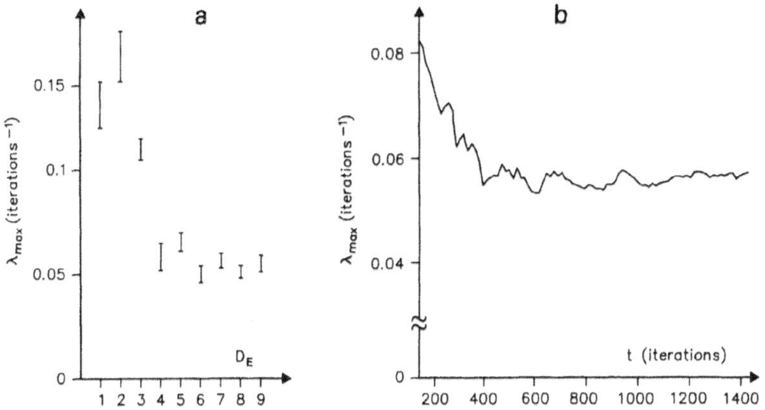

Figure 16. Maximum Lyapunov exponent λ_{max} for the aperiodicities developing in Fig. 15. (a): λ_{max} vs. the embedding dimension D_E. (b): λ_{max} as a function of time for D_e=8.

to these simulations, it is to be expected that other excitable media, e.g. those consisting of living cells, may well display chaotic behaviour if the control parameters are spatially inhomogeneous.

After suppressing the periodic inhomogeneity of n, we obtain reorganization into spirals, as shown in Fig. 17. Full reorganization is obtained for $t \geq 389$ (see last picture in Fig. 17), which is comparable to the reorganization of the BZ reagent after suppression of convection rolls [34,35].

For the quantification of spatial order and disorder, the minimum autodifference function has been successfully applied previously [34,35]. We determine here this measure as follows

$$D(d) = \frac{1}{R(d)} \sum_{\substack{i=1 \\ |\vec{x}_i - \vec{x}_j| = d}}^{N} \sum_{j=1}^{N} |S(\vec{x}_i) - S(\vec{x}_j)|,$$

where the $\vec{x}_i$ are the positions of the automaton points on the plane, $S(\vec{x}_i)$ their states, d the distance between two points, N the number of points in the medium, and R(d) the total number of point pairs. D_{min} is the minimum of D(d) over all d, the trivial minimum D=0 at d=0 being discarded. Order, resp. disorder, is indicated by low, resp. high, values of D_{min}. As shown in Fig. 18, D_{min} clearly indicates the disorganization and reorganization of the medium, as in the evaluation of the BZ experiments [34,35].

3.8. *Disruption of waves by obstacles*

In the BZ reagent, disruption and shrinking of waves caused by collision with CO_2 bubbles has been reported for the case of low excitability of the medium.

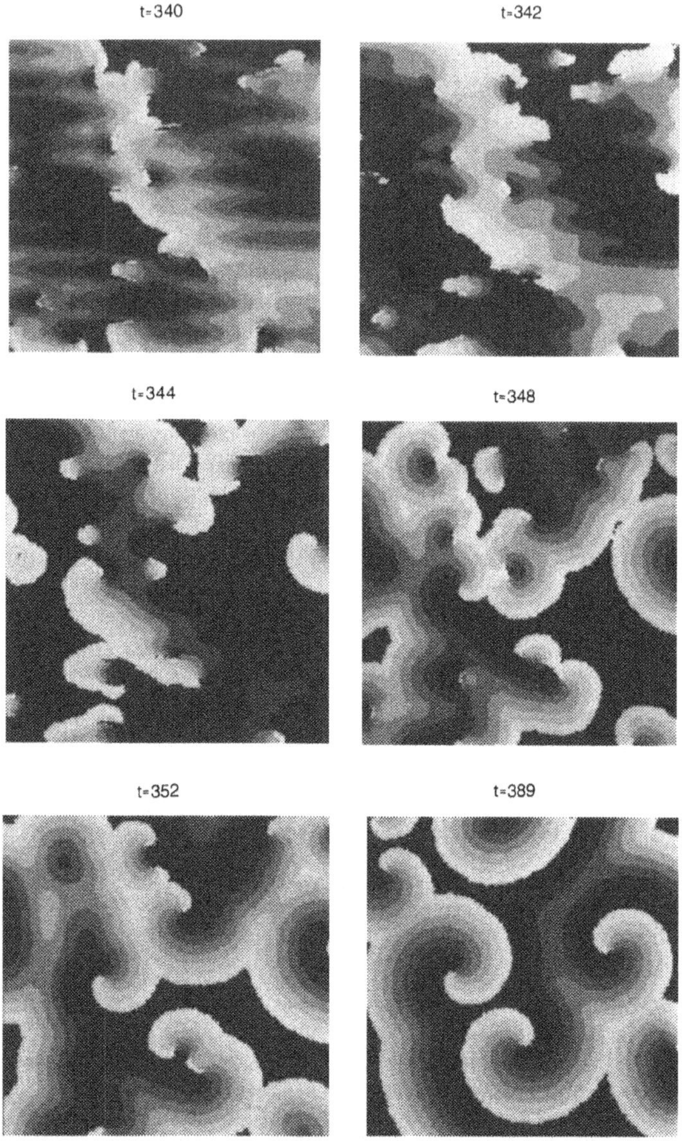

Figure 17. Simulation of reorganization of spatiotemporal chaos into spirals. For the first picture (t=340), which is the same as the last picture in Fig. 15, n varies periodically in space. For t ≥ 341, n is set to 6 everywhere. The other parameters are the same as in Fig. 15.

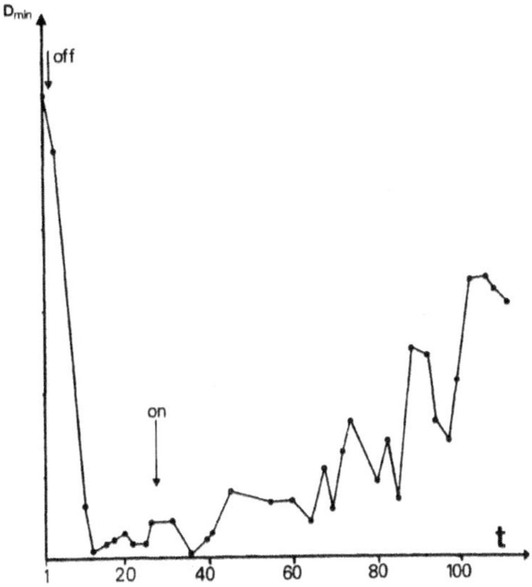

Figure 18. $\bar{D}_{min} = \hat{D} - D_{min}$ as function of time ($\hat{D}$ is a constant and D_{min} is the minimum autodifference function [34,35]). Ordinate units are arbitrary. (70x70 cells, model I, r=3, m=1). The arrows "off" and "on" correspond to those in BZ experiments [35]. Before the arrow "off", a spiral wave is obtained by setting n=3 everywhere. After the arrow "off", aperiodicities develop by varying n periodically as shown at the left of Fig. 14 with δ=10d, n_1=3 and n_2=7. After the arrow "on", reorganization into spirals is possible by setting again n=3 everywhere.

In contrast, waves travel across the bubbles and are self-repaired if the excitability is sufficiently high [39]. These effects have been simulated using differential equations of a modified FitzHugh type [40]. One can also simulate these effects using model I by correlating low excitability with high m. Results are shown in Fig. 19, where the obstacle is indicated by a white border. In Fig. 19a we assume high excitability, the obstacle causing a slight deformation of the wave, which rapidly repairs due to the curvature effects implicit in the model (see section 3.5). In Fig. 19b, which was obtained assuming low excitability, the wave is disrupted and the free ends shrink, as in experiments [39].

3.9. *Three-dimensional waves*

Experimental detection of spiral waves developing in three dimensions (the so-called scroll waves) was first reported by Winfree [41]. A recent review is given in [42]. An example of a scroll wave computed with model I, namely a so-called simple scroll ring, is shown in Fig. 20. The initial condition (t=0) of this wave is a closed ribbon, i.e. a cylinder segment, of thickness R buffered inside by closed refractory ribbons with the same thickness and with S decreasing (S=n,n-1,...,1) towards the cylinder axis z. The figure shows the configuration resulting from one of the cylinder edges. The excited state of the back half of the medium obtained by cutting with a plane parallel to the x-z-plane is displayed. Scroll waves similar to that shown in Fig. 20 have

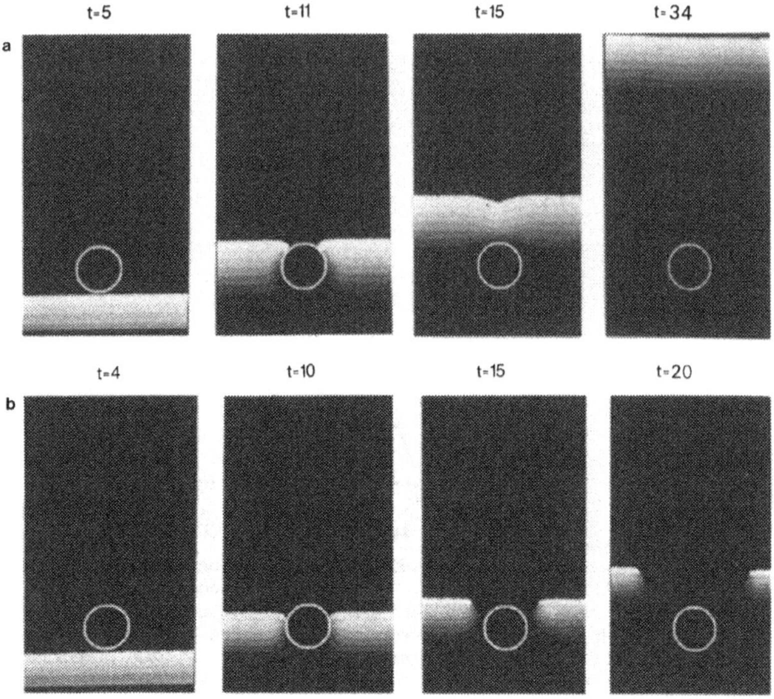

Figure 19. Simulation of the passage of a wave through an obstacle (indicated by a white circle), such as a CO_2-bubble in the BZ experiments reported in [39] (200x350 cells, model I, r=12, n=6). (a): High excitability (m=25) leads to self-repair. (b): Low excitability (m=86) leads to disruption and shrinking.

Figure 20. Simulated simple scroll ring (140x140x140 cells, model I, r=3, n=3, m=1, t=16).

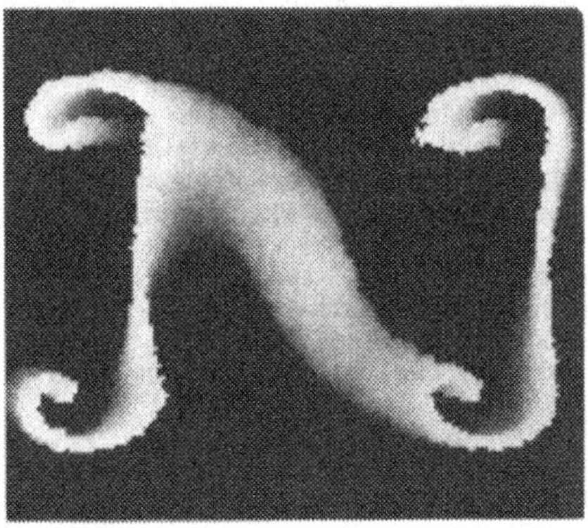

Figure 21. Simulated twisted scroll ring (200x200x200 cells, model I, n=3, r=3, m=1, t=4).

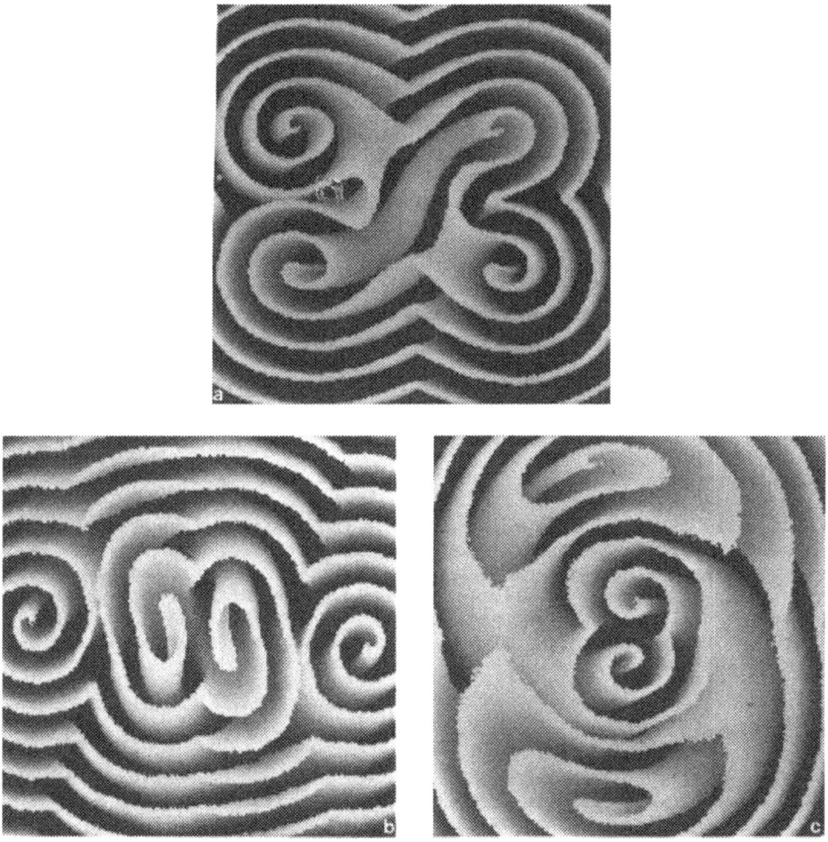

Figure 22. Same as Fig. 21, but for t=24, as seen from the front (a), from the right side (b) and from above (c). In each case, only the back half of the medium is shown, after cutting the medium with a plane perpendicular to the viewing direction.

been observed in thick heart muscle using the technique of multielectrode mapping [22,43] and in the BZ reagent [44].

More complex three-dimensional spiral waves are the so-called twisted scroll rings. This type of waves has been constructed geometrically (without dynamical simulations) [45] and by anisotropic automata (cubic cells) [3,4], but has not yet been observed experimentally. A simulation of a wave of this type was started here by ribbons such as those used to obtain Fig. 20 with the difference that they were given a 360° twist before their ends were closed into a ring. Figure 21 shows the result for t=4, i.e. shortly after initiation, and Fig. 22 shows the result for t=24, as seen from different sides. As in Fig. 20, only the excited state in the back half of the medium is shown in Figs. 21 and 22.

3.10. *Quantification*

The results presented so far and their comparisons with experiments were purely qualitative. In order to quantify our results, we had to determine our cell length and our time step in microns and in seconds. We do this here for the special case of the BZ reagent. Using model II and setting r=5, n=9, $m_o=2$, p=2 and $S_{max}=8$, we obtain a spiral period of 10.5 iterations (the spiral turns twice during 21 iterations) and a spiral propagation velocity of 3.8 cell lengths/iteration. Comparing these values with the measured spiral period (17.3 sec) and velocity (76 μm/sec) [27], we obtain for this case: 1 cell length = 33 μm and 1 iteration step = 1.65 sec. Using these transformations of units, we can now compare quantitatively our results with other measurements given in ref. [27]. First of all, we compare the velocity of planar waves c. Our calculations yield c=4.25 cell lengths/iteration. This corresponds to 85 μm/sec, in excellent agreement with the experimental value 82.5 μm/sec. Secondly, our critical radius $\rho_{crit} = K_{crit}^{-1}$ (obtained by extrapolation of N versus K to N=0) is 0.8 cell lengths, corresponding to 26.4 μm, as compared to the experimental value of 23.1 μm. Third, our diffusion coefficient, as determined by $D=c\ \rho_{crit}$ (see [28]) is D=3.4 (cell lengths)^2/iteration, corresponding to 2.24×10^{-5} cm^2/sec, which compares well with the experimental value of 2×10^{-5} cm^2/sec.

4. General conclusions

The simplicity of the present model suggests that while the underlying microscopic properties of an excitable medium may be very complex, the macroscopic manifestations are describable by a few rules and are common to a large variety of systems. Excellent quantitative agreement is obtained between automaton simulations and measurements with the BZ reagent. Work remains to be done to find parameters allowing us to match simulations to observations in other excitable media.

Acknowledgements

We thank the Stiftung Volkswagenwerk (Hannover) and the Commission of the European Communities (Brussels) for financial support. M.M. thanks John Tyson for fruitful suggestions. Also, we thank Gesine Schulte for the photographic work.

References

[1] Wiener, N. & Rosenbluth, A. (1946). The mathematical formulation of the problem of conduction of impulses in a network of connected excitable elements, specifically in cardiac muscle. *Arch. Inst. Cardiol. Mex.* **16**, 205-265.

[2] Madore, B.F. & Freedman, W.L. (1983). Computer simulation of the Belousov-Zhabotinskii reaction. *Science* **222**, 615-616 and cover.

[3] Winfree, A.T., Winfree, E.M. & Seifert, H. (1985). Organizing centers in a cellular excitable medium. *Physica* **17D**, 109-115.

[4] Panfilov, A.V. & Winfree, A.T. (1985). Dynamical simulations of twisted scroll rings in three-dimensional excitable media. *Physica* **17D**, 323-330.

[5] Plath, P.J. A new pattern of spreading excitations in cellular automata. In this volume.

[6] Hanusse, P., Perez-Muñuzuri, V. & Vidal, C. Phase dynamics and spatial patterns in oscillating and excitable media. In this volume.

[7] Kurrer, C. & Schulten, K. Propagation of chemical waves in discrete excitable media: anisotropic and isotropic wave fronts. In this volume.

[8] Markus, M. & Hess, B. (1988). An automaton with randomly distributed cells for the simulation of waves in excitable media. Abstract-book of the 16th Aharon Katzir-Katchalsky Conference on Dynamics in Molecular and Cellular Biology (Univ. Libre de Bruxelles), pp. 37-38.

[9] Markus, M. & Hess, B. (1990). Isotropic automata for simulations of excitable media: periodicity, chaos and reorganization. In *Dissipative Structures in Transport Processes and Combustion,* Meinköhn, D. (ed.). Springer-Verlag: Berlin (in press).

[10] Penrose, R. (1979). Pentaplexity. *Math. Intelligencer* **2**, 32-37.

[11] Nelson, D.R. & Halperin, B.I. (1985). Pentagonal and icosahedral order in rapidly cooled metals. *Science* **229**, 233-238.

[12] Hardy, J., Pomeau, Y. & dePazzis, O. (1973). The evolution of a two-dimensional model system. I: Invariant states and time correlation functions. *J. Math. Phys.* **14**, 1746-1759.

[13] Frisch, U., Hasslacher, B. & Pomeau, Y. (1986). Lattice gas automaton for the Navier-Stokes equation. *Phys. Rev. Lett.* **56**, 1505-1508.

[14] Nasilowski, R. An arbitrary-dimensional cellular automaton fluid model with simple rules. In *Dissipative Structures in Transport Processes and Combustion,* Meinköhn, D. (ed.). Springer-Verlag: Berlin (in press).

[15] Zykov, V.S. & Mikhailov, A.S. (1986). Rotating spiral waves in a simple model of an excitable medium. *Sov. Phys. Doklady* **31**, 51-52.

[16] Dorjsurangiyn, B. (1987). Autowave processes in active media formed by cellular automata. *Diploma Thesis.* M.V. Lomonosov Moscow State Univ. Dept. of Physics.

[17] Gerhardt, M. & Schuster, H. (1989). A cellular automaton describing the formation of spatially ordered structures in chemical systems. *Physica* **D36**, 209-221.

[18] Gerhardt, M., Schuster, H. & Tyson, J.J. A cellular automaton model of excitable media including curvature and dispersion. *Science*, in press.

[19] Keener, J.P. & Tyson, J.J. (1986). Spiral waves in the Belousov-Zhabotinskii reaction. *Physica* **21D**, 307-324.

[20] Müller, S.C., Plesser, Th. & Hess, B. (1987). Two-dimensional spectrophotometry of spiral wave formation in the Belousov-Zhabotinskii reaction. *Physica* **24D**, 71-86.

[21] Agladze, K.I. & Krinsky, V.I. (1982). Multi-armed vortices in an active chemical medium. *Nature* **296**, 424-426.

[22] Medvinsky, A.B., Panfilov, A.V. & Pertsov, A.M. (1984). Properties of rotating waves in three dimensions. Scroll waves in myocard. In *Self-Organization. Autowaves and Structures far from Equilibrium,* pp. 195-199, Krinsky, V.I. (ed.). Springer-Verlag: Berlin.

[23] Goldberger, A.L., Bhargava, V., West, B.J. & Mandell, A.J. (1986). Some observations on the question: Is ventricular fibrillation "chaos"? *Physica* **19D**, 282-289.

[24] Shibata, M. & Bures, J. (1974). Optimum topographical conditions for reverberating cortical spreading depression in rats. *J. Neurobiol.* **5**, 107-118.

[25] Krinsky, V.I. & Agladze, K.I. (1983). Interaction of rotating waves in a chemical active medium. *Physica* **8D**, 50-56.

[26] Foerster, P., Müller, S.C. & Hess, B. (1988). Curvature and propagation velocity of chemical waves. *Science* **241**, 685-687.

[27] Foerster, P., Müller, S.C. & Hess, B. (1989). Critical size and curvature of wave formation in an excitable chemical medium. *Proc. Natl. Acad. Sci. USA* **86**, 6831-6834.

[28] Tyson, J.J. & Keener, J.P. (1988). Singular perturbation theory of travelling waves in excitable media (a review). *Physica* **D32**, 327-361.

[29] Zykov, V.S. & Morozova, O.L. (1980). Speed of spread of excitation in a two-dimensional excitable medium. *Biophysics* **24**, 739-744.

[30] Zykov, V.S. (1980). Analytical evaluation of the dependence of the speed of an excitable wave in a two-dimensional excitable medium on the curvature of its front. *Biophysics* **25**, 906-911.

[31] Foerster, P. & Müller, S.C. Geometric parameters of aggregating waves in *Dictyostelium discoideum*. In this volume.

[32] Foerster, P., Müller, S.C. & Hess, B. Curvature and spiral geometry in aggregating patterns of *Dictyostelium discoideum*. *Development*, in press.

[33] Siegert, F. & Weijer, C. (1989). Digital image processing of optical density wave propagation in *Dictyostelium discoideum* and analysis of the effect of caffeine and ammonia. *J. Cell. Sci.* **93**, 325-335.

[34] Markus, M., Müller, S.C., Plesser, Th. & Hess, B. (1987). Characterization of order and disorder in two-dimensional patterns. In *Chaos in Biological Systems*, p. 295-303, Degn, H., Holden, A.V. & Olsen, L.F. (eds.). Plenum Press: New York.

[35] Markus, M., Müller, S.C., Plesser, Th. & Hess, B. (1987). On the recognition of order and disorder. *Biol. Cybern.* **57**, 187-195.

[36] Kurrer, C. (1989). Nichtlineare Oszillatoren und Selbstorganisation in biologischen Systemen. *Diplomarbeit*. Physik-Department, Technische Univ., München.

[37] Colli Franzone, P., Guerri, L., Rovida, S. & Tentoni, S. A model of excitation wavefronts spreading in the anisotropic cardiac tissue. In this volume.

[38] Wolf, A., Swift, J.B., Swinney, H.G. & Vastano, J.A. (1985). Determining Lyapunov exponents from a time series. *Physica* **16D**, 285-317.

[39] Nagy-Ungvarai, Zs. Chemical regulation of excitability in the Belousov-Zhabotinskii reaction. In this volume.

[40] Pertsov, A.M., Panfilov, A.V. & Medvedeva, F.U. (1983). Instability of autowaves in excitable media associated with the phenomenon of critical curvature. *Biophysics* **28**, 103-107.

[41] Winfree, A.T. (1973). Scroll-shaped waves of chemical activity in three dimensions. *Science* **181**, 937-939.

[42] Panfilov, A.V. Three dimensional vortices in active media. In this volume.

[43] Krinsky, V.I., Pertsov, A.M., Fast, V. & Biktashev, V.A. Study of the autowave mechanisms of cardia arrhythmias. In this volume.

[44] Welsh, B.J., Gomatam, J. & Burgess, A.E. (1983). Three-dimensional chemical waves in the Belousov-Zhabotinskii reaction. *Nature* **304**, 611-614 and cover.

[45] Winfree, A.T. & Strogatz, S.H. (1984). Organizing centres for three-dimensional chemical wavefronts. *Nature* **311**, 611-615.

18. CHEMICAL REGULATION OF EXCITABILITY IN THE BELOUSOV-ZHABOTINSKII REACTION

Zs. Nagy-Ungvarai

Max-Planck-Institut für Ernährungsphysiologie
Rheinlanddamm 201, D-4600 Dortmund 1, FRG

1. Introduction

In an excitable medium, there exists a spatially homogeneous resting state which is stable to small perturbations. If a small region of space of such a system is perturbed beyond threshold, self-propagating waves travel through the medium without damping and after their propagation the medium returns to its original resting state [1].

In two-dimensional cases, as the medium is in the same state before and after the wave has passed through it, propagating waves can be broken. During the further development of a broken wave front, free ends of the front either form spiral waves [2] or, as it was shown theoretically, they contract to wave pieces [3].

In three-dimensional cases several types of scroll waves - symmetric (involute) spirals, elongated spirals (tilted scrolls) and scroll rings (U-shaped scrolls) - can exist depending on the form of the central filament. The front surface of a scroll wave is almost always unstable. Scroll rings either shrink [4] or, as it was calculated, they sprout in time [3]. Theoretically, there exists a special case where stable scroll rings are possible. The properties of spatial patterns mentioned above strongly depend on the excitability of the medium. We have studied the dependence of spatial phenomena on excitability, and the possibilities of changing the excitability of a medium. Spatial patterns have been observed in various physical, chemical and biological systems. Among them, the Belousov-Zhabotinskii (BZ) reaction represents the most suitable system for studying spatial phenomena because of its extreme experimental ease. Therefore, we carried out our studies in thin layers of BZ reaction solutions or gels in a wide concentration range of the reagents: bromate, malonic acid, sulphuric acid and the catalysts ferroin, rutheniumbipyridyl complex, cerium and manganese. We studied the spatial phenomena by a 2D spectrophotometer system described elsewhere [5]. Every picture shown in this paper has an image area of 9 x 9 mm^2. The layer thickness is in every case 0.56 mm.

2. Dependence of pattern formation on excitability

2.1. *Highly excitable media*

Most types of chemical waves have been discovered in a BZ solution with reagent concentrations identical or close to the concentration set suggested by Winfree [6] which is a special case of highly excitable media. In this narrow concentration range wave fronts are always sharply shaped. That means, along the direction of propagation there are typically two regions: a narrow region in which there are sharp gradients and a broad region in which concentrations vary slowly in space [7]. In other words, the width of the wave, i.e. the size of the region where the medium is far from the state of rest (oxidized region), is smaller than the distance between consecutive wave fronts. There are theoretical descriptions which even consider the waves as oriented curves neglecting the width of the wave front [3]. We will call such solutions narrow wave excitable solutions, because the major part of the solution layer is in the reduced state.

In such highly excitable media concentric waves evolve in time regularly as it is commonly known (Fig. 1).

Wave fronts can be quite easily broken, for example mechanically or by hydrodynamic motions. Open ends of broken wave fronts curl up to stable spirals (Fig. 2).

A wave front travelling across a small obstacle, the size of which is comparable with that of the wave fronts, can be regarded as a special case of trying to break wave fronts. This phenomenon is easily observable in the BZ solution layers because CO_2 bubbles evolving during the course of the reaction behave as small mechanical obstacles in the way of a wave front. In highly excitable media a wave front running across an obstacle is quickly stabilized after a short rupture which can be explained by curvature effects [7]. Such a process is shown on the series of Fig. 3.

A propagating wave front represents a coupling between diffusion and autocatalytic reaction. If the autocatalytic species is produced in one volume element it then diffuses in the neighbouring volume elements and also triggers the autocatalytic reaction there. In a wave front, when arriving at a CO_2 bubble, diffusion is hindered into areas occupied by the bubble and achieves a higher level in the free directions. Therefore, on both sides of the bubble a large negative curvature is formed with respect to the direction of propagation, which leads to increased velocities. Having passed across half of

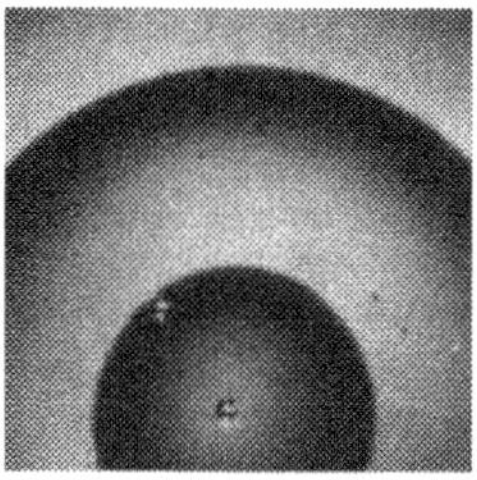

Figure 1. Concentric waves in a BZ solution with the initial composition of 0.52 M $NaBrO_3$, 0.4 M MA, 0.5 M H_2SO_4, 0.03 M Ce (IV).

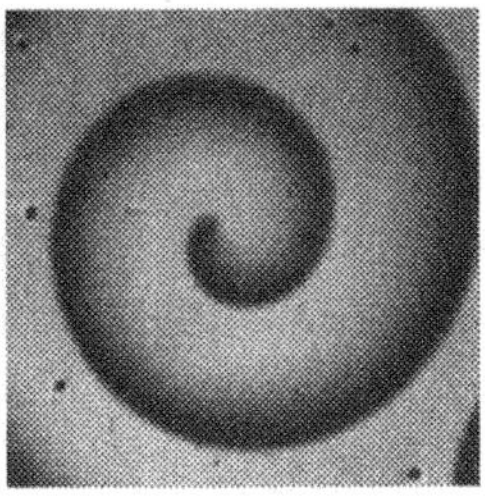

Figure 2. Concentric waves in a BZ reaction with the initial composition of 0.5 M $NaBrO_3$, 0.3 M MA, 0.46 M H_2SO_4, 0.012 M Ce (IV).

the bubble, as formerly closed directions of propagation become accessible again, the wave front attains a large positive curvature on both sides of the bubble causing a decrease in the velocities. In highly excitable media, if the slowest part of the wave front is still in an oxidized region, waves can grow together right after having detached from the bubble and the waveform is similar to the wall of a constricted cell (last picture of Fig. 3).

In highly excitable media three-dimensional elongated spirals tend to straighten up and approach a stable symmetric shape [4]. In the same medium, three-dimensional scrolls shrink and disappear as time proceeds [2,4,8].

There are chemically realizable cases of highly excitable media for which the consideration that waves are oriented curves neglecting the width of the wave front cannot be made. At very high $[BrO_3^-][H^+]/[MA]$ ratios oxidized wave fronts are broad and they occupy the major part of the reaction territory.

As is shown in Fig. 4 the whole solution layer consists of waves, and the width of the waves is much larger than the region behind. Therefore, we will call such solutions, opposite to the narrow wave excitable solutions, broad wave excitable solutions. Theoretical description of such solutions is still lacking.

2.2. *Poorly excitable media*

Below a critical excitability, pattern formation in the BZ reaction is considerably changed as compared to highly excitable media. In solutions with low excitability concentric rings evolve in time irregularly.

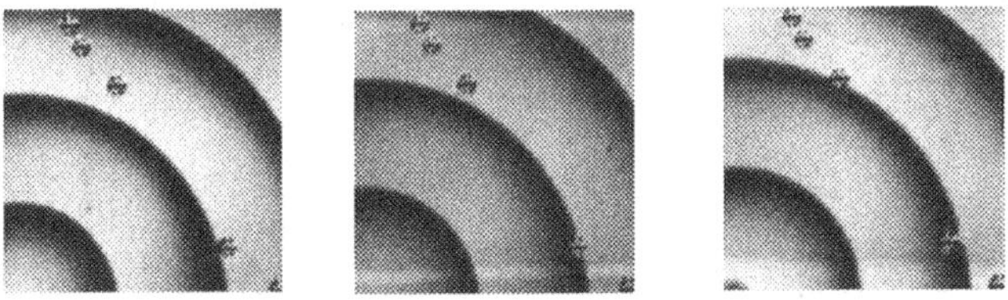

Figure 3. Travelling of waves across CO_2 bubbles in a BZ solution: 0.3 M $NaBrO_3$, 0.366 MA, 0.09 M BrMA, 0.41 M H_2SO_4, 0.006 Ce (IV).

Figure 4. Travelling waves in a solution with the initial composition of 0.3 M $NaBrO_3$, 0.01 M MA, 0.09 M BrMA, 0.41 M H_2SO_4, 0.006 M Ce (IV).

If we put in a poorly excitable solution a small mechanical obstacle (which can trigger waves) every second appearing wave will be unstable. That means, after every stable wave triggered (Fig. 5a) an unstable wave appears which disappears in pieces soon after its propagation (Fig. 5b).

Most of the spontaneously appearing waves in highly excitable solutions are formed at the wall of the Petri-dish. In poorly excitable solutions only unstable broken wave pieces appear at the wall of the Petri-dish (Fig. 6).

In a poorly excitable solution open ends of a broken wave front cannot curl up to spirals. Wave fronts can be this way parted into wave pieces which are moving in the direction determined by the former movement of the wave (Fig. 7).

The open ends are then shrinking. Small pieces tend to disappear this way, bigger ones exist for a long time until a new collision and division takes place.

In three-dimensional solutions of poorly excitable media scrolls are unstable, but they are not shrinking, they rather spread out in time.

3. **Regulation of excitability**

Pattern formation and the properties of different wave types have been

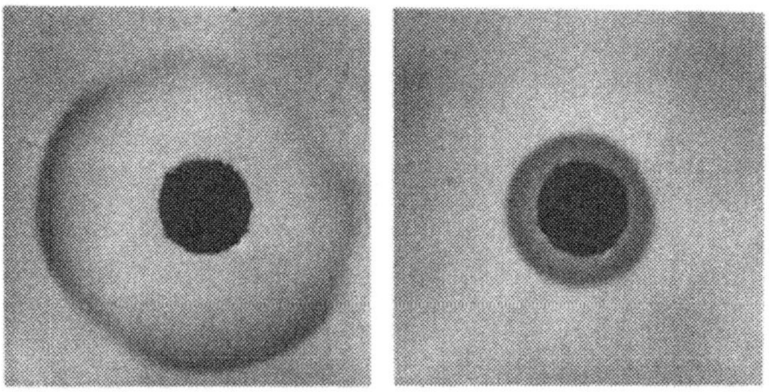

Figure 5. Wave formation in a BZ solution with the initial composition of 0.1 M $NaBrO_3$, 0.03 M MA, 0.09 BrMA, 0.1 M H_2SO_4, 0.006 M Ce (IV).

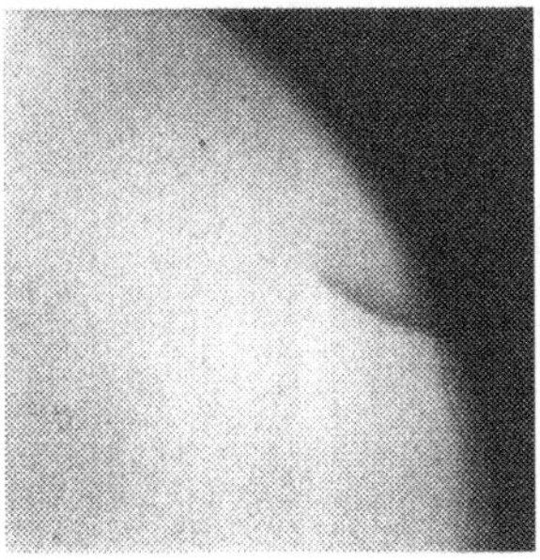

Figure 6. A piece of wave appearing at the wall of the Petri dish in the same solution as defined in Fig. 5.

experimentally extensively studied in highly excitable media (in the narrow wave case). Far less experimental work has been done in solutions with low excitability. The reason for it may be found in the fact mentioned above that most experimental work has been carried out in the highly excitable Winfree solution [6] and relatively little is known about the chemical modification of distributed BZ media [9] and its effect on the excitability of the medium [10,11].

In a large majority of the excitable media, the resting state of the solution corresponds to the reduced state of the medium in which oxidized waves can propagate. During excitation the autocatalytic oxidation step of the BZ reaction must be triggered. So, every change of the excitability of a BZ solution means a change in the autocatalytic reaction sequence:

$$HBrO_2 + BrO_3^- + H^+ = 2\ BrO_2\cdot + H_2O \tag{R5}$$

$$BrO_2\cdot + cat^{n+} + H^+ = cat^{(n+1)+} + HBrO_2 \tag{R6}$$

(We have not dealt with the special case of the reduced waves described earlier in a special version of the BZ reaction [12].)

We have found the following possibilities to vary the excitability of a BZ medium:

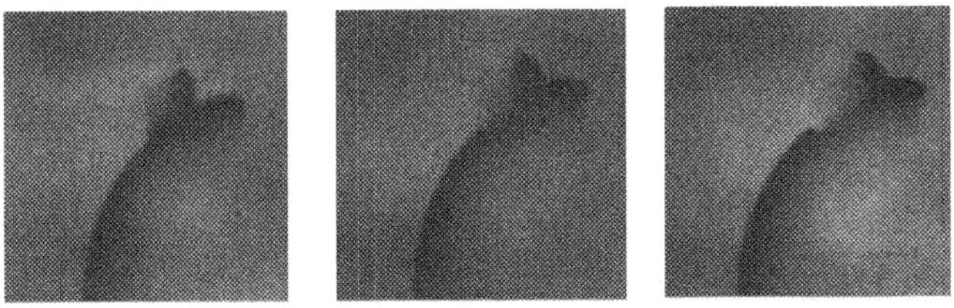

Figure 7. Formation and movement of wave pieces by collision with each other in a late stage (after 3 h) in a solution as defined in Fig. 1.

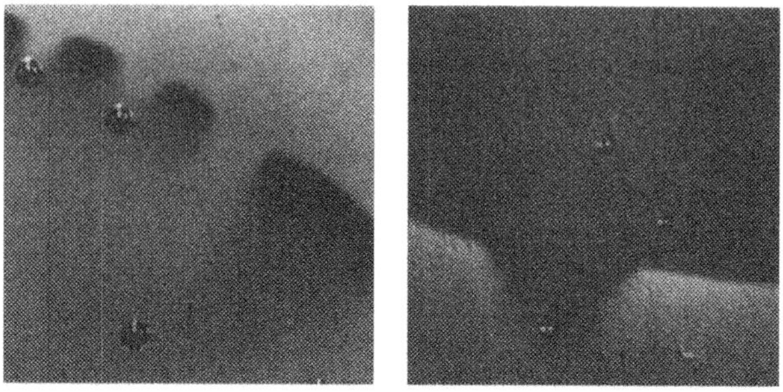

Figure 8. Disruption and shrinking of a wave front caused by collision with CO_2 bubbles in a Ce-catalyzed solution (late stage of the solution as described in Fig. 1) and in a ferroin gel system (0.15 M $NaBrO_3$, 0.1 M MA, 0.1 M H_2SO_4, 0.003 M ferroin fixed in a SiO_2 gel).

3.2.a. *With time.* The chemical properties of a BZ solution layer are changing in time. The assumption that the medium is in the same state before and after the wave has passed through it is an approximation. In every wavefront a certain small amount of initial substances is converted to products. As time passes initial substances are used up and the system becomes less and less excitable. During the temporal evolution of a highly excitable solution with a given initial concentration set we can observe continuously every stage of excitability.

As ferroin tends to decompose in acid solutions this observation is only possible in Ce^-, Mn^- or $Ru(bipy)_3^{2+}$-catalyzed solutions or in geled ferroin systems. Figure 8 shows the disruption and shrinking of a wave front in a Ce-catalyzed BZ solution layer and the same in a SiO_2 gel with fixed ferroin as a catalyst.

Depending on the excitability of a system, three-dimensional scrolls either shrink or spread out in time. Theoretically, we should find a stage with a critical excitability in which scrolls are stable. However, because of the temporal change of excitability of real solutions it is hardly possible to observe such a phenomenon. Until now, we have not succeeded in realizing it.

3.2.b. *Changing the initial concentrations.* The excitability of a solution is lowered with decreasing H^+-, BrO_3^-- and catalyst- concentrations. The effect of H^+ ions has already been described [10]. Typical for solutions with low $[H^+]$ or low $[BrO_3^-]$ at a medium excitability is the asymmetrical form of spirals formed in these systems (Fig. 9). The effect of MA is far less than that of the other components.

3.2.c. *Inhibiting the autocatalytic reaction.* Partly inhibiting the autocatalytic reaction step in a given highly excitable solution lowers the excitability of the system. It can be realized for example by adding different

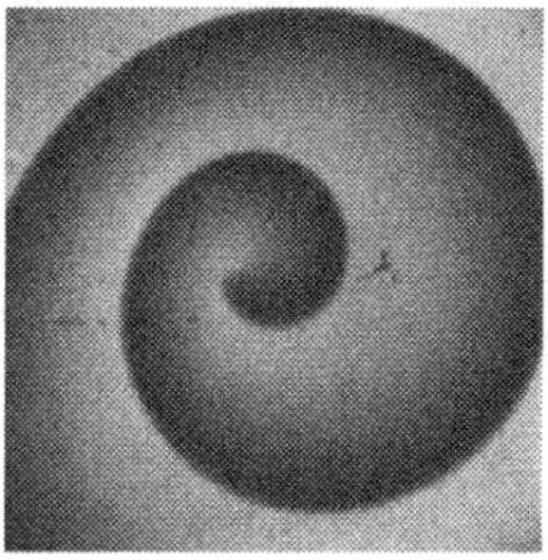

Figure 9. Asymmetrical spiral in a solution with the initial concentrations 0.1 M $NaBrO_3$, 0.45 M MA, 0.5 M H_2SO_4, 0.012 M Ce (IV).

amounts of ethanol to the solution. The aldehyde + $HBrO_2$ reaction competes with the catalyst + $HBrO_2$ (R5) reaction. The higher the concentration of ethanol, the lower the excitability [13].

3.2.d. *By light.* In the case of the Ru-catalyzed BZ reaction the inhibition of the autocatalytic reaction is also possible by light because of the light sensitivity of the Ru-complex [11].

Acknowledgement

The technical assistance of I. Beyer and B. Plettenberg is gratefully acknowledged.

References

[1] Tyson, J.J. (1977). *J. Chem. Phys.* **66**, 905.
[2] Winfree, A.T. (1973). *Science* **181**, 937.
[3] Davydov, V.A., Mikhailov, A.S. & Zykov, V.S. In *Nonlinear Waves in Active Media*, Crighton, D.G. & Engelbrecht, U. (eds.). Springer, in press.
[4] Keener, J.P. & Tyson, J.J. (1988). *Science* **239**, 1284.
[5] Müller, S.C., Plesser, Th. & Hess, B. (1987). *Physica* **24D**, 71.
[6] Winfree, A.T. (1972). *Science* **175**, 634.
[7] Keener, J.P. & Tyson, J.J. (1986). *Physica* **21D**, 307.
[8] Jahnke, W., Henze, C. & Winfree, A.T. (1988). *Nature* **336**, 662.
[9] Nagy-Ungvarai, Zs., Muller, S.C., Tyson, J.J. & Hess, B. (1989). *J. Phys. Chem.* **93**, 2760.
[10] Agladze, K.I., Panfilov, A.V. & Rudenko, A.N. (1988). *Physica* **29D**, 409.
[11] Kuhnert, L. (1986). *Nature* **319**, 393.
[12] Smoes, M.L. (1980). In *Dynamics of Synergetic Systmes*, p. 90-96, Haken, H. (ed.). Springer: Berlin.
[13] Nagy-Ungvarai, Zs. & Zimányi, I. (1986). *Reaction Kin. Catal. Lett.* **31**, 249.

19. DIFFUSIVE INSTABILITIES AND PATTERN FORMATION IN THE BELOUSOV-ZHABOTINSKY SYSTEM

A.B. Rovinsky

Institute of Biological Physics, USSR Academy of Science
Pushchino Moscow Region 142292, USSR

Complex dynamics of certain types of chemical systems like the Belousov-Zhabotinsky reaction has been attracting the attention of many researchers for nearly three decades [1,3,16]. This dynamics is governed by underlying nonlinear kinetics and manifests itself as multistability and periodic, multiperiodic or chaotic oscillations in perfectly stirred systems. At the same time, the investigations of spatial systems have mainly been confined to consideration of classical kinds of waves, or "autowaves" [1,4]. This term means that a local event, e.g. a pulse generated by a cell, is transferred by diffusion to neighbouring points. Such transport triggers the similar events there, and thus gives rise to a travelling wave. Diffusion plays rather a passive role here and dynamics of the whole medium is determined by dynamic properties of its local elements. Rashevsky [9] and Turing [15] found, however, that diffusion may be as important for a system's dynamics as its local kinetics. That means that behaviour of a spatial system may be far more complex than one can conceive from consideration of its local characteristics. Turing [15] was the first who clearly described the possible destabilizing role of diffusion.

This work shows that a variety of spatio-temporal patterns may arise as a result of the reaction-diffusion interplay in the extensively explored ferroin-catalyzed Belousov-Zhabotinsky system. The study is performed using the model of the system presented earlier [10,11,12]:

$$\varepsilon\frac{\partial x}{\partial\tau} = x(1-x) - \left(2q\alpha\frac{z}{1-z} + \beta\right)\frac{x-\mu}{x+\mu} + \Delta_\rho x$$
$$\frac{\partial z}{\partial\tau} = x - \alpha\frac{z}{1-z} + \delta\Delta_\rho z \qquad (1)$$

where x is scaled $HBrO_2$ concentration, z is scaled $Fe(phen)_3^{3+}$ concentration, τ is scaled time, ρ is scaled spatial coordinate, $\delta = D_{Fe(phen)_3^{3+}}/D_{HBrO_2}$, α, β, μ and ε are the scaled parameters of the model, as described by Rovinsky & Zhabotinsky [10] and Rovinsky [11,12]. This is a two variable model. So, the local system can exhibit only the two kinds of dynamics: steady state and periodic oscillations. Both may lose their stability in an extended reactor.

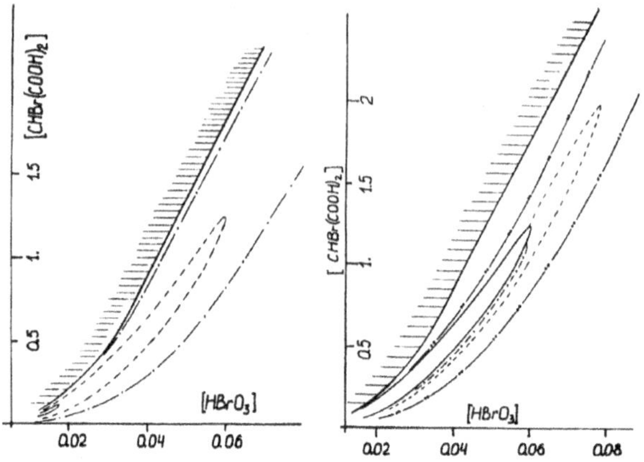

Figure 1. Left: The regions of homogeneous, stable stationary state (shaded) and bulk oscillations (bounded by ... for the catalyst concentration C-3·10^{-6}M, --- for C=10^{-5}M and -·-· for C=3·10^{-5}M). The Turing instability is possible on the part of the plane below the shaded region and beyond the bulk oscillations region. Right: The regions where the homogeneous state is unstable. An instability region is limited by the boundary of oscillation domain (solid line) and the line -·-· for δ = 1.5, - - - for δ = 2 and -··-·· for δ = 4. (Figures 3 and 4 of Rovinsky, A.B., J. Phys. Chem., 1987, 91, 4606-13).

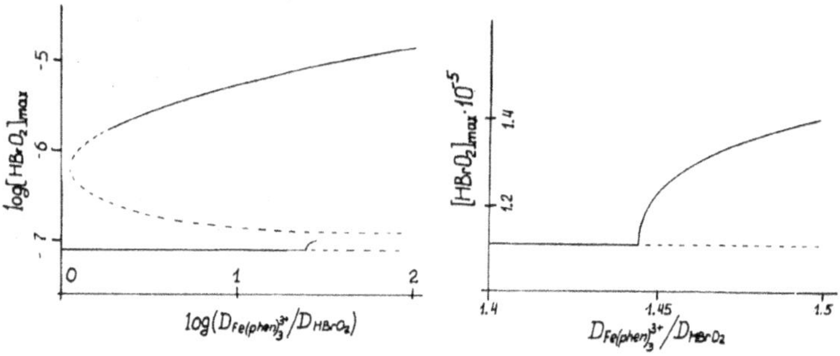

Figure 2. Stationary state bifurcation diagrams. Solid lines represent stable solutions while the dashed lines represent the unstable ones. The horizontal branch corresponds to the homogeneous steady state. Left: The subcritical Turing bifurcation. Right: The supercritical Turing bifurcation. (Figure 7 and 8 of Rovinsky, A.B., J. Phys. Chem., 1987, 91, 4606-13).

1. Symmetry breaking of a homogeneous stationary state in the continuous system

The analysis shows that the Turing bifurcation leading to the loss of stability of spatially uniform state can only occur when the diffusion coefficient of ferriin exceeds that of HBrO [11,12]. The domain of the parameter space where the uniform steady state is unstable expands with

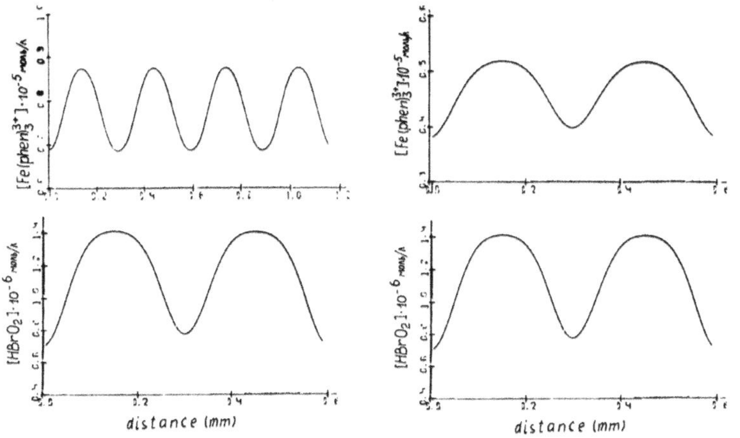

Figure 3. The stationary patterns emerging as a result of the subcritical (left, δ = 25) and supercritical (right, δ = 1.5) Turing bifurcation. (Figures 5 and 6 of Rovinsky, A.B., J. Phys. Chem., 1987, 91, 4606-13).

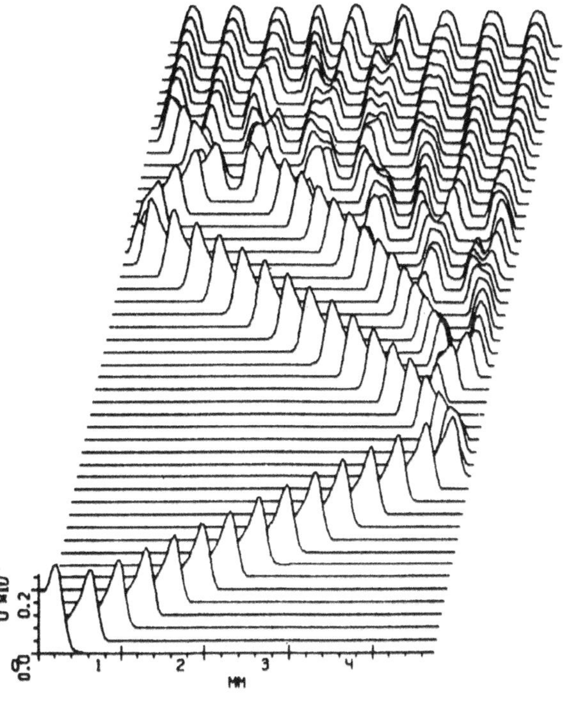

Figure 4. Reflection of a travelling pulse. In the wake of the reflected pulse a stationary periodic pattern appears (δ=5).

increasing $\delta = D_{Fe(phen)_3^{3+}}/D_{HBrO_2}$ and decreasing the total ferroin concentration (Fig. 1). The bifurcation can be of subcritical or supercritical type (Fig. 2), so a large or small-amplitude pattern can be formed as a result of instability (Fig. 3). The stability region of the large-amplitude structure may partly overlap the stability regions of the stationary homogeneous and/or small-amplitude inhomogeneous state. Thus, the system can be found in any state, depending on initial conditions, and can be transferred from one to another by an appropriate perturbation. For example, if the system rests in the homogeneous stationary state and is excited at one of its ends, a travelling pulse wave is generated and the pulse leaves behind the stationary periodic structure. Under certain conditions this process proceeds in a very exotic manner: the initial pulse travels like an ordinary pulse and the medium behind it returns to the initial homogeneous state; but as soon as it reaches the opposite wall, the pulse is reflected and, moving back, leaves behind the stationary periodic pattern (Fig. 4). It was also found that there can be two or three alternative stable, stationary inhomogeneous states differing by their spatial periods and profiles (Fig. 5).

With a changing parameter, the stationary periodic pattern may also lose its stability through a spatial analog of the Hopf bifurcation and give rise to oscillations of the inhomogeneous structure (Fig. 6). With the further change of the parameter, the oscillations may vanish through the inverse Hopf bifurcation (Fig. 5) reestablishing the stationary periodic structure, or disappear through the homoclinic infinite period bifurcation. The latter is usually of saddle-focus type and, under certain conditions, as the parameter approaches the critical value, the oscillatory state undergoes a series of bifurcations and becomes chaotic (Fig. 7). The parametric region of the chaotic oscillations also partly overlaps the stability region of the stationary homogeneous state. Thus, a certain symmetry breaking perturbation can cause chaos in the homogeneous system which has been in a steady state.

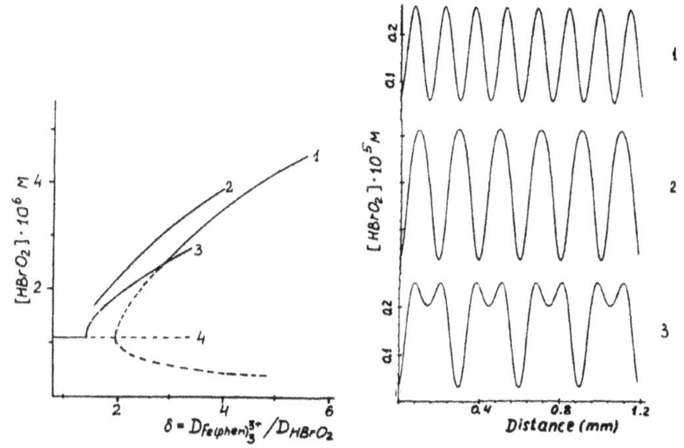

Figure 5. Multiplicity of periodic patterns. Left: The bifurcation diagram for three different inhomogeneous structures (solid lines represent stable solutions). Note that a short interval of branch 3 is unstable. Stability is lost and retrieved through the Hopf bifurcation. Right: The corresponding stable, alternative stationary patterns for δ=2.95.

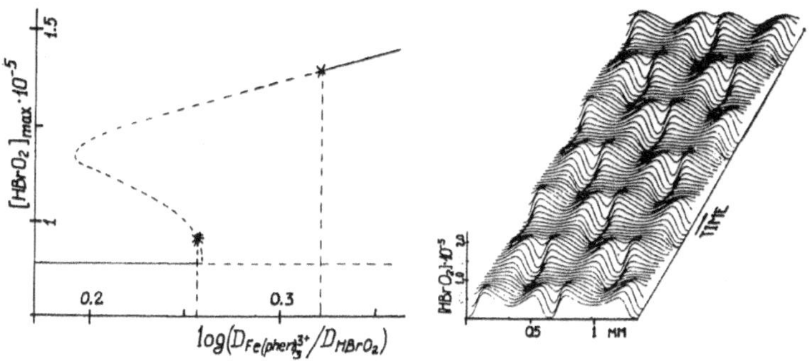

Figure 6. Right: The periodically oscillating inhomogeneous pattern which is formed due to the Hopf bifurcation of the stationary structure (δ=1.7). Left: The corresponding bifurcation diagram. The cross marks the point of the Hopf bifurcation. The asterisk marks the point of the homoclinic saddle-focus bifurcation. (Figures 3 and 4 of Rovinsky, A.B., J. Phys. Chem., 1987, 91, 5113-18).

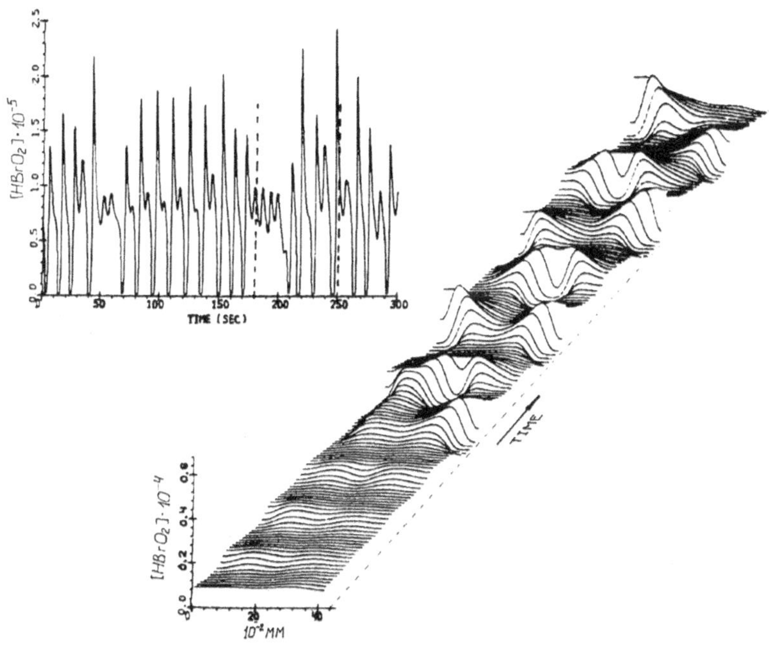

Figure 7. Chaotic oscillations of an inhomogeneous pattern. Top left: Oscillations at a fixed point of the pattern (δ=1.6). (Figure 15 of Rovinsky, A.B., J. Phys. Chem., 1989, 93, 2716-18).

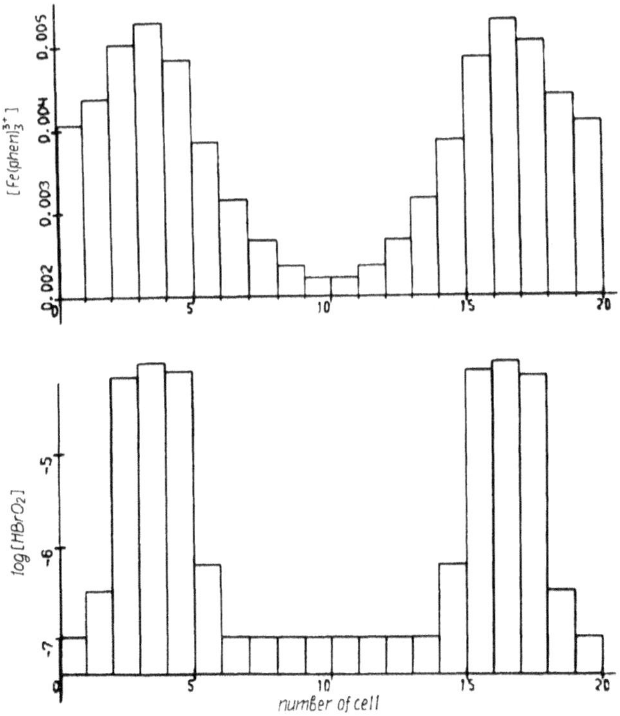

Figure 8. Stable stationary pattern in the discrete BZ-medium (δ=0.75). (Figure 1 of Rovinsky, A.B., J. Phys. Chem., 93, 2716-18).

2. Symmetry breaking of a homogeneous stationary state in the discrete system

The described phenomena were observed under condition $\delta = D_{Fe(phen)_3^{3+}}/D_{HBrO_2} > 1$. When, however, the model simulated a system of discrete, perfectly stirred and diffusively coupled cells it was found that the stability domain of the inhomogeneous stationary state extends into the region of $\delta < 1$ (Fig. 8). The inhomogeneous stationary state of the discrete ferroin catalyzed system was found at δ as small as 0.5 [13]. This fact is interesting because the diffusion coefficient of $HBrO_2$ most probably exceeds that of ferriin.

Moreover, the stable inhomogeneous state exists under conditions such that the Turing instability is impossible at any ratio of the diffusion coefficients and the homogeneous steady state is always stable.

3. Instability of homogeneous oscillations

The loss of stability of homogeneous oscillations was also studied. For this purpose, the model was reduced to the generalized Ginzburg-Landau equation and Kuramoto's stability criterion was applied [7].

The study reveals that when the system is close to the Hopf boundary the

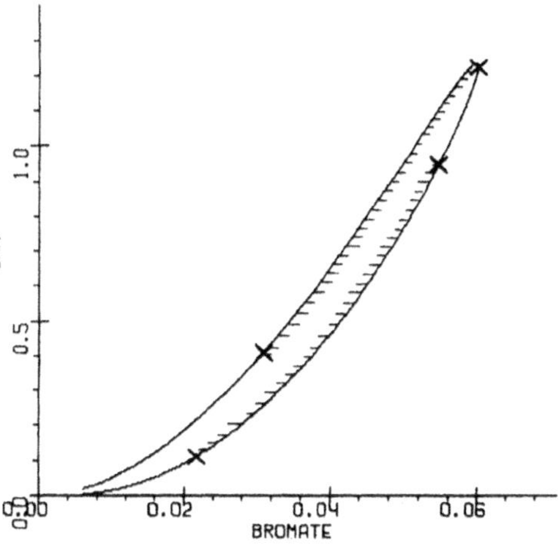

Figure 9. Regions of diffusive instability (shaded) within the oscillatory domain. The catalyst concentration is 10^{-5}M.

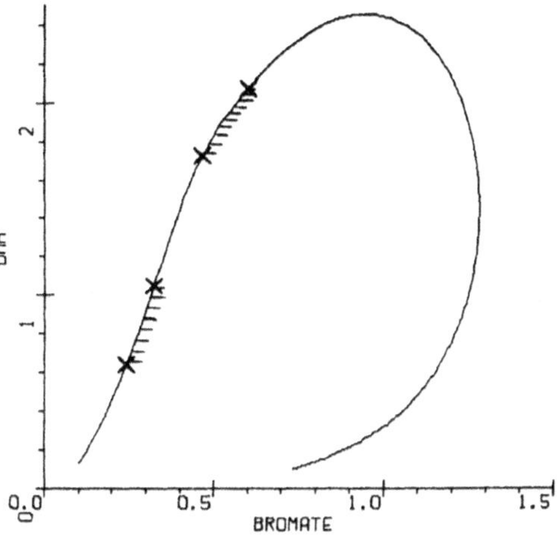

Figure 10. Diffusive instability regions (shaded) within the oscillatory domain. The catalyst concentration is 0.001M.

uniform oscillations are also unstable provided the diffusion coefficients of the species are sufficiently different. The result, however, strongly depends on whether diffusibility of the catalyst is larger or smaller than that of the other species. In the first case, again, we find stationary patterns. In contrast to the steady state case the homogeneous oscillations can also become unstable when the diffusion coefficient of the catalyst is small. Figures 9 and 10 show the instability regions within the oscillatory domains for different catalyst concentrations when $\delta < 1$. The approach applied allows us to treat the problem only near the Hopf boundary of the oscillatory region. Therefore, the internal boundaries of the instability regions are not found. Direct computations show that if the homogeneous oscillations become unstable

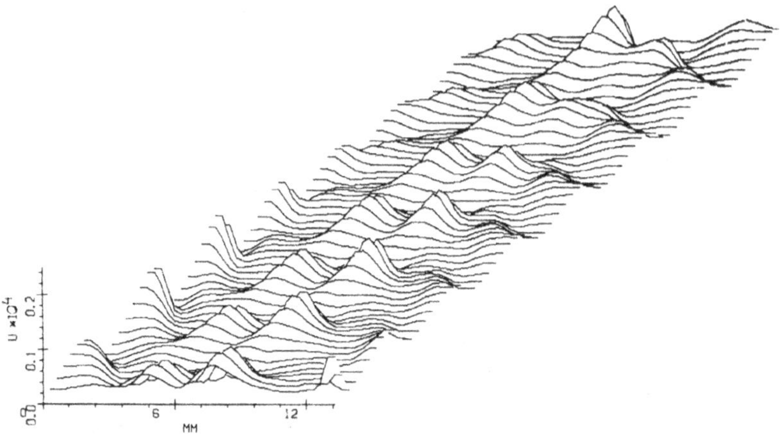

Figure 11. Chaotically oscillating pattern (δ=0, the catalyst concentration is 0.001M).

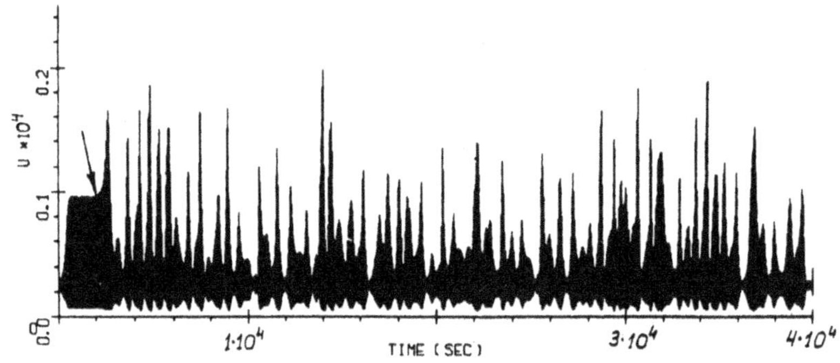

Figure 12. Oscillations at a fixed point of the pattern shown in Fig. 11.

for the ratio of the diffusion coefficients less than 1, the system forms a chaotically oscillating inhomogeneous structure (Figs. 11 and 12).

Efforts have been made to analyze what kind of attractor emerges here. Figure 13 presents three-dimensional reconstruction of the attractor from the time series with the time-lag technique [14]. Using the Grassberger-Procaccia method [2], the embedding and correlation dimensions were also estimated (Fig. 14). Apparently, the time series has turned out not sufficiently long for accurate calculations, though it includes about 1200 peaks. However, it is clear from Fig. 14 that the embedding dimension of the attractor is no less than 5. On the right part of the figure, one can discern something like a plateau for embedding dimensions N=5 and 6. The level of the plateau (which corresponds to the correlation dimension of the attractor) exceeds 3. This means that a higher chaos is present here.

When the system is deeper into the oscillatory region, small perturbations develop into periodic (and, perhaps, slightly modulated) travelling waves.

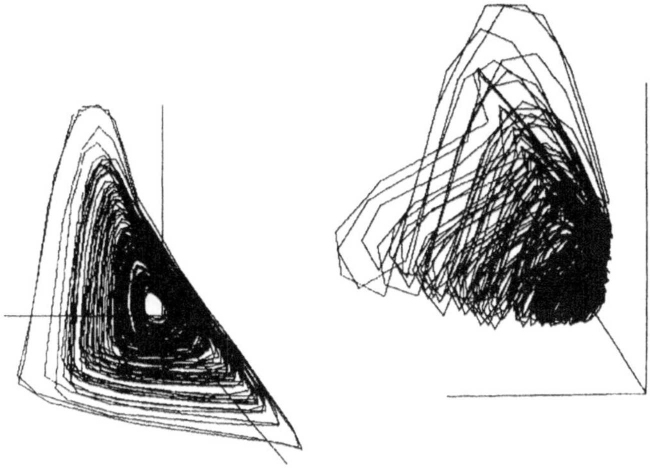

Figure 13. Three-dimensional reconstruction of the attractor for the spatio-temporal pattern shown in Figs. 11 and 12: two different views.

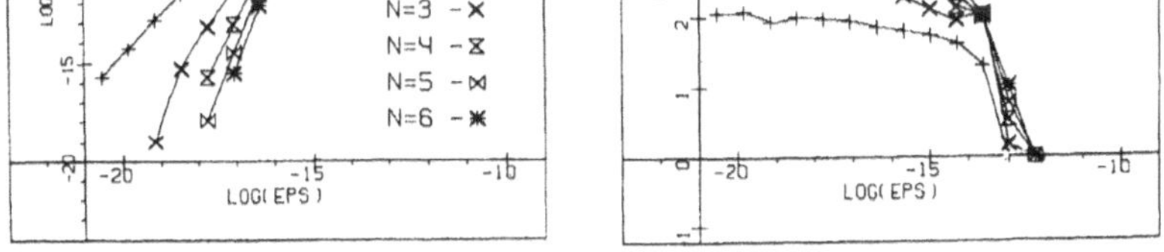

Figure 14. Dimensional characteristics of the attractor.

4. Experimental applications

As the analysis predicts, the emerging of the complex spatio-temporal patterns in the BZ system sensitively depends on the ratio of the diffusion coefficient of the catalyst to that of bromous acid. Though the precise value of the ratio is unknown, there is evidence that in a liquid phase it is less than 1. Therefore, to observe pattern formation from the homogeneous steady state, one should find a special medium where the molecular motions of the bromous acid are restricted.

On the other hand, the immobilization techniques for the catalyst of the BZ catalyst of the BZ reaction are rapidly developing now [5,6,8]. Therefore there is a good chance of discovering diffusive instabilities of the homogeneous oscillations and formation of complex spatio-temporal patterns in such acids.

Another application of the results may be that the diffusive instability of the homogeneous periodic oscillations could be responsible for complex oscillations in imperfectly stirred flow reactors.

References

[1] Field, R.J. & Burger, M. (eds.) (1985). *Oscillations and Traveling Waves in Chemical Systems*. Wiley Interscience: New York.

[2] Grassberger, P. & Procaccia, I. (1983). Measuring the Strangeness of strange attractors. *Physica* **D9**, 189.

[3] *J. Phys. Chem.* (1989), **93**, 7. The issue dedicated to Prof. Noyes' 70th birthday.

[4] Krinsky, V.I. (1984). *Self-organization. Autowaves and Structures far from Equilibrium*. Springer: Berlin.

[5] Kuhnert, L. (1983). Chemishe Structur Bildung in Festen Gelen auf der Basis der Belousov-Zhabotinsky reaction. *Naturwis.* **70**, 464.

[6] Kuhnert, L., Yamaguchi, T., Nagy-Ungvarai, Z., Muller, S.C. & Hess, B. (1989). Wave Propagation and Pattern Formation in Catalyst-Immobilized Gels. In *International Conference on Dynamics of Exotic Phenomena in Chemistry, Preprints of Lectures and Abstracts of Posters*, p. 289. Hajduszobozlo: Hungary.

[7] Kuramoto, Y. (1984). *Turbulence and Waves*. Springer: Berlin.

[8] Maselko, J., Reckley, J.S. & Showalter, K. (1989). Regular and Irregular Spatial Patterns in an Immobilized-Catalyst Belousov-Zhabotinsky Reaction. *J. Phys. Chem.* **93**, 2774.

[9] Rashevskky, N. (1940). An Approach to the Mathematical Biophysics of Biological Self-Regulation and of Cell Polarity. *Bull. Math. Biophys.* 2, 15.

[10] Rovinsky, A.B. & Zhabotinsky, A.M. (1984). Mechanism and Mathematical Model of the Oscillating Bromate-Ferroin- Bromomalonic Acid Reaction. *J. Phys. Chem.* **88**, 6084.

[11] Rovinsky, A.B. (1987). Turing Bifurcation and Stationary Patterns in the Ferroin-Catalyzed Belousov-Zhabotinsky Reaction. *J. Phys. Chem.* **91**, 4606.

[12] Rovinsky, A.B. (1987). Twinkling Patterns and Diffusion Induced Chaos in a Model of the Belousov-Zhabotinsky Chemical Medium. *J. Phys. Chem.* **91**, 5113.

[13] Rovinsky, A.B. (1989). Stationary Patterns in a Discrete Belousov-Zhabotinsky Medium with Small Catalyst Diffusibility. *J. Phys. Chem.* **93**, 2716.

[14] Takens, F. (1981). Detecting Strange Attractors in Turbulence. In *Lecture Notes in Mathematics*, 898,366. Springer: New York.

[15] Turing, A. (1952). The Chemical Basis of Morphogenesis. *Phil. Trans. R. Soc.* **237B**, 37.

[16] Zhabotinsky, A.M. (1974). *Concentration Autooscillations*. Nauka Publishing: Moscow (in Russian).

20. THREE-DIMENSIONAL WAVES IN EXCITABLE REACTION-DIFFUSION SYSTEMS: THE EIKONAL APPROXIMATION

J. Gomatam* and P. Grindrod**

*Department of Mathematics, Glasgow College
Cowcaddens Road, Glasgow G4 OBA
**INTERA-ECL, Highlands Farm, Greys Road
Henley-on-Thames, Oxon RG9 4PS

1. The need for analytically tractable models

There is a wealth of experimental observation on pattern formation in three dimensional Belousov-Zhabotinsky (BZ) reagent as documented by Welsh [1,2] and Winfree [3]. The diversity of wave forms evolving in the reagent highlights the complexity of the task faced by theoreticians working within the framework of the full reaction-diffusion (R-D) equations with excitable kinetics. This points to the need for developing analytically tractable, abbreviated models which will at least capture the geometry of the wave forms. An approach that offers considerable scope for generalization to three-dimensional R-D systems is a geometric theory (the eikonal method) proposed by Keener [4] and Zykov [5] in the context of spiral waves on a plane.

The main achievements of the investigation reported here are:

(i) The generalization of the eikonal method and its use in the analysis of waves on two-dimensional closed manifolds;

(ii) Precise testable predictions of wave propagation on spheres and tori;

(iii) The development of the eikonal equation in three-dimensional space; this equation provides a framework for demonstrating the existence of toroidal scroll waves and toroidal structures with twists.

2. R-D waves on two-dimensional closed surfaces

2.1. The effect of curvature of the wave-front on the velocity of propagation is easily derived in the case of two-dimensional closed surfaces, M, such as spheres or tori as discussed by Grindrod & Gomatam [6]. Consider an R-D system

$$\varepsilon u_t = \varepsilon^2 \nabla^2 u + F(u,w) \qquad (1)$$

$$\varepsilon w_t = G(u,w)$$

$$t \in [0,\infty) \tag{2}$$

$$\underline{r} \equiv (x,y,z) \in M$$

$$u_t = \frac{\partial u}{\partial t}$$

where F and G model excitable kinetics. Introduce the transformation

$$\underline{r} = \underline{r}\ (\mu,\eta,\lambda,t) \tag{3}$$

$$t = t \tag{4}$$

where $\lambda = 1$ defines the closed surface at any fixed time t. Further assume that

$$\nabla\mu . \nabla\eta = 0 \tag{5}$$

with the scaling

$$|\underline{r}_{\mu}(\mu,\eta,1,t)| = 1 \tag{6}$$

and

$$u = u(\mu) = V(\xi), \quad \mu = \varepsilon\xi \tag{7}$$

at the transition layer $\mu \approx 0$ one obtains

$$\varepsilon^2\nabla^2 u = V_{\xi\xi} + \varepsilon K V_{\xi} \tag{8}$$

$$\varepsilon u_t = -NV_{\xi} \tag{9}$$

Here K is twice the mean curvature of curve $\mu{\approx}0$ on the surface $\lambda = 1$, defined by

$$K \equiv - \frac{\underline{r}_{\eta\eta}\cdot\underline{r}_{\mu}}{|\underline{r}_{\eta}|^2} - \frac{r_{\lambda\lambda}\cdot\underline{r}_{\mu}}{|\underline{r}_{\lambda}|^2} \tag{10}$$

and N is the normal velocity of the curve $\mu \approx 0$ on M defined by

$$N \equiv \underline{r}_t\cdot\underline{r}_{\mu} \tag{11}$$

In the vicinity of the leading edge of the pulse solution, with the recovery variable $w = w_o$, a constant (1) simplifies to

$$V_{\xi\xi} + (N + \varepsilon K)V_{\xi} + F(V,w_o) = 0 \tag{12}$$

A comparison of (12) with the equation describing the travelling front solution in one dimension

$$V_{\xi\xi} + cV_{\xi} + F(V,w_o) = 0 \tag{13}$$

leads to the eikonal equation

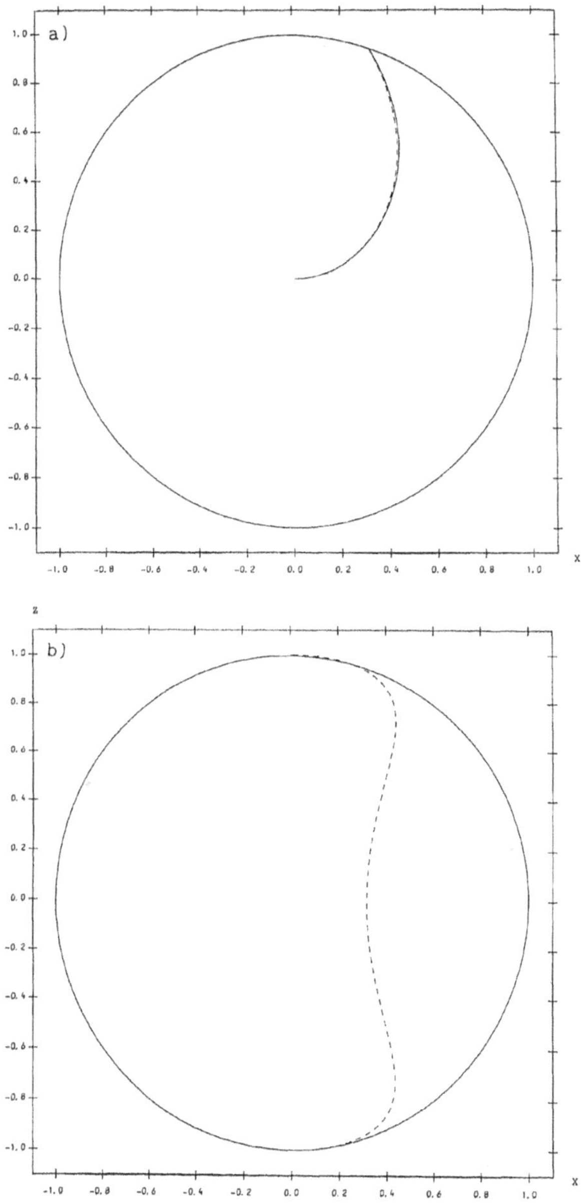

Figure 1. Double spiral on the unit sphere; dotted curve is in the rear:

$$\frac{c}{\varepsilon} = 3\ , \quad \frac{\omega}{\varepsilon} = -4.13$$

a) North polar view b) Equatorial view.

$$N + \varepsilon K = c \tag{14}$$

Conditions (5) and (6) permit the choice

$$\underline{r}_\mu = (\underline{r}_\eta \times \underline{r}_\lambda) / |\underline{r}_\eta||\underline{r}_\lambda| \tag{15}$$

By specifying the motion

$$\underline{r}(\mu,\eta,1,t) = \omega\hat{k} \times \hat{e}(\mu,\eta)$$

where ω is the angular velocity, equation (14) is numerically integrated to obtain solutions that start very near the north pole and spiral down to the south pole (Figs. 1 and 2). The significance of solutions depicted in Fig. 1 and Fig. 2 is that they are a pair of rotating spirals, analogues of a pair of spiral waves on the plane. A recent paper by Maselko & Showalter [7] on immobilized-catalyst B-Z reagent on a sphere (diameter 0.5 - 1.44 mm) demonstrated the existence of asymmetric spiral waves and attributed this to the differing chemical environments of the two hemispherical regions. Recently the emergence of almost symmetrical rotor waves as predicted in Figs. 1 and 2 has been observed by Showalter and co-workers [8]. The eikonal equation (14) can be represented in a toroidal co-ordinate system (ρ,ψ,ϕ) the major radius of the torus being R_o and solutions which rotate about the z-axis with speed ω obtained (Fig. 3).

3. The eikonal equation in three dimensions [3]

Introduce the moving coordinate system

$$\underline{r} = \underline{r}(\mu,\eta,\lambda,t), \tag{16}$$
$$t = t,$$

subject to the orthogonality and scaling restrictions

$$\underline{r}_\mu \cdot \underline{r}_\lambda = 0, \quad \underline{r}_\lambda \cdot \underline{r}_\eta = 0, \quad \underline{r}_\eta \cdot \underline{r}_\mu = 0 \tag{17}$$

$$|\underline{r}_\mu| = 1 \qquad h_2 \equiv |\underline{r}_\eta|, \qquad h_3 \equiv |r_\lambda| \tag{18}$$

In this case the normal velocity N and the mean curvature K are given by

$$N \equiv \frac{\underline{r}_t \cdot (\underline{r}_\eta \times \underline{r}_\lambda)}{h_2 h_3} \tag{19}$$

$$K \equiv - \frac{\underline{r}_{\eta\eta} \cdot (\underline{r}_\eta \times \underline{r}_\lambda)}{h_3^3 h_3} - \frac{\underline{r}_{\lambda\lambda} \cdot (\underline{r}_\eta \times r_\lambda)}{h_2 h_3^3} \tag{20}$$

3.1. *The expanding sphere and its stability*

A representation of the eikonal equation (14) with (3), (4) and (5) in the spherical coordinates (r,θ,ϕ) leads to an expanding spherical wave

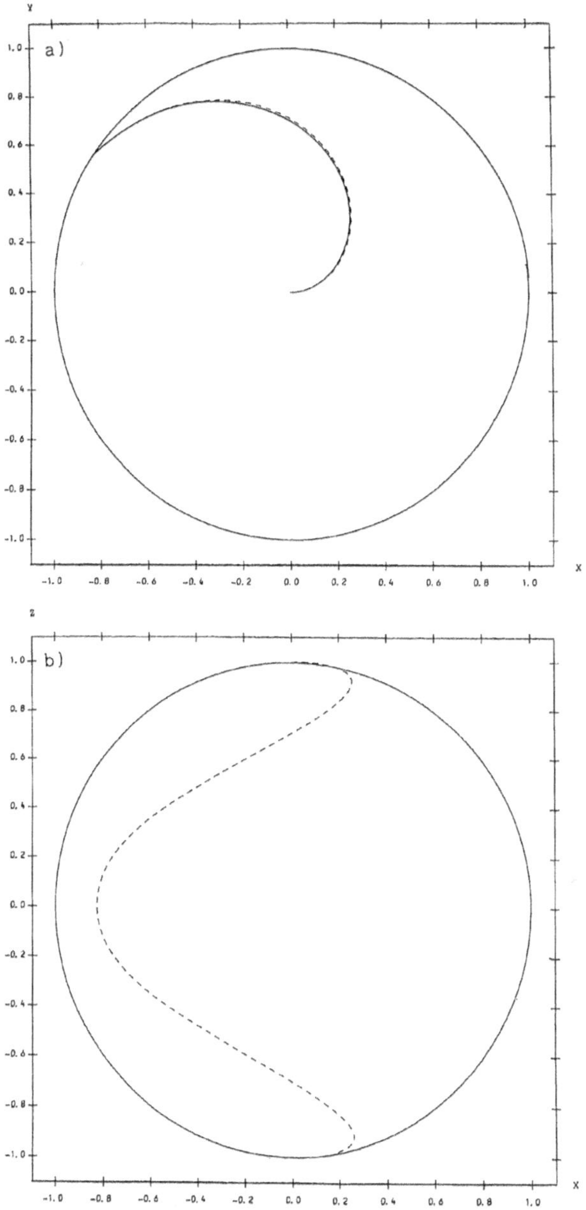

Figure 2. Double spiral on the unit sphere; dotted curve is on the rear:

$$\frac{c}{\varepsilon} = 5 \;, \quad \frac{\omega}{\varepsilon} = -9.0147$$

a) North polar view b) Equatorial view

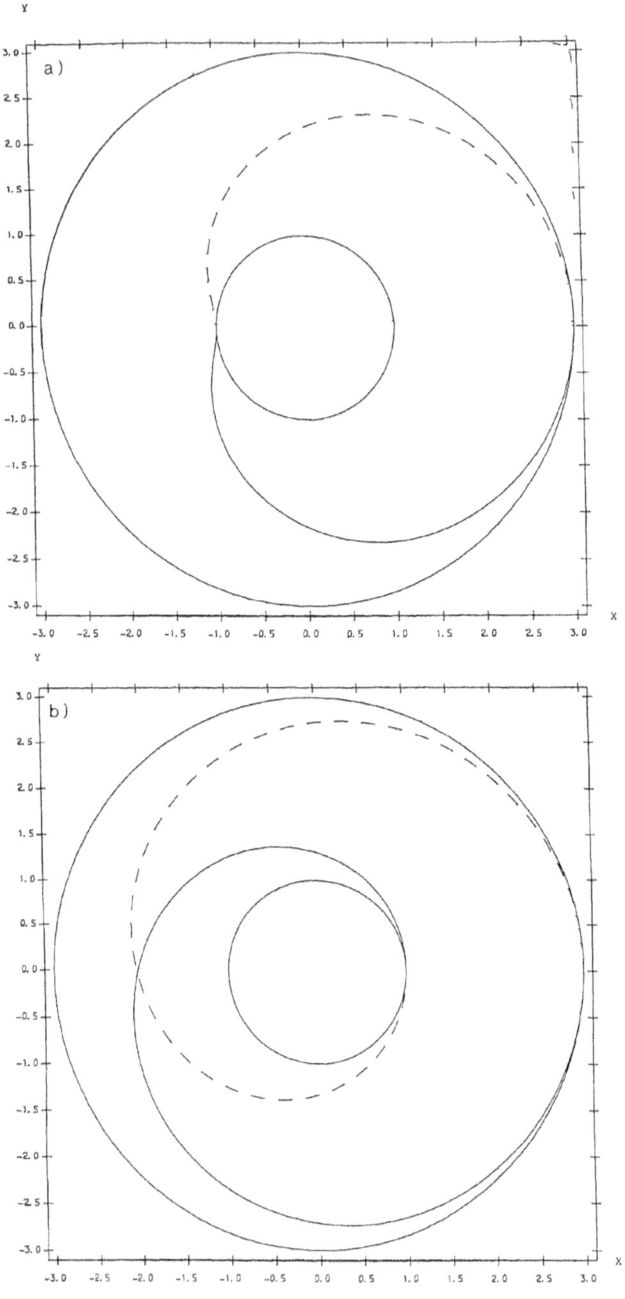

Figure 3. Solutions on the torus $R_o = 2$, $\rho = 1$; dotted curve is on the rear: a) $\gamma = 1.19$, $n = m = 1$, b) $\gamma = 2.10$, $n = 2$, $m = 1$.

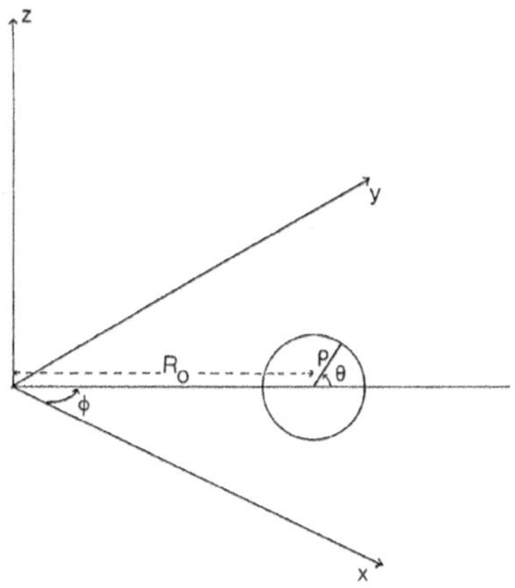

Figure 4. The toroidal coordinate system.

$$r_t = c - \frac{2\varepsilon}{r} \tag{21}$$

A linearized perturbation analysis leads to the conclusion that an expanding spherical wave is stable save for the freedom in the choice of the location of the source of the wave.

3.2. *Surfaces with axial symmetry*

The choice of variables $r = \lambda, \phi = \eta$ and $\theta = \theta(r,t)$ leads to axially symmetric structures supported by spherical coordinate systems. A simple solution of (14) for the case $\varepsilon = 0$ is the involute (on the plane ϕ = constant)

$$\theta = \omega \left\{ \left[r^2 - \frac{c^2}{\omega^2} \right]^{\frac{1}{2}} - \frac{c}{\omega} \operatorname{Cos}^{-1} \frac{c}{r\omega} \right\} + \omega t + \text{constant}$$

3.3. *Toroidal geometry*

The use of the toroidal coordinate system (Fig. 5) reduces (14) to a single partial differential equation of a rather complicated form. In what follows two cases of analytical interest will be presented.

3.3.a. *Toroidal scroll waves.* In this section the existence of toroidal scroll waves will be demonstrated under the assumption

$$\varepsilon/R_o \ll 1 \tag{22}$$

Setting

$$\lambda = \rho\ , \quad \eta = \phi\ , \quad \psi_\eta = 0 \tag{23}$$

equation (14) can be reduced to

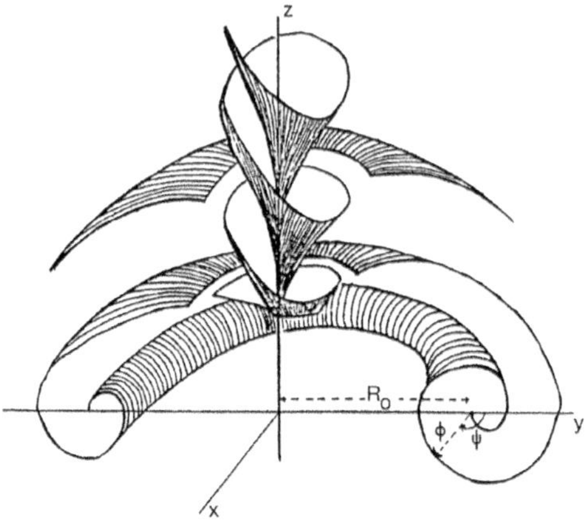

Figure 5. Illustration of a toroidal scroll with one uniform twist and a compatible helical surface near the z-axis.

$$\frac{\rho\psi_t}{M} - \varepsilon\left[\frac{\psi_\rho}{M} + \frac{\rho\psi_{\rho\rho} + \psi_\rho}{M^3}\right] - \varepsilon\,\frac{\text{Sin}\,\psi + \rho\psi_\rho\,\text{Cos}\,\psi}{LM} = c \tag{24}$$

where

$$M \equiv [1 + \rho^2\,\psi_\rho^2]^{\frac{1}{2}}, \tag{25}$$

and

$$L \equiv R_o + \rho\cos\psi. \tag{26}$$

The restriction on ψ is:

$$\psi(\rho,t) + 2\pi = \psi(\rho,t + 2\,\frac{\pi}{\omega}),\ \forall t \text{ and } \forall\rho. \tag{27}$$

For $R_o \gg \rho$ and $\frac{\varepsilon}{R_o} \ll 1$, the third term on the left-hand side of (24) can be neglected.

$$\frac{\rho\psi_t}{M} - \varepsilon\left[\frac{\psi_\rho}{M} + \frac{\rho\psi_{\rho\rho} + \psi_\rho}{M^3}\right] = c \tag{28}$$

If, in addition

$$\psi_t = \omega \tag{29}$$

(29) has solutions which are uniformly rotating spirals on any ϕ = constant plane.

The presence of the neglected term in (24) must perturb not only the functional form of the ϕ = constant cross section of the scroll wave, but also the uniformity of rotation [10] evident in the solution of the above simplified problem. To see this, suppose that ρ is very small so that in a neighbourhood of the toroidal axis, (24) becomes

$$\rho\psi_t - \varepsilon\left[2\psi_\rho + \rho\psi_{\rho\rho} + \frac{\sin\psi}{R_o}\right] = c$$

When $R_o \to \infty$, the desired solution of the above equation is

$$\psi = \psi_o = \omega t - \frac{c\rho}{2\varepsilon} + O(\rho^2)$$

which represents an expansion, for small ρ, of the spiral solution of (24).

Now substituting

$$\psi = \psi_o + \frac{1}{R_o}\psi_1 + O\left(\frac{1}{R_0^2}\right)$$

into equation for $\psi_{\rho\rho}$ above we obtain

$$\psi_1 = -\frac{\varepsilon}{c}\cos\omega t + O(\rho)$$

Thus

$$\psi = \omega t - \frac{c\rho}{2\varepsilon} + O(\rho^2) + \frac{1}{R_o}\left[-\frac{\varepsilon}{c}\cos\omega t + O(\rho)\right] + O\left(\frac{1}{R_0^2}\right)$$

which defines a surface for ρ small, which rotates in a non- uniform manner. (Nevertheless, it is still $2\frac{\pi}{\omega}$ periodic, since $\psi(\rho, t + 2\frac{\pi}{\omega}) = \psi(\rho, t) + 2\pi$.

3.3.b *Twisted toroidal scroll waves.* The existence of linked twisted toroidal scroll waves was demonstrated in the Debye limit of the singular solutions of linear reaction-diffusion equations [11]. It is not difficult to demonstrate the existence of twisted toroidal scroll waves using the eikonal equation (14) for $R_o \gg \varepsilon$. The existence of a helical wave in the neighbourhood of the z-axis emerges from the eikonal equation (14), but the continuous and differentiable transition of this helical surface to the twisted toroidal scroll surface obtained for $R_o \gg \varepsilon$ still remains to be demonstrated (Fig. 5).

Assumptions $R_o \gg \varepsilon$ and

$$\lambda \equiv \rho\,, \quad \eta \equiv \phi\,, \quad \psi_\eta = 1 \tag{30}$$

reduce (14) to

$$\frac{\rho\psi_t}{L} - \varepsilon\left[\frac{\rho\psi_{\rho\rho} + \psi_\rho}{L^3} + \frac{\psi_\rho}{L}\right] = c \tag{31}$$

The choice $\psi_t = \omega$, generates a solution (31) with one uniform twist.

Winfree & Strogatz [12] have argued on topological grounds that once twisted toroidal scroll waves exist in linked pairs. Equation (14) when represented in toroidal coordinate does not readily allow a demonstration of Winfree & Strogatz's result. As an intuitive compromise, (14) will be represented in cylindrical coordinates $[r,\phi,z]$ and solutions with helical symmetry will be sought.

$$\lambda = r\ ,\quad \eta = z \qquad \phi_\eta = 1 \tag{32}$$

Then (14) reduces to

$$\frac{r\phi_t}{PQ} + \varepsilon r\left[\frac{\phi_r}{PQ} + \frac{r\phi_{rr} + \phi_r}{PQ^3}\right] - \varepsilon\,\frac{r^2\phi_r}{P^3Q} = c$$

$$P \equiv [1 + r^2]^{\frac{1}{2}}\ ,\quad Q = [1 + r^2\phi_r^2]^{\frac{1}{2}} \tag{33}$$

For $\varepsilon=0$, and $\phi_t = \omega$, (22) has the solution

$$\phi = \int dr\ \frac{[r^2(\omega^2-c^2) - c^2]^{1/2}}{r[1+r^2]^{1/2}} + z + \omega t, \tag{34}$$

$$\omega^2 > c^2$$

Equation (34) represents a helical surface for $r > \frac{c}{[\omega^2-c^2]^{1/2}}$. It is easily seen that, due to the presence of $[1+r^2]^{1/2}$, z = 0 cross-sections are not involutes of the circle $r = \frac{c}{[\omega^2-c^2]^{1/2}}$.

Therefore the radius of the singular cylinder here is greater than the radius of the core region $[r = c/\omega]$ for the spiral wave. In this case the corrective role of the terms containing ε are the same as that encountered in the case of spiral waves on the plane [4].

4. The eikonal approach: conclusions and outlook

4.1. *Experimental implications of surface R-D wave*

The utility of the eikonal equation approach in a study of R-D wave propagation on closed two-dimensional manifold consists in the analytical control it provides on an otherwise stiff parabolic P.D.E. system. The predicted spiral waves on spheres have been observed and this should be of experimental interest to chemists and physiologists. An immediate experiment that suggests itself is a projected study of B-Z wave forms on closed manifolds. The major technical problem in setting up such an experiment is the selection of a proper surface coating capable of retaining B-Z reagent homogeneous for a substantial length of time.

4.2. *Three-dimensional R-D waves*

There are three conclusions reached here:

(i) The expanding spiral wave front is stable and small perturbations merely shift the source of the wave form.

(ii) Toroidal scroll waves are solutions to the eikonal equation for $R_o \gg \varepsilon$.

(iii) Twisted toroidal scroll wave forms are solutions to the eikonal equation for $R_o \gg \varepsilon$; the same equation allows the existence of helical wave fronts in the vicinity of the z-axis. A clear analytical demonstration of the precise nature of the transition region still remains.

(iv) A precise definition of the concept of stability and its analytical delineation for items (ii) and (iii) above within the framework of the eikonal equation is a substantial problem and needs to be addressed.

References

[1] Welsh, B.J. (1984). *Pattern Formation in the Belousov- Zhabotinsky Reaction.* Ph.D. thesis: Glasgow College.

[2] Welsh, B.J., Gomatam, J. & Burgess, A.E. (1983). Three- dimensional chemical waves in the Belousov-Zhabotinsky reaction. *Nature* **304**, 611-614.

[3] Winfree, A.T. (1987). *When time breaks down: the three- dimensional dynamics of electrochemical waves and cardiac arrhythmias.* Princeton University Press.

[4] Keener, J.P. (1986). A geometrical theory for spiral waves in excitable media. *SIAM J. Appl. Math.* **46**, 1039-1056.

[5] Zykov, V.S. (1988). *Simulation of wave processes in excitable media*; translation by A.T. Winfree, Manchester Univesity Press: Manchester.

[6] Grindrod, P. & Gomatam, J. (1987). The geometry and motion of reaction-diffusion waves on closed two-dimensional manifolds. *J. Math. Biol.* **25**, 597-610.

[7] Maselko, J. & Showalter, K. (1989). Chemical waves on spherical surfaces. *Nature* **339**, 609-611.

[8] Showalter, K. (1989). Private communication.

[9] Gomatam, J. & Grindrod, P. (1987). Three-dimensional waves in excitable reaction-diffusion systems. *J. Math. Biol.* **25**, 611-622.

[10] Grindrod, P. Patterns and waves in reaction diffusion. Oxford University Press, to appear.

[11] Gomatam, J. (1982). Pattern synthesis from singular solutions in the Debye limit; helical waves and twisted toroidal scroll structures. *J. Phys. A; Math. Gen.* **15**, 1463-1476.

[12] Winfree, A.T. & Strogatz, S.H. (1984). Singular filaments organize chemical waves in three-dimensions: 4 wave taxonomy. *Physica* **13D**, 221-233.

21. THE DYNAMICS OF FORCED EXCITABLE SYSTEMS

Hans. G. Othmer

Department of Mathematics
University of Utah
Salt Lake City, UT 84112, U.S.A.

1. **Introduction**

An excitable system is one whose dynamics have the following properties: (i) there is a rest point or a steady state that is globally attracting relative to some large set in phase space, and (ii) there is a region in state space that can be idealized as a surface of codimension one that locally partitions the phase space into two sets $\mathcal{D}$ and $\mathcal{A}$. The rest point lies in $\mathcal{D}$ (the **decaying set**) and all orbits through initial points in $\mathcal{D}$ return to the rest point without any substantial growth in any of the state variables. Thus an impulsive perturbation of the rest point that leaves the state in $\mathcal{D}$ decays without significant growth, and the responses are called **subthreshold**. By contrast, perturbations that carry the dynamics into $\mathcal{A}$ (the **amplifying set**) can lead to a large change in one or more of the state variables, even though the system eventually returns to the rest state. The surface that locally separates the amplifying and decaying sets is called the **threshold** surface, and perturbations that carry the state into $\mathcal{A}$ are called **superthreshold**. Excitable dynamics occur in many biological processes, including activation of contraction in cardiac tissue, nerve conduction, and cell signalling in development. In the Fitzhugh-Nagumo equations [10], the Hodgkin- Huxley equations [15], models of the cellular slime mold *Dictyostelium discoideum* [22,20,21], the Field-Noyes model of the Zhabotinskii-Belousov reaction [9] and other models, the parameters can be chosen so that the dynamics are excitable. Frequently these systems also have periodic solutions for parameters close to those that produce excitable behaviour.

It is often observed that an excitable system responds periodically when forced periodically with a superthreshold stimulus, in which case we say that entrainment or phase-locking occurs. This response is very similar to the behaviour found in forced oscillatory systems, and we call the regions in parameter space where phase-locking occurs the resonance zones, in analogy with forced oscillatory systems. Numerous experiments (cf. e.g. [13,18,6] and references therein) have been done in which excitable systems were forced periodically. The results of these experiments show much of the phase-locking structure found in forced oscillatory systems, including apparently chaotic behaviour at suitable combinations of the amplitude and frequency of the forcing. Numerical studies on Hodgkin-Huxley systems [16,19], on a Fitzhugh-Nagumo-like system [13], and on formal models of neurons [23,24], also show that the resonance structure can be quite similar to that in forced periodic systems. An example of the kind of experimental results that have been obtained is given in Fig. 1, and the results of some numerical computations for a model of *Dictyostelium discoideum* are shown in Fig. 2. Both of these examples deal with distributed systems, but as we shall see, the threshold behaviour and phase-locking structure in the forced ordinary differential equations we study is similar to that shown in the figures. This suggests that the critical step in each example is initiation of the wave at the point of stimulation. Of course in a nonuniform medium various types of wave blockage can occur "downstream" from the source of stimulation.

Here we review some of the recent work on forced excitable systems aimed at understanding the resonance structure [1,2,7]. The results we describe explain much of the coarse qualitative structure of the phase-locking zones seen experimentally and numerically in forced excitable systems, including the existence of threshold stimuli for phase-locking. However the fine structure in transition regions between resonance zones has not been examined in detail as yet, either numerically or analytically.

2. The reduction to a flow on a torus

Models of excitable systems often give rise to evolution equations of the form

$$\varepsilon\frac{dv}{dt} = f(v,w,\lambda)$$
$$\frac{dw}{dt} = g(v,w,\lambda) \qquad (1)$$

where $v \in \mathbb{R}^m$, $w \in \mathbb{R}^n$, $\lambda \in \mathbb{R}^p$ is a parameter vector, and ε is small. When $f(v,w,\lambda) \geq \mathcal{O}(1)$, v varies rapidly compared with w on the t scale, which implies that the v_i are the fast "voltage-like" variables, and the w_i are the "recovery-like" or "gating" variables. In general the forcing can enter both the fast and slow subsystems, and it can either enter parametrically, in which case the forcing is state dependent, or additively. In [1] we only

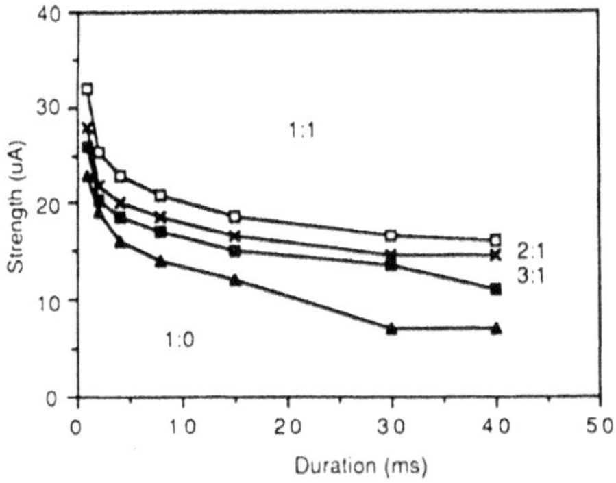

Figure 1. Strength-duration curves showing the phase-locking regions in a forced Purkinje fibre. The plot shows the ratio p:q of the number of forcing cycles (period = 700 ms) to the number of superthreshold responses for given combinations of the duration of the input pulse in milliseconds and the amplitude in microamps. From [6] with permission.

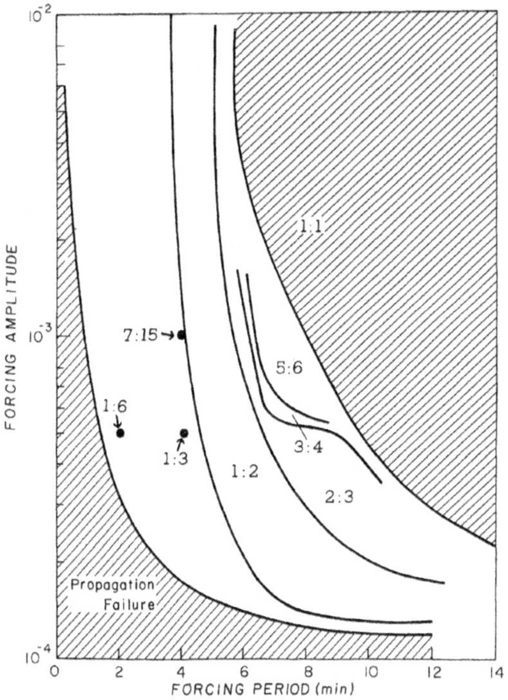

Figure 2. The phase-locking regions in frequency-amplitude space for a model of Dictyostelium discoideum. A field of excitable cells is stimulated at one point in space and the governing partial differential equations that arise from the Monk-Othmer model are solved numerically to determine the response far from the site of the stimulus. The cells in the aggregation field are excitable but not oscillatory. Note the similarity between the qualitative structure of the phase-locking regions in this figure and those in the preceding one. For details see [20,21].

consider the case in which the forcing enters additively in the slow subsystem. This case applies directly to a variety of chemical and biological systems, including for example, the Belousov-Zhabotinskii reaction in a continuous stirred tank reactor forced via periodic injection of the catalyst. Moreover, as we showed in [1], many T-periodic systems of the form

$$\varepsilon\frac{dv}{dt} = p(v) - q(w,\phi(t)),$$
$$\frac{dw}{dt} = r(v,w,\psi(t)), \qquad (2)$$

can be cast into the form

$$\varepsilon\frac{dv}{dt} = f(v,w,\lambda),$$
$$\frac{dw}{dt} = g(v,w,\lambda,t). \tag{3}$$

Thus forcing on the fast variables, as occurs for instance when the Belousov-Zhabotinskii reaction is forced by periodic addition of Br^- [17], can be treated within the framework developed in [1]. Since the phenomena of interest here already arise in planar systems, we only considered the case m = n = 1 in [1]. In the rigorous analysis there are a number of technical conditions that are needed, but we will not go into the details of these here. One of the main conditions is that the forcing on the slow variable moves the g = 0 curve in and out of the interval (v_l,v_r) in which the rest point is unstable (cf. Fig. 3). If the forcing is weak in an appropriate sense and the rest point of the unforced system is stable, then it is easy to prove that the forced system will only have small amplitude periodic solutions that are perturbations of the rest point. Such solutions correspond to subthreshold responses, i.e. responses in the region labelled 1:0 in Fig. 1.

On the other hand, if the amplitude of the forcing is too large there may be an intersection of g = 0 with f = 0 for $v > v_r$, and we exclude this at first because it complicates some of the analysis without introducing any significant new phenomena. Some remarks on this case are given later. Thus we assume that for some fraction of the time the forcing moves the rest point from the left branch of f = 0 into the region where it is unstable when regarded as a rest point of the system in which t is fixed in the function g, but does not move it to the right of v_r.

We can describe the qualitative dynamics under the conditions set forth in [1] as follows. Let $\tilde{t}$ and λ be fixed, and consider the autonomous system

$$\varepsilon\frac{dv}{dt} = f(v,w,\lambda),$$
$$\frac{dw}{dt} = g(v,w,\lambda,\tilde{t}) \tag{4}$$

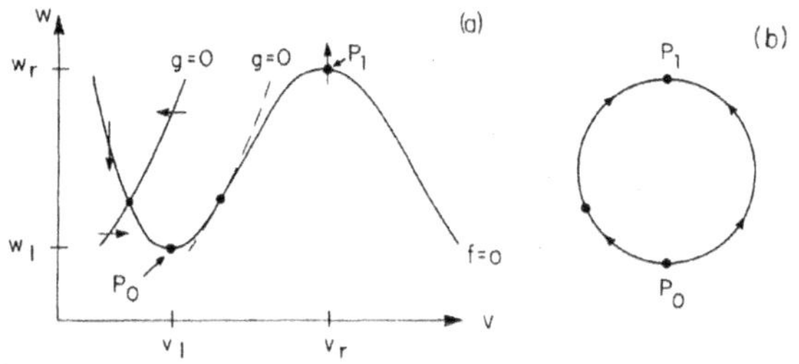

Figure 3. (a) The qualitative features of the isoclines f = 0 and g = 0. When g = 0 intersects f = 0 to the left of v_l the "steady" state is stable, while if g = 0 intersects f = 0 to the right of v_l there is a periodic solution. (b) The resulting state space S_1 in the limit $\varepsilon = 0$.

in which the forcing is frozen. Let $(v^0_{\tilde{t}}, w^0_{\tilde{t}})$ denote the rest point of this system. If $v^0_{\tilde{t}} < v_l$, then $P(\tilde{t}) \equiv (v^0_{\tilde{t}}, w^0_{\tilde{t}})$ is a globally attracting rest point. If $v_l < v^0_{\tilde{t}} < v_r$, the rest point is unstable, and the unique periodic solution attracts $\mathbb{R}^2 \setminus (v^0_{\tilde{t}}, w^0_{\tilde{t}})$. Thus if $\tilde{t}$ is regarded as a parameter of the frozen system, a periodic solution emerges or disappears as $\tilde{t}$ changes so that $v^0_{\tilde{t}}$ passes through v_l. We assume that for some $T_1 > 0$ and sufficiently small $\varepsilon > 0$ the rest point of (4) is in the decaying set $\mathcal{D}$ for $0 < \tilde{t} < T_1$, while if $T_1 < \tilde{t} < T$ it is in the amplifying set $\mathcal{A}$. Thus the effect of the forcing is to move the rest point of (4) in and out of the amplifying set.

Our approach to understanding resonance in forced excitable systems is based on first studying the singularly-perturbed system. It should be emphasized that it is necessary for this approach that the forcing varies on the slow time scale, rather than on the fast time scale. It is found numerically that when the frequency of the forcing is $\mathcal{O}(1/\varepsilon)$ new phenomena arise that cannot be described by the reduced system. On the other hand, when the forcing varies on the slow time scale any solution of (3) rapidly approaches a stable branch of the set where $f = 0$. For $v(t) > v_l$ (3) oscillates, and one can count the number of times N that the v-component crosses some value $v^* \geq v_l$ during one period of the forcing. This **firing number** gives the number of large excursions, or **spikes** in one period of the forcing. Evidently this firing number need not be an integer, since the number of large excursions may be different in successive forcing periods. In that case one averages the firing number over the period of the periodic solution.

Thus we first set $\varepsilon = 0$ and determine the behaviour of the resulting singular differential system. For the present we suppress the λ dependence of the vector field. When $\varepsilon = 0$ the equation $\varepsilon\dot{v} = f(v,w)$ reduces to the equation $f(v,w) = 0$ or $w = \gamma(v)$. For $v < v_l$ or $v > v_r$ the singular system is on a branch of the nullcline $w = \gamma(v)$. Let $v = \zeta_l(w)$ invert $w = \gamma(v)$ for $v < v_l$ and let $v = \zeta_r(w)$ invert $w = \gamma(v)$ for $v > v_r$. Then the system (3) reduces to the discontinuous scalar equation

$$\frac{dw}{dt} = \begin{cases} g(\zeta_l(w),w,t) & \text{for } v = \zeta_l(w) < v_l \\ g(\zeta_r(w),w,t) & \text{for } v = \zeta_r(w) > v_r. \end{cases} \tag{5}$$

If the solution lies on the left branch $v = \zeta_l(w)$, and w decreases until it reaches $w_l = \gamma(v_l)$, whereupon the solution jumps horizontally and instantaneously to the point $(v,w) = (\zeta_r(w_l),w_l)$ on the right branch. Similarly if the solution lies on the right branch $v = \zeta_r(w)$ and w reaches $w_r = \gamma(v_r)$, it jumps horizontally and instantaneously to the point $(v,w) = (\zeta_l(w_r),w_r)$ on the left branch. By identifying $(\zeta_l(w_l),w_l) \equiv (\zeta_r(w_l),w_l)$ and $(\zeta_l(w_r),w_r) \equiv (\zeta_r(w_r),w_r)$ the state space becomes a circle S^1, and thus the phase space for (5) is the cylinder $S^1 \times \mathbb{R}^+$. Without loss of generality we can rescale time so that the period T of the forcing is 1, and we identify the sections t = 0 and t = 1 of the cylinder. Then (5) gives rise to the discontinuous equation

$$\frac{du}{dt} = \begin{cases} T \cdot F_1(u,t) & \text{for } u \in (0,1/2), \\ T \cdot F_2(u,t) & \text{for } u \in (1/2,1) \end{cases} \tag{6}$$

on a torus. Here $u \in (0,1)$ is a coordinate for S^1, where $u = 0 (\equiv 1)$ corresponds to P_0 and $u = 1/2$ corresponds to P_1 (cf. Fig. 3). Further,

$$F_1(u,t) \equiv \frac{1}{2W} g(\zeta_r(2Wu + w_1),\ 2Wu + w_1, Tt)$$

$$F_2(u,t) \equiv \frac{-1}{2W} g(\zeta_1(-2Wu + W + w_r), -2Wu + W + w_r, Tt)$$

where $W \equiv w_r - w_1$. Both F_1 and F_2 are extended periodically with period 1 in both t and u. The associated flow is continuous except along the line $u = 0$ ($\equiv$ 1), and as was shown in [1], the classical theory of rotation numbers can be extended to the discontinuous flows that arise in the singular limit. The main features of these results are discussed in the following section.

3. Analytical results for the singularly-perturbed system

In the singular limit of a forced excitable system the vector fields are only piecewise continuous and orbits are not unique, so the classical theory does not apply directly. We suppose, as in [1], that the vector fields are piecewise smooth in the sense that F(u,t) is C^k for $k \geq 1$ except along certain lines. We assume that F(u,t) extends continuously along these lines and that the one-sided limits of the derivatives exist in the generalized sense that some of the limits may be infinite. These exceptional lines are classified as follows.

- *Space discontinuities.* There are a finite number (up to periodicity) of values $u = u_c$ such that

$$\Delta F(u_c,t) = \Delta_{u_c} F(u_c,t) = F(u_c^+,t) - F(u_c^-,t) \neq 0.$$

We assume that $F(u_c^+,t)$ and $F(u_c^-,t)$ are piecewise smooth in t (jumping only at points t_c).

- *Time discontinuities.* There are a finite number (up to periodicity) of values $t = t_c$ such that

$$\Delta F(u,t_c) = \Delta_{t_c} F(u,t_c) = F(u,t_c^+) - F(u,t_c^-) \neq 0.$$

We assume that $F(u,t_c^+)$ and $F(u,t_c^-)$ are piecewise smooth in u (jumping only at points u_c).

Thus (t,u)-space is partitioned into a collection of rectangles by the horizontal lines $u = u_c$ and the vertical lines $t = t_c$. We call these lines the **lines of discontinuity** of F(u,t). F(u,t) is smooth in the open rectangles and for any such rectangle R, F(u,t) can be extended to a continuous function on the closure of R.

The nature of the flow near the lines $u = u_c$ is determined by the sign of F(u,t) above and below such a line. One finds that every discontinuity is locally one of three types: (i) those at which the solution crosses the discontinuity transversely, (ii) those that are attracting, and (iii) those that are repelling. In the first case solutions are unique in both forward and reverse time, while in the second (resp., third) case the solution may not be unique in reverse (resp., forward) time. In all cases for which the orbit through a point on $u = u_c$ is not unique we regard the forward extension of the orbit to be the set of **all** orbits that emanate from this point. This includes orbits which may "travel along" the discontinuity $u = u_c$ for some finite interval of time. More precisely, for any point (t_0,u_0), let $S^+(t_0,u_0)$ denote the set of all forward orbits with (t_0,u_0) as initial point, and let $\bar{u}(t) = \sup\{u \in S^+(t_0,u_0)\}$ and $\underline{u}(t) = \inf\{u \in S^+(t_0,u_0)$. Since the supremum and infimum of orbits are orbits, $\bar{u}(t)$ and $\underline{u}(t)$ are orbits in $S^+(t_0,u_0)$, called the **maximal** and **minimal** orbits, respectively, [14] §III.2. We call the set of points in the plane between the minimal and maximal forward orbits through a point the **forward shadow** of that point. Since the horizontal component of the vector field is identically 1, the forward shadow of the point (t_0,u_0) is $S^+(t_0,u_0)$. If there is only one orbit through a point, the forward shadow consists of all points on the positive semi-orbit through that point.

In this way the discontinuous vector field under consideration defines a set-valued flow in the plane and, by factoring modulo the integer lattice, a set-valued flow on the torus T^2. However, in general a rotation number cannot be defined for these flows, and we must restrict the types of discontinuities admitted. This is done via the following hypothesis.

(H1) *There are no attracting space discontinuities and there is at most one orbit emanating in reverse time from any point at which $F = 0$ and F_u is infinite.*

Under this hypothesis uniqueness holds for the time-reversed flow, but not in general for the forward flow. Furthermore, one can prove that the properties of the rotation number given previously for smooth flows hold for this class of discontinuous systems. As a result, (6) has a periodic solution if and only if the rotation number is rational.

It is shown in [1] that the general theory applies to the type of equations that arise in the singular limit of planar forced excitable systems. In this case $u = 0$ is a repelling discontinuity for $0 \le t_1 (\equiv T_1/T) < 1$. We call the forward shadow of the point (0,0) the **critical shadow**, and any orbit through (0,0) is called a **critical orbit**. If the interior of the critical shadow contains a lattice point (m,n) (Fig. 4), one of the orbits in the critical shadow is periodic and the rotation number is n/m. For some of the analysis it is more convenient to use the circle map T induced by the Poincaré map, Ψ. T is set-valued at 0 because $T{:}0 \longrightarrow I^1 \equiv (\alpha_1,\beta_1) \subset S^1$, and is an

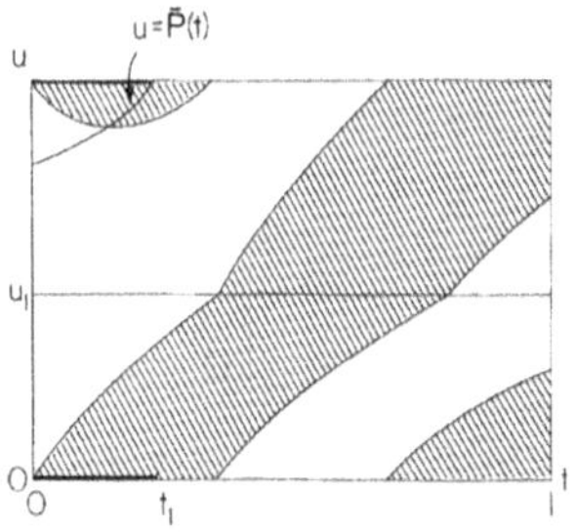

Figure 4. (a) The forward shadow of the origin at time 1, translated into the fundamental domain [0,1] x [0,1]. The critical shadow contains the lattice point (1,1).

orientation-preserving homeomorphism on (0,1) because orbits through points in (0,1) are unique. T^{-1} is point-valued but has a flat spot whenever the critical shadow at t = 1 is an interval.

In general it is difficult to characterize completely the dynamics of the iterates of Ψ, but when the mth power of Ψ is a contraction for some m a complete description can be given. The results are summarized in the following two theorems (cf. [1]).

Theorem 1. *Suppose that Ψ is an m-contraction for some m ∈Z.*

I. If the lattice point (m,n), m and n coprime, lies in the interior of the critical shadow then:

- *there is a unique periodic solution (modulo translation in t by k for k = 1,...,m-1) of (5) in the critical shadow and it is unstable,*
- *there is a unique periodic solution (modulo translation in t by k for k = 1,...,m-1) of (5) outside the critical shadow and it is asymptotically stable.*

II. If the critical shadow does not contain a periodic orbit then (5) has no periodic solution.

Theorem 2. *Suppose that the contraction hypothesis holds and that the rotation number is irrational. Then every noncritical orbit is dense in the complement of the critical shadow. The rotation number is stable under perturbations in (5) if and only if the critical shadow contains a lattice point in its interior. If (5) depends continuously on parameters, then on the boundaries in parameter space of a phase-locking region for rotation number n/m either the maximal orbit or the minimal orbit of the critical shadow at t = m passes through the lattice point (m,n).*

Thus when the rotation number ρ is rational the non-wandering set of the flow consists of periodic orbits. When ρ is irrational one can show [1] that the α- and ω-limit set of any point in (0,1) lies in the complement of the union of the forward and backward orbit through 0, where the forward orbit is set-valued. In this case the minimal set is a Cantor set. When ρ = n/m with m and n coprime, we say that the system is entrained in the ratio n:m or that it is in an n:m resonance, and we use $R_{n/m}$ to denote the n:m resonance zone in parameter space. In a generic one-parameter family of smooth circle maps T_λ the rotation number is constant on closed intervals where it is rational, strictly increasing where it is irrational, and the Lebesgue measure of the set Λ where it is irrational is positive. By contrast, for the set-valued circle map that arises here one can show that under reasonable hypotheses on the dependence of the map on λ the measure of Λ is zero.

4. The structure of the resonance zones for piecewise constant forcing

According to Theorem 1, when the contraction hypothesis holds (5) has a periodic solution if and only if the critical shadow contains a periodic orbit, which is true if and only if the rotation number is rational. Thus the parameter set for which the critical shadow contains a lattice point (m,n) coincides with $R_{n/m}$. As we stated earlier, if the vector field depends continuously on a parameter then one can show that the rotation number does also, and therefore it is constant on open intervals of the parameter. In this case Theorem 1 states that when the parameter lies on the boundary of $R_{n/m}$ either the maximal or the minimal orbit of the critical shadow passes through the lattice point (m,n), and a saddle-node bifurcation of periodic solutions occurs at this parameter value. Thus when the forcing is additive on the slow variable and piecewise constant the boundaries of the resonance zones can be determined analytically (up to solving certain transcendental equations) as follows.

Suppose that the contraction hypothesis is satisfied and suppose that (5) has the form

$$\frac{dw}{dt} = \begin{cases} g(\zeta_1(w),w) - f(t) & \text{for } v = \zeta_1(w) < v_1, \\ g(\zeta_r(w),w) - f(t) & \text{for } v = \zeta_r(w) > v_r \end{cases} \tag{7}$$

where f is piecewise constant and takes the values 0 and $\bar{A}$. Clearly there is threshold value $\bar{A}_0$ of $\bar{A}$, below which the solution never reaches the right branch even if the forcing is always on, and this value is

$$\bar{A}_0 = g(\zeta_1(w_1),w_1).$$

The minimum time t_0 needed to reach threshold from any point $w_0 \in (w_1,w_r)$ for any $\bar{A} > \bar{A}_0$ is given by

$$T_t = \int_{w_0}^{w_1} \frac{dw}{g(\zeta_1(w),w) - \bar{A}} \equiv \Theta(\bar{A}).$$

It is clear that $\Theta'(\bar{A}) < 0$, and thus the resulting strength- duration relationship $\bar{A} = \Theta^{-1}(T_t)$ is qualitatively similar to the boundary of the 0:1 region given in Fig. 1 for periodic forcing.

Now suppose that the forcing is given by the periodic extension of period T of the function

$$\bar{f}(t) = \begin{cases} 0 \text{ for } t \in (0,T_1), \\ \bar{A} \text{ for } t \in (T_1,T). \end{cases}$$

If $T_1 = 0$ then (7) has a periodic solution of period $T_0(\bar{A})$ (or frequency $\omega(\bar{A})$)

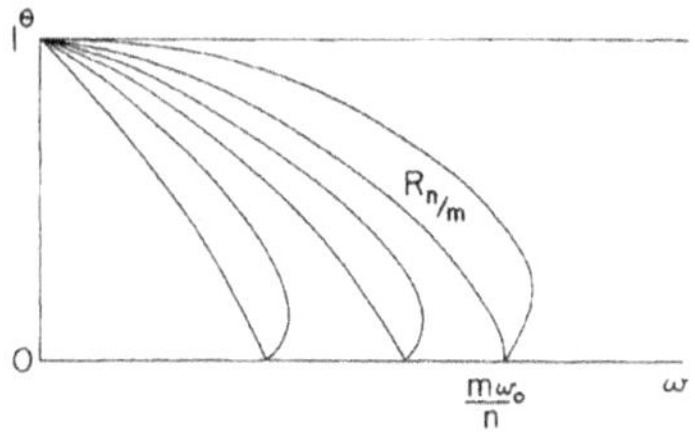

Figure 5. A sketch of the resonance zones in the (ω,θ) plane for a fixed amplitude of the forcing.

when regarded as an autonomous equation. However, this solution is only periodic as a solution of the nonautonomous equation if $T/T_0(\bar{A})$ is rational, since $\bar{f}$ is trivially periodic of period T.

Let $\theta \equiv 1 - T_1/T$ denote the fraction of time the forcing is off. After rewriting (7) in the scaled form given at (6) one is left with a three-parameter problem with parameters A, θ, and ω, where A is the scaled value of $\bar{A}$. Clearly there is a solution of period m for $\theta = 0$ if $T/T_0 = n/m$, or equivalently, if $\omega_0/\omega = n/m$, where $\omega = \frac{2\pi}{T}$. When this resonance condition is satisfied for $\theta = 0$ the flow on T^2 is parallel and the critical shadow reduces to a single orbit. If $\theta > 0$ and $A > A_0$ is fixed, $S^+(0,0)$ intersects t = 1 in an interval, and the boundaries of $R_{n/m}$ in the (ω,θ) plane correspond to the (ω,θ) pairs for which either the maximum orbit or the minimal orbit passes through (m,n). The analysis necessary to determine the qualitative structure of the resonance zones can be carried out explicitly [1]. The following proposition summarizes the results on the global (in (ω,θ)) disposition of these zones for any fixed $A \in (A_0, A_1)$.

Proposition 1.

- *$R_{n/m}$ does not intersect the line $\theta = 1$ for $\omega > 0$.*
- *$\rho = \infty$ along the line $\omega = 0$ for $\theta \in (0,1)$.*
- *If $r \neq km$ and $s \neq kn$ for some $k \in \mathbb{Z}$, then $R_{s/r} \cap R_{n/m} = \phi$.*
- *For every fixed (m,n) there exists an ω^* such that $R_{n/m} \subset \{(\omega,\theta) | \omega \leq \omega^*, \theta \in [0,1]\}$.*
- *The resonance zones that emanate from the line $\theta = 0$ all terminate at $(\omega,\theta) = (0,1)$.*

Figure 5 shows a sketch of the qualitative behaviour of the resonance zones in the (ω,θ)-plane for the general case. The structure of these zones is similar to that found in coupled oscillators [4] and in integrate-and-fire models [5,12]. The computations involved in determining the boundaries of the resonance zones can be carried out in detail for the piecewise-linear vector field

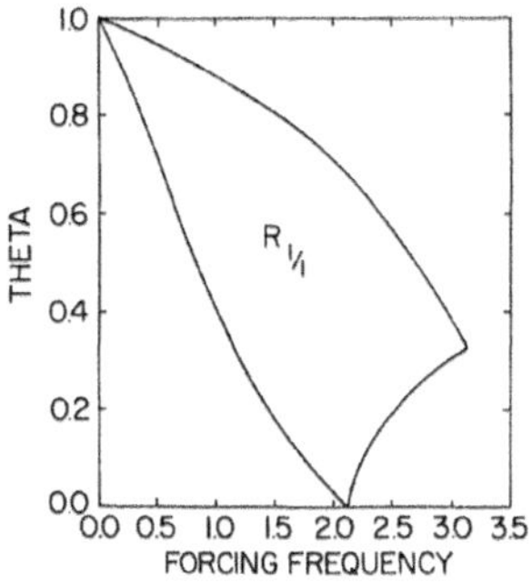

Figure 6. The 1:1 resonance zone for the piecewise linear system given by (8). $\alpha_1 = 2.0$, $\alpha_2 = \alpha_3 = 1$; $\beta_i = 1$, $i = 1,3$, $\gamma_1 = 0.0$, $\gamma_2 = 0.6$, $\gamma_3 = 1.8$; $\delta_1 = 1.0$, $\delta_2 = 0.5$. The forcing amplitude is $A = 0.75$.

$$\varepsilon\frac{dv}{dt} = \begin{cases} -\alpha_1 v - \beta_1 w + \gamma_1 & \text{for } v < v_1, \\ \alpha_2 v - \beta_2 w - \gamma_2 & \text{for } v \in (v_1, v_r), \\ -\alpha_3 v - \beta_3 w + \gamma_3 & \text{for } v > v_r, \end{cases} \tag{8}$$

$$\frac{dw}{dt} = \delta_1 v - \delta_2 w,$$

in the limit $\varepsilon = 0$. the 1:1 zone is shown in Fig. 6 for a particular choice of the coefficients. The details of how the boundaries of the resonance zones are determined are given in [1].

Next we consider the (θ, A) section for fixed $\omega > 0$. The functional relationship between A, ω and θ that determines the boundary of the resonance zones for fixed A can also be used at fixed ω, and from this relation one can show that $\frac{dA}{d\theta} > 0$ for $A - A_0$ positive and sufficiently small [cf. [1]]. Thus the forcing amplitudes corresponding to the upper and lower boundary of $R_{1/1}$ are decreasing functions of the fraction of time $1 - \theta$ that the forcing is on.

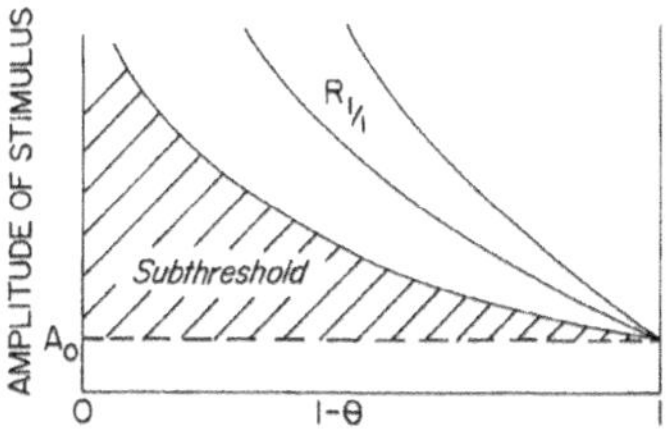

Figure 7. A qualitative sketch of the amplitude-duration relations for the piecewise linear vector-field, showing the threshold behaviour and the 1:1 locking region.

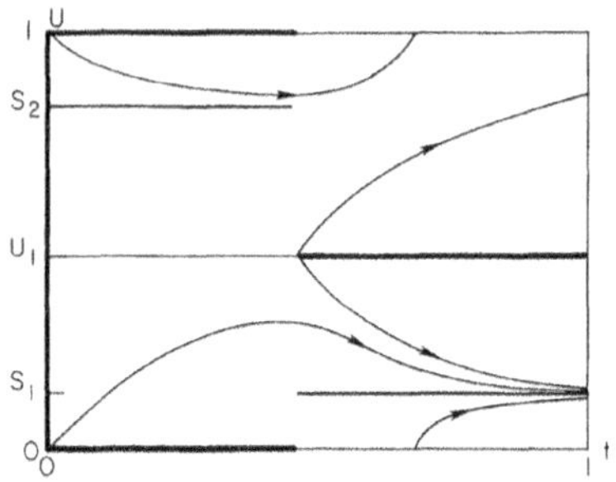

Figure 8. A sketch of the flow in the fundamental domain [0,1] x [0,1] when the forcing amplitude is greater than A_1. Both u = 0 and u_1 = 1/2 are repelling discontinuities during part of the forcing cycle, the former when the forcing is off and the latter when the forcing is on. S_1 and S_2 represent the rest points during the corresponding parts of the forcing cycle.

Earlier we showed that the time to threshold is also a decreasing function of the time the forcing is on, and these two results lead to the qualitative sketch of the amplitude-duration plane shown in Fig. 7. How the resonance zones terminate at $\theta = 0$ depends on the frequency of the forcing. This diagram is in qualitative agreement with the experimental results shown in Fig. 4, except that the latter does not show the upper boundary of the 1:1 zone.

Even though we earlier restricted A to be less than $A_1 \equiv g(\zeta_r(w_r),w_r)$ in order to preclude a rest point on the right-hand branch during part of the cycle, this case can be handled similarly. In fact, when this restriction on A is lifted and there is a rest point on the right branch for part of the cycle,

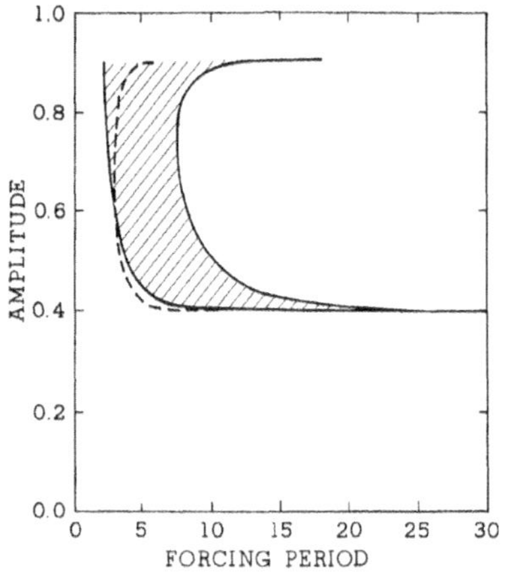

Figure 9. The 1:1 resonance zone in the period-amplitude plane for fixed θ. This region is shown hatched for θ = 1/2. This zone collapses to the dashed curve as $\theta \to 0$.

both $u = 0$ and $u_1 = 1/2$ are repelling discontinuities for part of the forcing cycle, and the flow on the torus is as shown in Fig. 8. It is easy to see that the resonance zones extend into the region $A > A_1$ and that some of them terminate along the line $\{(\theta,A)|\theta = 0, A \in (A_0,A_1)\}$. Again the details depend on the frequency of the forcing.

Finally, in Fig. 9 we show the 1:1 locking region in the period-amplitude plane for fixed θ. The dashed line, which corresponds to $\theta = 0$, represents the locus in this section of the vertex of the 1:1 region shown in Fig. 6. The hatched region between the solid curves is the 1:1 zone for $\theta = 1/2$.

5. The dynamics of a piecewise linear system for $\varepsilon > 0$

As we showed in [1] and summarized in previous sections, the dynamical behaviour of the system (3) in the limit $\varepsilon = 0$ can be characterized by whether or not the system is in resonance with the forcing. The rotation or firing number ρ varies continuously as a parameter λ in the equations is varied, and for a generic system the domain of λ decomposes into a disjoint set of resonance zones $R_{n/m}$, on each of which the rotation number is constant and equal to n/m. In this section we consider the case $\varepsilon > 0$ in which v is not infinitely fast. It is to be expected that for small ε the dynamics approximates that of the singular limit when the parameter point lies in the interior of a resonance zone, and this fact is proven in [1]. On the other hand, one also expects new phenomena to arise in the transition regions, and some of these will be described here for the piecewise linear system considered in the previous section.

When studying a T-periodically forced oscillatory system such as the van der Pol oscillator one can, for sufficiently small $\varepsilon > 0$, construct a cylindrical shell containing the periodic orbit that is invariant under the flow. The time T map associated with the flow maps the annular section of this shell to itself and the dynamics can be studied via this map. However, this does not work in forced excitable systems. One can still construct an attracting cylinder that captures the long-term dynamics, but because of the existence of subthreshold responses the time T map, which maps a section of this cylinder to itself, does not simplify for small ε. Here we define a different type of section, and as we shall see, the map of this section to itself is usually discontinuous. We consider the intersections of a trajectory for a system like (3) or (8) with the section S defined by

$$S = \{(v,w,t)|\quad v = v_1, \quad w \le w_1 \quad t \in \mathcal{R}\}.$$

(cf. Fig. 10). When $\varepsilon = 0$ such intersections occur at the point P_0 at which the system jumps to the right branch of the curve $f = 0$ and fires (cf. Fig. 3(b)). As in the previous section, we consider piecewise-constant forcing, and without loss of generality we define the origin of time to coincide with the onset of forcing in a fixed cycle. Thus the forcing is **on** in the first portion of the cycle and **off** in the second portion, which is the opposite convention to that used earlier. Since the forcing is periodic of period 1 we can identify successive time slices separated by one unit. As a result, we obtain a mapping of a cylinder to itself, and we call the time (mod 1) the phase.

Suppose that a trajectory crosses S at the sequence of phases $\phi_1, \phi_2, \ldots \phi_n, \ldots$. This sequence is generated by the phase return map $\phi_n \mapsto$

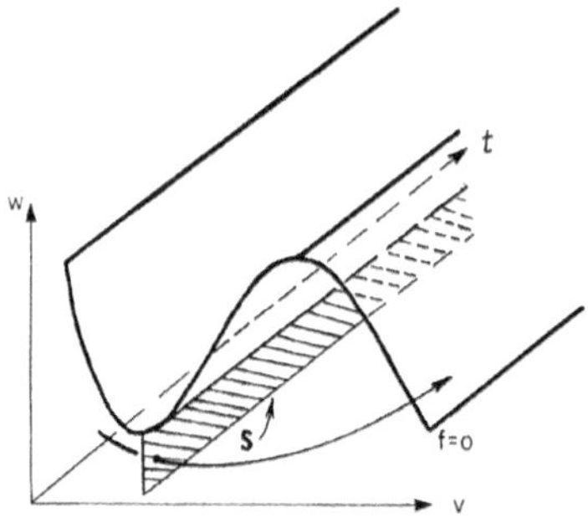

Figure 10. A schematic of the section S used for $\varepsilon > 0$.

ϕ_{n+1}. Of course this map does not capture the entire dynamics for $\varepsilon > 0$ because the w-component varies as well, but one expects that the phase component is most important when ε is small. In our earlier discussion of the singular limit we considered the time 1 map, which maps u(n) into u(n + 1). When $\varepsilon = 0$ the phase return map, which maps $\phi(u = 0)$ into $\phi(u = 1)$, is just the map between the other pair of boundaries of the fundamental domain. The flow for three copies of the fundamental domain in the limit $\varepsilon = 0$ is shown in Fig. 11(a), and the associated phase return map is shown in Fig. 11(b). Evidently the phase return map is both multiple-valued and discontinuous. On

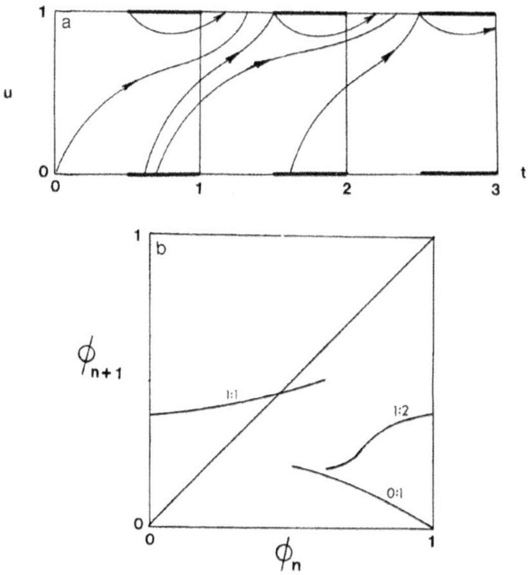

Figure 11. (a) A sketch of the flow for $\varepsilon = 0$. Heavy lines indicate repelling discontinuities in the flow. (b) The phase return map for the flow in (a). In the labels p:q, q denotes the number of cycles within which the trajectory returns to the section, and p indicates the number of times that it has crossed u = 1/2.

the branch marked 1:1 in Fig. 11(b) the trajectory returns to the section within one period of the forcing and "fires", i.e. it traverses the right branch of $\dot{v} = 0$. On the branch marked 0:1 the trajectory returns to the section within one period, but does so without firing. The leftmost point on this branch corresponds to a phase equal to $1-\theta$. Finally, on the branch marked 1:2 the trajectory returns within two periods of the forcing, having fired once. The three distinct cases can easily be understood by tracing selected trajectories in Fig. 11(a).

The fact that the phase return map captures most of the dynamics for small $\varepsilon > 0$ is illustrated in Fig. 12, where we show the return map for the piecewise linear system (8) with $\varepsilon = 0.001$. For these values of the parameters $(v_1, w_1) = (0.2 - 0.4)$. As in the previous section, we use piecewise constant forcing. The amplitude of the forcing is 0.7567, the period is 2.399315, and the fraction of the time the forcing is on is 0.5. Figure 12(a) is a composite of the phase components for eight values of $w \in (-.50, -.4001)$. The contraction in the w-component at this value of ε is so strong that the entire range of w is mapped onto (essentially) a curve (Fig. 12(b)), even though there is some dispersion in the phase values. However at larger ε the image of the

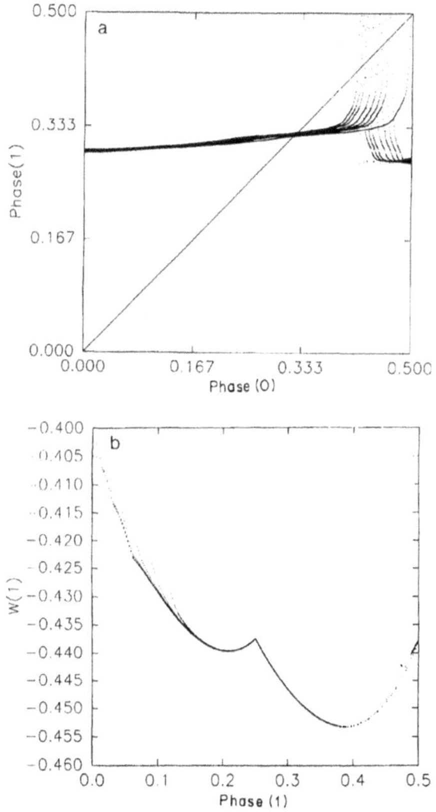

Figure 12. (a) The phase return map for $\varepsilon = 0.001$. (b) The image of the cylindrical section $\{(w,\phi) | w \in (-.50,-.40)\ \phi\ \varepsilon\ (0,1)\}$.

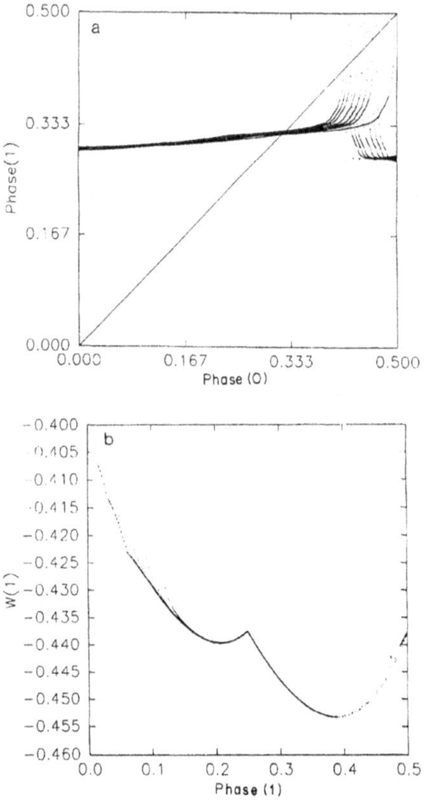

Figure 13. As in Fig. 12, but for $\varepsilon = 0.2$.

cylindrical section is genuinely two-dimensional, as is shown in Fig. 13.

It is clear that for the present choice of the forcing the invariant set for the flow must be contained in the set $\{w,\phi)| \ \phi \le 0.5\}$, and so we can ignore the initial points on the complementary interval when we analyze the long-term dynamics. The graphs of the phase return map for five different values of ε are shown in Fig. 14.

For each ε there are two types of results superimposed in this figure. First is the phase return map for fixed $w = -.411$ and $\phi \in (0.25, 0.50)$, which is the interval that contains the essential information. For ε small the major component corresponds to a 1:1 response in which the trajectory makes one cycle during each forcing cycle. There is also a 0:1 component in the lower right corner that corresponds to an aborted spike. Here the forcing turns off soon after the trajectory crosses back into the region $v < v_1$ and does not make a full spike until the forcing turns on again. This component, which for $\varepsilon = 0$ does not extend into the interval where the forcing is on, does so here because for $\varepsilon > 0$ the trajectory does not instantaneously jump across to another branch of the nullcline. Rather it requires a finite amount of time

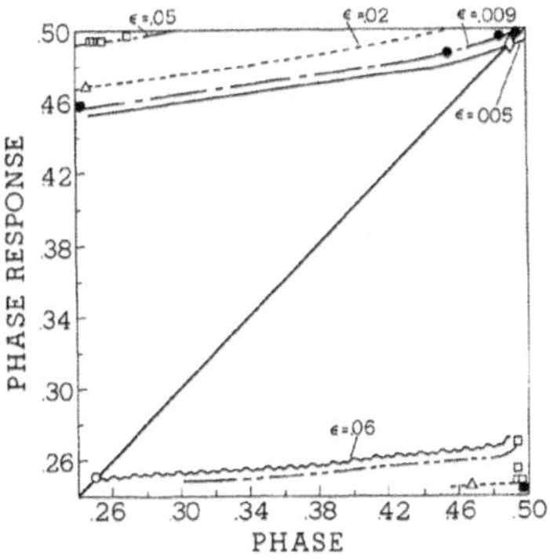

Figure 14. A composite of the phase return graphs for five values of ε. The symbols denote points of the invariant set for the corresponding value of ε.

(proportional to ε for small ε) to move to another branch of the nullcline.

In addition to scanning the phase return for a range of phases, stable periodic orbits were computed by iterating the phase return map for many cycles. These orbits are shown as small symbols on the graphs. For ε = .005 the 1:1 branch of the phase return map crosses the diagonal at a stable fixed point of the dynamics, and indeed there is a single stable periodic orbit with phase return at this fixed point. In fact, at this value of ε the dynamics is qualitatively the same as in the singular limit.

The first-order effect of increasing ε is to move the graph of the phase return map upwards, which means that the trajectory returns to the section at a later phase. At ε ~ .00705 the right hand tip of the 1:1 component touches the diagonal and a new unstable fixed point appears. As ε increases further the pair of fixed points approach and finally disappear via a tangent bifurcation at ε ~ .00785. Beyond this value the graph no longer crosses the diagonal and there is no fixed point. Rather there is a stable orbit with several returns on the 1:1 branch and then one on the lower 0:1 branch (cf. the curve for ε = .009 in Fig. 14). As ε varies the number of spikes, i.e. the number of returns on the upper branch, changes by an arithmetic progression. This is essentially the known phenomenon of period adding via intermittency.

For larger ε the reverse period-adding cascade continues, in the sense that there are fewer full spikes between abortive spikes as ε increases. At ε = .02 the period-adding cascade has collapsed to a period 2 orbit with successive phase returns on the 1:1 branch and the 0:1 branch. At ε = .05 a new phenomenon occurs: there are two coexisting stable period 4 orbits. One has successive phase returns on the branches 1:1,0:1,1:1,0:1 and the other on the branches 1:1,1:2,1:1,0:1. There are other dynamics for ε near this value. For example for ε = .502, there are coexisting period 4 and period 2 orbits. The analysis of these dynamics will clearly be more complicated, and the singular limit ε = 0 has no relevance in these cases.

Acknowledgements

Supported in part by NIH Grant # GM29123.

References

[1] Alexander, J.C., Doedel, E.J. & Othmer, H.G. On the resonance structure in a forced excitable system. To appear in *SIAM J. Appl. Math.*, 1990.

[2] Alexander, J.C., Doedel, E.J. & Othmer, H.G. (1989). Resonance and phase-locking in excitable systems. *Lects. on Math. Life Sci.* **21**, 1-37.

[3] Aronson, D., Doedel, E.J. & Othmer, H.G. (1986). Bistable behavior in coupled oscillators. In *Nonlinear Oscillations in Biology and Chemistry*, pp. 221-231, Othmer, H.G. (ed.). Lecture Notes in Biomathematics **66**, Springer-Verlag.

[4] Aronson, D., Doedel, E.J. & Othmer, H.G. (1987). An analytical and numerical study of the bifurcations in a system of linearly-coupled oscillators. *Physica* **25D**, 20-104.

[5] Builder, G. & Roberts, N.F. (1939). The synchronization of a simple relaxation oscillator. *AWA Tech. Rev.* **4**, 164-180.

[6] Chialvo, D.R. & Jalife, J. (1987). Non-linear dynamics of cardiac excitation and impulse propagation. *Nature* **330**, 749-752.

[7] DeYoung, G. & Othmer, H.G. Resonance in oscillatory and excitable systems. To appear in *Ann. N.Y. Acad. Sci.*, 1990.

[8] Doedel, E.J. (1981). AUTO: A program for the automatic bifurcation and analysis of autonomous systems. In *Proc. 10th Manitoba Conf. Num. Anal. and Comp.*, pp. 265-284.

[9] Field, R.J. & Noyes, R.M. (1974). Oscillations in chemical systems, IV. Limit cycle behavior in a model of a real chemical reaction. *J. Chem. Physics* **60**, 1877-1884.

[10] Fitzhugh, R. (1969). Mathematical models of excitation and propagation in nerve. In *Biological Engineering*, pp. 1-85, Schwan, H.P. (ed.). McGraw-Hill.

[11] Glass, L., Guevara, M.R., Belair, J. & Shrier, A. (1984). Global bifurcations of a periodically forced biological oscillator. *Phys. Rev.* **A29**, 1348-1357.

[12] Glass, L. & Belair, J. (1986). Continuation of Arnold tongues in mathematical models of periodically forced biological oscillators. In *Nonlinear Oscillations in Biology and Chemistry*, pp. 232-243, Othmer, H.G. (ed.). Lecture Notes in Biomathematics **66**.

[13] Guttman, R., Feldman, L. & Jakobsson, E. (1980). Frequency entrainment of squid axon membrane. *J. Membrane Biol.* **56**, 9-18.

[14] Hartman, P. (1964). *Ordinary Differential Equations.* John Wiley and Sons.

[15] Hodgkin, A.L. & Huxley, A.F. (1952). A quantitative description of membrane current and application to conduction and excitation in nerve. *J. Physiol.* **117**, 500-544.

[16] Holden, A.V. (1976). The response of excitable membrane models to a cyclic input. *Biol. Cybernetics* **21**, 1-7.

[17] Hudson, J.L., Lamba, P. & Mankin, J.C. (1986). Experiments on low-amplitude forcing of a chemical oscillator. *J. Phys. Chem.* **90**, 3430-3434.

[18] Matsumoto, G., Aihara, K., Hanyu, Y., Takahashi, N., Yoshizawa, S. & Nagumo, J.-I. (1987). Chaos and phase locking in normal squid axons. *Phys. Rev. Lett. A* **123(4)**, 162-166.

[19] Markevich, N.I. & Sel'kov, E.E. (1989). Parametric resonance and amplification in excitable membranes. The Hodgkin-Huxley model. *J. Theor. Biol.* **140**, 27-38.

[20] Monk, P.B. & Othmer, H.G. (1989). Cyclic AMP oscillations in suspensions of *Dictyostelium discoideum*. *Phil. Trans. R. Soc. Lond.* **323(1215)**, 185-224.

[21] Monk, P.B. & Othmer, H.G. Wave propagation in aggregation fields of the cellular slime mode *Dictyostelium discoideum.* Submitted.

[22] Othmer, H.G. & Monk, P.B. (1988). Concentration waves in aggregation fields of a cellular slime mold. In *Biomathematics and Related Computational Problems,* pp. 381-398, Ricciardi, L. (ed.). Kluwer Academic Publishers: Dordrecht.
[23] Nagumo, J.-I. & Sato, S. (1972). On a response characteristic of a mathematical neuron model. *Kybernetik* **10**, 155-164.
[24] Sato, S., Hatta, M. & Nagumo, J.-I. (1974). Response characteristics of a neuron model to a periodic input. *Kybernetik* **16**, 1-8.

22. PATTERN FORMATION ON THE SHELLS OF MOLLUSCS BY TRAVELLING WAVES WITH UNUSUAL PROPERTIES

Hans Meinhardt and Martin Klingler

Max-Planck-Insitut für Entwicklungsbiologie
Spemannstr. 35, D-74 Tübingen, FRG

1. **Shell patterns - a space-time plot**

The shells of many molluscs are decorated with either a relief-like or a pigmentation pattern. These patterns are of great diversity and frequently of great beauty. The formation of these patterns proceeds in most species in a strictly linear manner since new pattern elements are added only along a marginal zone, the growing edge of the shell. The second dimension is a protocol of what happens as a function of time. The shell is, so to speak, a space-time plot. The shells provide a unique situation in that the complete history of a very dynamical process is preserved. We have proposed a model for shell patterning based on reaction-diffusion mechanisms. As shown by computer simulations, this model can account for apparently very different shell patterns if small variations of the parameters or minor changes in the underlying mechanism are assumed [5,8]. Details of computations can be found in [7]. A model based on similar principles but underlining the possible role of the nervous system has been proposed by Ermentrout et al. [1].

Basic elements of shell patterns are lines perpendicular, parallel and oblique to the growing edge. Keeping in mind the space-time character of the shell pattern, lines perpendicular to the growing edge indicate the formation of a spatially periodic pattern along the growing edge that is stable in time. Pigment production or bulging occur at particular positions interspersed by regions without such events. The latter two patterns indicate that pigment deposition oscillates. A particular cell produces pigment only during a certain time interval and enters then into an inactive (refractory) period until the next pigment production takes place. A synchronous oscillation in pigment production leads to lines that are essentially parallel to the shell margin. Oblique lines originate from travelling waves of pigment production. Such waves arise if pigment-producing cells trigger their neighbouring cells such that these cells also start to produce pigment after a certain delay. More complex patterns require, as a rule, the superposition of at least two pattern forming reactions.

2. **Pattern formation by autocatalysis and lateral inhibition**

In his pioneering work, Turing [13] has shown that stable patterns can be generated by a system of two interacting substances of different diffusion rates. Prigogine and Lefever [3,10] have introduced a non-linear reaction, later called the Brusselator, which can generate oscillations as well as stable spatial patterns. Gierer and Meinhardt [2,4,6] have proposed that pattern formation requires local self-enhancement and long ranging inhibition. The following interaction between an autocatalytic activator a(x) and its long-ranging antagonist, the inhibitor h(x) provides an example.

$$\frac{\partial a}{\partial t} = \frac{\rho a^2}{h} - \mu a + D_a \frac{\partial^2 a}{\partial x^2} + \rho_0 \tag{1a}$$

$$\frac{\partial h}{\partial t} = \rho a^2 - \nu h + D_h \frac{\partial^2 h}{\partial x^2} + \rho_1 \tag{1b}$$

The antagonistic effect can also result from the depletion of a substrate s(x) which is consumed during the autocatalytic activator production.

$$\frac{\partial s}{\partial t} = \rho s a^2 - \mu a + D_a \frac{\partial^2 a}{\partial x^2} + \rho_0 \tag{2a}$$

$$\frac{\partial s}{\partial t} = \sigma - \rho s a^2 - \nu s + D_s \frac{\partial^2 s}{\partial x^2} \tag{2b}$$

Whether stable patterns, oscillations or travelling waves are generated, depend on the parameter involved. Stable patterns require an antagonistic reaction with a shorter time constant compared to that of the autocatalytic substance ($\mu < \nu$ or $\mu < \sigma$) as well as a higher diffusion range of the antagonist ($D_h \gg D_a$ or $D_s \gg D_a$). A local activator increase will grow further due to the autocatalysis while the surrounding region will become de-activated either due to the inhibitor spreading out from the region with elevated activator concentration or due to the depletion of the substrate. Periodic patterns arise if the range of the antagonistic substance is smaller then the total length of the growing edge. Since particular regions remain steadily activated and since high activation is assumed to trigger pigment production, a periodic pattern leads in the course of time to stripes of pigmentation or bulging perpendicular to the growing edge.

If the antagonistic substance adapts too slowly to a change of the activator concentration, i. e. if $\mu < \nu$ or $\mu < \sigma$, oscillations do occur. If the activator diffuses more rapidly than the antagonist, along the growing edge the oscillations become sychronized. This leads to stripes parallel to the edge (for simulations see [8]). In contrast, if the activator has a small diffusion range but the antagonistic substance is nearly non-diffusible, travelling waves of high activator concentration will result. An activated region can "infect" its neighbouring region such that it becomes fully activated after a certain time delay, infecting its neigbouring region, and so on. In each cell, after a certain time, the activation breaks down due to the action of the antagonistic substance. In the time record, such wandering regions of high activation lead to oblique lines of pigmentation or bulging.

3. Travelling waves generate oblique lines

In view of the general subject of this volume, patterns containing oblique lines are of special interest since, as already mentioned, they result from travelling waves. Characteristic for many such shell patterns is a W-like arrangement of the oblique lines. The Λ-like pattern element results from a spontaneous trigger of a cell followed by an "infection" of both neighbours, and so on. In other words, a spontaneous trigger of a cell gives rise to two diverging oblique lines. If two such lines merge, both lines become extinct because all neighbouring cells are still in the refractory period, leading to the v-like pattern element. Many shell patterns indicate that the travelling waves can have unusual properties. For instance, some shells show crossings or branchings of pigmentation lines (see Fig. 4 and 5). A crossing indicates that a collision of two waves without the mutual annihilation mentioned above has occurred. The formation of oblique lines with branches indicates the sudden formation of backwards waves. These patterns require more elaborate reactions, as will be discussed in further detail below.

The pattern on the bivalved mollusc *Stragella carnea* [11,12] is an example for a pattern generated by simple travelling waves. The specimen shown in Fig. 1 shows a shell in which pattern formation has been perturbed at a particular time. As shown in the simulation, the simple interactions mentioned above are able to account not only for the pattern formation itself but also for fine details of pattern regulation that occurs after a perturbation.

4. Superposition of a spatially stable and a periodic pattern

Many shell patterns can be explained by assuming a superposition of two patterns, one that is stable in time and one that oscillates. The first

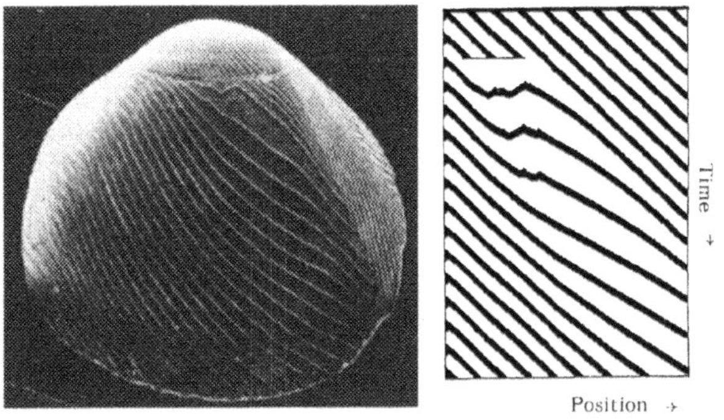

Figure 1. Oblique lines resulting from travelling waves and pattern regulation. The oblique rib-like pattern of *Strigilla* [see 11,12] is assumed to be generated by travelling waves of high activator concentrations. After an interruption of the normal pattern (caused presumably by an external event), regulatory processes take place. After the interruption, lines from one side only bend towards the gap. In addition, an initiation point with two diverging lines emerges. Model: an activator-subtrate model (eq. 2) is assumed. The external event causes a sudden lowering of the substrate concentration to 40 % (horizontal line). The chain of trigger events is interrupted. Due to the absence of activation for a prolonged period, the substrate concentration (not shown) increases to such a level that a spontaneous activation becomes possible (Λ-element). One of these waves annihilates with a wave that survived the external event (V-element). The bending of lines indicates a higher speed of the travelling waves. Due to the accumulating substrate, the cells become more and more susceptible to the trigger from neighbouring cells. This leads to a speeding up of the travelling waves. The bending is unilateral since the waves on the other side of the gap are, of course, not retarded by events that will take place in the future (after [8]).

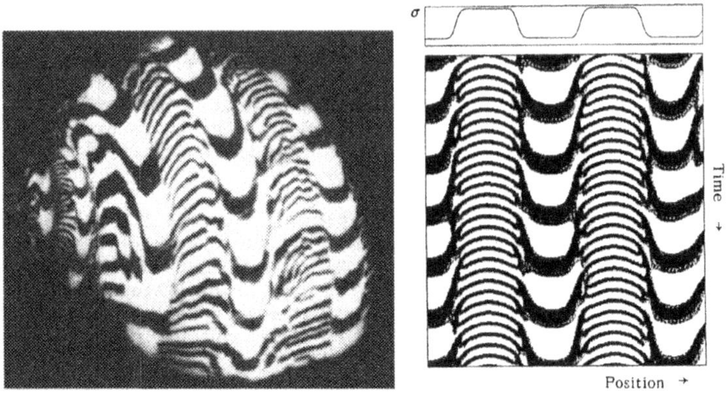

Figure 2. Superposition of stable and periodic patterns. (a) Shell of *Palmadusta diluculum*. The oscillation frequency is much higher in some broad bands in comparison to the intervening regions, indicating the existence of a spatially stable pattern which influences the oscillation frequency. Model: the substrate production σ (eq. 2) as well as the decay rate of the activator μ and the source density ρ is assumed to be a function of position. These parameters have the same space dependency as that shown for σ. Regions at which these parameters are high oscillate more rapidly. This leads to bands in which the pigment lines have a smaller spacing. At the border between these bands, the pigment lines partially merge (after [8]).

pattern controls the parameter of the second. In the simulation of *Palmadusta diluculum* an activator-substrate model (eq. 2) has been assumed. The spatially stable pattern modulates substrate production σ, the activator lifetime μ as well as the source density ρ. This has the consequence that regions of high and low oscillation frequencies alternate. The resulting pattern consists of broad stripes perpendicular to the growing edge that contain pigmentation lines of different spacings. Bands with large and small spacings alternate. The regions of the highest oscillation frequencies form pace maker regions from which waves spread out. In regions of low oscillation frequencies, the cells fail to follow such rapid triggering. Some pulses are skipped. This has the consequence that some pigmentation lines merge at the border between two bands.

Fig. 3 shows the shell of the snail *Voluta musica*. The pattern is characterized by fine lines perpendicular to the growing edge interspersed by periodic patterns of dots and fine lines parallel to the growing edge. These pattern elements are reproduced in the simulation. Assumed is an activator-substrate model in which, in contrast to the simulation shown in Fig. 2, only the substrate production is a function of position, but, in addition, the substrate diffuses. The diffusion has the consequence that pattern elements of the lateral inhibition type can be formed in regions of high substrate production. Moreover, if the parameters of the pattern forming system are such that the system is still in the oscillating regime, the diffusion of the antagonistic substance provides a tendency to switch into the non-oscillating mode. Diffusion of substrate into a depleted region mimics a higher production rate or the the diffusion of inhibitior away from an activated region mimics a shorter life time since the inhibitor disappears more rapidly. As outlined above, higher production rate of the substrate or shorter lifetime of the inhibitor favours patterns that are stable in time. In the simulation in Fig. 3, this effect leads to the fine lines stable in time.

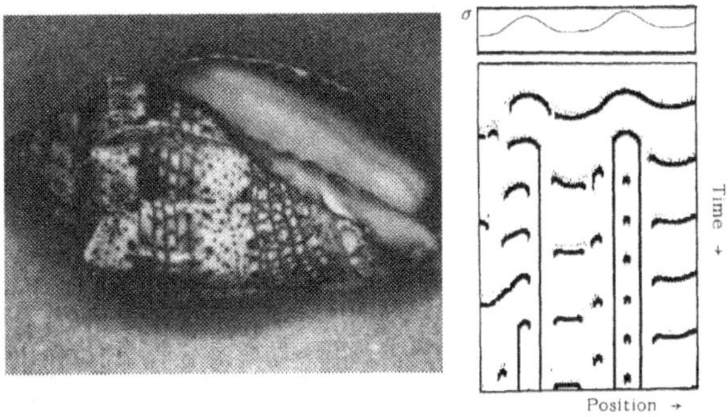

Figure 3. The shell of *Voluta musica* and simulation. The pattern is characterized by several fine lines perpendicular to the growing edge, interspersed by periodic pattern elements. In the simulation with eq. 2, the time constants have been chosen in such a way that in regions of high substrate production (high σ), the system would be close to the transition from the oscillating to the non-oscillating mode. Nevertheless, steady state activations become possible due to the diffusion of the substrate since this diffusion provides an additional supply of substrate for depleted regions from neighbouring non-activated regions with high substrate concentrations. In this way the steady state activation becomes stabilized. The steadily activated regions become so sharp and separated from each other due to the effect of lateral inhibition.

In the remaining field, oscillations remain. The activated regions are separated from each other due to substrate diffusion.

5. Formation of branches: the sudden formation of backwards waves

Many species form patterns with oblique lines that branch. Figure 4 shows as an example a portion of the shell of the snail *Olivia porphyria*. Keeping in mind the space-time character of the shell pattern a branch indicates the sudden formation of a backwards wave. Its initiation requires a modification of the normal serial trigger mechanism that leads to the oblique lines. From a particular moment onwards and for a short time interval, the activated cells must remain in the activated state until the refractory period of the neigbouring, still oscillating cells is over. Then, these cells can be re-infected and the backwards wave is initiated. For instance, in a reaction according to eq. 1 the temporary shift from the oscillating into a steady state activation can result from a change in the decay rate of the inhibitor, from $\nu < \mu$ to $\nu > \mu$ and later on back to $\nu < \mu$. Of course, this requires a signal as to where and when this transition should occur.

As mentioned, as a rule, travelling waves annihilate each other during a collision. Branching is one of the possible mechanisms to maintain the number of travelling waves. In the simulation in Fig. 4, this has been achieved by the assumption of a hormone-like, homogeneously distributed substance that is produced by the activated cells. The hormone concentration is thus a measure of the total number of cells activated. It provides a signal that is available in each cell along the growing edge. The hormone has an inhibiting influence on the inhibitor decay. Thus, whenever the number of activated cells becomes too small, the inhibitor life time becomes shortened such that the cells are shifted from the oscillatory mode into a steady state. After the refractory period of a neighbouring still oscillating cell is over, a backwards wave is

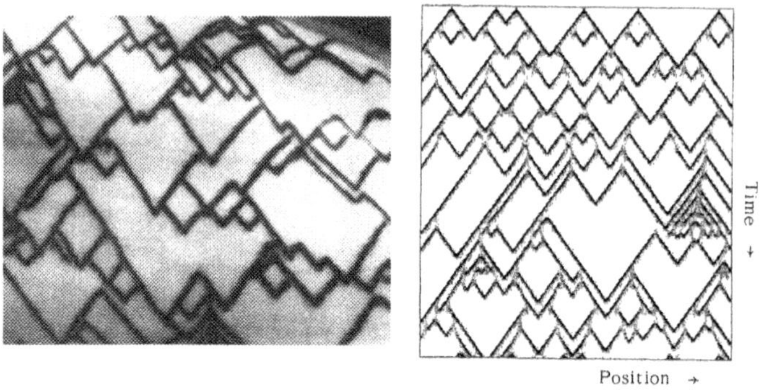

Figure 4. Formation of branches. At the left, a shell of *Olivia porphyria* is shown. A branch indicates the sudden formation of a backwards wave. Model: branching occurs whenever the number of travelling waves drop below a certain threshold value. The controlling agent is a hormone-like substance which is rapidly distributed within the organism and which is produced at a rate proportional to the local activator concentration. The hormone provides thus a measure of how many travelling waves are present. In the simulation shown it is assumed that the hormone inhibits inhibitor decay. Below a certain hormone concentration (not shown in the simulation), i.e. whenever the inhibitor life time becomes sufficiently short, the just activated cells switch from the oscillatory mode into a steady state activator production. Groups of cells remain activated during the refractory period of their still oscillating neighbours. After this refractory period is over, backwards waves are initiated by re-infection (after [8]).

initiated. This causes an increase of the number of travelling waves and thus an increase in the inhibitor lifetime. All cells return to the oscillating mode and the formation of the branch is completed. A further condition for such pattern formation is that the refractory period is much shorter than the time interval after which a spontaneous activation can occur, a property that can be adjusted, for example, in a reaction according to equation 1 by an appropriate choice of ρ_0 and ρ_1.

The simulation reproduces many details of the natural pattern. Branchings are simultaneously initiated at many positions. In the model, this results from the global control due to the homogeneous distribution of the hormone. In the final pattern, this leads to the V-like elements with the same distance between the tip and the points of branch initiation. While the original wave proceeds in an unperturbed way, a branch appears frequently only loosely connected to the original line. In the simulation, this effect results from the passage through the temporary steady state activation, which is lower than that achieved during the pulse-like activation in the oscillatory regime. Sometimes, a little hook appears close to the point of branch initiation. This results from an incipient initiation of a wave parallel to the primary wave. Usually this wave does not survive since the medium is not yet exitable enough. If this wave survives, however, a line close to the original line is formed together with the branch.

6. Formation of crossings by an additional diffusible inhibitor

Pigmentation lines that cross each other can be found on several shells. Figures 5 and 6 provide examples. Crossings are unusual for non-linear waves in exitable media since, as a rule, two colliding waves annihilate each other. Crossings of oblique lines can be regarded as the initiation of two new diverging waves at the position where two waves collide. Similar to the case of branch formation, the new waves can result from cells which remain activated for a prolonged period due to a temporary shift into the steady state. In the simulation in Fig. 5, an activator-substrate mechanism has been assumed with parameters such that a cell, once activated, would remain in a steady state. An additional diffusible inhibitor is produced by the activated cells. A travelling wave results since each newly activated cell extinguishes the activation of the preceding cell. However, if two waves collide, the situation is very different. No newly activated cell is available to extinguish the activation of the cells at the point of collision. These cells would remain in the steady state until the refractory period of the neighbouring region is over such that these cells can become re-infected. The newly activated cells extinguish via the diffusible inhibitor the steady state activation of the cells at the point of collision.

7. Formation of crossings by coupling of two pattern forming reactions

The mechanism for crossings described above is very parameter-sensitive. If the inhibition is too strong, the activation does not survive at the point of collision, if too weak, the steady state activation cannot be extinguished. A more robust mechanism results if the extinction of the activation is accomplished by a separate pattern forming system. Equation 3 provides a set of coupled reactions with such a property, Fig. 6 a simulation. As it will be shown further below, the very same mechanism can also be adapted to generate a very different pattern (see Fig. 7).

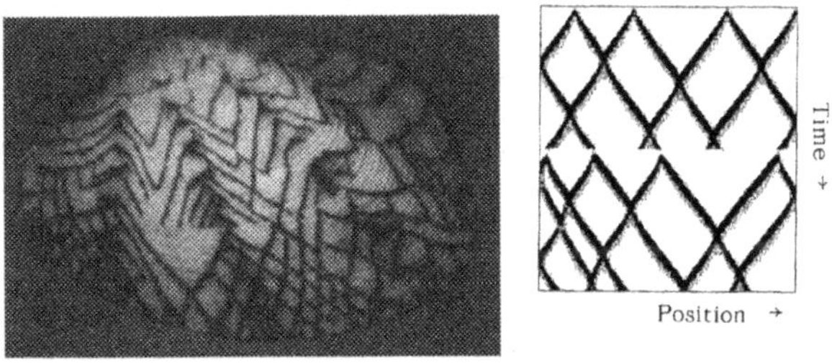

Figure 5. Crossings of pigmentation lines. Left: Pattern on *Tapes* spec. Crossings indicate that two waves do not annihilate each other but penetrate each other at a collision. Model: assumed is a three-component system; an activator-substrate mechanism tuned to produce a steady state activation, and, in addition, a diffusible inhibitor. Travelling waves result from the extinguishing of the activation in each cell by their subsequently triggered neighbouring cells. At the point of collision, no new neighbours are available which could provide this inhibition. The cells remain in the steady state until two backwards waves are initiated. A sudden lowering of the activator concentration can lead either to the termination of a line, to branch formation or to the continuation of a line. The same features can be seen in the perturbation of the natural pattern (for details of the calculation see [8])

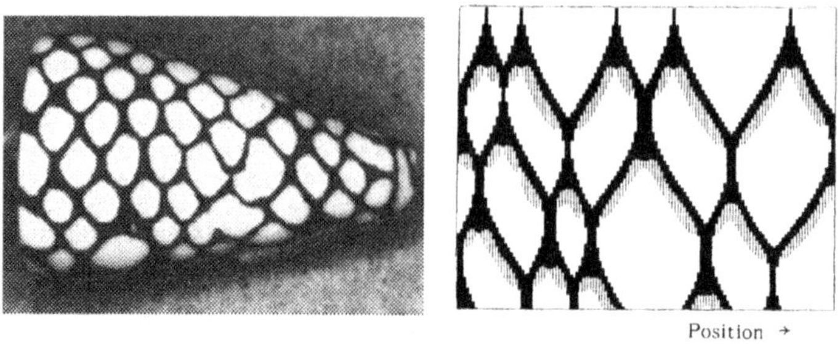

Figure 6. An alternative model for branch formation. Left: Patterns on *Conus marchionatus*. The asymmetry of the crossing along the time axis and the long time spent in the steady state after collision is clearly visible. Model: assumed are two pattern forming systems (Eq. 3a-d), one initiates pigment formation and a second one (shaded) that is initiated by the first and that extinguishes first. Due to the diffusible antagonist **d** of the second reaction, the extinguishing wave does not collide. Therefore, the activation of the first system survives at the point of collision. Calculated with Eq 3 and $\rho_a = 0.01$, $\kappa_a = 0.05$, $\mu_a = 0.01$, $D_a = 0.005$, $\gamma = 0.08$, $\sigma_b = 0.06$, $\nu_b = 0$, $D_b = 0$, $\rho_c = 0.05$, $\kappa_c = 0.05$, $\mu_c = 0.05$, $D_c = 0.15$, $\rho_{ac} = 0.002$, σ_d, $d = 0$, $\sigma_{ad} = 0.02$, $\rho_0 = 0$, $\nu_d = 0$, $D_d = 0.1$.

The reaction of the activator-depleted substrate type that generates the signal for pigmentation:

$$\frac{\partial a}{\partial t} = \frac{\rho_a b a^2}{1 + \kappa_a a^2} - \mu_a a + D_a \frac{\partial^2 a}{\partial x^2} - \gamma c a \tag{3a}$$

$$\frac{\partial b}{\partial t} = \sigma_b - \frac{\rho_a b a^2}{1 + \kappa_a a^2} - \nu_b b + D_b \frac{\partial^2 b}{\partial x^2} \tag{3b}$$

The extinguishing reaction:

$$\frac{\partial c}{\partial t} = \frac{\rho_c d c^2}{1 + \kappa_c c^2} - \mu_c c + D_c \frac{\partial^2 c}{\partial x^2} + \rho_{ac} a + \rho_0 \tag{3c}$$

$$\frac{\partial d}{\partial t} = \sigma + \sigma_{ad} a - \frac{\rho_c d c^2}{1 + \kappa_c c^2} - \nu_d d + D_d \frac{\partial^2 d}{\partial x^2} \tag{3d}$$

The **a-b** system provides as usual the signal for pigment production. It is tuned to remain in the steady state after an activation. It initiates the activation of the **c-d** system via the term $\rho_{ac} a$ in eq. 3c. The substance **c**, in turn, extinguishes the **a-b** - system via an additional destruction term $-\gamma ca$. The result is a travelling **a**-wave that is followed with some delay by a travelling **c**-wave. In order to achieve survival of the activator at the point of collision of the **a**-wave, a high diffusion of the **d**-substance is assumed.

This has the consequence that two **c**-waves keep a distance from each other. They become extinguished even before they collide due to the entry into the zone of a high inhibitor concentration of the wave that runs in the opposite direction. Therefore, at the point of collision of two **a**-waves, no extinguishing **c**-wave is present and the **a**-system remains in the steady state until the onset of wave formation due to the restoration of excitability. At this stage, the extinguishing **c**-wave becomes initiated again and a normal-looking travelling wave remains.

The crossings of lines in some species are characterized by a high degree of symmetry along the time axis (see Fig. 5). The waves move towards each other with about the same speed as they move away after the collision. Other species show a pronounced asymmetry (Fig. 6). Waves move towards each other with increasing speed, as indicated by the rounded, U-shaped pattern before the collision takes place, followed by a relatively long period in the steady state with little if any movement. The latter phase generates a stem-like pattern below the U-shaped element. After a certain period, movement starts again and accelerates in the course of time. It is easy to see how such a pattern can result. With increasing time after the last activation, the medium becomes more exitable, therefore speeding up the waves. After the burst of activator production, some time is required before the excitability of the medium becomes restored.

As shown in Fig. 6, this simulation reproduces the asymmetry of branch formation more closely although not perfectly. This problem results from the long-ranging inhibiton assumed in both mechanism. It slows down the speed of the waves shortly before collision.

8. Formation of drop-like, unpigmented regions

The pattern on the shell of *Conus marmoreus* (Fig. 7) is characterized by a pigmented background with drop-like unpigmented regions. The beginning of these drops indicates an almost simultaneous break down of the activation over a larger extent of the field. Due to the spread of activation from those regions in which activation has survived, the region becomes pigmented again in the course of time while in other regions, new pigment-free regions emerge, and so on. One can find an almost continous transition from the pattern of asymmetric crossings (Fig. 6) to the white drop-like pattern, suggesting that both patterns are generated by closely related mechanisms. In the simulation Fig 7, eq. 3 has been used again with diiferent parameters. The **a-b**-system has been tuned so that a steady state activation is possible. From a single activated point, the activation spreads out to both sides, generating in the course of time a growing triangle. In contrast to the simulation in Fig. 6 the production of the substrate **d** of the extinguishing **c-d**-system is assumed to be the under control of the **a**-substance. Whenever the **a**-system was active for a sufficient amount of time, i.e. if the pigmented triangle has achieved a certain size, a burst-like **c**-activation is generated over a larger region and the **a**-activation is switched off. New triangles emerge from the activation surviving in the marginal zone of that triangle.

9. Oblique and parallel lines in the same pattern

Several species produce lines oblique and parallel to the growing edge simultaneously. Figure 8 provides an example. As outlined above, both pattern elements require different parameters. Oblique lines, i.e. travelling waves, require a small diffusion while parallel lines, i.e. a synchronous oscillation, require a high diffusion of the activator. Since parallel lines emerge at positions at which somewhat earlier an oblique line has been formed,

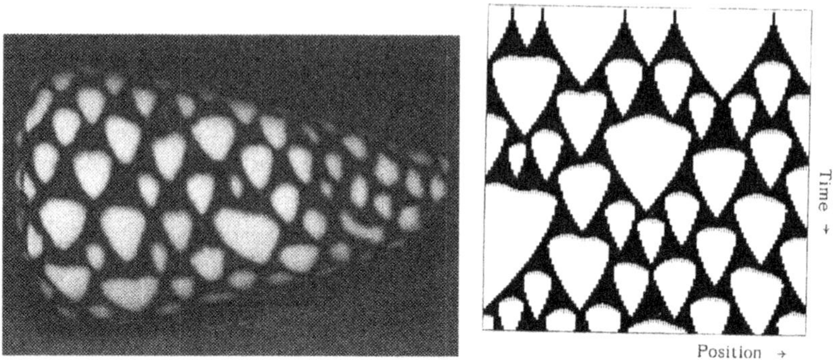

Figure 7. Unpigmented drop-like pattern on a pigmented background. Left: The shell of *Conus marmoreus*. Model: Assumed is the same model as used in Fig. 6. Whenever cells have remained activated for a sufficient time, a second system is fired which extinguishes the first (shaded regions at the top of the white drops). Therefore, pigment production stops almost simultaneously over larger regions which become filled up in the course of time by the spread of activation from regions in which the activation of the first system has survived. Calculated with Eq 3 and $\rho_a = 0.01$, $\kappa_a = 0.05$, $\mu_a = 0.01$, $D_a = 0.005$, $\gamma = 0.07$, $\sigma_b = 0.1$, $\nu_b = 0$, $D_b = 0$, $\rho_c = 0.1$, $\kappa_c = 0.05$, $\mu_c = 0.1$, $D_c = 0.1$, $\rho_{ac} = 0$, $\rho_0 = 0.015$, $\sigma_d = 0.01$, $\sigma = 0.02$, $\nu_d = 0$, $D_d = 0.1$.

this cannot result from a superposition of a spatially stable pattern that controls the parameters of the pigmentation reaction (as was assumed in Figs. 2 and 3).

In the simulation in Fig 8, similar to that of Figs. 6 and 7, the superposition of two pattern-forming reactions has been assumed. The activation on one system, the **c-d** system, is the precondition of an activation of the **a-b**-system since the **b**-production depends on a high **c**-concentration. The **a-b**-system controls pigmentation. If the diffusion of the **c**-molecules is small, the spread of the **c**-activation is necessarily small. With the onset of the **c**-activation, an **a** activation will immediately occur. Thus, the wave propagation of the **a**-system is not determined by the large **a**- but by the small **c**-diffusion. Within an activated **c** region, oscillations of the **a**-system are possible. They become synchronized due to the high **a**-diffusion. The following equation (4) has been used for this simulation:

The reaction generating the signal for pigmentation:

$$\frac{\partial a}{\partial t} = \frac{\rho_a b a^2}{1 + \kappa_a a^2} - \mu_a a + D_a \frac{\partial^2 a}{\partial x^2} + \rho_0 \qquad (4a)$$

$$\frac{\partial b}{\partial t} = c\sigma_b - \frac{\rho_a b a^2}{1 + \kappa_a a^2} - \nu_b b + D_b \frac{\partial^2 b}{\partial x^2} \qquad (4b)$$

The system generating the precondition for the **a-b** system:

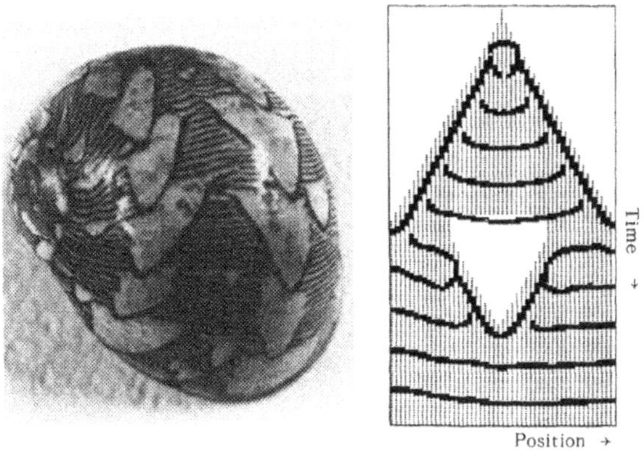

Figure 8. Pattern containing lines parallel and oblique to the growing edge. Left: Pattern on *Clithon ovalanensis* (see [9]). The parallel lines are enframed by oblique lines. Both patterns would require different parameters, a high and a low diffusion of the activator. Model: Assumed are two pattern forming reactions, one which controls pigment production and a second one (shaded in the simulation) that is required as a precondition that the pigment-controlling reaction can be activated (Eq.4). Thus, the oblique lines are not determined by the spread of the activation of the pigment-controlling but by that of the precondition reaction. Pattern repair after a drop-out of the precondition reaction is also simulated. Calculated with Eq 4 and $\rho_a = 0.08$, $\kappa_a = 0.2$, $\mu_a = 0.08$, $D_a = 0.01$, $\rho_0 = 0.01$, $\sigma_b = 0.02$, $\nu_b = 0$, $D_b = 0.03$, $\rho_c = 0.01$, $\kappa_c = 0.4$, $\kappa_d = 0.02$, $\mu_c = 0.01$, $D_c = 0.001$, $\nu_d = 0.02$, $D_d = 0$.

$$\frac{\partial c}{\partial t} = \frac{\rho_c c^2}{(d + \kappa_d)(1 + \kappa_c c^2)} - \mu_c c + D_c \frac{\partial^2 c}{\partial x^2} \tag{4c}$$

$$\frac{\partial d}{\partial t} = \frac{\rho_c c^2}{1 + \kappa_c c^2} - \nu_d d + D_d \frac{\partial^2 d}{\partial x^2} \tag{4d}$$

The simulation in Fig. 8 reproduces several features of the natural pattern. For instance, the parallel lines are only loosely connected with the oblique lines. Occasionally, the activation of the precondition reaction disappears in some region. This has been introduced artificially in the simulation, but can, of course, be accomplished by an extinguishing reaction such as that shown in Fig. 7. After such an event, the gap in the **c**-activation disappears due to the spread of **c**-activator from regions that still carry a high **c**-activation. The resulting pigmentation resembles closely the natural pattern. However, the simulation is certainly incomplete. In addition a pattern must be superimposed that is stable in time, which restricts that reaction to a particular part of the shell.

10. Conclusion

The patterns on the shells of molluscs can be simulated with simple or coupled reaction-diffusion mechanisms that have been successfully applied for the simulation of pattern formation in the development of higher organisms. Thus, the shell patterns seem to be more a special adaptation of a general pattern forming mechanism than a unique invention for that special purpose during evolution. The diversity of patterns in different species provides an inroad for the investigation of the range of possible patterns which can be generated by modifications of a basic mechanism. Superposition of several pattern-forming systems in a hierarchical way are the rule in all higher organisms since their structure is by far too complex to be, for instance, under the control of a single gradient. Thus, the complex patterns provide an extraordinary tool for the investigation of superpositions due to the preservation of the complete history.

References

[1] Ermentrout, B., Campbell, J. & Oster, G. (1986). A model for shell patterns based on neural activity. *The Veliger* **28**, 369-338.

[2] Gierer, A. & Meinhardt, H. (1972). A theory of biological pattern formation. *Kybernetik* **12**, 30-39.

[3] Lefever, R. (1968). Dissipative structures in chemical systems. *J.Chem.Phys.* **49**, 4977-4978.

[4] Meinhardt, H. (1982). *Models of biological pattern formation*. Academic Press:London.

[5] Meinhardt, H. (1984). Models for positional signalling, the threefold subdivision of segments and the pigmentation pattern of molluscs. *J. Embryol. exp. Morph.* **83** (Supplement), 289-311.

[6] Meinhardt, H. & Gierer, A. (1974). Applications of a theory of biological pattern formation based on lateral inhibition. *J. Cell Sci.* **15**, 321-346.

[7] Meinhardt, H. & Klingler, M. (1986). Pattern formation by coupled oscillations: The pigmentation pattern on shells of molluscs. *Lecture Notes in Biomath.* **71**, 184-198.

[8] Meinhardt, H. & Klingler, M. (1987). A model for pattern formation on the shells of molluscs. *J. theor. Biol.* **126**, 63-69.

[9] Neumann, D. (1958). Morphologische und experimentelle Untersuchungen über die Variabilität der Farbmuster auf der Schale von Theodoxus fluviatilis L. *Z. Morph. Ökol. Tiere* **48**, 349-411.

[10] Prigogine, I. and Lefever, R. (1948). Symmetry breaking instabilities in dissipative systems. II. *J. chem. Phys.* **48**, 1695-1700.

[11] Seilacher, A. (1972). Divaricate patterns in pelecypod shells. *Lethaia* **5**, 325-343.

[12] Seilacher, A. (1973). Fabricational noise in adaptive morphology. *Systematic Zool.* **22**, 451-465.

[13] Turing, A. (1952). The chemical basis of morphogenesis. *Phil. Trans. B.* **237**, 37-72.

23. TWO DIMENSIONAL WAVE PROPAGATION IN A MODEL OF DICTYOSTELIUM DISCOIDEUM

P.B. Monk

Department of Mathematics
University of Delaware
Newark, DE 19716, U.S.A.

We summarize a continuum model for cyclic AMP wave propagation in *Dictyostelium discoideum*, and show that this model predicts the correct amplitude and period for spiral waves. Particular attention is given to the numerical method used in our simulations.

1. **Aggregation in** *Dictyostelium*

Under favourable conditions, amoebae of the cellular slime mold *Dictyostelium discoideum* exist as free ranging individuals. However, if the food supply is exhausted, the amoebae change behaviour and begin to aggregate in response to intercellular signals to form mobile multicellular slugs. These slugs develop further into a spore mass supported by a stalk of dead cells. Germination of the spores completes the life cycle. It is the combination of single cell simplicity, together with elaborate developmental and signalling behaviour that explains the biologists' interest in *Dictyostelium*. For a more detailed description of the life cycle of *Dictyostelium* and references to the experimental literature the reader can consult Durston [15].

In a typical experiment to study aggregation fields, a uniform monolayer of cells is spread on an agar plate in a thin film of fluid [7]. A typical cell density is of the order of 10^5 cell per square centimetre. If the cells are starved, the monolayer at first exhibits no concerted motion. However, after about 8 hours, randomly located spontaneous pacemaker sites appear. These sites periodically release the chemical cyclic AMP, and if the density of cells is high enough these pulses of cyclic AMP diffusing away from the pacemaker sites evoke two responses from surrounding cells. First the surrounding cells transiently synthesize and release cyclic AMP reinforcing or **relaying** the cyclic AMP signal away from the pacemaker. Second, the cells orient themselves by the cyclic AMP wave and move for a short time towards the source of the signal. In this way, a wave of elevated cyclic AMP concentration moves out from the pacemaker across the aggregation field triggering cell movement [7]. Each pacemaker centre organizes an aggregation territory (which may change as other pacemakers appear or are extinguished) of cells that will ultimately form a slug. If the density of cells is too low the cyclic AMP signal is not relayed, but local cell movement and aggregation into small slugs can still occur [17].

We should note that the nature of the pacemaker centre is not well understood. It is thought to consist of a single cell or small group of cells that attain the ability to signal periodically before their peers in the surrounding field (cf. [14]). In [9] it is shown that the model for cyclic AMP synthesis to be presented in this paper predicts that a single cell can act as a pacemaker. The cells surrounding the pacemaker region are not thought to be capable of periodic cyclic AMP synthesis [13]. Instead, they are termed excitable or relay competent, since they can transiently produce a large pulse of cyclic AMP in response to a cyclic AMP signal so relaying the cyclic AMP signal across the aggregation field.

From our description of pacemakers it is clear that one wave form that is expected in aggregation fields is axisymmetric wave propagation centred on a pacemaker. However, a second wave form is also encountered in which the cyclic AMP wave forms a rotating spiral structure. This structure does not require an autonomous pacemaker at its centre [13]. The proportion of spiral waves to axisymmetric waves in an aggregation territory varies with investigators, but in some aggregation fields a large proportion of wave forms present may be spiral waves [7]. Durston [14] reports that the first aggregation territories have axisymmetric waves but that later territories with spiral or axisymmetric waves may appear. In addition he observes that aggregation territories revert to axisymmetric wave propagation before aggregation ceases [14].

Typical parameters for axisymmetric wave propagation are a wave speed of about 350 μm/min and a period of about 3-10 minutes although there is a great deal of variability in the results [1]. Spiral waves are generally reported to be slower with speeds from around 300 μm/minute for spirals soon after formation dropping to as low as 100 μm/minute for well developed spirals [18,32]. The total maximum concentration of cyclic AMP (intra and extracellular) is estimated to be about 10^{-6} M and to grow along the spiral [7]. In this paper we show that our model for cyclic AMP relay developed in [30,27,22,25] can predict spiral waves of the correct speed, period and geometry. Since cells move at about 20 μm/minute [32], which is slow compared to the speed of cyclic AMP propagation, we shall neglect cell movement in this paper, and will focus on the intercellular signalling mechanism active during aggregation.

In the next section, we give a brief presentation of the model for cyclic AMP synthesis and secretion. It has been shown previously that this model can account in a quantitative fashion for most of the experimental results concerning cyclic AMP relay obtained in perfusion and suspension experiments [22]. Then in section 3 we discuss in detail our numerical method for integrating the system of equations arising from the model. Finally, in section 4 we give some results from the model. Firstly we show that our numerical method is superior to the standard Euler method when applied to the cyclic AMP relay model. Secondly we show that the model supports spiral wave solutions that have characteristics very close to those of experimentally observed spirals.

2. The continuum model for an aggregation field

The model for cyclic AMP synthesis and signal transduction in *Dictyostelium* has been developed and investigated in several previous papers [30,27,22,26,9,21,23]. The model explains cyclic AMP relay as follows.

Extracellular cyclic AMP binds to receptors on the cell surface, thereby stimulating adenylate cyclase, the enzyme that catalyzes the synthesis of cyclic AMP, which leads to a rapid increase in the intracellular cyclic AMP concentration and the secretion of cyclic AMP. Concurrently, the binding of cyclic AMP to cell surface receptors causes an increase in the influx of calcium into the cytoplasm, and it is the resulting slower rise in the cytoplasmic calcium concentration that leads to a decrease to basal levels of cyclic AMP production and secretion via inhibition of adenylate cyclase. The combination of stimulation of adenylate-cyclase by external cyclic AMP and inhibition mediated via calcium results in a short burst of cyclic AMP secretion. For a more complete discussion of the biochemistry of the model the reader can consult [30,27,22].

The network of intracellular and extracellular reaction steps on which the upcoming equations are based is shown in Fig. 1 of [22]. The equations governing the dynamics of the intracellular network are derived in [27,22] on a per cell basis using singular perturbation reduction techniques to reduce the dimensionality of the system. Since the time scales used in this and previous papers are the same, the reductions carried out in [27,22] also hold here. Following the reduction there are three intracellular variables, x_2, x_3 and x_4, which represent the dimensionless concentrations of intracellular cyclic AMP, cytoplasmic calcium, and sequestered calcium, respectively. The intracellular dynamics are governed by the equations

$$\frac{dx_2}{d\tau} = V(x_1,x_3) + H(x_1 - R_H x_2) - f_2(x_2) - V_{PI}\left(\frac{x_2}{1 + x_2}\right)$$

$$\frac{dx_3}{d\tau} = R_1(x_1)\left(\frac{x_6}{K_F + x_6}\right) - V_{PU}\left(\frac{x_3}{K_{PU} + x_3}\right)$$
$$+ V_S\left(\frac{R_S + K_S x_2}{1 + K_S x_2}\right)(x_4 - x_3) - V_{SA}\left(\frac{x_3}{1 + x_3}\right) \qquad (1)$$

$$\frac{dx_4}{d\tau} = \frac{R_c}{1 - R_c}\left\{V_{SA}\left(\frac{x_3}{1 + x_3}\right) - V_S\left(\frac{R_S + K_S x_2}{1 + K_S x_2}\right)(x_4 - x_3)\right\},$$

where x_1 represents the dimensionless cyclic AMP concentration at the external surface of the cell. As in [22] we take the dimensionless secretion rate of cyclic AMP to be

$$f_2(x_2) = \begin{cases} V_{VP}\left(\dfrac{(x_2 - x_{sw})}{K_{VP} + (x_2 - x_{sw})}\right) + V_{sw} & \text{if } x_2 \geq x_{sw} \\ \left(\dfrac{V_{sw}}{x_{sw}}\right)x_2 & \text{if } x_2 < x_{sw} \end{cases} \qquad (2)$$

For computational purposes we use a smoothed version of this function, as was done in [22]. Finally, the dimensionless function that governs the cyclic AMP controlled calcium influx is given by

$$R_1(x_1) = V_I\left(\frac{R_I + K_I x_1}{1 + K_I x_1}\right)\left(\frac{R_{I1} + K_{I1}x_1}{1 + K_{I1}x_1}\right), \tag{3}$$

and the dimensionless rate of production of cyclic AMP is given by

$$V(x_1,x_3) = V_1\left(\frac{ax_1^2 + bx_1 + c}{(dx_1^2 + ex_1 + f)x_3 + gx_1^2 + hx_1 + 1}\right). \tag{4}$$

From the requirement that the steady state secretion rate of cyclic AMP be independent of the external cyclic AMP concentration (as demonstrated experimentally by Devreotes, et al. [6,8,10, 11,12]) we obtain a relationship between the parameters in the adenylate cyclase velocity V given by (4) and the parameters in the cytoplasmic calcium equation. To first order in H there are only four free parameters in (4), which we choose to be the scale factor V_1 and f, g and h. To simplify the expression for these relationships let

$$\omega = \frac{V_I K_{PU}\tilde{x}_6}{V_{PU}} \text{ and } \chi = \frac{(K_{PU} - \tilde{x}_6)V_I}{V_{PU}},$$

let

$$\alpha = \omega R_I R_{I1} \qquad \delta = 1 + \chi R_I R_{I1}$$

$$\beta = \omega(R_I K_{I1} + R_{I1}K_I) \qquad \mu = K_I + K_{I1} + \chi(R_I K_{I1} + R_{I1}K_I)$$

$$\gamma = \omega K_I K_{I1} \qquad \rho = (1 + \chi)K_I K_{I1},$$

and let

$$R_1 = \beta\delta - \alpha\mu \qquad R_4 = R_2^2 - R_1 R_3$$

$$R_2 = \gamma\delta - \rho\alpha \qquad R_5 = R_2 R_3 - \mu R_1 R_3/\delta + \rho R_1 R_2/\delta$$

$$R_3 = \gamma\mu - \beta\rho$$

Then the constants a,...,e are given in terms of f, g and h as follows:

$$a = (\gamma/\rho)(\rho\delta R_2 R_5/R_4 - \rho^2 R_1)f/(R_3\delta^2) + g$$

$$b = ((\alpha/\delta)(R_5/R_4) + (R_1/\delta^2))f + h$$

$$c = 1 + (\alpha/\delta)f$$

$$d = (\rho\delta R_2 R_5/R_4 - \rho^2 R_1)f/(R_3\delta^2)$$

$$e = (R_5/R_4)f$$

The numerical values of f,g,h and the other constants in (1) are given in

Table 1: Values of the dimensionless constants

α	1.0	ρ	1.0×10^{-1}	V_{PE}	5.0×10^{3}
f	1.5×10^{3}	R_c	9.8×10^{-1}	V_{PO}	1.0×10^{2}
g	3.0×10^{1}	R_I	1.5×10^{-1}	V_{SA}	1.2×10^{4}
h	3.0×10^{2}	V_{PU}	1.25	V_S	3.0×10^{2}
H	1.5×10^{-6}	R_{I1}	3.3×10^{-2}	V_{sw}	3.0×10^{-3}
K_F	5.0×10^{2}	R_S	2.0×10^{-1}	V_{VP}	5.0
K_I	2.0×10^{2}	R_e	2.0×10^{3}	V_1	1.2×10^{-1}
K_{I1}	1.1	R_H	2.0×10^{1}	$\bar{x}_6$	1.0
K_{PU}	4.5	V_I	2.2×10^{2}	$\tilde{x}_6$	1.0
K_S	3.3×10^{3}	V_{PI}	6.0×10^{-1}	x_{sw}	2.5×10^{-1}

Table 1. For more details of this derivation, and a general discussion of adaptation see [26].

The choice of parameter values is discussed in [27,22] and these values are given in Table 1. In simulating spiral waves, we use the same parameters as for simulations of axisymmetric waves, suspension and perfusion experiments, with the exception of V_1 which is allowed to vary in accordance with experimental evidence [22] (the value of V_1 is the same as is used when simulating perfusion experiments).

The equations governing extracellular variables in an aggregtion field were derived in [21,23]. This derivation entails numerous assumptions, the most important of which are as follows:

- There are sufficient cells present and the scale of phenomena of interest are such that the field of cells can be treated as a continuum.
- The calcium concentration in the region exterior to the cells is uniform in space and constant in time.
- The concentration of free extracellular phosphodiesterase is uniform in space and time.

Thus the aggregation field is treated as a distributed two-phase mixture in the sense of continuum mechanics. Under these hypotheses the equation governing extracellular cyclic AMP x_1 is

$$\frac{\partial x_1}{\partial \tau} = D\Delta x_1 + R_c\left(\frac{\rho}{1-\rho}\right)\left(R_H f_2(x_2) - V_{PO}\left(\frac{x_1^{\alpha}}{1+x_1^{\alpha}}\right)\right.$$
$$\left. + H R_H(R_H x_2 - x_1) - V_{PE}\left(\frac{x_1}{R_e + x_1}\right)\right), \qquad (5)$$

where D is the dimensionless diffusion constant for cyclic AMP and Δ is the Laplacian operator. The remaining equations are as given at (1). The constancy assumption for external calcium implies $x_6 = \bar{x}_6$ for some constant $\bar{x}_6$.

To model spiral wave propagation, we solve (1) and (5) on a square [-a,a] x [-a,a] with no flux boundary conditions on x_1. Parameters used are given in Table 1. The dimensional diffusion constant for cyclic AMP is taken to be 5 x 10^{-6} cm^2/sec [4,3], and the square is typically either 1 cm x 1 cm or 0.5 cm x 0.5 cm. Initial conditions for $x_1, \ldots, x_4$ are piecewise constant, and will be discussed more later. In order to relate dimensionless time τ to dimensional time t we have $t = 0.5\tau$ minutes.

3. Numerical methods

The model given by (1) and (5) cannot be solved analytically, and so we must resort to numerical techniques to investigate the solution. The *Dictyostelium* model has the general structure of a Hodgkin-Huxley system which is a coupled system of parabolic partial differential equations and ordinary differential equations of the following form. Let the dependent variables be represented by vector functions **u** and **v** such that $\mathbf{u}(\mathbf{x},t) \in \mathbb{R}^{m_1}$ and $\mathbf{v}(\mathbf{x},t) \in \mathbb{R}^{m_2}$. Let $\Omega \subset \mathbb{R}^2$ be a bounded domain with outward normal **n** and boundary Γ, then **u** and **v** satisfy

$$\mathbf{u}_t = D\Delta\mathbf{u} + \mathbf{f}(\mathbf{u},\mathbf{v}) \quad \text{in } \Omega \times (0,T] \tag{6}$$

$$\mathbf{v}_t = \mathbf{g}(\mathbf{u},\mathbf{v}) \quad \text{in } \Omega \times (0,T] \tag{7}$$

$$\mathbf{u}_n = 0 \quad \text{on } \Gamma \times (0,T]. \tag{8}$$

Here **f** and **g** are specified functions and D is a diagonal matrix of positive diffusion coefficients. In addition the initial conditions $\mathbf{u}(\cdot,0)$ and $\mathbf{v}(\cdot,0)$ are assumed known.

The problems of biological interest modelled by (6)-(8) tend to have similar characteristics. The numerical analysis is made easier by the fact that the spatial domain Ω is often simple (in our case a square) which allows the use of finite difference methods. Furthermore, the right hand sides **f** and **g** are smooth (often rational functions), and the Jacobian is usually easy to compute. The numerical analysis is complicated by the highly nonlinear nature of the problems. But the main complication is that we shall seek wave-like solutions that travel over the domain. These waves typically have sharp fronts in space which move through the domain in a way that is not a priori known. It is thus necessary either to have a very fine grid throughout the domain, or to use a space adaptive method at every time step. The design and programming of space adaptive schemes for parabolic problems is still a research topic, and so we elect to use a simple fine grid. The stiffness of the kinetics also implies stiff behaviour in time, however a time adaptive scheme is inappropriate here since the presence of all phases of a wave somewhere in the domain at each time implies that the time adaptive scheme would always select the same minimum time step. It is better to select the time step once and for all by experimentation than pay for the recurring overhead of a time adaptive method.

In the multidimensional case, extensive computations in two and three space dimensions have been carried out by Winfree and his co-workers [24,20] and by workers in Russia [38,29,35,37,36,28,34] on a particular case of equations (6)-(8). These investigations concern a version of a simplified model of nerve conduction called the Fitzhugh-Nagumo equations in which **u** and **v** are both scalar functions of space and time. In addition Tyson et al. [33] have investigated a model of chemical signalling in *Dictyostelium* (see [22,23] for a detailed review of other models for *Dictyostelium*). In all cases the Euler method is used to advance a finite difference version of (6)-(8). In our model, the equations (7) are stiff if **u** is held constant, and so we need to use a method with improved stability on this portion of the model. However, to avoid large matrix problems we prefer to keep the method for (6) explicit in the nonlinearity, while using a method that controls instability due to diffusion. Finally, to improve the speed of our computations, we will use a locally second order in time method.

Our method is as follows. We use a uniform mesh and a finite difference scheme to discretize (6)-(8) in space. For simplicity we use the five-point difference approximation of the Laplacian modified by a term of $O(\Delta t^2)$. Let $\Omega = [-a,a] \times [-a,a]$, let

$$\Delta x = \frac{2a}{N_x} \qquad \Delta y = \frac{2a}{N_y}$$

and let

$$x_j = -a + j\Delta x \quad 0 \le j \le N_x, \qquad y_k = -a + k\Delta y \quad 0 \le k \le N_y.$$

We seek a grid function $(\mathbf{u}^n_{j,k}, \mathbf{v}^n_{j,k})$, $0 \le j \le N_x$, $0 \le k \le N_y$, $n \le 0$ which approximates $(\mathbf{u}(x_j,y_k,t_n), \mathbf{v}(x_j,y_k,t_n))$. We denote the ith component of $\mathbf{u}^n_{j,k}$ by $\mathbf{u}^n_{(i)j,k}$.

In order to improve stability compared to the Euler method and provide a higher order time stepping method we combine the Alternating-Direction-Implicit or ADI method with the trapezoidal rule and modified Euler method [31]. Let I_{N_x} and I_{N_y} denote the $N_x \times N_x$ and $N_y \times N_y$ identity matrices. Let L_x be the $N_x \times N_x$ matrix

$$L_x = \frac{D}{(\Delta x)^2}\begin{pmatrix} -2 & 2 & 0 & \cdots & 0 \\ 1 & -2 & 1 & 0 & \cdots \\ 0 & \ddots & \ddots & \ddots & 0 \\ \cdots & 0 & 1 & -2 & 1 \\ 0 & \cdots & 0 & 2 & -2 \end{pmatrix}$$

and let L_y be the corresponding $N_y \times N_y$ matrix with Δx replaced by Δy. Now let us enumerate the unknowns in the natural grid ordering by columns so that the $(N_x + 1)(N_y + 1)$ dimensional vector $\mathbf{U}^n_{(i)}$, $1 \le i \le m_1$ is defined by

$$\left[\mathbf{U}^n_{(i)}\right]_{(j-1)(N_y+1)+k} = \mathbf{u}^n_{(i)j,k} \qquad 0 \le j \le N_x, \quad 0 \le k \le N_y,$$

where $\left[\mathbf{U}^n_{(i)}\right]_\ell$ denotes the ℓth component of $\mathbf{U}^n_{(i)}$. If A and B are matrices, let $A \otimes B$ denote the standard tensor product of A and B (cf. [5] for a discussion of the tensor product and its properties).

Now combining the ADI method, the trapezoidal and the modified Euler method we first compute an approximation to $\mathbf{u}^{n+1/2}$. For $1 \le i \le m_1$, $\mathbf{U}^{n+1/2}_{(i)}$ satisfies:

$$((I_{N_x} - \frac{\Delta t}{4} L_x) \otimes I_{N_y})(I_{N_x} \otimes (I_{N_y} - \frac{\Delta t}{4} L_y))\mathbf{U}^{n+1/2}_{(i)}$$

$$= ((I_{N_x} + \frac{\Delta t}{4} L_x) \otimes I_{N_y})(I_{N_x} \otimes (I_{N_y} + \frac{\Delta t}{4} L_y))\mathbf{U}^n_{(i)} + \frac{\Delta t}{2} \mathbf{F}^n_{(i)} \tag{9}$$

where the vector $\mathbf{F}^n_{(i)}$ is defined for $1 \le i \le m_1$ by

$$\left[\mathbf{F}^n_{(i)}\right]_{(j-1)(N_y+1)+k} = f_{(i)}(\mathbf{u}^n_{j,k}, \mathbf{v}^n_{j,k}) \qquad 0 \le j \le N_x, \quad 0 \le k \le N_y.$$

Then we advance $\mathbf{v}$ by a full time step using the trapezoidal rule for $0 \le i \le N_x$ and $0 \le j \le N_y$

$$\mathbf{v}^{n+1}_{i,j} = \mathbf{v}^n_{i,j} + \Delta t \mathbf{g}\left(\mathbf{u}^{n+1/2}_{i,j}, \frac{\mathbf{v}^{n+1}_{i,j} + \mathbf{v}^n_{i,j}}{2} \right). \tag{10}$$

Now let $\mathbf{F}^{n+1/2}_{(i)}$ be defined by

$$\left[\mathbf{F}^{n+1/2}_{(i)}\right]_{(j-1)N_y+k} = f_{(i)}\left(\mathbf{u}^{n+1/2}_{j,k}, \frac{\mathbf{v}^{n+1}_{j,k} + \mathbf{v}^n_{j,k}}{2} \right) \quad 0 \le j \le N_x,$$
$$0 \le k \le N_y.$$

Finally, we apply the modified Euler ADI method to obtain $\mathbf{U}^{n+1}_{(i)}$. For $1 \le i \le m_1$

$$((I_{N_x} - \frac{\Delta t}{2} L_x) \otimes I_{N_y})(I_{N_x} \otimes (I_{N_y} - \frac{\Delta t}{2} L_y))\mathbf{U}^{n+1}_{(i)}$$

$$= ((I_{N_x} + \frac{\Delta t}{2} L_x) \otimes I_{N_y})(I_{N_x} \otimes (I_{N_y} + \frac{\Delta t}{2} L_y))\mathbf{U}^n_{(i)} + \Delta t \mathbf{F}^{n+1/2}_{(i)} . \tag{11}$$

This method has an improved local order of accuracy in time compared to Euler's method, since by using Taylor's theorem, we can show that the local truncation error is second order in time and space. The global accuracy is likely to be second order in space and time at least for linear problems. For nonlinear problems there may be a loss of order as is encountered for example in the Peaceman-Rachford ADI procedure [19].

The stability of the method is not known. The ADI procedure should stabilize the method to diffusion. For the kinetic terms, the stability in general should be dictated by the modified Euler term, but in our case the **v** equations are stiff and the trapezoidal rule should give improved stability and allow a long time step.

The solution of (9) or (11) is a two step procedure. Since the steps are identical in each case we shall only discuss (9) in detail. At each time step, we introduce two auxiliary vector variables $\mathbf{U}^*$ and $\tilde{\mathbf{U}}$ of dimension $(N_x + 1)(N_y + 1)$. Successively for each i, $1 \le i \le m_1$, we compute $\mathbf{U}^*$ and $\tilde{\mathbf{U}}$ by solving

$$(I_{N_x} \otimes (I_{N_y} - \frac{\Delta t}{4} L_y))\mathbf{U}^* = ((I_{N_x} + \frac{\Delta t}{4} L_x) \otimes I_{N_y})\mathbf{U}^n_{(i)} \qquad (12)$$

$$(I_{N_x} \otimes (I_{N_y} - \frac{\Delta t}{4} L_y))\tilde{\mathbf{U}} = \frac{\Delta t}{2}\mathbf{F}^n_{(i)}. \qquad (13)$$

We can compute the solution of each of (12) and (13) by solving N_x independent linear systems in which the matrix is $(I_{N_y} - \Delta t\ L_y/4)$. This matrix is tridiagonal, and hence each tridiagonal system can be solved in $O(N_y)$ operations. Once $\mathbf{U}^*$ and $\tilde{\mathbf{U}}$ have been computed, we can compute $\mathbf{U}^{n+1/2}_{(i)}$ by solving the following system:

$$((I_{N_x} - \frac{\Delta t}{4} L_x) \otimes I_{N_y})\mathbf{U}^{n+1/2}_{(i)} = (I_{N_y} + \frac{\Delta t}{4} L_y))\mathbf{U}^* + \tilde{\mathbf{U}}. \qquad (14)$$

To solve (14) involves solving N_y independent linear systems in which the matrix is the tridiagonal system $(I_{N_x} - \Delta t L_x/4)$. Thus the solution of (9) and (11) requires $O(N_x N_y)$ operations once $\mathbf{F}^n_{(i)}$ and $\mathbf{F}^{n+1/2}_{(i)}$ are computed. The computation of these right hand side vectors requires two evaluations of **f** at all the mesh points, compared to a single evaluation at all the grid points for Euler's method. In addition, the trapezoidal rule in (10) requires at least one evaluation of **g** at all the mesh points, and at least one evaluation of the Jacobian $\partial\mathbf{g}/\partial\mathbf{v}$ at all mesh points together with the solution of at least one $m_2 \times m_2$ matrix problem at each mesh point. For the time steps we use with our model, the Newton method usually does at least two iterations at each grid point, and rarely more than three. However the solution of the nonlinear equations in (10) is often a very time-consuming part of the overall stepping algorithm consuming about 60% of the overall computing time for our model. Of course the Euler method requires the evaluation of **g** at all mesh points once per iteration.

Recently, some papers on related methods have appeared in the literature. The Peaceman-Rachford ADI scheme for nonlinear equations is analyzed in [19]. This method involves the solution of implicit equations along each x and y grid line. Thus at each time step one must solve $(m_1 + m_2)N_x$ nonlinear problems each with $(m_1 + m_2)N_y$ equations and unknowns, and $(m_1 + m_2)N_y$ problems each with $(m_1 + m_2)N_x$ equations and unknowns. In each case the

Table 2: A table of relative ℓ_2 error and relative maximum norm error (in parentheses) in u against time step size for (1)-(5) with excitable kinetics and D = 0.

Δt(minutes)	Euler	Trap.-Mod.-Euler
0.05	> 0.9	4.55×10^{-2} (5.06×10^{-2})
0.025	> 0.9	9.51×10^{-3} (1.10×10^{-2})
0.01	> 0.9	1.93×10^{-3} (2.13×10^{-3})
0.005	> 0.9	3.49×10^{-4} (4.01×10^{-4})
0.00005	2.40×10^{-4} (2.85×10^{-4})	5.34×10^{-8} (8.51×10^{-8})

Jacobian is banded with band width $O(m_1 + m_2)$. The method is unconditionally stable and can be second order (although on nonlinear problems the method may lose precision [19]). Since our method is semi-explicit, we have to solve N_xN_y nonlinear problems, each of dimension m_2^2. This is less work per step, but the stability properties of our method are not likely to be as good as for Peaceman-Rachford. A second fully implicit method multistep method is proposed by Bramble et al. [2] and suggests using an ADI scheme as a preconditioner in the nonlinear solver. Their method concerns a more complex problem than ours (a nonlinear diffusion coefficient is used), but specializing their method to our case would result in an interesting method that could be as high as third order in time. This method would seem to be a good candidate for a more advanced technique for handling reaction diffusion equations.

4. Numerical results

4.1. *A comparison of methods when D = 0*

We compare the Euler and trapezoidal-modified Euler scheme (our scheme with D = 0) applied to the ordinary differential equations obtained from (1) and (5) with D = 0. The initial condition is chosen to excite a large response from the system and we integrate until T_{max} = 2 minutes by which time $u = x_1$ has decayed below one per cent of its maximum value. In this case the initial condition is

$$u = x_1 = 5.\ \times 10^{-2} \qquad v_1 = x_2 = 2.424 \times 10^{-1}$$

$$v_2 = x_3 = 1.264 \times 10^{-2} \qquad v_3 = x_4 = 1.904 \times 10^{-2}$$

Each of the variables v_1, v_2 and v_3 are at their steady state values, but u is elevated from its steady state value. The results for each method are compared to a solution computed via Gear's stiffly stable method with a very low error criterion [16].

As can be seen from the results in Table 2, the stiffness of the kinetic equations forces an extremely small step size to be used for Euler's method. A detailed examination of the solution shows that when Δt ≥ 0.0005, Euler's

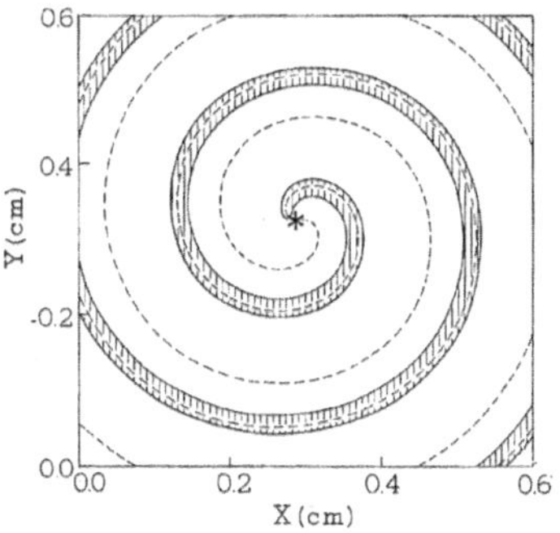

Figure 1. A single spiral wave rotating in a 0.6 cm x 0.6 cm region. This plot shows both cytoplasmic calcium and external cyclic AMP concentrations. The hatched area shows the region where the external cyclic AMP concentration is above 1×10^{-8} M. The dashed line marks the contour on which the cytoplasmic calcium concentration is 0.1×10^{-6} M. The spiral is rotating counter-clockwise around the point marked with a *. (Reproduced from [21]).

method entirely fails to integrate the transient excitable response. The Euler solution returns directly to steady state, and the physically important transient amplification aspect of the response is missed. By contrast, the semi-implicit trapezoidal-modified Euler method always approximates the transients (at least up to $\Delta t = 0.05$). This indicates that the use of a partially implicit method has stabilized the marching scheme when applied to our model.

Since the solution of the ordinary differential equation problem ($D = 0$) is a special uniform solution of the full partial differential equation model ($D \neq 0$), we know that the data in Table 2 give an upper limit on the time step necessary for integrating the partial differential equation to a given accuracy. Since the time step necessary for Euler's method is two orders of magnitude lower than for the ADI-trapezoidal-modified Euler method, we judge Euler's method inappropriate for our problem and have not tried Euler's method on the partial differential equation model.

4.2. *Spiral waves in the Dictyostelium model*

To compute spirals we use (1) and (5) with $D \neq 0$. Since spiral waves do not require a periodic pacemaker centre [13], we use parameters that result in an excitable but not oscillatory field of cells. All the parameters have the values given in Table 1 and are independent of time and position. Note that in this case there is a uniform steady state for (1) and (5) that corresponds to the steady state of the excitable system in the absence of diffusion. This steady state is stable, but the medium is also excitable so that a sufficient perturbation may trigger a wave moving across the field.

For the spiral shown in Fig. 1, we discretize a 1 cm x 1 cm domain using a step size of 0.005 cm and a time step of 0.05 minutes. These values were arrived at by observing the effects of halving the space and time steps until the computed solution (particularly x_1) stopped changing appreciably.

A simple way to obtain spiral waves numerically is to use as initial data a single detached plain wave with a cross-section that approximates a travelling pulse. A spiral will grow from these intial conditions and settle down to a characteristic rotation period and spatial structure as in Fig. 1. The period of this spiral wave is about 6.1 minutes, the wave speed far from the centre is 225 μm/minute, and the spatial wavelength is about 1400 μm. These figures are for a spiral wave that has settled to a steady rotation rate and form, and thus should be compared to the late time figures measured by Gross et al. [18] discussed in the introduction. Clearly the periods and wavelengths computed by us are in reasonable agreement with experimental results. Our results (not shown here) show that the amplitude of the total cyclic AMP concentration predicted by the model is also in good agreement with measured results, but that the predicted half-width of the wave is not as large as in Devreotes et al.'s [7] experiments. This may be because the experimental medium has slight non-uniformities in the density of the cells as well as developmental age of the cells. Numerical experiments suggest that these non-uniformities tend to broaden the peak of the cyclic AMP wave.

Much more complex, yet stable patterns of cyclic AMP waves can be generated using a suitable choice of initial data. For instance in [23] we show a persistent two-armed spiral that alternately connects and disconnects the two arms as it rotates about a central point with a period of 12.2 minutes. Limited numerical experimentation suggests that the single spiral discussed above has a unique geometry and period since perturbations of the spiral are damped out and the spiral returns to its pre-perturbation form. The same may not be true of the two-armed spiral. It seems that a variety of closely separated co-rotating centres will produce double-armed spirals.

References

[1] Alcantara, F. & Monk, M. (1974). Signal propagation during aggregation in the slime mold *Dictyostelium discoideum*. *J. Gen. Microbiol.* **85**, 321-334.

[2] Bramble, J.H., Ewing, R.E. & Li, G. (1989). Alternating direction multistep methods for parabolic problems - iterative stabilization. *SIAM J. Numer. Anal.* **26**, 904-919.

[3] Cohen, M.H. & Robertson, A. (1971). Chemotaxis and the early stages of aggregation in cellular slime molds. *J. Theor. Biol.* **31**, 119-130.

[4] Cohen, M.H. & Robertson, A. (1971). Wave propagation in the early stages of aggregation of cellular slime molds. *J. Theor. Biol.* **31**, 101-118.

[5] Curtis, C.W. & Reiner, I. (1966). *Representation theory of finite groups and associative algebras.* John Wiley, New York.

[6] Devreotes, P.N., Derstine, P.L. & Steck, T.L. (1979). Cyclic 3',5' AMP relay in *Dictyostelium discoideum* I. A technique to monitor responses to controlled stimuli. *J. Cell. Biol.* **80**, 291-299.

[7] Devreotes, P.N., Potel, M.J. & MacKay, S.A. (1983). Quantitative analysis of cyclic AMP waves mediating aggregation in *Dictyostelium discoideum*. *Dev. Biol.* **96**, 405-415.

[8] Devreotes, P.N. & Steck, T.L. (1979). Cyclic 3',5' AMP relay in *Dictyostelium discoideum* II. Requirements for the initiation and termination of the response. *J. Cell. Biol.* **80**, 300-309.

[9] De Young, G., Monk, P.B. & Othmer, H.G. (1988). Pacemakers in aggregation fields of *Dictyostelium discoideum*. Does a single cell suffice? *J. Math. Biol.* **26**, 486-517.

[10] Dinauer, M.C., MacKay, S.A. & Devreotes, P.N. (1980). Cyclic 3',5' AMP relay in *Dictyostelium discoideum* III. The relationship of cAMP synthesis and secretion during the cAMP signaling response. *J. Cell. Biol.* **86**, 537-544.

[11] Dinauer, M.C., Steck, T.L. & Devreotes, P.N. (1980). Cyclic 3',5' AMP relay in *Dictyostelium discoideum* IV. Recovery of the cAMP signaling response after adaptation to cAMP. *J. Cell. Biol.* **86**, 545-553.

[12] Dinauer, M.C., Steck, T.L. & Devreotes, P.N. (1980). Cyclic 3',5' AMP relay in *Dictyostelium discoideum* V. Adaptation of the cAMP signaling response during cAMP stimulation. *J. Cell. Biol.* **86**, 554-561.

[13] Durston, A.J. (1973). *Dictyostelium discoideum* aggregation fields as excitable media. *J. Theor. Biol.* **42**, 483-504.

[14] Durston, A.J. (1974). Pacemaker mutants of *Dictyostelium discoideum*. *Dev. Biol.* **38**, 308-319.

[15] Durston, A.J. (1977). The control of morphogenesis in *Dictyostelium discoideum*. In Horgan, P.A. & O'Day, D.H. (eds.): *Eucaryotic Microbes as Model Developmental Systems*, pp. 294-321. Marcel Dekker, New York.

[16] Gear, C.W. (1971). *Numerical initial value problems in ordinary differential equations*. Prentice Hall, Englewood Cliffs.

[17] Gingle, A.R. (1976). Critical density for relaying in *Dictyostelium discoideum* and its relation to phosphodiesterase secretion into the extracellular medium. *J. Cell. Sci.* **20**, 1-20.

[18] Gross, J.D., Peacey, M.J. & Trevan, D.J. (1976). Signal emission and signal propagation during early aggregation in *Dictyostelium discoideum*. *J. Cell. Sci.* **22**, 645-656.

[19] Hundsdorfer, W.H. & Verwer, J.G. (1989). Stability and convergence of the Peaceman-Rachford ADI method for initial- boundary value problems. *Math. Comp.* **53**, 81-101.

[20] Lugosi, E. & Winfree, A.T. (1988). Simulation of wave- propagation in three dimensions using Fortran on the CYBER 205. *J. Comp. Chem.* **9**, 689-701.

[21] Monk, P.B. & Othmer, H.G. (1989). Wave propagation in aggregation fields of *Dictyostelium discoideum*. In Othmer, H.G. (ed.): *Lects. on Math. in Life Sci.* American Math. Soc., Providence, R.I.

[22] Monk, P.B.& Othmer, H.G. (1989). Cyclic AMP oscillations in suspensions of *Dictyostelium discoideum*. *Phil. Trans. R. Soc. Lond.* **323(1215)**, 185-224.

[23] Monk, P.B. & Othmer, H.G. Wave propagation in aggregation fields of the cellular slime mold *Dictyostelium discoideum*. In preparation.

[24] Nandpurkar, P.J. & Winfree, A.T. (1987). A computational study of twisted linked scroll waves in excitable media. *Physica D* **29**, 69-83.

[25] Othmer, H.G., Dunbar, S.R. & Alt, W. (1988). Models of dispersal in biological systems. *J. Math. Biol.* **26**, 263-298.

[26] Othmer, H.G. & Monk, P.B. (1988). Concentration waves in aggregation fields of a cellular slime mold. In Ricciardi, L. (ed.): *Biomathematics and Related Computational Problems*, pp. 381-398. Kluwer Academic Publishers, Dordrecht.

[27] Othmer, H.G., Monk, P.B. & Rapp, P.E. (1985). A model for signal relay and adaptation in *Dictyostelium discoideum* Part II. Analytical and numerical results. *Math. Biosciences* **77**, 77-139.

[28] Panfilov, A.V., Rudenko, A.N. & Krinsky, V.I. (1986). Vortical rings in three dimensional active media with diffusion by two components. *Biofizika* **31(5)**, 850-854.

[29] Pertsov, A.M., Ermakova, A. & Panfilov, A.V. (1984). Rotating spiral waves in a modified Fitzhugh-Nagumo model. *Physica D* **14**, 117-124.

[30] Rapp, P.E., Monk, P.B. & Othmer, H.G. (1985). A model for signal relay and adaptation in *Dictyostelium discoideum* Part I. Biological processes and the model network. *Math. Biosciences* **77**, 35-78.

[31] Sod, G. (1985). *Numerical Methods for Fluid Dynamics: Initial and Initial Boundary-value Problems.* Cambridge University Press, New York.

[32] Tomchik, K.J. & Devreotes, P.N. (1981). Adenosine 3',5'- monophosphate waves in *Dictyostelium discoideum*: A demonstration by isotope dilution-fluorography. *Science* **212**, 443-446.

[33] Tyson, J.J. & Keener, J.P. (1988). Singular perturbation theory of traveling waves in excitable media (a review). *Physica D* **32**, 327-361.

[34] Yermakova, Y.A. & Pertsov, A.M. (1987). Interaction of rotating spiral waves with a boundary. *Biophysics* **31(5)**, 932-940.

[35] Yermakova, Y.A., Krinskii, V.I., Panfilov, A.V. & Pertsov, A.M. (1986). Interaction of helical and flat periodic autowaves in an active medium. *Biophysics* **31(2)**, 348-354.

[36] Zykov, V.S. (1986). Cycloid circulation of spiral waves in an excitable medium. *Biophysics* **31(5)**, 940-944.

[37] Zykov, V.S. (1987). Kinematics of the non-steady circulation of helical waves in an excitable medium. *Biophysics* **32(2)**, 365-369.

[38] Zykov, V.S. & Petrov, A.A. (1977). Role of the inhomogeneity of an excitable medium in the mechanisms of self-sustained activity. *Biofizika* **22(2)**, 300-306.

24. PATTERNS OF SPORE FORMATION IN NEUROSPORA CRASSA AND THEIR SIMULATION WITH A CELLULAR AUTOMATON

A. Deutsch*, L. Rensing* and A. Dress**

*Fachbereich Biologie der Universität
Postfach 330440, D-2800 Bremen, FRG

**Fakultät für Mathematik der Universität
Universitätsstrasse, D-4800 Bielefeld, FRG

1. Introduction

We have analyzed the spatial pattern formation of macroconidia in *Neurospora crassa* (bd-strain). Macroconidia are the vegetative spores produced during vegetative growth of a mycelium as shown in Fig. 1. A fungal mycelium is initiated by a spore (or a small portion of mycelium) in the centre of a nutrient agar plate. Hyphae grow and branch in radial directions from this centre (Fig. 2). They can differentiate into aerial hyphae which give rise to spores (macroconidia). The well known concentric ring pattern of spore distribution is due to an internal clock mechanism (circadian rhythm) controlling spore differentiation (see [5,6] for review). Various facts support the hypothesis that each hyphal filament contains an internal clock. Thus the fungal mycelium may represent an ensemble of coupled or noncoupled oscillators [2,14].

Surprisingly this distinct pattern of spore distribution has not been examined systematically so far. We think that the differentiation pattern of *Neurospora* is worth studying with the aim of a better understanding of pattern formation in higher organisms.

Neurospora has no genetically determined form or size, instead it is able to fill a given "medium space". This flexible form is probably an adaptation to the varying habitat structures (e.g. soil). The flexibility of form and size relies on a strategy of a stepwise growth and branching process. At each time step the fungus responds to its environmental conditions - for example medium composition and the "influence" (e.g. inhibiting influence) of already existing hyphae. On the basis of this response the fungus grows or branches at suitable sites along the mycelium. The growth and branching response seems to be governed by some "rules". The "new" mycelium and the changed medium composition (e.g. reduced contents of nutrients due to the preceding growth process) constitute the basis of the next response. One of the rules probably defines a threshold of the nutrient concentration for hyphal growth. The growth process is continued until a stationary state is reached at an overall nutrient concentration below this threshold. This process (the creation of a

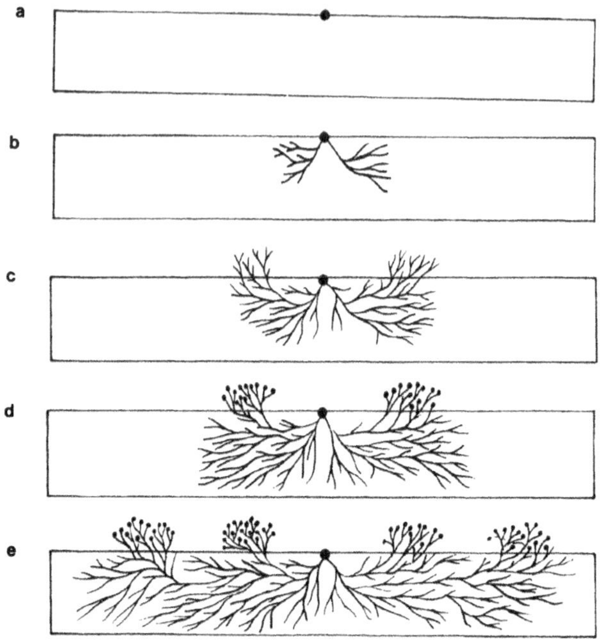

Figure 1. The development of *Neurospora crassa*. Schematic section through a glass Petri dish containing "growth medium" in agar (experimental conditions see "Materials and Methods").
a. Inoculation in the centre of the Petri dish.
b. Vegetative hyphae grow and branch inside the agar medium.
c. After a characteristic time interval, vegetative hyphae at the growing front start to branch into aerial hyphae that grow out of the medium.
d. The aerial hyphae differentiate into macroconidia (spores). After a typical time interval branching into aerial hyphae stops. Vegetative growth and branching continues exclusively inside the agar medium.
e. *Neurospora* culture after two sporulation periods; the period length is approximately 22 hours, characteristic for the circadian rhythmicity.

new state from an old state by some rules, definition of the new state as old, application of the same rules to this state, reaching a new state, and so on) corresponds exactly to a mathematical recursion process.

The remarkable spatial order expressed by concentric ring formation has long been known from studies of other fungi, e.g. *Nectria cinnabarina* [11,12]. Under different growth conditions (especially concerning medium composition) *Neurospora* shows other patterns of spore distribution apart from this ring pattern: mainly "radial" structures (Fig. 3a). Two or more growth initiation sites lead to various "inhibition patterns" (Fig. 3b). Apparently, *Neurospora* does not show spiral patterns of spore distribution as they are known from other fungi [1].

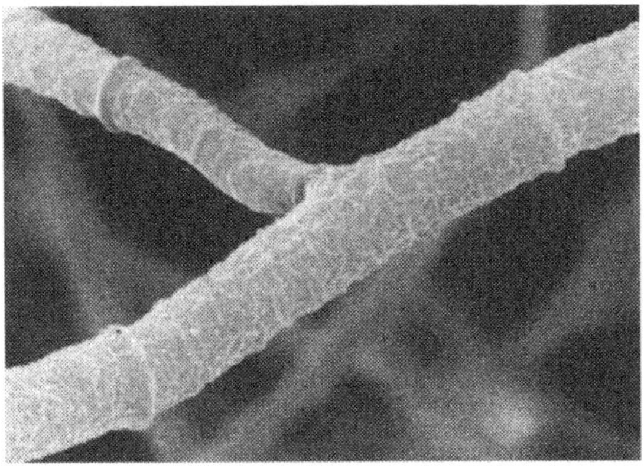

Figure 2. Scanning electron micrograph of hyphal ramification (hyphal diameter approximately 2 μm).

In order to explain the different patterns of spore formation in *Neurospora*, we considered various hypotheses. These hypotheses were modelled by using a cellular automaton that is able to simulate growth and branching of the *Neurospora* mycelium. States of the automaton represent patterns observed in the biological experiments. The essential assumptions of the model are a limited "source of nutrient" (resource) and an "inhibitory" substance. The concentration of nutrients decreases whereas the inhibitor is produced during growth and correlates negatively with further development.

2. Materials and Methods

We used the bd-A strain of *Neurospora crassa* exclusively. The term bd refers to the band mutation causing very distinct conidial bands. These banding patterns mark the circadian rhythm.

After three to seven days of growth on Horowitz-slants [7] in constant light at 25°C, the spores were suspended in sterile H_2O and collected after filtration through glasswool. 100-200 μl of the conidial suspension (≈ 13 x 10^5 spores) were deposited in a plastic Petri dish (90 mm diameter) containing 25 ml growth medium consisting of 2% Vogel salts [10] and 2% saccharose. The pH was adjusted to 5.8.

After 38 hours of growth in constant light, discs (4 mm diameter) were cut from the mycelium by using a cork bore. Each disc was transferred to a glass Petri dish (20 cm diameter) containing 200 ml growth medium (0.4% saccharose and varying amounts of Vogel salts and agar). The Petri dishes were held in constant dark and 25°C thereafter. After 3-4 days, the pattern of spore distribution in the Petri dish was photographed.

3. Patterns

When grown on Petri dishes as described above, spores emerge on top of the vegetative mycelium exactly at those sites where vegetative hyphae have

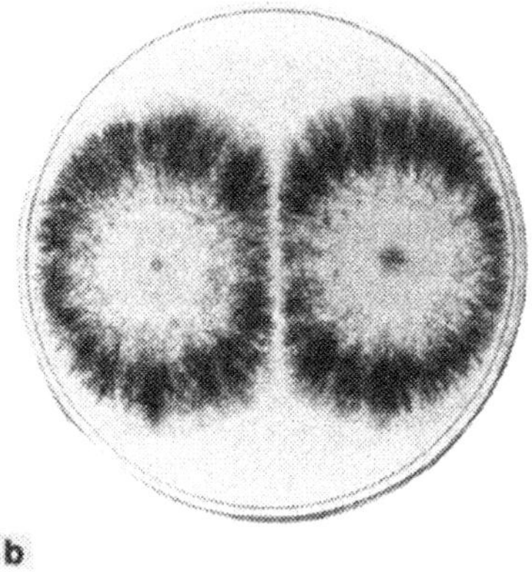

Figure 3. a. "Radial patterns" in the spore distribution of *Neurospora*. Experimental conditions as described under "Material and Methods" with 6% Vogel salts and 1.5% agar in the growth medium.
b."Inhibition structures" in the spore distribution of *Neurospora*. Experimental conditions as described under "Materials and Methods" with 2% Vogel salts and 1.5% agar in the growth medium.

branched into aerial hyphae. The spore distribution thus reflects the underlying mycelial structure. Consequently, the key to the understanding of the pattern of spore formation can be found on the level of aerial hyphae development. This implies that we must examine processes that determine growth and branching of the hyphae.

The mycelial development is the result of a frequent organism-environment interaction. We believe that the particular environment of a developing hypha in a mycelium is mainly determined by the neighbouring hyphae. In order to test this assumption a series of experiments on the influence of additional hyphae on the development of a mycelium was performed. Inoculating a Petri dish with two (or more) discs is giving rise to two (or more) mycelia growing towards and eventually meeting each other. The contact zone is characterized by a lack of spores (**"inhibition pattern"**, Fig. 3b). The suppressed sporulation is probably due to competition over nutrients and to inhibitors (see below). Similar patterns are known from the bacterium *Proteus* [3,4,9].

Under certain growth conditions **"radial structures"** in the distribution of spores appear combined with concentric rings (Fig. 3a). The radial structure is characterized by two features that can be found in many other biological structures and even in chemical and physical systems. One feature is the "start of the pattern", namely the splitting of an initially homogeneous structure into some "bunches". In the *Neurospora* mycelium, the bunches correspond to bundles of hyphae: visible in the darker patches in each of the single spore rings. The other feature is the continued growth of the bunches. The bundles of hyphae can be followed by connecting corresponding darker patches of two successive rings.

4. Modelling fungal development

Our model of *Neurospora* spore formation follows the lines of a cellular automaton approach introduced by John v. Neumann in 1966 [8]. The idea to use a cellular automaton for modelling fungal development was fostered by the hope that the macroscopic order (i.e. the spore distribution patterns) could be derived from local interactions between hyphae in microscopic surroundings. Particularly the following arguments and assumptions seemed to justify that hope:

- The spatial structure of the fungus results from a combination of just three different "modules" - vegetative hypha, aerial hypha and spore.
- Each module seems to contain an "internal clock". The clock is responsible for the periodical branching of vegetative hyphae into aerial hyphae. All clocks have the same period length and are nearly synchronous (a small phase gradient between new and old parts of a mycelium was observed by Dharmananda & Feldman [2]).
- The decision for differentiation is made at the site of differentiation. The decision is not genetically fixed, but rather determined by the oscillator's phase at the decision- point.
- *Neurospora*'s development is successive in a sense that corresponds to the state development of a cellular automaton. Assuming discrete time steps ($t = 0,1,2,3,\ldots$) only the mycelium (together with the medium) at time t ("generation t") defines the boundary conditions for the development of "generation t+11 ($t = 1,2,3,\ldots$).
- The medium's influence on growth and branching of the culture is represented by two parameters, nutrient concentration (r) and growth probability (ε). The nutrient concentration decreases during growth (or branching). At the same time, an inhibitory substance (i) is produced. This correlates negatively with further growth processes.

5. Definition of the cellular automaton

The definition of the (vector-) cellular automaton can be given here only in a very abbreviated form. A structogram is shown in Fig. 4.

(i) "Space consists of a two dimensional quadratic grid with n^2 cells (n = no. of lines = no. of columns) that are marked by (k,1), (k,1 = 1,2,3,...,n); typically we take n = 400.

(ii) For each cell (k,1) a vector (s,Φ,r,i) is defined; s is the state of (k,1) (see (iii)), Φ the phase of the local oscillator, r the nutrient concentration and i the inhibitor concentration in the cell.

(iii) The state set consists of the elements U, V and A. U symbolizes

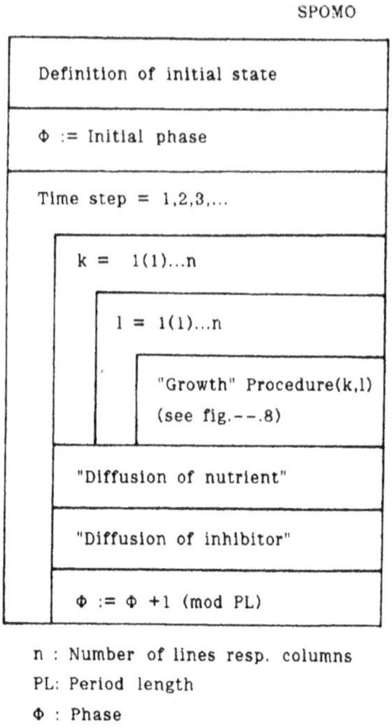

Figure 4. Structogram of the cellular automaton (SPOMO) described in the text.

"unoccupied", V "vegetative" and A "aerial". Please note that this definition makes possible the "projection" of the three dimensional system "spore pattern formation" into a two dimensional space.

(iv) A cell's neighbours are defined by the Moore- neighbourhood.

(v) The state development of the cellular automaton is governed by two processes, a "growth" and a "diffusion" process. At each timestep the growth process occurs and the "diffusion" takes place later.

The growth algorithm is shown in Fig. 5. The diffusion processes of nutrients and inhibitor are simulated by simple "averaging procedures" in the model: the concentration of nutrients and inhibitor respectively are substituted by the average of the corresponding concentrations of their Moore neighbours. At each step the diffusion processes are repeated for a fixed number of times which may be different for nutrients and inhibitor corresponding to different diffusion rates of the two.

6. Results and discussion

Some states of the cellular automaton show typical patterns: concentric rings (Fig. 6a) and "inhibitory" structures (Fig. 6b) that correspond to suppressed development of aerial hyphae (and hence sporulation) in the "contact areas" of the two mycelia. Although decreasing nutrient concentration has inhibitory effects, the assumption of an additional inhibitory substance seems to be

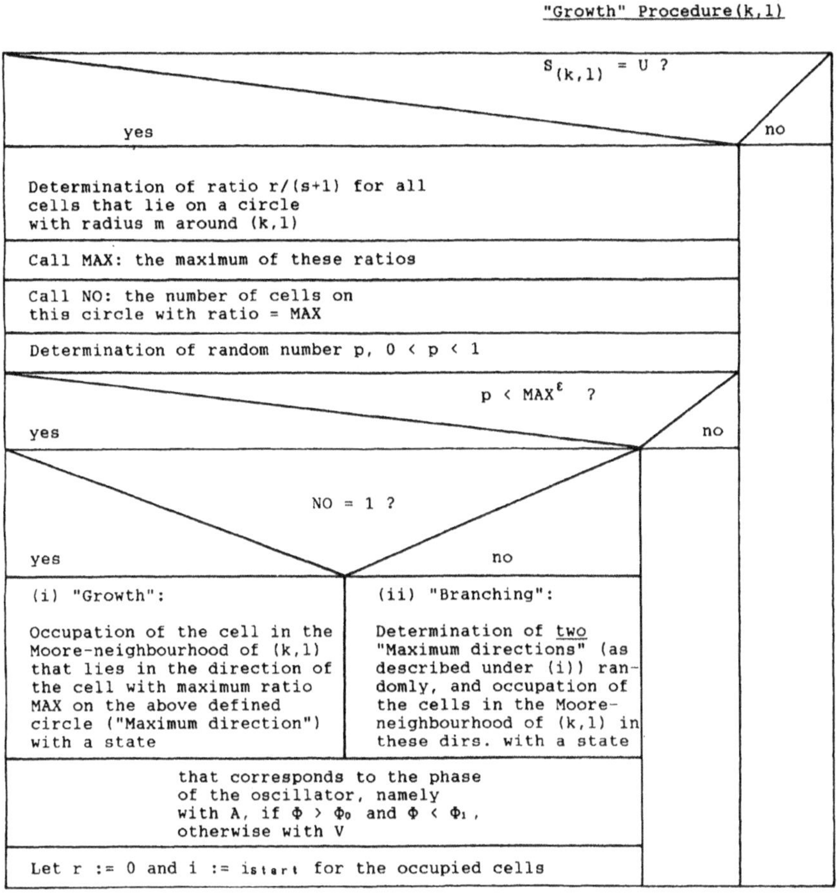

Explanation of symbols

S: state variable (can be either U, V or A)
r: concentration of nutrient
i: concentration of inhibitor
ε: growth probability
Φ: phase of the oscillator
($\Phi > \Phi_0$ and $\Phi < \Phi_1$: time of "branching into aerial hyphae")

Figure 5. Structogram of the "growth procedure" of the cellular automaton SPOMO.

essential for the realization or simulation of the patterns. At present, we are working on parameter configurations that might enable the automaton to produce radial structures as reported for the *Neurospora* spore distribution patterns.

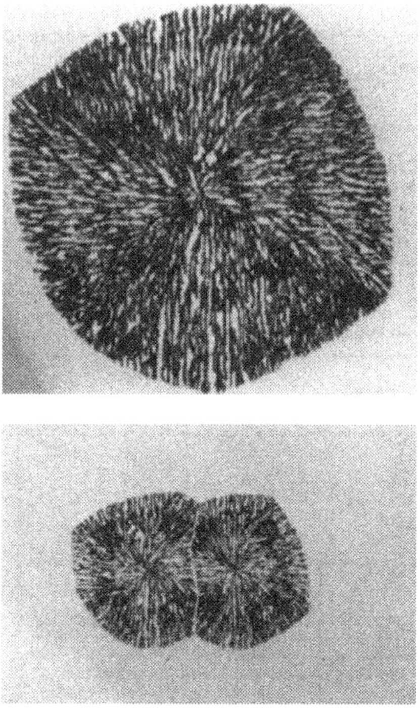

Figure 6. (Top) State of the cellular automaton SPOMO showing concentric ring patterns. The cells of the automaton are represented by pixels (light pixel: vegetative hypha; dark pixel: aerial hypha). The initial state corresponds to "one inoculum in the biological experiment of *Neurospora* growth" (parameters ε = 1.5, initial nutrient concentration = 1).
(Bottom) State of the cellular automaton SPOMO showing "inhibition structures" that means suppression of branching into aerial hyphae in the "contact zone". Representation as described in legend to Fig. 6a. The initial state corresponds to "two inocula in the biological experiment of *Neurospora* growth" (parameters ε = 1.5, initial nutrient concentration = 1).

Acknowledgements

The work of A. Deutsch is funded by Stiftung Volkswagenwerk, Hannover, FRG.

References

[1] Bourret, R.A., Lincoln, R.G. & Carpenter, B.B. (1969). Fungal endogenous rhythms expressed by spiral figures. *Science* **166**, 763.

[2] Dharmananda, S. & Feldman, J.F. (1979). Spatial distribution of circadian rhythm in aging cultures of *Neurospora Crassa*. *Plant Physiol.* **63**, 1049.

[3] Dienes, L. (1946). Reproductive processes in *Proteus* cultures. *Proc. Soc. exp. Biol. N.Y.* **63**, 265.

[4] Dienes, L. (1947). Further observations on the reproduction of bacilli from large bodies in *Proteus* cultures. *Proc. Soc. exp. Biol. N.Y.* **66**, 97.

[5] Edmunds, L.N. (1988). *Cellular and molecular bases of biological clocks*. Springer: New York, Berlin, Heidelberg.

[6] Feldman, J.F. & Dunlap, J.C. (1983). *Neurospora Crassa*: A unique system for studying circadian rhythms. *Photochem. Photobiol. Rev.* **7**, 319.

[7] Horowitz, N.H. (1947). Methionine synthesis in *Neurospora*. *JBC*, 255.

[8] Neumann, J.v. (1966). *Theory of self-reproducing automata*. University of Illinois Press.

[9] Smith, D.G. (1972). The *Proteus* swarming phenomenon. *Sc. Prog. (Oxford)* **60**, 487.

[10] Vogel, H.J. (1964). Distribution of lysine pathways among fungi: evolutionary implications. *Amer. Naturalist* **98**, 430.

[11] Winfree, A.T. (1970). The oscillatory control of cell differentiation in *Nectria*. *Proc. IEEE Symp. on Adaptive Processes* **XXIII**, 4.1-4.7.

[12] Winfree, A.T. (1973). Polymorphic pattern formation in the fungus *Nectria*. *J. theor. Biol.* **38**, 363.

[13] Winfree, A.T. (1980). *The geometry of biological time*. Springer: New York, Berlin, Heidelberg.

[14] Winfree, A.T. & Twaddle, G.M. (1981). The *Neurospora* mycelium as a two-dimensional continuum of coupled circadian clocks. In *Mathematical Biology*, Burton, T.A. (ed.). Pergamon Press.

25. A MODEL FOR GLIDING AND AGGREGATION OF MYXOBACTERIA

Angela Stevens

SFB 123, Institut für Angewandte Mathematik
Im Neuenheimer Feld 294, D-6900 Heidelberg

1. **Introduction**

In morphogenesis, cells recognize and find each other to build complex structures. Up to now, the mechanism for this phenomenon is not fully understood. Because of their cooperative gliding and aggregation, the myxobacteria serve as a model problem for investigations in this direction. To get a better insight into these problems, a theoretical approach is chosen and represented in a cellular automaton model.

2. **Biology**

The myxobacteria show the cell cycle illustrated in Fig. 1 (compare the picture in [11]).

The model is dealing with the vegetative state of the myxobacteria where they form different patterns during their communal gliding.

The bacteria produce extracellular slime which they prefer to glide on. Once on the slime trail, they increase their gliding velocity. They build tracks or swarms of cells which merge to form larger ones or split into smaller ones. They glide around open circles and in spirals. Just before aggregation, patterns of rhythmic pulsating waves occur. Different biological hypotheses and models for this behaviour are described in [1-6,9,13,15]. These hypotheses are discussed in [14].

3. **Monte-Carlo-Simulations**

The modelling features for the Monte-Carlo-Simulations presented in this paper are:

(a) the following of the slime trails;
(b) the introduction of a chemoattractant which is produced by the bacteria to reach the final aggregation.

The simulations are realized with a 'Four Nearest Neighbours Cellular Automaton Model'. The bacteria (length: 8 units; width: 1 unit) are distributed randomly over an area of $(100 \text{ units})^2$ with periodic boundary conditions. Each of them glides to one of the four nearest neighbours of its

Nonlinear Wave Processes in Excitable Media
Edited by A. V. Holden *et al.*, Plenum Press, New York

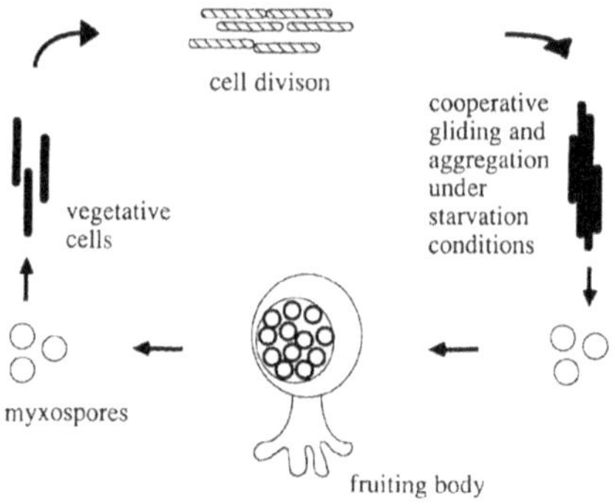

Figure 1.

head (Fig. 2) without crossing itself.

Define $G:=\{1,\dots,100\}$, let

$$S:G^2 \times \mathbf{N} \longrightarrow \mathbf{R}_+$$

describe the density of slime in space and time, and

$$D:G^2 \times \mathbf{N} \longrightarrow \mathbf{R}_+$$

describe the density of the diffusing chemoattractant, let w_S, $w_D \in \mathbf{R}_+$ be weightfactors for the slime and the chemoattractant. Let N_y be the set of the four nearest neighbours of $y \in G^2$. Now looking at one bacterium, let $h(t)$ describe the orientation of the head of the bacterium at time t (see Fig. 2). Define $B:=\{x \in G^2 | x \text{ is part of the bacterium}\}$, then the probability for a bacterium with its head positioned at y at time t to glide to $x_0 \in N_y$, $x_0 \notin B$ is:

$$\frac{w_D \cdot D(x_o,t) + (w_S \cdot S(x_0,t) + 1)\cdot c(y,x_0,h(t))}{\sum_{x \in N_y, x \notin B} (w_D \cdot D(x,t) + (w_S \cdot S(x,t) + 1)\cdot c(y,x,h(t))} \tag{1}$$

where

$$c(y,x,h(t)) = \begin{cases} 10.0 & \text{if } x \text{ is neighbour in direction of the} \\ & \text{orientation } h(t) \text{ of } y \\ 1.0 & \text{else} \end{cases}$$

The probability is 0 for $x_0 \in B$.

The parameters w_S and w_D are chosen so that $w_D \gg w_S > c(\cdot,\cdot,\cdot)$. This means that the bacteria react with the following preference:

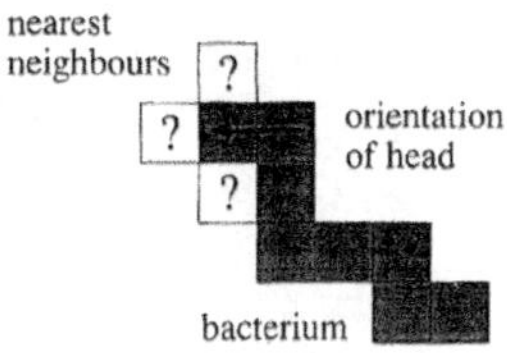

Figure 2.

(i) density of diffusing chemoattractant;
(ii) density of slime;
(iii) preference of gliding straight ahead.

In every time step, slime is produced under the whole cell length of the bacterium, and the old slime on the field, which serves as memory for the bacteria, loses 10% of its effect. The gliding velocity increases up to tenfold if the slime density is high enough and the bacterium is not gliding in a cell track. When during gliding a certain threshold density of bacteria is reached at a point in the field, those bacteria which are positioned at that point start to produce the diffusing chemoattractant. Every bacterium which senses a certain threshold concentration of the chemoattractant starts to produce it too and stops if this concentration is not attained. Diffusion is discretized as usual, and the diffusion coefficient is 0.05. As a second assumption, 2% of the chemoattractant dissolves every time step. There exists an upper bound for both the density of slime and of the chemoattractant.

These assumptions have the following effect: the bacteria are gliding in rectangles over each other until the production of the diffusing chemoattractant starts. Then they glide over the slime trails into the aggregation centres which are stabilized because the bacteria stop to produce the chemoattractant when sensing a density too low.

A modification is to simulate cells gliding parallel and in long cell tracks. To reach this aim, a structure like that in Fig. 3 is constructed around and behind each bacterium. All bacteria are gliding with the same velocity, and the probability for gliding from position y at time t to $x_0 \in N_y$, $x_0 \notin B$ t is:

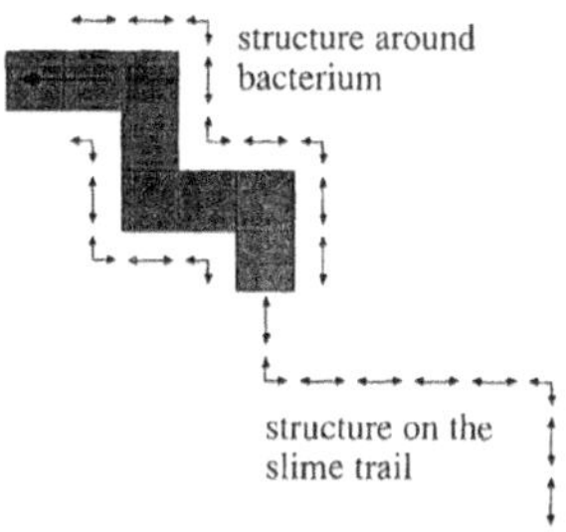

Figure 3.

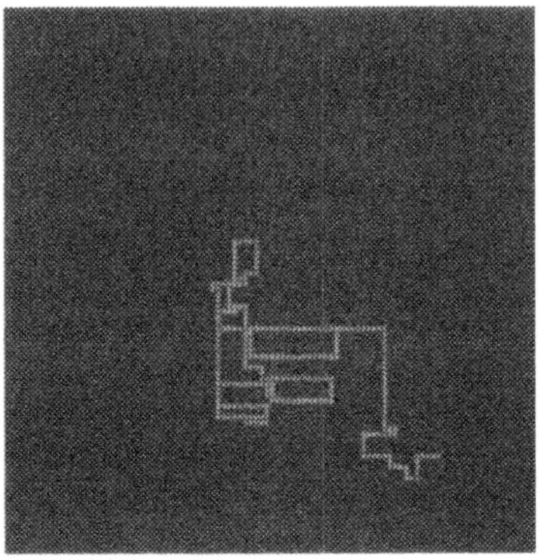

Figure 4.

$$\frac{w_S \cdot S(x_0,t) + a(y,x_0)}{\sum\limits_{x\in N_y,\, x\notin B} (w_S \cdot S(x,t) + a(y,x)} \qquad (2)$$

where

$$a(y,x) = \begin{cases} w_a \in R_+ & \text{if there exists a structure on } y \text{ in direction of } x \text{ produced by another bacterium} \\ 1.0 & \text{else} \end{cases}$$

The probability is 0 for $x_0 \in B$.

The parameters have to fulfil the relation $w_a \gg w_S$ to keep the bacteria together. Otherwise, they would break away from cell tracks every time they sense a slime trail. So no longer tracks would develop.

In Fig. 4 a typical random walk of a single bacterium (black) is shown. The slime trail is represented by the white dots. The rectangles marked by the slime depend on the preference for gliding straight ahead and the memory effect of the slime.

In Fig. 5 1000 bacteria are randomly distributed over the field (light grey for one bacterium to black for more than nine bacteria). This is an average cell density for the inspected area [12].

In Fig. 6 the bacteria are running around open rectangles (over each other). These rectangles are stable for a while but then disappear and appear at other places again.

To stabilize these 'preaggregations', a diffusing chemoattractant (chequered) is introduced. Its effect can be seen in Fig. 7. Once the bacteria are in the aggregation centres, they stay there because of the strong but spatially restricted effect of the chemoattractant. In areas without a

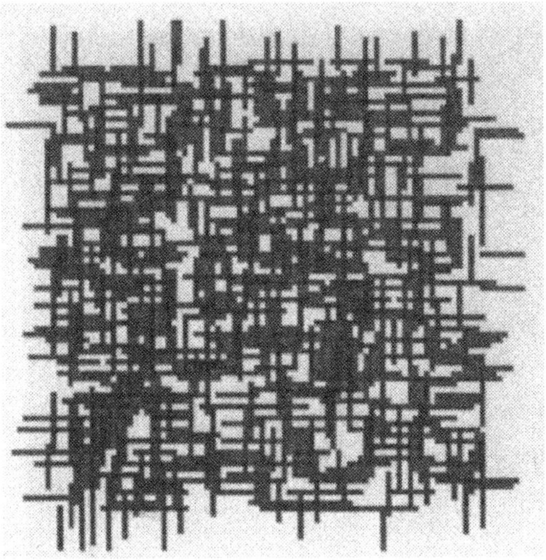

Figure 5.

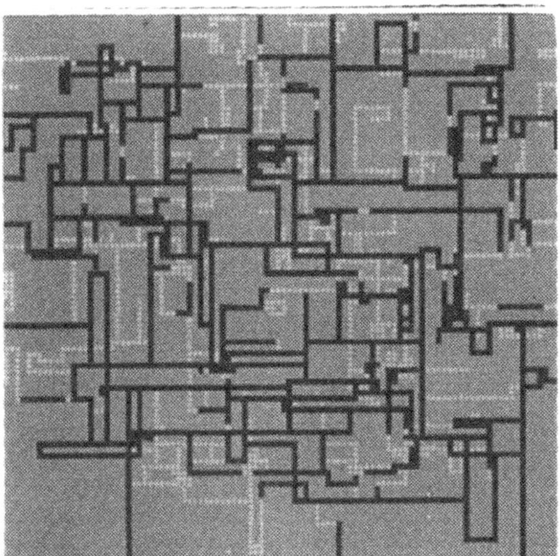

Figure 6.

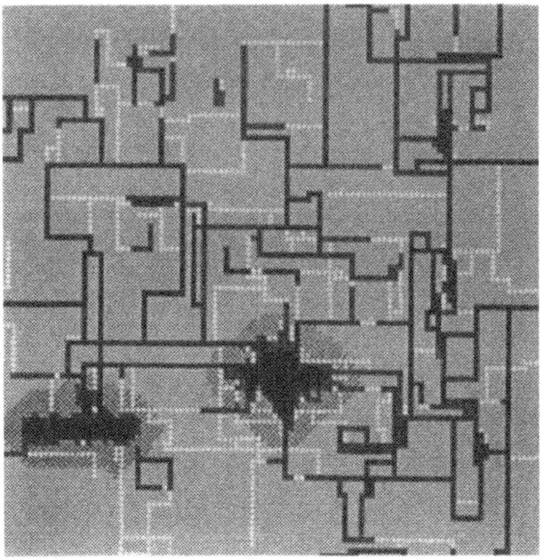

Figure 7.

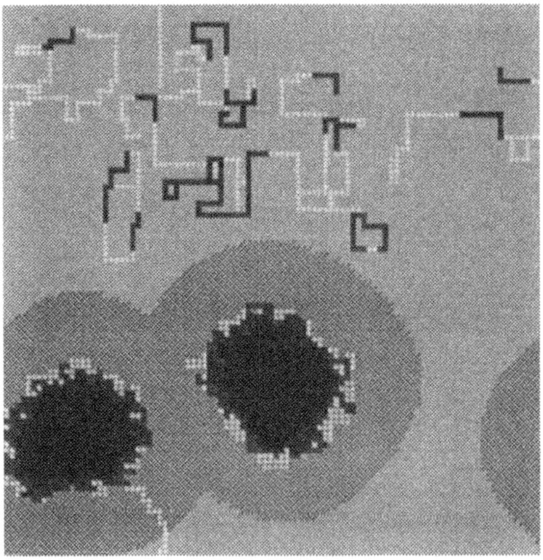

Figure 8.

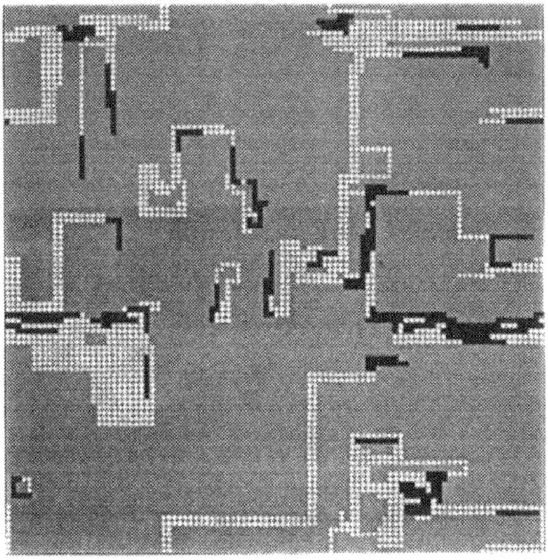

Figure 9.

developing fruiting body, the bacteria are still gliding around open rectangles.

In Fig. 8 the final aggregation is shown.

A more refined approach is the parallel gliding which is presented in Fig. 9. Long cell tracks develop and stay together for a realistic time. If there are too many bacteria on the field, these tracks do not develop.

4. Discussion and further research

The following of the slime trails is not enough to reach aggregation but seems to produce the patterns which can be found before aggregation. The production of the density dependent chemoattractant is necessary for the formation of fruiting bodies but its effect has to be explained biologically. These results are taken into account for an interacting stochastic many-particle system which models the gliding of the myxobacteria. A system of reaction-diffusion equations can be derived at least heuristically (compare [7],[8]). This system has to be analysed in detail to prove that the model is reasonable.

Acknowledgement

This work has been supported by the Deutsche Forschungsgemeinschaft.

References

[1] Dworkin, M. & Eide, D. (1983). Myxococcus xanthus does not respond chemotactically to moderate concentration gradients. *J. Bacteriol.* **154**, 437-442.

[2] Dworkin, M., Keller, K.H. & Weisberg, D. (1983). Experimental observations consistent with surface tension model of gliding motility of Myxococcus xanthus. *J. Bacteriol.* **155**, 1367-1371.

[3] Hodgekin, J. & Kaiser, D. (1979). Genetics of gliding motility in Myxococcus xanthus (Myxobacterales): two gene systems control movement. *Mol. Gen. Genet.* **171**, 177-191.

[4] Keller, K.H., Grady, M. & Dworkin, M. (1983). Surface tension gradients: Feasible models for gliding motility in Myxococcus xanthus. *J. Bacteriol.* **155**, 1358-1366.

[5] Kuner, J.M. & Kaiser, D. (1982). Fruiting body morphogenesis in submerged cultures of Myxococcus xanthus. *J. Bacteriol.* **151**, 458-461.

[6] Lauffenburger, D. (1984). An hypothesis for approaching swarms of Myxobacteria. *J. theor. Biol.* **110**, 257-274.

[7] Oelschläger, K. (1987). A fluctuation theorem for moderately interacting diffusion processes. *Probab. Th. Rel. Fields* **74**, 591-616.

[8] Oelschläger, K. On the derivation of reaction-diffusion equations as limit dynamics of systems of moderately interacting stochastic processes. To appear in Probab. Th. Rel. Fields.

[9] Reichenbach, H. (1965). Rhythmische Vorgänge bei der Schwarmentfaltung von Myxobakterien. *Berichte der Deutschen Botanischen Gesellschaft* **78**, 102-105.

[10] Reichenbach, H. (1965). Untersuchungen an Archangium violaceum. Ein Beitrag zur Kenntnis der Myxobakterien. *Berichte der Deutschen Botanischen Gesellschaft. Arch. Microbiol.* **52**, 376-403.

[11] Reichenbach, H. (1986). The myxobacteria: common organisms with uncommon behaviour. *Microbiol. Sc.* **3 (No. 9)**, 268-274.

[12] Shimkets, L.J. & Dworkin, M. (1981). Excreted adenosine is a cell density signal for the initiation of fruiting body formation in Myxococcus xanthus. *Dev. Biol.* **84**, 51-60.

[13] Stanier, R.Y. (1942). A note on elasticotaxis in myxobacteria. *J. Bacteriol.* **44**, 405-412.

[14] Stevens, A. (1990). Simulations of the aggregation and gliding behaviour of myxobacteria. To appear in Lecture Notes in Biomathematics.

[15] White, D. (1987). Cell interactions and the control of development in myxobacteria populations. *Int. Rev. Cytol.* **71**, 203-227.

26. GEOMETRIC PARAMETERS OF AGGREGATION WAVES IN DICTYOSTELIUM DISCOIDEUM

P. Foerster and S.C. Müller

Max-Planck Institut für Ernährungsphysiologie
Rheinlanddamm 201, D-4600 Dortmund 1, FRG.

The *dictyostelium discoideum* amoebae have been studied as a model system for pattern formation in a biological excitable medium. During the development cycle of this system from an assembly of single cells to a multicellular fruiting body, there is an aggregation phase when the cells start to undergo coherent motion, stimulated by intercellular wave-like propagation of the signal transmitter cyclic AMP. During this phase the majority of the cell population rest in an excitable state but there are groups of cells serving as pacemakers emitting the cAMP pulses which excite neighbouring cells. The response of the excited cells is chemotactic motion along the cAMP gradients. Because the moving cells have a slightly elongated shape while during rest the shape is circular, the light scattering behaviour of the cells depends on their actual state of motion. With a suitable optical set-up these differences can be visualized.

In this study we used dark-field illumination of cells spread out on an agar surface in a petri dish and combined it with image recording by a sensitive video camera and computerized evaluation techniques of the digitally stored images. This allows a quantitative calculation of the geometric shape of the patterns by extracting contour maps from the dark-field images. We focused on the temporal evolution of the shape and curvature of cusps formed by the collision of wave fronts and on the geometry of spiral- shaped aggregation waves.

A typical dark-field pattern is shown in the Figure. It consists of spirals and several cusp structures. An analysis of the evolution of the cusps in accordance with previous investigations on chemical waves [1] yields a linear relationship between the curvature and the normal velocity of the waves, so the theoretically predicted proportionality between these two quantities [2] was verified experimentally for negative curvatures. Analysis of positive curvatures as found for small circular waves is on the way. A fit of splines to spiral-shaped isointensity contour maps shows that the pitch of the spiral is nearly constant (2 mm), except for an area close to the spiral core, where the pitch is about 10% larger. The core can be identified at high magnification as a small dark spot into which the bright and dark stripes along the spiral edge merge (see Figure). Its diameter is about 300 μm. The measurement of the spiral shape indicates that it is well represented by the involute of a circle.

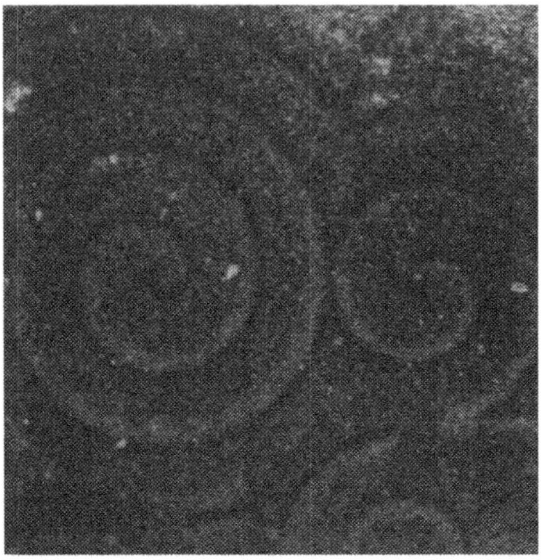

Figure 1. Dark-field image of a pattern of colliding spiral waves in *dictyostelium discoideum*.

The main result of our study is the experimental verification of the predicted "eikonal equation" [2]

$$N = c - D \cdot K$$

where N is the normal velocity of the aggregation waves, c the velocity of plane waves, D the diffusion constant of the autocatalytic species, and K the front curvature. From the slope of the line relating N with K we obtained the diffusion coefficient of cyclic AMP: $(0.66 \pm 0.05) \times 10^{5}\ cm^{2}/sec$. In addition, a critical radius R_{crit} equal to D/c was derived, below which propagation of a circular wave cannot be initiated [3]. Its value was found to be about 200 μm and provides an estimate of the size of pacemaker centres.

Our investigations emphasize in quantitative detail that the system of slime mould aggregation shows properties which appear to be quite general for excitable systems. They are in good agreement with experimental results obtained in the Belousov-Zhabotinskii reaction. Beyond the phenomenological similarity between the patterns of the *dictyostelium discoideum* system and the Belousov-Zhabotinskii reaction, we thus provide evidence that the predictions from the formally almost identical reaction-diffusion models [4] are quantitatively correct. This strongly supports the concept of developing a general theory for pattern formation in excitable media, that can be applied also to biological systems evolving on a higher level of complexity.

References

[1] Foerster, P., Müller, S.C. & Hess, B. (1988). Curvature and propagation velocity of chemical waves. *Science* **241**, 685-687.

[2] Tyson, J.J., Alexander, K.A., Manoranjan, V.S. & Murray, J.D. (1989). Spiral waves of cyclic AMP in a model of slime mould aggregation. *Physica* **34D**, 193-207.

[3] Martiel, J.-L. & Goldbeter, A. (1987). A model based on receptor desensitization for cyclic AMP signalling in *dictyostelium* cells. *Biophys. J.* **52**, 8076-828.

[4] Foerster, P., Müller, S.C. & Hess, B. (1989). Critical size and curvature of wave formation in an excitable chemical medium. *Proc. Natl. Acad. Sci. USA* **86**, 6831-6834.

27. ELECTRICAL ROTORS IN THE HEART

P.A. Guse, D.W. Frazier, N. Shibata, P.-S. Chen
and R. E. Ideker

Departments of Medicine and Pathology
Duke University Medical Center
Duram, NC, U.S.A., and
The School of Engineering
Duke University, Durham NC, U.S.A.

Electrophysiological processes in the heart are responsible for impulse formation and conduction. The normal action of these processes is necessary for the production of an efficient contraction. Interruption of impulse formation or conduction [3,18] by either biological, pathological, or toxicological mechanisms can cause a disruption of the normal rhythm, which is called an arrhythmia. Two types of abnormal impulse formation can cause cardiac arrhythmias. The first type involves those cells that have the ability to depolarize spontaneously [19] and generate their own action potentials. This includes cells found in the sinoatrial (SA) node, atrioventricular (AV) node, His bundle, and ventricular Purkinje system. By suppressing or enhancing the rate of spontaneous depolarization, an arrhythmia can develop.

The second type of abnormal impulse formation, called triggered activity, occurs in cells that have become injured or exposed to drugs [10]. Triggered activity occurs when a cell partially or completely depolarizes during or immediately after normal repolarization. If an after depolarization reaches threshold, an impulse may be generated.

When changes occur in cardiac conduction, another type of arrhythmia may form, called a reentrant arrhythmia [18,20]. In reentry, an impulse blocks in one region while continuing to conduct through other regions. If the conducted activation fronts circulate around to enter the region of block from another direction, and if the area proximal to the region of block has had time to recover, then this area may be reentered by the activation front [12] (Figs. 1 and 2). It is generally believed that reentry mechanisms are responsible for generating the majority of arrhythmias that are seen clinically.

There are at least two types of reentry arrhythmias: those that require an anatomical obstacle about which the reentry circuit rotates (Fig. 1) and those that do not require such an obstacle (Fig. 2). In the first type, reentry circuits can be produced around normal anatomical areas such as the vena cava, branches of the Purkinje system, through the A-V node, S-A node, the tricuspid valve or in abnormal areas such as regions of myocardial infarction. The pathway for this type of reentry is fixed. The second type of reentry, not involving an anatomical barrier, has been called "leading circle"

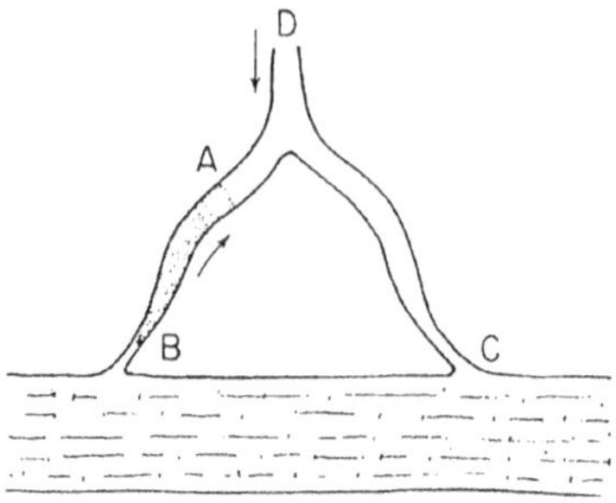

Figure 1. Reentry circuit around an anatomic obstacle. The diagram depicts a branch of the Purkinje system that terminates on a strip of ventricular muscle. The shaded (A-B) area represents an area of slow conduction and delayed recovery. An impulse conducts normally from area D to areas A and C. However, because of delayed recovery, unidirectional conduction block occurs at A. The impulse instead conducts to C and then B. If the conduction from B to A is sufficiently slow such that the area proximal to A has had time to recover, the impulse can reenter and conduct back to D and C. If the electrophysiological conditions remain the same, the impulse can continuously circle around the pathway resulting in a sustained arrhythmia. (From Schmidt & Erlanger [12], with permission.)

if it involves a single rotor of reentry in a two dimensional thin layer of myocardium [1], or "figure of eight" if it involves two mirror image rotors of reentry [5]. In contrast to the reentry circuit around an anatomical barrier, the length of the circuit for this second type of reentry is determined by the functional electrophysiological properties of the cells, such as their refractory periods [1].

Both types of reentry are thought to require areas of unidirectional block and slow conduction. Unidirectional block means that the impulse is blocked when it reaches the tissue from one direction, but will later conduct the impulse in the opposite direction. Several mechanisms have been proposed to explain the formation of the unidirectional block. Cranefield [3] demonstrated that the block could be caused by prolonged refractoriness brought on by ischaemic heart disease. During the absolute refractory period, cardiac cells are transiently inactivated and cannot depolarize. Thus, an impulse that reaches an area during its absolute refractory period will be blocked. Should this area recover its excitability by the time the impulse has circulated around it, a reentry pathway will be generated if the blocked area can now support conduction in the opposite direction.

The other prerequisite commonly accepted for the generation of a reentry arrhythmia is that the cardiac impulse be conducted slowly during some portion of its course. Slow impulse conduction allows previously excited tissue time to repolarize. In normal hearts, a conduction pathway may have to be as long as 7.5 cm for reentry to occur [20]. By slowing conduction velocity by a factor of 10, the length of the pathway need only be 7.5 mm for reentry to occur.

Ventricular fibrillation is a relatively disorganized arrhythmia that for the first half of this century was thought to be caused by both enhanced

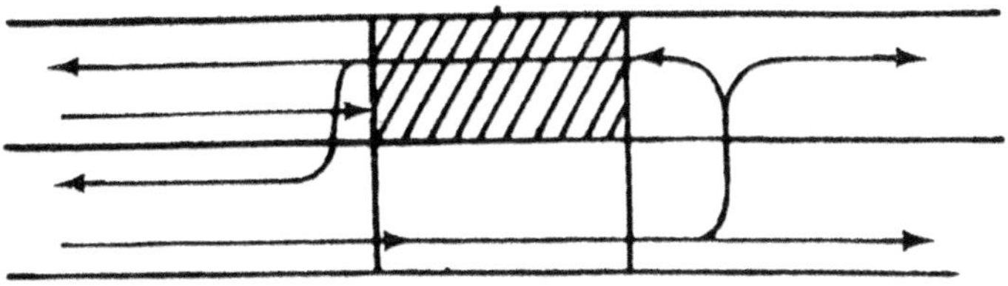

Figure 2. Reentry circuit without an anatomic obstacle.
The diagram shows a similar set of pathways as in Fig. 1 except that the anatomic obstacle is not present. The bottom pathway is normal and the upper pathway contains an area of slow conduction and delayed recovery. A normal impulse, moving from left to right down both fibres, blocks at the shaded area. After the impulse has traversed the normal lower pathway, it spreads to the upper pathway where it divides to conduct to the right and back through the shaded area. If the area proximal to the shaded region has had sufficient time to repolarize, the impulse will reenter the normal tissue to the left. (From Schmidt & Erlanger [12], with permission.)

automaticity and reentry mechanisms [10]. Based on computer simulations and cardiac mapping studies, for the last 30 years, most investigators have favoured reentry mechanisms. Until recently, it has generally been considered that nonuniform dispersion of refractoriness of cardiac tissue is responsible for electrically initiating ventricular fibrillation [8]. According to this hypothesis, an impulse is conducted away from the site of premature electrical stimulation during the "vulnerable period" of the cardiac cycle until it reaches a nonhomogeneous distribution of refractoriness. At this site a temporary unidirectional conduction block occurs in tissue that is more refractory than neighbouring tissue. Reentry occurs if this neighbouring tissue has recovered by the time the impulse reaches it after conducting backwards through the regions in which unidirectional block occurred.

Recently, with the aid of computer assisted methods of mapping cardiac impulses [14], other possible mechanisms for the initiation of reentry have been found. One such mechanism was put forward by Spach et al. [15], who suggest that the anisotropy of conduction caused by fibre orientation may be important for the induction of reentry. They showed experimentally that the safety factor for conduction was less when the impulse conducted along the fibres than across them. Thus, a prematurely conducted impulse could block unidirectionally along fibres but continue to propagate across fibres, leading to reentry.

Our research group has performed three studies which show that the nonuniform dispersion of refractoriness is not necessary for the induction of reentry by electrical stimuli during the vulnerable period of the cardiac cycle. In these studies, reentry rotors were created by generating a potential gadient field from a large premature stimulus that interacted with a region of relatively refractory tissue. The distributions of gradient and refractoriness were dispersed uniformly and at angles to each other. The reentry circuit was centred around a certain critical point formed by the intersection of a critical value of potential gradient created by the stimulus with a critical degree of tissue refractoriness.

In one study [7], activation complexes and potentials generated by

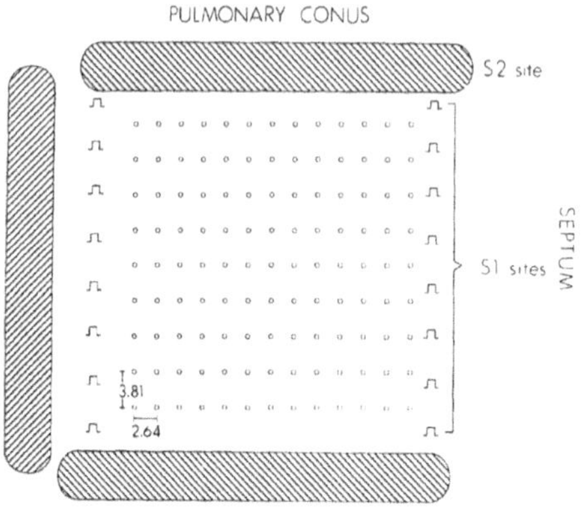

Figure 3. Epicardial electrode array.
Epicardial recordings were obtained from an array of 117 surface bipolar electrodes (open circles) held within an epicardial array placed on the anterior right ventricle. The top of the array was located within 2 cm of the pulmonary conus and the right side parallel to and within 1 to 2 cm ventricular septum. A row of eight epicardial pacing wires was sutured along the right (labelled "S1 sites") and left sides of the array. S1 stimuli were delivered from one of the two rows of pacing wires. Stainless steel mesh electrodes, 4.5 x 1 cm, were sutured to the epicardial surface along the top (labelled "S2 site"), bottom, and/or left sides of the array. S2 stimuli were delivered from one of the three mesh electrodes. In the figures that follow, the sites of the pacing wires and mesh electrodes are designated as S1 and S2, respectively. (From Frazier et al. [7], with permission.)

premature stimuli were recorded from either 120 transmural plunge electrodes or 117 epicardial electrodes (Fig. 3) in a holder sutured to the right ventricle of anaesthetized dogs. During pacing at a cycle length of 300-350 msec (S1 pacing), recordings were made simultaneously from the electrodes with a computer-assisted mapping system and activation times were determined at each electrode [14]. The recovery times at 24 - 44 of these recording sites were also determined by delivering a 2 mA cathodal stimulus after every 10 regularly paced S1 beats to each site. This 2 mA stimulus was made progressively more premature, in 2 ms steps, until it no longer captured the heart. The interval between the last of the 10 S1 stimuli and the earliest 2 mA stimulus that still captured the heart was called the recovery time. The activation times and recovery times were found to be dispersed uniformly across the array of recording electrodes. The isolines of activation and recovery were parallel to each other and to one side of the array of recording electrodes (Fig. 4).

Next, large shocks (S2 = 25-250 V) were delivered via a wire mesh electrode that was sutured at a right angle to the S1 pacing electrodes [7]. These potentials were also recorded from the 117 or 120 electrodes and the potential gradient was calculated by a finite element method that also took into account the anisotropy in conductivity introduced by the orientation of the myocardial fibres (Fig. 5). The potential gradients were highest close to the S2 mesh electrode and diminished with distance from the mesh electrode.

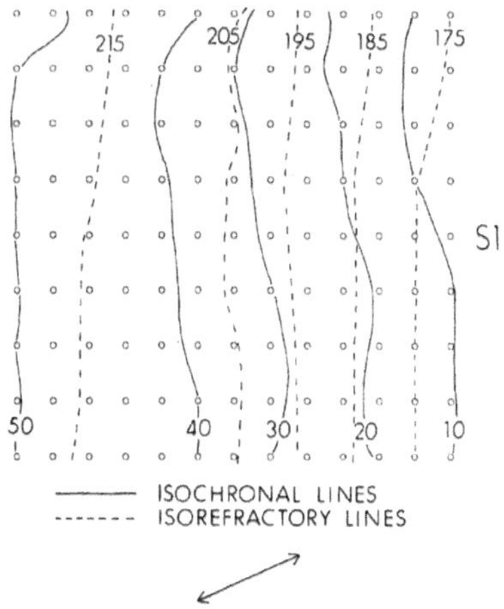

Figure 4. S1 activation and isorecovery patterns. The row of eight epicardial pacing wires was tied together as a single source (S1) for pacing at 10 mA. Approximately parallel isochronal lines (solid lines) are created by S1 pacing, with conduction velocities of 0.5 to 0.7 m/sec. The wider isochronal spacing to the left of the array, distant from the S1 site, probably indicates a sizable endocardial to epicardial component to the spread of activation in that region [6]. The recovery periods (dashed lines) were calculated at 32 electrode sites evenly spaced across the array for this example. The refractory periods were similar at all electrode sites (166±3 ms), indicating uniform recovery. The double-headed arrow at the bottom represents the mean epicardial fibre orientation of the tissue under the array, in this case 24±5 degrees with respect to the horizontal axis. Figures 4 through 6 are taken from the same animal. (From Frazier et al. [7], with permission.)

Thus, the potential gradients were dispersed perpendicularly to the activation-recovery isolines caused by S1 stimulation.

The ventricles were paced for 10 S1 cycles followed by a 25 to 250 V premature 3 ms S2 stimulus for which the S1-S2 interval was varied in 3 to 5 ms steps to find a coupling interval that initiated ventricular fibrillation. Fibrillation was induced in all animals. Activation maps revealed a single rotor of reentry (Fig. 6). Surprisingly, the earliest activation after the S2 shock was not directly adjacent to the S2 electrode where the potential gradient was strongest. Rather, the earliest post-shock activation was observed several cm distant from the S2 electrode, conducting away from the border of a region that was assumed to be directly excited by the potential gradient field of the S2 shock. Where the S2 potential gradient was greater than a critical value of approximately 5-6 V/cm, activation did not conduct away from the border of the directly excited region after the S2 shock. This unidirectional block induced by the shock allowed establishment of the reentry rotor. The rotor circled round a critical point formed by the intersection of the critical value of potential gradient with tissue in a critical stage of

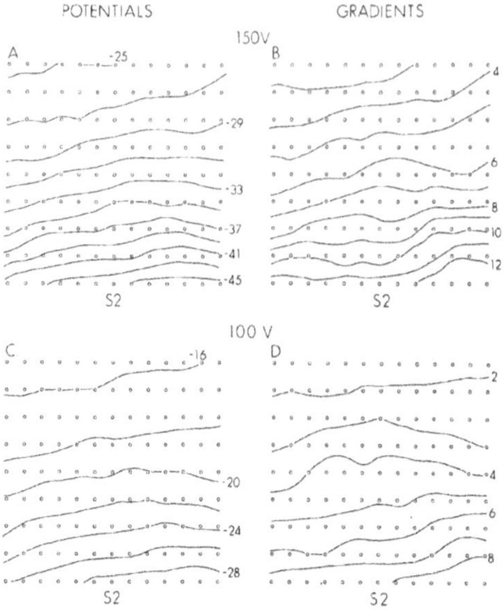

Figure 5. Stimulus potentials and potential gradients. Panels A and B show the epicardial stimulus potentials and potential gradients, respectively, for a unipolar cathodal shock of 150 v from the bottom mesh electrode. Isopotential and isogradient lines approximately parallel to the bottom of the array are created, with gradients of 6 V/cm occurring near the middle of the array. The isopotential values are given in volts (V), and the isogradient values in V/cm. Panels C and D show the potentials and gradients for a 100 V S2. Gradients of 6 V/cm occur closer to the S2 site than for the 150 V S2. (From Frazier et al. [7], with permission.)

refractoriness, i.e. the recovery time. The location of this critical point and hence the location of the centre of the reentry rotor could be changed by changing either the S2 strength or the timing interval between S1 and S2.

The direction of rotation of the rotor was determined by the relationship of the dispersion of refractoriness to the dispersion of the potential gradient. When the S1 electrodes were to the right of the mapped region and the S2 mesh electrode was to the bottom, the direction of rotation was counter clockwise (Fig. 6). When the S1 electrodes were to the right and the S2 electrode was at the top, the direction of rotation of reentry was clockwise. When the S1 electrodes were to the left and the S2 electrode was at the top, the direction of rotation was counter-clockwise. Finally, when the S1 electrodes were to the left and the S2 electrode was at the bottom, the direction of rotation was clockwise.

Because a change in S1-induced recovery patterns or in the S2 potential gradient distribution produced a change in the location and direction of the reentry circuits, the rotors were formed by the interaction of these two phases at a critical point rather than by inhomogeneous electrical properties of the tissue. Thus, a nonuniform dispersion of refractoriness is not necessary for the electrical induction of arrhythmias by a large premature S2

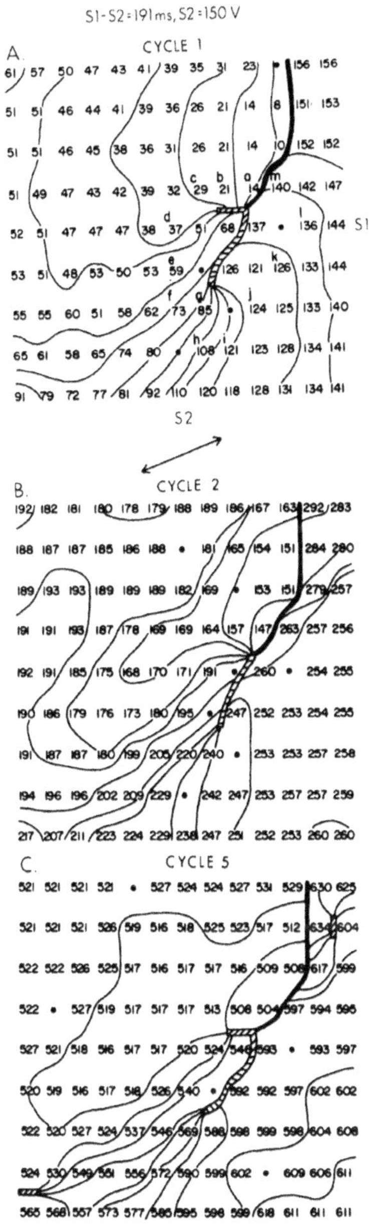

Figure 6. Activation patterns for perpendicular isorefractory and isogradient lines.
The patterns of activation following S1 pacing from the right and S2 shock from the bottom initiating fibrillation are shown in Panels A (cycle 1), B (cycle 2) and C (cycle 5). Activation times, shown in ms, are measured from the start of the 3 ms S2 shock. Solid dots represent sites of deficient recordings. The

(continued)

stimulus delivered to a different site than the S1 stimuli. This finding does not negate the possible importance of the nonuniform dispersion of refractoriness for the induction of other types of electrically or spontaneously induced arrhythmias. In addition, inhomogeneous electrophysiological properties of the tissue may add to the complexity of arrhythmias electrically initiated by the critical point mechanism.

In the experiment described above, it is the nonlinear response of the tissue induced by the S2 potential gradient field that causes a nonuniform dispersion of refractoriness and unidirectional block; activation conducts away from the border of the zone of tissue directly activated by the S2 field where this field is less than a certain critical value (5-6 V/cm for the waveform used in this experiment, but probably a different value for other waveforms). The activation does not conduct away from the border of the directly excited zone where the potential gradient is greater than this critical value. This concept suggests that a reentry rotor should be formed wherever a critical point is formed by the intersection of the critical potential gradient with tissue in a critical degree of relative refractoriness. Further evidence for this conclusion is given by the second and third experiments.

In the second experiment [2], 120 plunge electrodes in an acetal block were used to record transmurally from the anterior right ventricular free wall (Fig. 7). However, the large, premature S2 stimulus was given from a single point electrode in the centre of the mapped region rather than from a long mesh electrode along the edge. This study was performed before the previously described study, and we did not yet have the ability to measure the S2 potentials. Presumably, the potential gradient decreased with distance from the S2 site with secondary effects because of the anisotropy of conductivity along and across the long axis of the myocardial fibres. Therefore, the critical gradient level probably formed an ellipsoidal surface. Since the S2

---

solid line represents the transition between successive activation maps and is called a "frame line". The hatched line represents a zone of conduction block and is called a "block line". Isochrones are at 10 ms intervals. The double-headed arrow for this figure represents the mean epicardial fibre orientation in the area of conduction block, in this case 21 degrees with respect to the horizontal. Panel A shows the initial activation pattern following the S2 shock at an S1-S2 interval of 191 ms and an Ss strength of 150 V. The hatched area indicates the region assumed to be directly excited by the S2 shock field. Earliest post-shock activation occurs distant from the S2 site. No early activation wavefronts conduct away from the directly excited region located between the S2 site and the critical point, i.e. the point where the activation front blindly ends at the junction between the frame line and the block line. A counter-clockwise reentrant circuit is formed around the region containing the critical point and the line of block. The potential gradient equals 5.8 V/cm. The region at the critical point recovered beyond its effective refractory period (ERP) 2 ms before the S2 conduction occurs at the frame line between the first and second reentrant cycles. The activation pattern is similar to the first cycle except that the circuit time is decreased to 122 ms. Panel C represents the fifth cycle of the reentrant pattern. The reentrant pattern is similar to the previous cycles; however, the total circuit time has further decreased to 101 ms. (From Frazier et al. [7], with permission.)

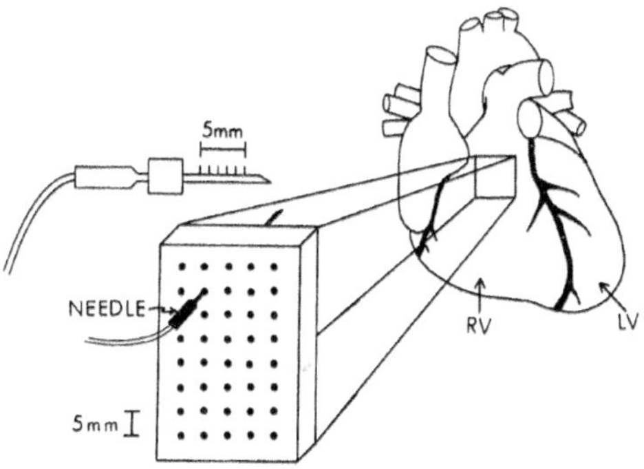

Figure 7. Transmural electrode recording array.
Recordings were made from 40 plunge needles held within a plate which is made of parallel acetal bars. Semicircles along the bars have diameters equal to that of the plunge needles. The semicircles on each bar match to form holes spaced equal distances apart for holding the needles. A plunge needle is shown inserted through the plate. At the top left is a side view of a needle within the plate. The needles are 21-gauge and contain 6 electrodes 1 mm apart. Recordings are made from six electrodes wired as three bipolar pairs, one to record from the subepicardium, one from the midmyocardium and one from the subendocardium. Thus, 120 electrodes from 40 needles in a 5 x 8 array were used to acquire the data. These electrodes were used to make the recordings shown in Fig. 8. (From Frazier et al. [7], with permission.)

stimuli were large (20 to 80 mA) and the right ventricular free wall is thin, the elliptical surface was probably almost perpendicular to the epicardium and thus was open at the endocardium.

S1 stimuli were given from a point electrode at one edge of the mapped volume of tissue. Activation fronts and, hence, isorecovery surfaces also were elliptical and approximately perpendicular to the epicardium in the centre of the mapped volume. Thus, the critical refractoriness level and the critical potential gradient level formed two ellipses that intersected at two critical points, one above and the other below an imaginary line connecting the S1 and S2 electrodes. The area of earliest activation following S2 stimulation occurred between the S1 and S2 electrodes. This activation front initiated fibrillation by forming two mirror image rotors, one reentrant rotor around each of the two critical points. One rotated clockwise, the other counter-clockwise (Fig. 8). This activation pattern has been observed in other types of arrhythmias and has been called a figure of eight configuration [5].

The ventricular refractory period was determined at 54 of the recording sites. Those areas that had the greatest degree of nonuniform refractoriness did not correspond to the regions of unidirectional block after S2. Thus, nonuniform dispersion of refractoriness was not necessary for initiating ventricular fibrillation in this experimental model either.

Since other investigators [15] have noted that the orientation of fibres may be important for creating areas of slow conduction or block, myocardial

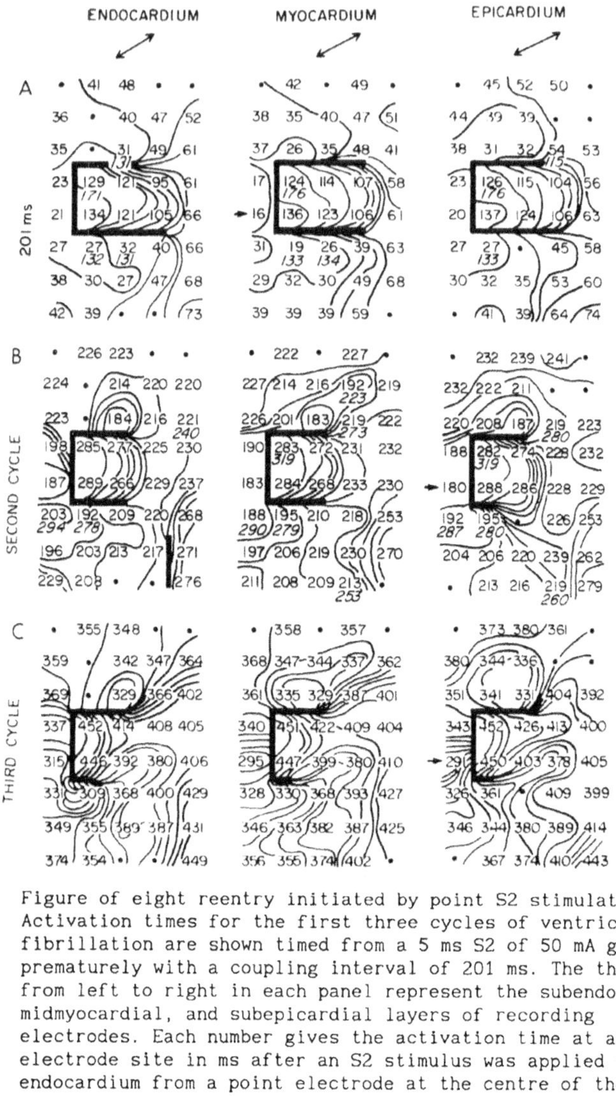

Figure 8. Figure of eight reentry initiated by point S2 stimulation. Activation times for the first three cycles of ventricular fibrillation are shown timed from a 5 ms S2 of 50 mA given prematurely with a coupling interval of 201 ms. The three maps from left to right in each panel represent the subendocardial, midmyocardial, and subepicardial layers of recording electrodes. Each number gives the activation time at an electrode site in ms after an S2 stimulus was applied to the endocardium from a point electrode at the centre of the array. S1 pacing was from the left side of the array. Small closed circles represent damaged recording electrodes. (From Chen et al. [2], with permission.)

fibre orientation was determined. In both this experiment and the first one, histological examination revealed that neither activation fronts nor recovery isochrones correlated with the fibre orientation. Furthermore, the lines of conduction block also were not determined by fibre orientation. It is still

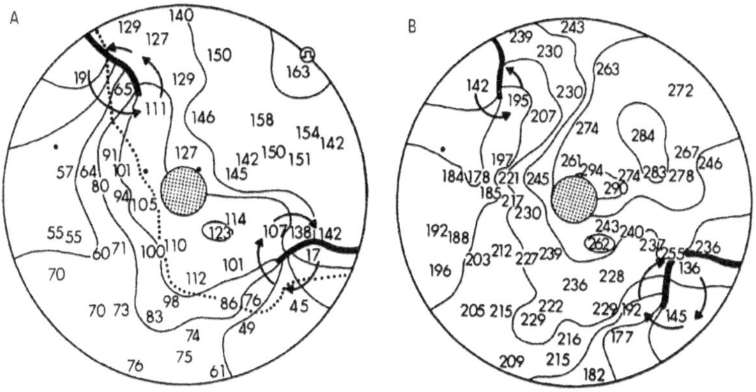

Figure 9. Mirror image reentrant rotors widely distant from S2 site. A polar projection of the ventricles is shown with the stippled apical S2 electrode in the centre and the AV groove circumferentially at the periphery. Numbers represent locations of electrodes with satisfactory recordings and give time of activation for the first (Panel A) and the second (Panel B) cycles of ventricular fibrillation timed from a 1 joule premature S2 shock. Two rotors are present on opposite sides of the ventricles, distant from the S2 apical electrode. (From Shibata et al. [13], with permission.)

possible that fibre orientation was important, but that the approximately 90 degree rotation of fibre orientation transmurally through the anterior right ventricular free wall in the mapped volume obscured this effect.

The critical point concept predicts that as the S2 stimulus strength is increased, the distance between the two rotors forming the figure of eight pattern should move apart. This prediction was tested in the last study [13], in which a 1 Joule premature shock was applied to defibrillation-type electrodes on the apex of the left ventricle and on the right atrium. This shock strength is several orders of magnitude larger than the 20 to 80 mA stimuli used in the previous study. This marked increase in the shock potential field would be expected to move the two critical points several cm away from the S2 electrode at the left ventricular apex. S1 pacing was performed from the base of the right or left ventricle. Again two rotors were created that rotated in opposite directions (Fig. 9). The two mirror image rotors were both at the base of the ventricles, the assumed location of the two critical points. As in the first experiment, the location of the rotors along the base of the ventricles was dependent upon the coupling interval of S1-S2.

The mechanism of unidirectional block and reentry in these studies has been postulated to be production of a graded response by the S2 stimulus [7]. A graded response is defined as a local response that prolongs action potential duration and refractoriness but does not initiate a propagated impulse [9]. Graded responses are thought to occur when a stimulus is applied at a time when most but not all sodium channels are inactivated. The stimulus is thought to activate those few sodium channels that are not inactivated, causing a small local response. Optical recording techniques have demonstrated

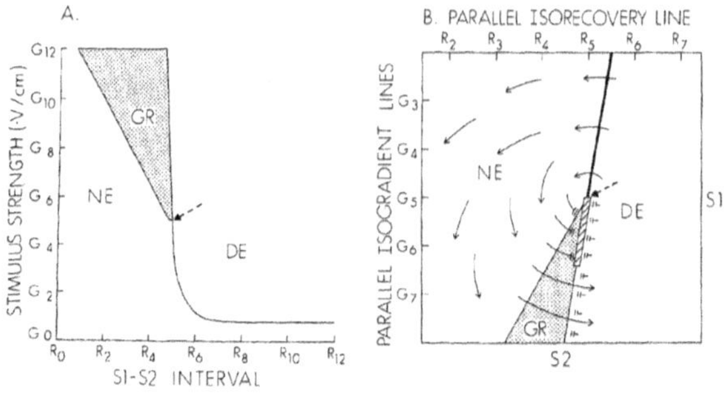

Figure 10. Graded Response Theory for Reentry.
Panel A - Strength Interval Curve: A modified strength-interval curve is shown, with stimulus strength in V/cm on the y-axis and S1-S2 interval, or refractoriness, on the x-axis. On the y-axis, G_{12} represents the largest potential gradient and G_0 the smallest. On the x-axis, R_{12} represents the most repolarized tissue and R_0 the most refractory. R_0 does not represent phase 0 of the action potential, but rather tissue refractory to a premature stimulus. Interaction of the isorecovery and isogradient lines produces either direct excitation (DE), graded response (GR), or no effect (NE). The potential gradient at G_5 is approximately 5 V/cm while the potential gradient for fully recovered tissue is approximately 0.8 V/cm [6]. The critical refractory period (for the 2 mA stimuli) equals the preshock interval at R_5. The dashed arrow represents the position of the critical point in this panel and in panel B.
Panel B - Perpendicular Isorecovery and Isogradient Lines: A diagrammatic representation of the results of Fig. 6A is shown, with gradient and refractory values corresponding to the diagram in Panel A. The row of pacing wires (S1) on the right creates parallel isorecovery lines (R_7 through R_2), with R_7 the least refractory and R_2 the most refractory. The S2 from the bottom creates parallel isogradient lines (G_7 through G_3), with G_7 the largest gradient and G_3 the weakest. The solid and hatched lines are the frame and block lines, respectively. The ╫ represent areas of temporary unidirectional conduction block. For this idealized diagram, the S1-S2 interval equals the recovery period (preshock interval equals the critical refractory period) at R_5, and the S2 voltage creates a potential gradient of 5 V/cm at G_5. The S2 shock directly excites the area near the S1 site (DE) and produces a region of graded response (GR). Activation fronts only propagate away from part of the directly excited area. Activation fronts do not propagate away from the directly excited area abutting regions of graded response, thus forming an area of temporary unidirectional conduction block. An activation front conducting towards the S2 site from the early sites of activation produces another region of conduction

(continued)

block (hatched line) when it enters tissue insufficiently repolarized from graded response or direct excitation. When the myocardium has recovered, activation can conduct through this area and form a reentrant circuit, as shown by the arrows spanning the GR zone and entering the DE zone. (Modified from Frazier et al. [7], with permission.)

---

that shock fields of defibrillation strength can prolong action potential duration when given during the latter half of the action potential [4]. As shown in the strength-interval plot in Fig. 10A, the interaction of different strengths of S2 potential gradient with tissue in various stages of recovery is hypothesized to evoke three responses: (a) direct excitation (DE) in which the tissue is directly activated by the premature S2 stimulus; (b) graded response (GR) in which action potential duration and refractoriness are prolongated without initiating a propagated activation front in adjacent tissue; and (c) little or no effect (NE). The study illustrated in Figs. 3 to 6 created a strength- interval relationship spatially in a small portion of myocardium. Figure 10B represents the type of response thought to occur in the experiment whose results are shown in Fig. 6. Where the S2 potential gradient field is less than the critical value of approximately 5 V/cm (arrow), an activation front propagates away from the border of the area that was directly activated by the S2 stimulus field. Where the potential gradient is greater than the critical value, activation does not propagate away from the border of the directly excited area after the S2 shock. This is the site of unidirectional block. The activation front rotates around the critical point through the NE region, giving the graded response area time to recover. The activation front traverses the region of graded response and then the directly excited region, forming the first cycle of the reentry rotor.

The above explanation is based in terms of classical cardiac electrophysiology. These experimental results have also been predicted and explained in terms of nonlinear dynamics of excitable media [11,17]. Similar behaviour has been observed in Belousov-Zhabotinsky reagent [21], spontaneously beating cells [16], and FitzHugh-Nagumo simulations [17].

Acknowledgements

Supported in part by the National Institutes of Health research grants HL-28429, HL-33637, and HL-17670, and the National Science Foundation Engineering Research Centre grant CDR-8622201.

References

[1] Allessie, M.A., Bonke, F.I.M. & Schopman, F.J.G. (1977). Circus movement in rabbit atrial muscle as a mechanism of tachycardia. III. The "leading circle" concept: A new model of circus movement in cardiac tissue without the involvement of an anatomical obstacle. *Circ. Res.* **41**, 9-18.

[2] Chen, P.-S., Wolf, P.D., Dixon, E.G., Danieley, N.D., Frazier, D.W., Smith, W.M. & Ideker, R.E. (1988). Mechanism of ventricular vulnerability to single premature stimuli in open chest dogs. *Circ. Res.* **62**, 1191-1209.

[3] Cranefield, P.F., Wit, A.L. & Hoffman, B.F. (1973). Genesis of cardiac arrhythmias. *Circulation* **47**, 190-204.

[4] Dillon, S. & Morad, M. (1981). A new laser scanning system for measuring action potential propagation in the heart. *Science* **214**, 453-456.

[5] El-Sherif, N., Mehar, R., Gough, W.B. & Zeiler, R.H. (1982). Ventricular activation patterns of spontaneous and induced ventricular rhythms in canine one-day-old myocardial infarction: Evidence for focal and reentrant mechanisms. *Circ. Res.* **51**, 152-166.
[6] Frazier, D.W., Krassowska, W., Chen, P.-S., Wolf, P.D., Danieley, N.D., Smith, W.M. & Ideker, R.E. (1988). Transmural activations and stimulus potentials in three- dimensional anisotropic canine myocardium. *Circ. Res.* **63**, 135-146.
[7] Frazier, D.W., Wolf, P.D., Wharton, J.M., Tang, A.S.L., Smith, W.M. & Ideker, R.E. (1989). Stimulus-induced critical point: Mechanism fro the electrical initiation of reentry in normal canine myocardium. *J. Clin. Invest.* **83**, 1039-1052.
[8] Han, J. & Me, G.K. (1964). Nonuniform recovery of excitability in ventricular muscle. *Circ. Res.* **14**, 4-60.
[9] Kao, C.Y. & Hoffman, B.F. (1958). Graded and decremental response in heart muscle fibres. *Am. J. Physiol.* **194**, 187-196.
[10] Kisch, B. (1950). The mechanics of flutter and fibrillation. A short review of a century of studies. *Cardiologia* **17**, 244-250.
[11] Krinsky, V.I. (ed.) (1984). *Self Organization: Autowaves and Structures far from equilibrium.* Springer-Verlag: Berlin.
[12] Schmidt, F.O. & Erlanger, J. (1928-29). Directional differences in the conduction of the impulse through heart muscle and their possible relation to extrasystoles and fibrillary contractions. *Am. J. Physiol.* **87**, 3236-347.
[13] Shibata, N., Chen, P.-S., Dixon, E.G., Wolf, P.D., Danieley, N.D., Smith, W.M. & Ideker, R.E. (1988). Influence of shock strength and timing on induction of ventricular arrhythmias in dogs. *Am. J. Physiol.* **255**, H8891-H901.
[14] Smith, W.M., Funk, A.L., Ideker, R.E., Bartram, F.R. & Talbert, P.V. (1982). A microcomputer-based multichannel data acquisition system for the study of complex arrhythmias. In *Proc. Computers in Cardiology*, pp. 131-134.
[15] Spach, M.S., Miller, W.T.,III, Geselowitz, D.B., Barr, R.C., Kootsey, J.M. & Johnson, E.A. (1981). The discontinuous nature of propagation in normal canine cardiac muscle. Evidence for recurrent discontinuities of intracellular resistance that affect the membrane currents. *Circ. Res.* **48**, 39-54.
[16] Winfree, A.T. (1983). Sudden cardiac death. *Sci. Am.* **248**, 144-161.
[17] Winfree, A.T. (1987). *When time breaks down: The three- dimensional dynamics of electrochemical waves and caridac arrhythmias.* Princeton University Press: Princeton, NJ.
[18] Wit, A.L., Rosen, M.R. & Hoffman, B.F. (1974). Electrophysiology and pharmacology of cardiac arrhythmias. II. Relationship of normal and abnormal electrical activity of cardiac fibers to the genesis of arrhythmias B. Reentry. *Am. Heart J.* **88**, 664-670.
[19] Wit, A.L. & Rosen, M.R. (1981). Cellular electrophysiology of cardiac arrhythmias. Part I. Arrhythmias caused by abnormal impulse generation. *Mod. Concepts Cardiovas. Dis.* **50**, 1-6.
[20] Wit, A.L. & Rosen, M.R. (1981). Cellular electrophysiology of cardiac arrhythmias. Part II. Arrhythmias caused by abnormal impulse conduction. *Mod. Concepts Cardiovas. Dis.* **50**, 7-12.
[21] Zaikin, A.N. & Zhabotinsky, A.M. (1970). Concentration wave propagation in two-dimensional liquid-phase self-oscillating systems. *Nature* **225**, 535-537.

28. NONLINEAR DYNAMICS AND IONIC MECHANISMS OF EXCITATION PATTERNS IN MODELS OF THE CARDIAC MYOCYTE

Alain Vinet, Dante R. Chialvo, Donald C. Michaels and
Jose Jalife

State University of New York Health Science Center
Syracuse, NY, USA.

Ventricular arrhythmias are spatial phenomena which result from one or more activation fronts travelling at a given time across the cardiac muscle. Two basic mechanisms have been postulated for the genesis and/or perpetuation of such depolarization wavefronts: (i) the activity of local pacemakers, referred to as ectopic foci [23,24]; and (ii) the formation of reentrant circuits, whereby the wavefront circles around an electrically excitable ventricular pathway [24]. Both mechanisms require that individual excitable cells involved in the arrhythmias be able to sustain high frequency pacing, and both mechanisms demand a description of the phase-locking behaviour of such cells when subjected to external driving.

Experiments in quiescent isolated nerve fibres subject to sinusoidal current stimulation, as well as numerical simulations using mathematical analogs of such fibres, have illustrated much of the phase-locking structure typical of forced oscillatory systems [1,4,16,18,26,28]. Apparently chaotic behaviour has also been demonstrated for some parameter combinations (amplitude, frequency, bias current, temperature), as well as for a model of single cardiac myocytes [20]. Recently, a formal approach was developed which provides an analytical basis to the bifurcation structure in such models [10-12]. Similar results have also been obtained in nerve and cardiac fibres driven repetitively with square depolarizing current pulses [21,22].

Recently, the behaviour of isolated cardiac Purkinje fibres stimulated repetitively with square current pulses has been investigated [7,8]. The bifurcation structure and phase-locking sequences have been obtained as functions of the basic cycle length (BCL) and amplitude (I_{st}) of the pulses. The bifurcation structure, expressed as activation ratio (AR : N:M, N = number of stimuli, M = number of active responses), was shown to change drastically when the potassium concentration of the bathing solution was changed. At high potassium concentrations, typical period adding sequences [8] were found; i.e. successive N:I patterns of activation were detected when the BCL was decreased. At low potassium concentrations, the bifurcation structure was highly dependent on I_{st}. For example, at intermediate values of I_{st}, regions of presumably chaotic irregular dynamics were encountered after sequences of period doubling as the BCL was progressively abbreviated.

The modification of the bifurcation structure with potassium concentration was shown to be primarily a consequence of the change of the threshold properties of the preparation; recovery of excitability being monotonic at high potassium concentrations, but highly nonlinear and with a supernormal phase at low concentrations. A difference-equation (DE) model was devised on the basis of electrophysiological parameters and shown to reproduce accurately the experimental results [8].

We have more recently compared the predictions of the DE model with those of numerical simulations using a modified highly dimensional ionic model [31,32] of the space-clamped cardiac myocyte. Iteration of the difference equation model reproduced all important features of the ionic model results, including a wide variety of frequency-dependent locking patterns, period doubling and chaotic dynamics. From those studies, we concluded that the bifurcation structure of the ionic model under repetitive stimulation may be understood to a large extent on the basis of two functions: excitability and action potential duration. We have extended those results and in the present study we demonstrate that, in the parameter plane, both models predict that the behaviour of the cardiac cell should change appreciably as a function of the stimulus duration. The overall results in our single cell models suggest that the pattern of propagation during repetitive stimulation in an extended structure comprised of many such cell models should depend on the degree of intercellular communication.

1. **Models of the cardiac myocyte**

1.1. *The "difference-equation" model*

The DE model predicts whether or not, following an action potential, a new stimulation will be successful in inducing a new active response. Briefly, the independent variable of the model is the diastolic interval (DI); i.e. the time elapsed from the end of the previous action potential to the beginning of the next stimulus. Three functions must also be defined: Thr(DI), which gives the minimum amount of current necessary to induce an action potential; Lat(DI), which defines the delay between the onset of a successful simulation and the beginning of the resulting action potential; and APD(DI), which is the duration of the action potential. Let DI_i be the diastolic interval preceding the i'th stimulus of intensity I_{st}; then if $I_{st} < Thr(DI_i)$, the stimulus is subthreshold, no action potential occurs, and

$$DI_{i+1} = DI_i + T_{i,i+1} \tag{1}$$

where $T_{i,i+1}$ is the time to the next stimulus. If $I_{st} \geq Thr(DI_i)$, then an action potential of duration $APD(DI_i)$ is induced after a delay $Lat(DI_i)$, and

$$DI_{i+1} = T_{i,i+1} - Lat(DI_i) - APD(DI_i) \tag{2}$$

The original DE model also includes a "memory" function [8]. Such a function modifies the action potential duration depending on previous activity, and multiplies the APD function. When the memory factor is removed and repetitive stimulation is applied to the model at fixed BCL and I_{st}, a unique function g(DI) = Lat(DI) + APD(DI) may be defined, and eq. (1,2) becomes

$$DI_{i+1} = BCL - g(DI_i), \quad I_{st} \geq Thr(DI_i)$$

$$DI_{i+1} = (BCL + DI_i), \quad I_{st} < Thr\ (DI_i) \qquad (3)$$

The DE model is thus reduced to the iteration of a one- dimensional function, a context in which numerous concepts from nonlinear dynamics are directly applicable [for review, see 19].

1.2. *The ionic model*

As a first approximation to providing an ionic basis to complex excitation patterns in cardiac cells, we have used the Beeler- Reuter model (MBR) [5] of the space-clamped myocyte with a modification of the sodium current [11]. MBR is a seven- dimensional Hodgkin-Huxley (H-H) type model [17] which includes the potential (V), the reversal potential of the calcium current (E_{si}) and five gating variables (Y_i, i=1,5). The model equations are as follows:

$$\frac{dV}{dt} = -I_{ion}(V, E_{si}, Y_i, i=1,5) + I_{st} \qquad (4)$$

$$I_{ion} = I_{Na}(V, Y_1, Y_2) + I_{si}(v, Y_3, Y_4, E_{si}) + I_k(V, Y_5)$$

$$\frac{dY_i}{dt} = \frac{1}{\tau_i(V)}(Y_{i\infty}(V) - Y_i), i = 1,5$$

$$\frac{dCa}{dt} = k_1 Y_3 Y_4 (V - E_{si}) - k_2 Ca$$

$$E_{si} = k_3 \ln(Ca)$$

All functions and parameters are the same as those described in [5], except for the sodium current (I_{Na}) which was taken from [11]. The inclusion of I_{si}, the slowly activating inward current, is the main difference between the H-H squid axon model [17] and the MBR description of cardiac myocyte. Under normal conditions, I_{si} is responsible for the long action potential plateau (~ 75 ms) which follows the upstroke and maintains the cell in a state of absolute refractoriness. Thus, physiologically, I_{si} acts as a protection mechanism against premature activation of the ventricle. Mathematically, I_{si} provides an additional time scale (~ 40 ms) which is intermediate between the time constant of activation for the fast inward I_{Na} responsible for the upstroke (< 1 ms) and that of the slow potassium repolarizing current (>400 ms).

1.3. *Calculating the Lyapunov exponent*

In the one-dimensional case

$$APD_{i+1} = g(DI_i) = g(n*BCL - APD_i) = F(APD_i)$$

where n is the number of stimuli preceding a successful response. Thus, the Lyapunov exponent [29] is:

$$e^{\lambda} = \lim_{n \to \infty} \left\| \prod_{i=0}^{n-1} \frac{d\ APD_i}{d\ APD} \right\|^{1/n}$$

or

$$\lambda = \lim_{n \to \infty} \frac{1}{n} \ln \left\| \prod_{i=0}^{n-1} \frac{d\ APD_i}{d\ APD} \right\|$$

2. Functions, parameters and bifurcation structures

The major goal of our study was to obtain the bifurcation structure of the ionic model as a function of stimulus frequency, intensity and duration and to relate such a structure to the dynamics of the ionic currents. The bifurcation structures of the MBR and DE models were also compared. Unfortunately, the high dimensionality of the MBR model and the nonlinearity of its functions preclude any analytical work concerning the solution of the periodically driven system. Moreover, two dimensional Fitzhugh-Nagumo [12,13,25] type approximations cannot be directly applied in the context of external driving since three significant time scales are involved in the dynamics. However, since the one-dimensional DE model accurately reproduces the important features of the full ionic model [31,32], it can be used to complement the MBR model, and to stress its important dynamical features. The ionic bases of such dynamics may then be investigated by a close scrutiny of individual current systems.

2.1. *Thr(DI) and g(DI) in the MBR model*

Experimentally, the Thr(DI), APD(DI) and Lat(DI) functions are obtained through the so-called S_1-S_2 protocol [8]. Briefly, the preparation is first paced using trains of 10 to 20 basic stimuli (S_1) of fixed magnitude (≥ 1.5 threshold from rest) and long period (S_1-S_1 = 1.5 to 2 s) until a stable 1:1 response is established. Premature test pulses (S_2) are then applied at various coupling intervals (S_1-S_2). Threshold and APD restitution curves for barely suprathreshold stimuli are then obtained as functions of S_1-S_2. DI is computed from S_1-S_2 by subtracting the APD and latency of the last action potential in the S_1 train. In order to define Thr and the beginning of the absolute refractory period (ARP, i.e. the shortest DI at which an action potential can be obtained), the protocol must be supplemented by specific criteria which provide a definition of "active response" [32]. The DE model assumes that APD(DI) is independent of the intensity of stimulation. This is not the case for latency, which obviously decreases as the intensity of stimulation is increased. Hence, for any fixed I_{st}, the Lat(DI) function provides an indirect measure of Thr(DI). As an approximation, Lat(DI) is set to be equal to k*Thr(DI); where k is a constant which is selected to obtain a value of latency typical of the I_{st} range under study.

The S_1-S_2 protocol was conducted in the MBR model, with a set of criteria for "active response" as described in [31,32]. Results are illustrated in

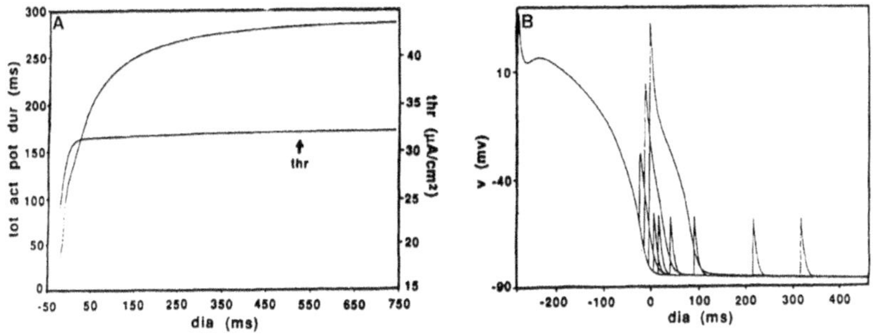

Figure 1. Thr(DI) and g(DI) in the MBR model for pulses of 1 ms. The system was paced with S_1 stimuli of 37.5 $\mu A/cm^2$, 1 ms duration and a BCL of 1.5ms. S_2 premature pulses of 1 ms were applied after the tenth beat. A: Left scale: total action potential duration, g = APD + k Thr (ms) as a function of the diastolic interval (ms), k is fixed at 0.1. Right scale: Threshold ($\mu A/cm^2$). B: Superimposed tracings showing last response to S_1 and responses of premature S_2 stimuli of 30 $\mu A/cm^2$ at various diastolic intervals.

Figs. 1 to 3. In all cases, the pacing S_1-S_1 period was 1.5 s. Figures 1 and 2 were obtained with square pulses S_1 of 37.5 $\mu A/cm^2$ and 1 ms duration, while square pulses of 2.75 $\mu A/cm^2$ and 100 ms duration were used in Fig. 3. Each figure was obtained with S_2 pulses that were of different duration; namely 1 ms (Fig. 1), 25 ms (Fig. 2) and 100 ms (Fig. 3). Panel A of each figure gives the corresponding Thr(DI) function (curve labelled thr; right vertical scale) and total action potential duration (left vertical scale; g(DI) = APD(DI) + k*Thr(DI), k being set to 0.1, 7.0 and 25.0, respectively, in Figs. 1, 2 and 3).

It may be readily seen that the shape of Thr(DI) changes dramatically with the duration of S_2. When S_2 is relatively brief (1 ms in Fig. 1A), a monotonically increasing Thr function is obtained. Such a function has an initial steep rise near the end of the ARP and a much slower progression toward the saturation value of the threshold at very long DI's. In this case, a brief stimulus whose amplitude is between the minimum of Thr(DI) and its saturation value (Thr(∞)) will induce an action potential only when applied at an extremely brief DI; i.e. immediately after the ARP. This phenomenon, which we have designated as "type-1 absolute supernormality" [8], is illustrated in Fig. 1B. An S_2 stimulus of 30 $\mu A/cm^2$, corresponding to 92% of Thr(∞), was applied at various diastolic intervals after the last S_1 response. The superimposed tracings show the last S_1 response, together with the depolarizations induced by S_2 at the various DI's. An active response, with a full-blown action potential, occurred only at a brief DI. All other responses were passive (i.e. electrotonic). Note that the premature response is

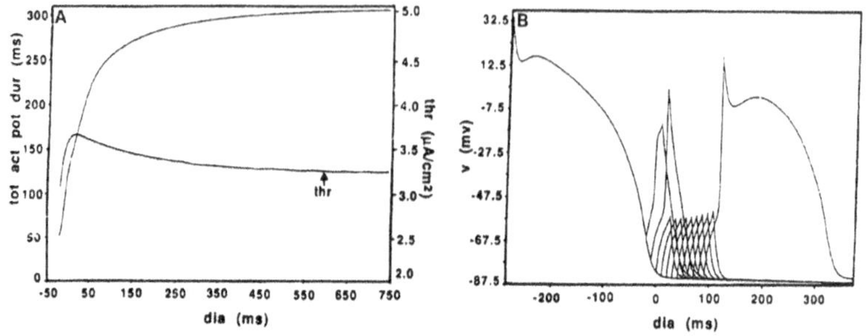

Figure 2. Thr(DI) and g(DI) functions for 25 ms. Same pacing protocol as in Figure 1; S_2 duration of 25 ms.
A: same as in Figure 1A, with k = 7.0. B: same as Figure 1B, but with pulse intensity = 3.5 $\mu A/cm^2$ and S_2 duration = 25 ms.

characterized by a steep rise of the potential and a maximum amplitude which are almost identical to that of the S_1 response, thus indicating that I_{Na} did contribute to the upstroke of this early event. Its duration, however, is much reduced, in accordance with the g(DI) function of Figure 1A. Note that the slope of the g(DI) function is not monotonically increasing; at short diastolic intervals, g(DI) shows a small segment whose slope is not steep.

When the duration of S_2 is 25 ms (Fig. 2A), a maximum appears in the Thr curve after an initial steep increase, which gives rise to the phenomenon which we have labelled as "type-2 absolute supernormality" [8]. In this case, $Thr_{min} < Thr(\infty) < Thr_{max}$. Consequences of type-2 supernormality are illustrated by the superimposed tracings presented in Fig. 2B. An S_2 stimulus of 3.5 $\mu A/cm^2$ was applied at various DI's. Since $Thr(\infty) < 3.5 < Thr_{min}$, active responses can occur at short as well as long DI's. However, the morphology of the action potential varies markedly, depending on the position of the test pulse. Near the ARP, action potentials have a very slow upstroke and low amplitude and duration, which indicates that these responses are mediated primarily by the slow inward current (I_{si}). On the other hand, the g(DI) function remains quite similar to that in Fig. 1A, retaining its gradual slope segment at short DI's.

As illustrated in Fig. 3, when the duration of S_2 is much longer (100 ms), the Thr curve displays only "relative supernormality" ($Thr(\infty) < Thr_{min} < Thr_{max}$), with a very short ascending portion before the maximum is reached. In Fig. 3B, a 100 ms test stimulus of 3.22 $\mu A/cm^2$ gives rise to premature responses at two different DI's, which are separated by a subthreshold (silent) region. In this case, the g(DI) function includes an appreciable contribution of latency (25*Thr = 75 ms) which is responsible for the shape of the curve at short DI's.

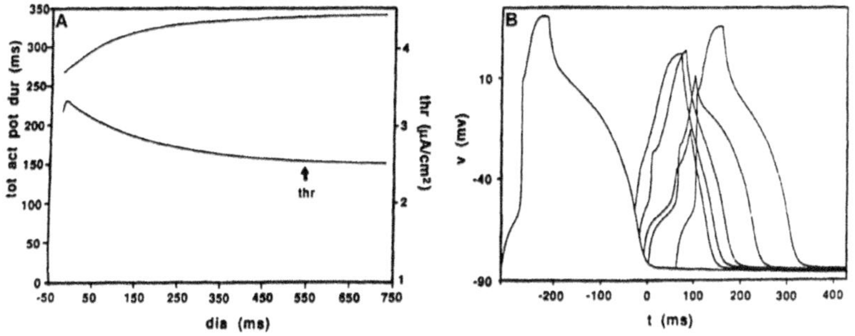

Figure 3. Same as in Figure 1; S_1 = 100 ms in duration and 2.75 $\mu A/cm^2$ in intensity; and S_2 duration = 100 ms. A: same as in Figure 1A, with k = 25.0. B: same as in Figure 1B, with S_2 intensity = 3.22 $\mu A/cm^2$ and duration = 100 ms.

2.2. *Ionic mechanisms*

Absolute supernormality has also been observed in the H-H model [30]. The mechanism responsible for type-2 absolute supernormality, and for the occurrence of a local maximum (Thr_{max}) in the threshold curve of the H-H model, has been attributed to the occurrence of a delayed after-hyperpolarization at the end of the action potential. When such an after-hyperpolarization is prevented from occurring, the threshold curve becomes similar to the monotonically increasing Thr function of Fig. 1A (i.e. type-1 supernormality). After-hyperpolarization does not exist in the MBR model (see paced action potentials, Figs. 1 to 3B). Yet, if the duration of S_2 is brief, the MBR threshold curve is similar to the H-H threshold function with suppressed hyperpolarization. In both models, the fast inward sodium current must be responsible for the particular shape of the threshold function. Indeed, in both models, I_{Na} activates so rapidly and its conductance becomes so large relative to the outward current that both have literally a fixed voltage threshold. Such a threshold corresponds to the potential at which the activation gate of I_{Na} begins to open. The threshold curve thus gives a measure of the distance between threshold voltage and the potential at which the onset of stimulation occurs. As the stimulation becomes more premature, the starting voltage increases and Thr diminishes. This occurs only if the inactivation variable of I_{Na}, which closes during the preceding upstroke, has recovered to a minimal value (~0.01). If the stimulation is too premature, the inactivation variable is still closed, the system is in the state of absolute refractoriness, and no active response may be initiated.

At longer S_2 durations, the inactivation gating variable of the inward sodium current closes partially during the course of stimulation. This reduces the availability of I_{Na} and hence moves the threshold voltage to less negative values. This effect becomes more important if the initial value of the inactivation gate is relatively low, as happens at brief DI's. As the threshold potential moves up, the amount of current needed to reach that threshold also increases, as seen in Fig. 2A. However, as DI is further reduced, the slow inward current begins to contribute significantly to the response. At short DI's, the cell is still depolarized at the beginning of the stimulation and the activation variable of the slow inward current overshoots its resting value. This initial overshoot, together with the long stimulus duration, allows the inward current to induce an action potential with a slow upstroke (see Fig. 2B) at short diastolic intervals and explains the early drop in the Thr function. As S_2 is increased even further (Fig. 3), inactivation of the calcium current and activation of the outward potassium current are allowed to occur during the stimulus, which leads to a reduction in the DI range of supernormal responses.

APD(DI) curves were computed for all S_2 pulse durations shown in Figs. 1 to 3. For each S_2 duration, the APD curve was practically independent of the period or pulse duration used in the conditioning S_1-S_1 train, which confirms that long-term "memory effects" do not exist in the MBR model. A comparison of the results of Fig. 1B with those of Fig. 3B reveals that the action potential upstrokes are very different from each other, while at the end of repolarization, particularly below -40 mv, the responses look very much the same. This comparison also applies for the time course of the currents and the dynamic variables.

As discussed above, the mechanism of the upstroke and, consequently the threshold curve, are very much dependent on the relation between the characteristics of the S_2 stimulus and the time constants of activation of the inward currents. In contrast, the repolarization phase of the action potential is controlled by the slow inactivation of I_{si} and by the activation of the potassium current, I_k ($t > 400$ ms). In the phase plane, all action potentials converge toward the same trajectory before the end of repolarization and reach together the arbitrary voltage value which is set to define the end of the action potential and the beginning of the diastolic interval. DI is thus a good measure of the state of the system, independent of its past history. This fulfils a first condition necessary for the difference equation model to be applicable in the analysis of the dynamics of MBR. In addition, if, for a given stimulus, the slow variables are not much affected during the time elapsed from the onset of stimulation to the resulting upstroke, the following APD will still be primarily dependent on the starting DI, which is a second condition necessary for the relevance of the DE model. This condition was shown to be reasonably fulfilled for the case of the MBR model, even for the S_2 duration of 100 ms.

Finally, the duration of S_2 is the major determinant of the latency and thus of the total action potential duration (Lat(DI) + APD(DI)). If the intention is to use the DE model for one particular stimulus amplitude, the APD and Lat functions may be constructed for that specific current intensity. However, our aim was to study a wide range of stimulus current amplitudes and, in the range of current studied, APD functions were shown to be reasonably independent from the intensity of stimulation. This was obviously not the case for the Lat(DI) functions. Latency is highly dependent on the stimulus duration as well as on the current intensity above the threshold. However, the shape of Lat(DI) curves obtained with fixed suprathreshold stimuli is very similar to that of the corresponding Thr(DI) function. In our study, the decrease in latency with increase of stimulus current intensity was not uniform at any given DI. In fact, the latency decreases rapidly for a short range of currents above the threshold. Subsequently, with further increases in stimulus amplitude, the decrease becomes quasi-linear. Thus, as a compromise in our use of the DE model, Lat(DI) was expressed in all cases as k*Thr(DI), k being a scaling factor such that the resulting values corresponded to the observed range of latencies for the specific current interval under consideration. Since memory effects were excluded, the DE model is then reduced to the iteration of the single functions shown in panel A of Figs. 1, 2 and 3.

2.3. *Bifurcation structure for DE model*

To obtain the bifurcation structure of the DE model, numerical simulations were performed for repetitive stimulation using the various Thr(DI) and g(DI) functions of Figs. 1A, 2A and 3A. Wide ranges of I_{st}'s and BCL's were explored for each set of curves. For each I_{st} and BCL value, the system was set at rest (DI ⇒ ∞) before the onset of repetitive stimulation. At a given BCL, the presence of stimulus:response locking and periodicity was assessed following a procedure described in [32], with a maximum of 1000 repetitive stimuli in each run. When periodicity was detected, AR of the stable sequence was computed. Otherwise the mean AR of 1000 beats was retained. Results are presented as parameter spaces, where the loci of AR changes are indicated. Detailed bifurcation structures are also presented for specific values of I_{st} (see Figs. 5A and 8A to C). The latter are plots of g(DI) values of the stable periodic sequences, or of 50 successive beats in non-periodic cases. g(DI) of skipped beats was set to 0.

As shown in Fig. 4A, for S_2 of 1 ms, a zone of irregular dynamics (I.D.), with no detected periodicity, is observed as the BCL is abbreviated progressively. The zone of irregular dynamics occurs after the sequence of bifurcation 1:1 ⇒ 2:2 ⇒ 2:1 ⇒ 4:2, and ends at briefer BCL with the appearance of a stable 3:1 pattern. The transition from 4:2 to the irregular patterns is not direct, but is preceded by a sequence of 2n:n+1 AR, in which n decreases from 32 to 4. The detailed bifurcation structure for 1 ms current pulses of 37.5 $\mu A/cm^2$ is presented in Fig. 5A. The BCL interval range with 2n:n+1 AR

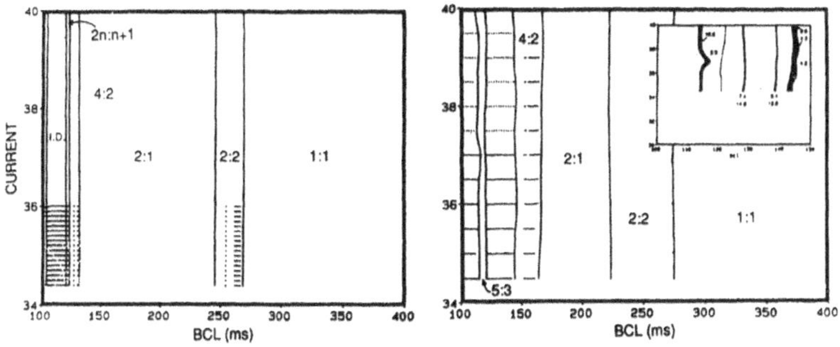

Figure 4. Parameter spaces of the DE (A) and MBR (B) models, for repetitive stimuli of 1 ms. Lines indicate where changes in stimulus:response pattern occur. No stable period was detected in the regions labelled I.D. (Irregular Dynamics). The insert in panel B details the transitions after the 4:2 region.

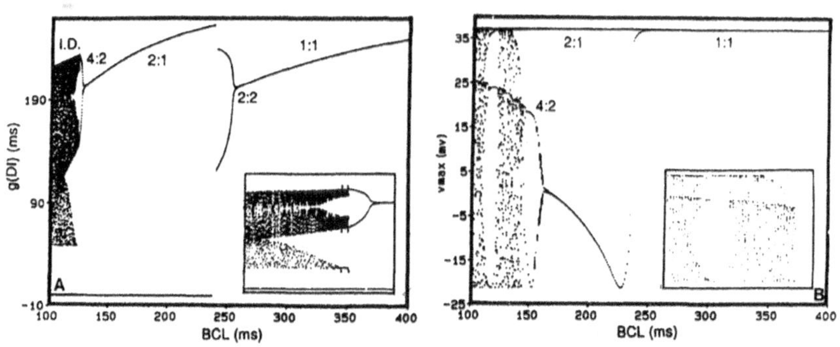

Figure 5. Detailed bifurcation diagram for I_{st} = 37.5 $\mu A/cm^2$ and 1 ms pulses. For each BCL, plots show stable values or 50 successive beats if no periodicity was detected. A: DE model, g(Di) (ms) as a function of BCL (ms). In the region between the two marks in the insert, successive 2n:n+1 ratios are detected before the onset of irregular dynamics (I.D). B: MBR model; the maximum potential reached after each stimulation, V_{max} (mv), is plotted in the vertical scale. The insert provides details of the region near the 4:2 stimulus:response pattern.

appears in the inset. This peculiar transition sequence also appears for the stimulus duration of 25 ms, and is discussed in detail in [32]. Such a transition seems to result from the non-monotonicity of dg(DI)/dDI which has a minimum at a short diastolic interval. In both cases, the zones of irregular dynamics are characterized by a Lyapunov exponent greater than 0 (see Fig. 9), indicating that the dynamics there are truly chaotic [29].

In the parameter plane of Fig. 4A, transition lines begin to appear near the current amplitude of 34.3 $\mu A/cm^2$. Beneath this point, only n:0 AR's were detected. In this case, any decrease in current to a level below the maximum threshold (Thr(∞) = 34.34 $\mu A/cm^2$) would abruptly reduce the g(DI) function to a short branch near the beginning of the ARP (Fig. 1A). This branch is too short to sustain any stable pattern of activation.

It is readily seen that the DE model's parameter plane associated with stimulus duration of 25 ms (Fig. 6 A,B) has the most complicated structure. In this case, the bifurcation structure is very dependent upon stimulus intensity as could be predicted by the examination of the corresponding Thr function (Fig. 2A). In fact, the results with 25 ms pulses provided one of the most convincing pieces of evidence for the applicability of the DE model to real physiological preparations [7,8]. Detailed description of the parameter plane under these conditions has been presented elsewhere [32]. Here we will discuss only the major features of these data.

As shown in panel A of Fig. 6, the parameter plane for 25 ms pulses may be divided into three major regions of current intensity (horizontal broken lines). In region III, the current intensity is always higher than Thr_{max}, and the complete g(DI) function is used for iteration. The sequence of transition is exactly similar to that of Fig. 4A (1 ms pulses), except for the structure

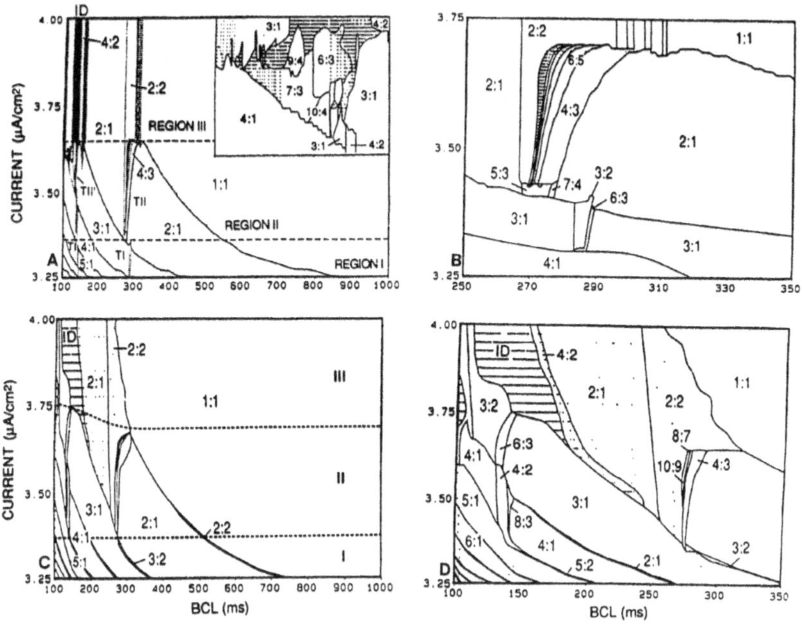

Figure 6. Parameter spaces for stimuli of 25 ms. A,B: DE model. The plane could be divided into three regions of current intensity, each with a typical bifurcation structure. The inset of panel A details the region labelled TII'. C,D: MBR model.

of the zone of irregular dynamics. In the 1 ms case, the dynamics were uniformly chaotic throughout the entire range of BCL, as confirmed by analysis of the Lyapunov exponent (see Fig. 9A). On the other hand, for stimuli of 25 ms, the I.D. zone is interrupted by short bursts of stable periodicities, as may be seen from the Lyapunov exponent for $I_{st} = 4\ \mu A/cm^2$ (see Fig. 9B).

In the 25 ms stimulus case, as the current intensity is decreased, an enlarging gap is introduced in the g(DI) function (see Fig. 2B). Iterations falling in this gap correspond to subthreshold stimuli with failed responses. The position of this gap in relation to solutions obtained with the complete g function in region III explains the occurrence of different activation ratios in regions I and II. Details for these transitions are given in panel B of Fig. 6 and in the inset of panel A which presents an enlarged representation of the zone labelled TII'. In region I, with very low current intensity, the section of g(DI) which is accessible at low DI's is very small. The accessible branch of g(DI) extending towards DI = ∞ becomes the main determinant of the bifurcation structure whose major feature is a succession of n:1 ratios with the characteristic Arnold tongue organization [14].

Examination of g(DI) obtained for S_2 = 100 ms (Fig. 3A) shows only a very small decrease at short diastolic intervals. When the complete function is available for iteration (i.e. at high current intensities), the fixed point of

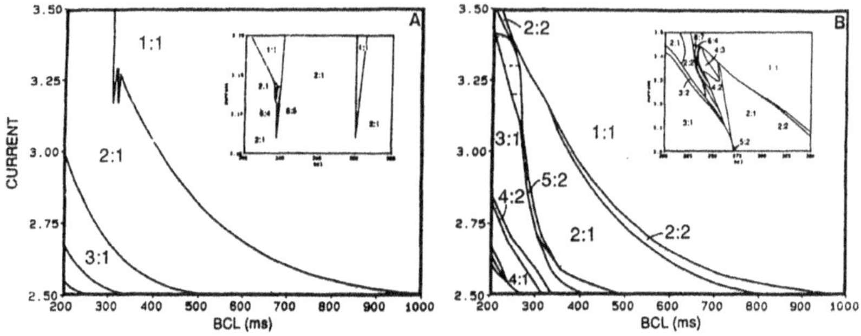

Figure 7. Parameter spaces for stimuli of 100 ms. A: DE model. The insert details the peaks in the full diagram. B: MBR model. The insert gives details of the closed non-labelled region near the 3:1 zone. Note the absence of any irregular activity in this case.

the first iterate (DI = BCL - g(DI)) remains stable as long as it exists. When it no longer falls on the function, the second iterate takes its place, and successive n:1 ratios are detected (see Figs. 7A and 8A to C). A similar bifurcation structure was demonstrated for the 25 ms case at low current intensities (see region I in Fig. 6A), the only difference being that, in the latter case, short transitional zones (labelled TI) become manifest. Such transitions are demonstrable at pulse durations of 25 ms because, in that case, the g(DI) function has two branches which are separated from each other by a gap at low I_{st}. The shape of the main branch at high DI's is similar to that of the g function at 100 ms, and results in the same bifurcation structure. Consequently one may conclude that the bifurcation structure in the 25 ms case is somewhat generic since it contains the characteristics associated with pulse duration of 1 and 100 ms.

Returning to the 100 ms threshold function (see right scale in Fig. 3A), one may see that, except for a short interval of current near the maximum threshold, reduction of the current intensity should result in a truncated g(DI) function extending to DI = ∞. In this case g(DI) will also have a very short branch near the minimum DI. The influence of this short branch may be appreciated by analysing the results presented in the insert of Figs. 7A and 8B.

2.4. *Bifurcation structure of the MBR model*

The bifurcation structure of the MBR model was investigated for stimulus durations of 1, 25 and 100 ms. Resting state was always used as the initial condition. As discussed in [32], at any given BCL, we searched for periodicity in a maximum of 400 stimulations. AR was then computed. Any depolarization with a net inward current lasting for more than 0.5 ms ($-I_{ion} > 0$) was considered to be an active response. The resulting parameter planes are presented in Figs. 4B, 6C and 7B. Detailed bifurcation structures are presented for specific I_{st} values (see Figs. 5B and 8D to 8F). In panels D to F of Fig. 8, the maximum voltage (V_{max}) reached after each stimulation is

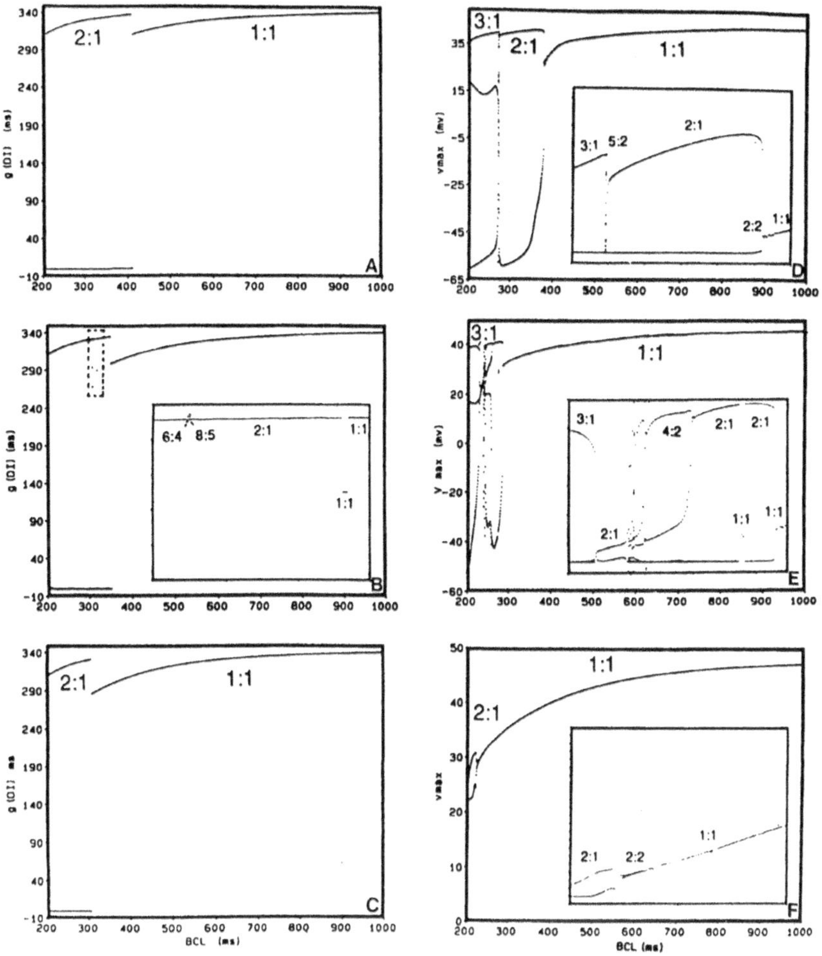

Figure 8. Detailed bifurcation structure for 100 ms pulses. Panels A, B and C are from the DE model. Panels D, E and F are from the MBR model. I_{st} = 3 (A,D), 3.17 (B), 3.3 (E) and 3.5 (C,F) $\mu A/cm^2$. Inserts in panels D to F give $-I_{ionmax}$.

plotted as a function of BCL, to illustrate the stable periodic sequences as well as 50 successive beats for each case in which periodicity was not detected.

As illustrated in panel B of Fig. 4, the parameter plane of the MBR model for a pulse duration of 1 ms is very similar to that of the corresponding DE model (panel A). In both models, as current intensity is increased there is an abrupt change from n:0 to the characteristic 1:1 $\Rightarrow$ 2:2 $\Rightarrow$ 2:1 $\Rightarrow$ 4:2 $\Rightarrow$ I.D. $\Rightarrow$

3:1 sequence. 3:1 patterns start at a slightly lower BCL than 100 ms (not shown in figure). However, there are some differences between the two models for the zone enclosed by the 4:2 and 3:1 regions. Firstly, the transition from 4:2 to irregular dynamics in the MBR model occurs through a classical sequence of period doubling bifurcations [9]. Periods 8, 16 and 32 were successively observed at decreasing BCL, in place of the decreasing 2n:n+1 cascade of the DE model. Secondly, the zone of irregular dynamics is not uniform, as in the DE model, but is interrupted by the periodic patterns depicted in the insert of Fig. 4B. These can also be seen in Fig. 5B which was obtained for a current of 37.5 $\mu A/cm^2$.

For pulse duration of 25 ms, the global correspondence between the parameter plane of the DE (Fig. 6, panels A and B) and MBR (Fig. 6, panels C and D) models is impressive, particularly in relation to the changes of the bifurcation structure as the amplitude of stimulation is changed. As in the DE model at high current intensity, in region III, the MBR model also displays the sequence of bifurcation 1:1 ⇒ 2:2 ⇒ 2:1 ⇒ 4:2 ⇒ ID ⇒ 3:1. In the irregular dynamics zone, short bursts of stable patterns (ΔBCL =.1 to .3 ms) with different periods (4, 8, 12, 16) are intermingled with regions in which no stable period could be detected. A similar situation is illustrated for the DE model by the Lyapunov exponent of Fig. 9B. The only difference is in the transitional zone from 4:2 to irregular dynamics. For the MBR model, this region was scanned using BCL steps of 0.1 ms. Both period and AR were very irregular. Points with no detected periodicity were interposed between periodic regions (periods 4, 8, 12, 16) without period adding or doubling organization.

Two major types of transition are also present in both models in region II, corresponding to intermediate current intensities. Differences are mainly related to the zone labelled TII' in Fig. 6A (also in the inset). In this region, the DE model gives a complicated family of ratios, with a large number of dropped beats (i.e. subthreshold events). In the DE model, dropped beats are assumed to have no influence on the dynamics, while in the MBR model, subthreshold events can slightly prolong the previous action potential, particularly when they occur at short BCL's. As discussed elsewhere [32], we

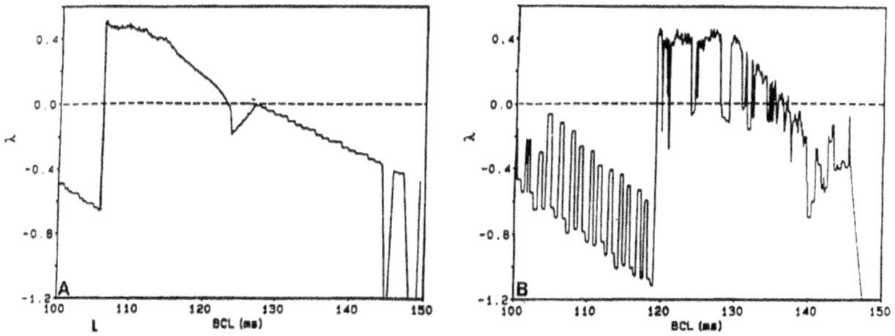

Figure 9. Lyapunov exponent of the DE model for BCL intervals including regions where irregular dynamics were detected. A: I_{st} = 37.5 $\mu A/cm^2$, S2 = 1 ms, corresponding to Figures 3 and 4. B: I_{st} = 4 $\mu A/cm^2$, S2 = 25 ms, corresponding to Fig. 5. Lyapunov exponent > 0 indicates chaotic activity.

have proposed the effect of subthreshold events to be the main source of difference between the two models. Finally, in region I, the MBR model has the same successive n:1 Arnold tongues, but it does not present the short TI transition seen in the parameter plane of the DE model. Such a difference can also be attributed to the effect of subthreshold stimuli and to the increase of the latency for nearly threshold stimuli in the MBR model. The latter effect was removed from the DE model since only one latency function was used for all intensities of stimulation.

For pulse duration of 100 ms, the correspondence between the DE and the MBR models is also very good (Figs. 7 and 8). It should be noted that in this case the maximum ionic current used as a criterion for active upstroke was often very low (see insert Fig. 8D to F), making the ratio markedly unstable in some regions. In Fig. 7B, the occurrence of a 5:2 region between 2:1 and 3:1 could be ascribed to a prolongation of the latency for barely suprathreshold stimuli. The complicated region for intermediate current intensity (see insert Fig. 8B) recorded for the MBR model is somewhat similar to the DE model's results presented in the insert of Fig. 8A, although the detailed structures of the two maps are different.

3. Discussion

3.1. *The DE vs. MBR models*

The DE model is a remarkably good predictor of the bifurcation structure of the MBR model, although some differences exist within restricted BCL and I_{st} intervals. The major achievement of using the DE model is that it is possible to predict correctly the effects of supernormality, and specifically to demonstrate that absolute supernormality yields a bifurcation structure that is highly dependent on the intensity of the stimulus.

The agreement between the two types of model results depends on the existence of a strongly attractive post-upstroke trajectory in the MBR model. Such a trajectory is controlled by slow variables that, for a wide range of stimulus intensity, duration and frequency, are not much influenced by the exact nature of the upstroke. The slow inward current, I_{si}, slows down the repolarization process and transiently holds the membrane potential at a depolarized level after an upstroke, thus creating a time window during which the evolution of the slow variables becomes almost autonomous. This allows the system to return to its nearly invariant trajectory of repolarization and explains the predictive power of the DE model. I_{si} is also responsible for the appearance of supernormality when its time constant of activation matches the stimulus duration.

We have identified two factors as contributors to the difference between the two models: latency and effects of subthreshold stimuli. In the MBR model, latency was shown to vary in a non-monotonic fashion. Barely suprathreshold stimuli lead to responses which are preceded by long latencies. Such latencies are longer than the stimulus duration. Latency decreases, first rapidly, and then more gradually as current intensity is increased. This could be easily incorporated into the DE model by constructing latency curves for specific sets of I_{st}, and using those curves in an interpolation procedure for any given range of I_{st} under study.

A second and more profound difference between the two models is the effect of subthreshold stimuli occurring during repolarization. Such stimuli were shown to prolong slightly the action potential in the MBR model with a consequent shortening of the following diastolic interval. This shortening of DI may have different effects, depending on the context in which it happens. In some instances, it suppresses a beat predicted by the DE model. In other cases, it allows the occurrence of an additional upstroke; changing for example a 4:2 into a 4:3 as is seen near the I.D. region of the parameter plane for a duration of 1 ms (Fig. 4B). In this particular case, the failed response allows shortening of the following DI which in turn leads to a reduction of the duration of the following action potential, and allows the immediate occurrence of an additional action potential. The DE model could be extended to include the effects of subthreshold stimuli. However this modification would be at the expense of the simplicity of the model, which is a major asset since it gives it the power of prediction at a low computational cost, and allows access to all the resources associated with nonlinear dynamics tools.

We will close the discussion by outlining a few problems that are currently under investigation in our laboratory. Calculation of the Lyapunov exponent has confirmed that the regions of irregular dynamics in the DE model are truly chaotic (Fig. 9). Although such a test has not been performed as yet for the MBR model (for sinusoidal input, see 9), return maps for 25 ms pulse durations, have been analysed [32]. The structure of such maps was shown to be similar to the corresponding return map of the DE model, strongly suggesting that chaotic activity was also present in the MBR model. However, definite diagnosis of chaotic activity in this model, as well as evaluation of its dimensionality, require computation of the Lyapunov exponents.

In the DE model, transition to chaotic activity happens through a cascade of 2n:n+1 AR, in which n decreases as BCL is reduced. Such a transition mode has never been reported for published cases in which g functions with monotonic slope decrease were used [21]. This suggests that non-monotonicity in dg(DI)/dDI might be required for this bifurcation scenario to occur. However, there is no formal proof that this is indeed the case.

Finally, it should be noted that repetitive 100 ms stimuli with BCL around 200 ms represents a kind of limiting case. Constant application of depolarizing current in the range of intensities of Figs. 7 and 8 may lead to automaticity in the MBR model (A. Vinet, unpublished observations). In such a limit, the situation of applying a square depolarizing stimulus for 100 ms at a BCL < 200 ms is equivalent to applying a repetitive hyperpolarizing perturbation with duration < 100 ms to an autonomous oscillatory system. In the latter case of the perturbed oscillator, iteration of a phase resetting curve (PRC) becomes the appropriate low dimensional analog for the MBR model [6,15]. There might be a continuous transition from DE to PRC models as the pulse duration increases. Understanding the nature of the transition should prove to be an interesting and challenging problem.

3.2. *Pathophysiologic implications*

The major goal of our research work is to determine the cellular substrate underlying the initiation and maintenance of cardiac arrhythmias. In this article, we have shown that models of individual cells may sustain high frequency pacing and, under certain circumstances, may undergo chaotic activity. The most complex families of behaviour are those associated with supernormality when depolarizing pulses of intermediate duration are used as repetitive stimuli. The implications of our results in terms of cardiac arrhythmias may be readily appreciated in the context of conduction delay between two regions of extended tissues which are separated by a zone of depressed excitability [19]. Such zones are often though of as direct or indirect sources of arrhythmias [3]. If one thinks of the DE or MBR models as representing a cell or group of cells distal to the zone of depression, then one can also think of the depolarizing current pulse as representing the electrotonic image of an active response initiated proximally and being blocked in the depressed zone. Under the conditions in which the degree of depression is relatively low, repetitive stimulation of the proximal tissue at a slow BCL should lead to 1:1 activation of the distal end. This would be analagous to application of brief (e.g. 1 ms) depolarizing pulses of high intensity at a slow BCL. As predicted by the DE and MBR models, abbreviation of the BCL in the extended one-dimensional cable should lead to a sequence of propagation of the form 1:1 ⇒ 2:2 ⇒ 2:1 ⇒ 4:2 ⇒ I.D. across the depressed zone. The sequence should end at brief BCL's with the appearance of a stable 3:1 pattern. On the other hand, at intermediate ranges of depression, the situation at the distal end would become more analogous to that presented for the case in which pulses of 25 ms were used (see Fig. 6). In such a case, extremely rich dynamics of propagation would be expected which will depend not only on the BCL but on the specific degree of block. Experimental evidence in this regard has already been published [7]. Thus, as such, the present results provide additional insight into some of the dynamics of excitation and cell-to-cell communication which may be involved in the genesis of arrhythmias.

References

[1] Aihara, K. & Matsumoto, G. (1987). Forced oscillations and route to chaos in the Hodgkin-Huxley axons and squid giant axons. In *Chaos in Biological Systems,* Holden, A.V. & Olsen, L.F. (eds.), pp. 121-131. NATO ASI series, Series A: Life Sciences, Vol. 138. Plenum Press: New York.

[2] Alexander, J.C., Doedel, E.J. & Othmer, H.G. On the resonance structure in forced excitable systems. *Siam J. App. Math.,* in press.

[3] Antzelevitch, C., Jalife, J. & Moe, G.K. (1980). Characteristics of reflection as a mechanism of re-entrant arrhythmias and its relation to parasystole. *Circulation* **61**, 182-191.

[4] Arrigo, P., Marconi, L., Morgavi, G., Ridella, S., Rolando, C. & Scalia, F. (1987). High sensitivity chaotic behaviour in sinusoidally driven Hodgkin-Huxley equations. In *Chaos in Biological Systems,* Holden, A.V. & Olsen, L.F. (eds.), pp. 113-119. NATO ASI series, Series A: Life Sciences, Vol. 138. Plenum Press: New York.

[5] Beeler, G.W. & Reuter, H. (1977). Reconstruction of the action potential of ventricular myocardial fibres. *J. Physiol.* **268**, 177-210.

[6] Chay, T.R. & Lee, Y.S. (1985). Phase resetting and bifurcation in the ventricular myocardium. *Biophys. J.* **47**, 641-651.

[7] Chialvo, R.D. & Jalife, J. (1987). Non-linear dynamics of cardiac excitation and impulse propagation. *Nature* **330**, 749-752.

[8] Chialvo, R.D., Michaels, D.C. & Jalife, J. Supernormal excitability as a mechanism of chaotic dynamics of activation in cardiac Purkinje fibers. *Circ. Res.,* in press.

[9] Devaney, R.L. (1989). *An introduction to chaotic dynamical systems,* pp. 1-147. Addison-Wesley: New York.

[10] DeYoung, G. & Othmer, H.G. Resonance in oscillatory and excitable systems. *N.Y. Acad. Sci.,* in press.

[11] Drouhard, J.P. & Roberge, F.A. (1987). Revised formulation of the Hodgkin-Huxley representation of the sodium current in cardiac cells. *Comput. Biomed. Res.* **20**, 333-350.

[12] Fitzhugh, R. (1960). Threshold and plateaus in the Hodgkin- Huxley nerves equations. *J. Gen. Physiol.* **43**, 867-897.
[13] Fitzhugh, R. (1961). Impulses and physiological states in theoretical models of nerve membrane. *Biophys. J.* **1**, 445-461.
[14] Glass, L. & Belair, J. (1986). Continuation of Arnold tongues in mathematical models of periodically forced biological oscillators. In *Nonlinear oscillations in biology and chemistry,* Othmer, H.G. (ed.), pp. 232-243. Vol. 66 of Lecture Notes in Biomath. Springer-Verlag: New York.
[15] Glass, L. & Mackey, M.C. (1988). *From clocks to chaos*, p. 99-143. Princeton Univ. Press: Princeton.
[16] Hayashi, H. & Ishizuka, S. (1987). Chaos in molluscan neuron. In *Chaos in Biological Systems,* Holden, A.V. & Olsen, L.F. (eds.), pp. 157-166. NATO ASI series, Series A: Life Sciences, Vol. 138. Plenum Press: New York.
[17] Hodgkin, A.L. & Huxley, A.F. (1952). A quantitative description of membrane current and its application to conduction and excitation in nerve. *J. Physiol.* **177**, 500-544.
[18] Holden, A.V. (1976). The response of excitable membrane models to a cyclic input. *Biol. Cybernetics* **21**, 1-7.
[19] Jalife, J. & Moe, G.K. (1981). Excitation, conduction and reflection of impulses in isolated bovine and canine cardiac Purkinje fibers. *Circ. Res.* **49**, 233-247.
[20] Jensen, J.H., Christiansen, P.L., Scott, A.C. & Skovgaard, O. (1984). Chaos in the Beeler-Reuter system for the action potential of ventricular myocardial fibres. *Physica* **D13**, 269-277.
[21] Lewis, T.J. & Guevara, M.R. Chaotic dynamics in an ionic model of the propagated action potential. (Submitted)
[22] Matsumoto, G., Takahashi, N. & Hanyu, Y. (1987). Chaos, phase locking and bifurcation in normal squid axon. In *Chaos in Biological Systems,* Holden, A.V. & Olsen, L.F. (eds.), pp. 143-165. NATO ASI series, Series A: Life Sciences, Vol. 138. Plenum Press: New York.
[23] Moe, G.K. & Mendez, C. (1973) Physiologic basis of premature beats and sustained tachycardia. *N. Eng. J. Med.* **288**, 250-254.
[24] Moe, G.K. (1975). Evidence for reentry as a mechanism of cardiac arrhythmias. *Rev. Physiol. Biochem. Pharmacol.* **72**, 55-81.
[25] Nagumo, J., Arimoto, S. & Yoshizawa, S. (1962). An active pulse transmission line simulating nerve axon. *Proc. IRE* **50**, 2061-2070.
[26] Nagumo, J.I. & Sato, S. (1972). On a response characteristic of a mathematical neuron model. *Kybernetik* **10**, 155-164.
[27] Othmer, H.G. The dynamics of forced excitable systems. In *Nonlinear wave processes,* Holden, A.V., Markus, M. & Othmer, H.G. (eds.). Plenum Press, in press.
[28] Sato, S., Masaaki, H. & Nagumo, J.I. (1974). Response characteristics of a neuron model to a periodic input. *Kybernetik* **16**, 1-8.
[29] Schuster, H.G. (1984). *Determinstic chaos: an introduction,* p. 101. Springer-Verlag: New York.
[30] Stockbridge, N. (1988). Etiology for the supernormal period. *Biophys. J.* **54**, 77-78.
[31] Vinet, A., Chialvo, D.R. & Jalife, J. Irregular dynamics of excitation in biological and mathematical models of cardiac cells. In *Electrocardiography: Past and future,* Garfein, O.B. & Coumel, P. (eds.). *N.Y. Acad. Sci.*, in press.
[32] Vinet, A., Chialvo, D.R., Michaels, D.C. & Jalife, J. Nonlinear dynamics of rate-dependent activation in models of single cardiac cells. (Submitted)

29. A MODEL OF EXCITATION WAVEFRONTS SPREADING IN THE ANISOTROPIC CARDIAC TISSUE

P. Colli Franzone*, L Guerri*, S. Rovida** and S. Tentoni**

*Dipartimento di Informatica e Sistemistica
dell'Università di Pavia, Pavia, Italy
**Istituto di Analisi Numerica del C.N.R.
Corso C. Alberto 5, Pavia, Italy

1. Modelling the excitation wavefront spreading in the cardiac tissue

The bioelectric activity of the myocardium during the excitation process, also called **depolarization phase**, is associated with a propagating front-like variation of the transmembrane potential which is the jump of the potential across the cellular membrane. In this phase the transmembrane potential undergoes an abrupt change from a resting to an excited value occurring in a moving layer about 1 mm thick and lasting about 2 msec.

To describe the excitation process we use a macroscopic model of the cardiac tissue called **bidomain** or **syncytium** of electrically coupled media. In this model the cardiac tissue is defined by two interpenetrating continuous conducting media occupying the same volume: the **intracellular** and the **extracellular** (or interstitial) medium which will be referred to as (i) and (e). These two media are connected at each point by the cardiac cell membrane. Due to the fibre structure of the myocardium these two media are anisotropic; concerning the electric conductivity we assume axial symmetry around the fibre direction. Moreover the conductivity tensors of these two media M_i, M_e have the same principal axes, one parallel and two transverse to the fibre axis. Assuming a **regular** fibre structure, we denote by $\underline{a}_l = \underline{a}_l(\underline{x})$ a unit vector parallel to the fibre direction at a point $\underline{x}$. We introduce the matrix $A(\underline{x}) = [\underline{a}_1(\underline{x}), \underline{a}_2(\underline{x}), \underline{a}_3(\underline{x})]$ where $\underline{a}_1$, $\underline{a}_2$, $\underline{a}_3$ are three mutually orthogonal unit vectors with $\underline{a}_3$ parallel to $\underline{a}_l$ and $\underline{a}_1$, $\underline{a}_2$ defined apart from a rotation.

Setting $M^*_{i,e}(\underline{x}) = \text{diag}\,(\sigma_t^{i,e}(\underline{x}), \sigma_t^{i,e}(\underline{x}), \sigma_l^{i,e}(\underline{x}))$ where $\sigma_l^{i,e}$, $\sigma_t^{i,e}$ are the conductivity coefficients of the intra-, extra- cellular medium measured respectively along a direction parallel or perpendicular to the fibre axis at the point $\underline{x}$, the conductivity tensors M_i, M_e in the global (cartesian) reference system are given by $M_{i,e}(\underline{x}) = AM^*_{i,e}A^T$. Hence at any point M_i, M_e are symmetric positive definite matrices with at most two distinct eigenvalues. The ventricular heart volume (the bisyncytium (i) + (e)) is denoted by H and

$\underline{a}_l$ is assumed regular in H. In the media (i) and (e) the flow of current is described by

$$\underline{j}_i = -M_i \nabla U_i \; , \quad \underline{j}_e = -M_e \nabla U_e \tag{1}$$

where $\underline{j}_i$, $\underline{j}_e$ and U_i, U_e are respectively the intra-, extra-cellular current density and the electric potential. Under the usual assumption that the cardiac electric fields are quasi-static, the current conservation law implies div $(\underline{j}_i + \underline{j}_e) = 0$. The effective cardiac source of the electric field is the transmembrane current J_m per unit volume:

$$J_m = \text{div}\ \underline{j}_e = -\text{div}\ \underline{j}_i \tag{2}$$

which is also the sum of capacitive and ionic currents (see [13],[17]). Denoting by

χ : the membrane surface area per unit volume of the tissue
C_m : the membrane capacitance per unit area of the membrane surface
I_{ion} : the ionic current per unit area of the membrane surface which is carried by the flow of ions across the membrane
V : the transmembrane potential given by $V = U_i - U_e$

then

$$J_m = (C_m V_t + I_{ion})\chi \tag{3}$$

We recall that the so-called fast sodium current is the main cause of the depolarization of the cardiac cells during the excitation phase of the transmembrane action potential. Experimental advances have been recently made in measuring the fast sodium current in different cardiac preparations by means of voltage-clamp techniques (see [10] and for a survey [9]); these results justify the modelling of the sodium current mechanism by means of gating equations of the Hodgkin-Huxley type (see [16]).

The aim of this investigation is the macroscopic description of the propagation of the excitation wavefront using an approximation of the ionic current model given by the following nonlinear current-voltage law

$$I_{ion}(V) = I_{Na}(V) + I_L(V)$$

with $I_{Na} = \bar{G}_{Na}\, m_\infty(V)^3(V - V_{Na})$, $I_L = \bar{G}_L(V - V_L)$ denoting respectively the sodium ion turn on and the early repolarizing current. For a biophysical interpretation we recall that the membrane is crossed by an inward depolarizing current whenever $I_{ion}(V) < 0$. Using as $\bar{G}_{Na}$, V_{Na} and $m_\infty(V)$ the fitted expression for voltage data of the heart muscle obtained by Ebihara-Johnson in [10], choosing $\bar{G}_L = 0.05$ mS/cm^2, $V_L = -80$ mV and plotting $I_{ion}(V)$ one sees that I_{ion} is a cubic-like function of V, i.e. it has the following properties:

$$\begin{cases} I_{ion} \in C^1(\mathbf{R}) \text{ and there exist only three zeros:} \\ I_{ion}(V_r) = I_{ion}(V_{th}) = I_{ion}(V_p) = 0 \\ \text{with } V_r < V_{th} < V_p,\ I'_{ion}(V_r) > 0,\ I^t_{ion}(V_p) > 0 \end{cases}$$

where V_r, V_{th}, V_p are respectively the resting, threshold and plateau values of V. Moreover $\int_{V_r}^{V_p} I_{ion}(v)dv < 0$, so that a depolarizing current is available for the impulse propagation.

Without loss of generality in the following we set $V = (U_i - U_e) - V_r$, i.e. V is the displacement of the transmembrane potential from its constant resting value V_r and correspondingly we shift the I_{ion} curve taking the origin at V_r.

Combining equations (1)-(3) and setting $M = M_i + M_e$ we arrive at the following **reaction-diffusion system**:

$$\begin{cases} c_m V_t + I(V) = - \operatorname{div} M_e \nabla U_e \quad \text{in } H \\ \\ \operatorname{div} M_i \nabla V = - \operatorname{div} M \nabla U_e \quad \text{in } H \end{cases} \tag{4}$$

where $c_m = \chi\, C_m$, $I(V) = \chi\, I_{ion}(V)$.

The system (4) must be supplemented with boundary conditions related to the intra-, extra-cellular and cardiac current fluxes through ∂H. In the **ideal** bidomain model, the **bulk** tissue is described by the tensor M of the composite medium (i) + (e) and the current flowing in this tissue is given by $-M\nabla U_e$; denoting by $\underline{j}_0$ the extracardiac current density, the tissue current flux through ∂H must equal the flux entering the extracardiac medium, i.e. $-\underline{n}^T M\nabla U_e = \underline{n}^T \underline{j}_0$, where $\underline{n}$ is the unit normal to ∂H oriented toward the extra cardiac volume. On the other hand the conservation of currents implies $\underline{n}^T \underline{j}_0 = \underline{n}^T(\underline{j}_i + \underline{j}_e) = -\underline{n}^T M\nabla U_e - \underline{n}^T M_i \nabla V$. Combining the last last relations it follows that $\underline{n}^T M_i \nabla V = 0$ on ∂H.

The approximate kinematic properties of travelling wavefront solutions of the system (4) are investigated by means of a singular perturbation analysis of an appropriate scaled form of the system (4). We shall consider the **stationary** or **regular** propagation of the excitation wavefronts, i.e. without taking into account the initial phase related to the starting stimuli or front-front and front-boundary collisions.

In the **stationary** propagation we can scale the $\underline{x}$ and t variables taking spatial (L) and temporal (T) scale factors in such a way that the corresponding mean wavefront velocity results in unitary order. In this phase of the excitation process the velocity varies between 0.3 ÷ 0.6 mm/msec, according to whether it is measured longitudinally or tranversally to the fibre direction. Using an average value of 0.45 mm/msec the ($\underline{x}$,t) variables are rescaled with L = 13.5 and T = 30. This rescaling can be interpreted as reducing to unit a characteristic wall thickness of 13.5 mm.

Setting $f(V) = I(V)/\bar{g}_{Na}$, $\bar{g}_{Na} = \chi\bar{G}_{Na}$, $\hat{x} = x/L$, $\hat{t} = t/T$, and using the following typical parameter values:

$C_m \simeq 1\ \mu F\ cm^{-2}$, $\kappa^{-1} \simeq 2\cdot 10^{-4}cm$, $\bar{G}_{Na} \simeq 20\cdot 10^{-3}\ \Omega^{-1}cm^{-2}$, $\sigma \simeq 2\cdot 10^{-3} \div 10^{-4}$ $\Omega^{-1}cm^{-1}$ results in $\dfrac{c_m}{\bar{g}_{Na}T} = \varepsilon\tau_m$, $\varepsilon^2 D_i = \dfrac{1}{\bar{g}_{Na}L^2} M_i$, $\varepsilon^2 D_e = \dfrac{1}{\bar{g}_{Na}L^2} M_e$, $D = D_i + D_e$ where τ_m, D_i, D_e denote scaled quantities of unit order of magnitude and ε is a dimensionless small parameter of order 10^{-3}.

After rescaling and dropping the superscript ^ system (4) transforms into the following singular perturbation problem:

$$\begin{cases} \varepsilon\tau_m V_t + f(V) = -\varepsilon^2 \operatorname{div} D_e\nabla U_e & \text{in } H \\ \operatorname{div} D_i\nabla V = -\operatorname{div} D\nabla U_e & \text{in } H \end{cases} \tag{5}$$

Systems of reaction-diffusion equations involving a small parameter have been studied, e.g. in [2],[11-12],[18-19].

We recall that the excitation process of the tissue is associated with a travelling wavefront solution of (5); for a fixed point $\underline{x}_o \in H$ the passage of the wavefront solution is characterized by an upstroke of the transmembrane action potential $V(\underline{x}_o,t)$. Away from the stimulating regions we define the **activation** or **excitation time** $t = \psi(\underline{x})$ as the unique time instant for which $V(\underline{x},\psi(\underline{x})) = V_p/2$ and for fixed t we consider the surface $S_t = \{\underline{x} \in H : \psi(\underline{x}) = t\}$, called the **excitation wavefront**.

To describe the wavefront motion during the **stationary** propagation we carry out a formal perturbation analysis of (5) looking for solutions with a special structure.

We consider then the following change of variables: $\underline{\xi} = \underline{x}$, $\tau = \varepsilon^{-1}J(t - \psi(\underline{\xi}))$, with $J = (\alpha_i\alpha_e/\alpha)^{-\frac{1}{2}}$, $\alpha_{i,e} = \underline{p}^T D_{i,e}\underline{p}$, $\alpha = \underline{p}^T D\underline{p} = \alpha_i + \alpha_e$, and $\underline{p} = \nabla\psi$.

Seeking solutions of (5) with $V = v(\tau)$, i.e. independent of $\underline{\xi}$, and assuming $\underline{p}(\underline{x}) \neq 0$, neglecting $O(\varepsilon)$ or $O(\varepsilon^2)$ terms according to whether we consider zero or first order approximations, we are led to the following nonlinear eigenvalue problem:

$$\begin{cases} \text{Find } K > 0 \text{ and } v(\tau) \text{ bounded on } \mathbf{R} \text{ such that:} \\ Kv_\tau + f(v) = v_{\tau\tau},\quad v(\pm\infty) = v^\pm,\quad v(0) = (v^+ + v^-)/2 \end{cases} \tag{6}$$

$v^{\mp}$ being the resting and plateau values of v. K is unique and the corresponding v is monotonically increasing. The following zero and first order eikonal equations hold:

$$\tau_m (\underline{q}^T \underline{p})^{-\frac{1}{2}} = K, \quad (\tau_m + \varepsilon \operatorname{div} \underline{q}) (\underline{q}^T \underline{p})^{-\frac{1}{2}} = K \tag{7}$$

with $\underline{q} = (\beta_i^2 D_e + \beta_e^2 D_i)\underline{p}$, $\beta_i = \alpha_i/\alpha$, $\beta_e = \alpha_e/\alpha$. From the definition of $\alpha_{i,e}$, $f_{i,e}$ it follows easily that $\alpha_i \alpha_e/\alpha = \underline{p}^T\underline{q}$ hence $\tau = \varepsilon^{-1}(\underline{p}^T\underline{q})^{-\frac{1}{2}}(t - \psi(\underline{x}))$. As a byproduct we have the following zero and first order approximation of the transmembrane potential corresponding to the eikonal equations (7):

$$V(\underline{x}, t) = v(\tau) + O(\varepsilon^s) = v\left(\frac{t - \psi(\underline{x})}{\varepsilon\sqrt{\underline{q}^T\underline{p}}}\right) + O(\varepsilon^s), \quad s = 1,2$$

which combined with the second equation of (5) yields

$$- \operatorname{div} D\nabla U_e(\underline{x}, t) = \operatorname{div} D_i \nabla v \left(\frac{t - \psi(\underline{x})}{\varepsilon\sqrt{\underline{q}^T\underline{p}}}\right) + O(e^s), \quad s = 1,2 \tag{8}$$

The unscaled version of (7) is:

$$\frac{K\sqrt{\bar{g}_{Na}}}{c_m}\sqrt{\underline{q}^T\underline{p}} = 1, \quad -\frac{1}{c_m}\operatorname{div} \underline{q} + \frac{K\sqrt{\bar{g}_{Na}}}{c_m}(\underline{q}^T\underline{p})^{\frac{1}{2}} = 1 \tag{9}$$

Neglecting $O(\varepsilon)$ or $O(\varepsilon^2)$ terms for the transmembrane potential we have:

$$V(\underline{x}, t) = v\left(\frac{\sqrt{\bar{g}_{Na}}}{\sqrt{\underline{q}^T\underline{p}}}(t - \psi(\underline{x}))\right) \tag{10}$$

and from equation (8) it follows

$$- \operatorname{div} M\nabla U_e(\underline{x}, t) = \operatorname{div} M_i \nabla v \left(\frac{\sqrt{\bar{g}_{Na}}}{\sqrt{\underline{q}^T\underline{p}}}(t - \psi(\underline{x}))\right)$$

Given the excitation time $t = \psi(\underline{x})$, determined for instance using the zero or first order eikonal equation in (7), the above equation provides a local approximation of the extracellular potential field in H and in general it is coupled by means of transmission conditions on ∂H to the equation describing the extracardiac potential field.

The range of applicability of the zero order equation depends on the magnitude of the term div $\underline{q}$ appearing in the first order eikonal equation and accounting for the influence of the wavefront shape on the propagation velocity in anisotropic media. In [15,19,22-24] a different approach was followed to develop eikonal equations in isotropic excitable media with velocity dependent on the front curvature. For the connections between the eikonal equations proposed in this paper and those of [19,22-23] and for a numerical study of propagation in an isotropic medium see [5,6].

At the same order of approximation, the eikonal equation (9) is associated with the transmembrane action potential profile (10) which, in the case of the first order equation, depends on the point $\underline{x}$ and on the way the front reaches $\underline{x}$.

Related to the wavefront propagation on the epi-endocardial surface, we remark that for $t = \psi(\underline{x})$ we have $V(\underline{x},\psi(\underline{x})) = V_p/2$ and $\nabla V(\underline{x},\psi(\underline{x})) = -\underline{p}V_t(\underline{x},\psi(\underline{x}))$; hence if $\underline{x} \in \partial H$ and $\underline{p} \neq 0$, condition $\underline{n}^T M_i \nabla V = 0$ becomes

$$\underline{n}^T M_i \underline{p} = 0 \quad \text{for} \quad \underline{x} \in \partial H \text{ and } \nabla\psi(\underline{x}) \neq 0 \tag{11}$$

This relation provides a boundary condition for the eikonal equations setting also a constraint on the wavefront motion on the epi-endocardial surface.

To build a far-distance approximation of the extracellular potential we consider the so-called outer solution of the singular perturbation problem (5). We look for solutions of the following form

$$V(\underline{x},t) = \bar{V}(\underline{x},t) + \hat{V}(\underline{\xi},\tau) + O(\varepsilon), \quad U_e(\underline{x},t) = \bar{U}_e(\underline{x},t) + \hat{U}_e(\underline{\xi},\tau) + O(\varepsilon)$$

with $\underline{\xi} = \underline{x}$ and $\tau = (t - \psi(\underline{x}))/\varepsilon$ and $\hat{V},\hat{U}_e \longrightarrow 0$ as $\tau \longrightarrow \pm\infty$.

Away from the wavefront S_t we obtain

$$\text{div } M\nabla U_e = 0 \text{ in } H - \bar{S}_t \tag{12}$$

$$[U_e]_{S_t} = V_p \underline{n}^T M_i \underline{n}/\underline{n}^T M \underline{n}, \qquad [\underline{n}^T M \nabla U_e]_{S_t} = V_p \, \text{div}_{S_t} \underline{\omega} \tag{13}$$

where, since $M = M_i + M_e$, the tangential vector field is given by

$$\underline{\omega} = (\beta_e M_i - \beta_i M_e)\underline{n}, \text{ with } \beta_{i,e} = \underline{n}^T M_{i,e} \underline{n}/\underline{n}^T M \underline{n}$$

The equations implicitly provide a macroscopic representation of the electric cardiac sources on S_t and define a reduced potential problem suitable for far-field simulations, given the excitation wavefront S_t and the tensors M_i, M_e. Using the following splitting

$$M_i = \frac{\sigma_t^i}{\sigma_t} M + \left(\sigma_l^i - \frac{\sigma_l}{\sigma_t}\sigma_t^i\right) \underline{a}_l \underline{a}_l^T$$

the solution of (12)(13) can be written as

$$U_e(\underline{x},t) = V_p[U_\nu(\underline{x},t) + U_a(\underline{x},t)]$$

where U_ν and U_a satisfy respectively

$$\text{div } M\nabla U_\nu = 0, \quad [U_\nu]_{S_t} = \frac{\sigma_t^i}{\sigma_t}, \quad [\underline{n}^T M\nabla U_\nu]_{S_t} = 0$$

$$\text{div } M\nabla U_a = 0, \quad [U_a]_{S_t} = \left(\sigma_l^i - \frac{\sigma_l}{\sigma_t}\sigma_t^i\right) \frac{(\underline{a}_l^T \underline{n})^2}{\underline{n}^T M\underline{n}}, \quad [\underline{n}^T M\nabla U_a]_{S_t} = \text{div}_{S_t} \underline{\omega}$$

The split form shows that the potential U_e is the superposition of the potential fields due to a **conormal** and an **axial dipole layer** on S_t. These layers are characterized respectively by the dipole directions $\underline{\nu} = M\underline{n}/|M\underline{n}|$, $\underline{a}_l$ and by the moment densities $\frac{\sigma_t^i}{\sigma_t}|M\underline{n}|$, $\left(\sigma_l^i - \frac{\sigma_l}{\sigma_t}\sigma_t^i\right) (\underline{a}_l^T \underline{n})^2/\underline{n}^T M\underline{n}$. In the particular case of $M = \sigma I$ the conormal direction $\underline{n}$ becomes normal, i.e. $\underline{\nu} = \underline{n}$.

To take into account the extracardiac potential, i.e. for H imbedded in a bounded and insulated conducting medium Ω, homogeneous and isotropic with constant conductivity σ_0, we must add to (12)(13) the following equations for the extracardiac potential U_0:

$$\sigma_0 \text{div } \nabla U_0 = 0 \text{ in } \Omega, \qquad \sigma_0 \underline{n}^T \nabla U_0 = 0 \text{ on } \Gamma$$

$$U_0 = U_e, \qquad \sigma_0 \underline{n}^T \nabla U_0 = \underline{n}^T M\nabla U_e \text{ on } \partial H$$

where Γ is the insulated boundary of Ω.

This mathematical model, obtained considering the transmembrane potential V discontinuous through the excitation wavefront, should describe adequately the far-field extracardiac potential U_0 for distances greater than 0.5 cm from the epicardial surface, i.e. distances at least five times greater than the thickness of the real moving boundary layer.

For greater details on the mathematical development of the model see [6-7].

2. Numerical approximation of the zero order eikonal equation and of the reduced potential problem

If the excitation wavefront, spreading in the bounded domain H representing the myocardium, is described approximately by the zero order eikonal equation supplemented by the boundary condition (11) then setting

$$F = F(\underline{x},\underline{p}) = \frac{1}{2}\left[\rho^2 \frac{(\underline{p}^T M_i \underline{p})(\underline{p}^T M_e \underline{p})}{\underline{p}^T M \underline{p}} - 1\right], \text{ with } \rho = \frac{K\sqrt{g_{Na}}}{c_m}$$

where the dependence on $\underline{x}$ is through M_i, M_e and $M = M_i + M_e$, we have the problem:

$$F(\underline{x},\underline{p}) = 0 \text{ in } H, \; \underline{n}^T M_i \underline{p} = 0 \text{ on } \partial H \tag{14}$$

where $\underline{n}$ is the outward normal to ∂H.

Since $F = 0$ is a partial differential equation of first order, the boundary value problem (14) is generally incompatible in the whole domain H. However, since a boundary affects the wavefront only in its proximity, the condition $\underline{n}^T M_i \underline{p} = 0$ can be neglected for those parts of the boundary which do not interact with or are not close to the front. This situation occurs, for instance, for small fronts generated by a point stimulation. To be more definite, we shall consider a small portion of the myocardium with epi- and endocardial surfaces flat and parallel and the volume in which the wavefront propagates is a parallelepiped.

We choose a local system of coordinates x_1, x_2, x_3 with the axes' origin taken at the centre of the volume and the x_3-axis perpendicular to the epi- and endocardial surfaces. We shall also assume a particular fibre structure (see [21]), i.e. the myocardium is made up of layers parallel to the epicardium; the fibre direction is constant on each layer and rotates with depth. In this model $\underline{a}_l^T = (\cos\gamma, \sin\gamma, 0)$ where $\gamma = \gamma(x_3)$ is the fibre rotation function. This assumption is reasonable for the study of a local stimulation delivered at point $\underline{x}_0$. Moreover we assume uniform anisotropy, i.e. constant conductivity coefficients $\sigma^i_{l,t}, \sigma^e_{l,t}$, in H. If $\psi(\underline{x})$ is the solution of the eikonal equation $F = 0$, corresponding to a stimulation delivered at point $\underline{x}_0$ at the initial time, we have $\psi(\underline{x}_0) = 0$ and $\psi(\underline{x}) = t$ is the wavefront at time t. If $\underline{x}$ is on the epi- or endocardial surface it must be $\underline{n}^T \underline{p} = 0$ on this surface.

As is well known from the Hamilton-Jacobi theory (see e.g. [8]) the solution of the equation $F = 0$ is equivalent to the solution of the system of ordinary differential equations:

$$\dot{\underline{x}} = F_p \;, \quad \dot{\underline{p}} = -F_x \tag{15}$$

with initial condition $\underline{x}(0) = \underline{x}_0$ and $\underline{p}(0) = \underline{p}_0$ where $\underline{p}_0$ must satisfy the condition $F(\underline{x}_0, \underline{p}_0) = 0$.

The local solvability of (15) is assured if and only if ∂H is a so-called basic characteristic manifold (see [8]); this condition is satisfied if $\left.\frac{d\gamma}{dx_3}\right|_{\partial H} = 0$.

The differential system (15), given the intial data $(\underline{x}_0, \underline{p}_0)$, is solved numerically by means of a classical fourth-order Runge-Kutta scheme monitoring the zero value of the Hamiltonian function. For a given set of $\underline{r}_0$ defining a triangular mesh on the unit sphere (or hemisphere) the corresponding set of points $\underline{x}(t;\underline{p}_0)$ defines a polyhedral surface S_t^h with triangular elements approximating the wavefront S_t.

Assuming $M = \sigma_0 I$, the computation of the extracellular potential generated by the excitation wavefront S_t, based on an integral representation of U_e, is carried out by means of a numerical approximation using the boundary integral element method (see [1], and for details [4],[7]). The influence of transmural fibre rotation on the wavefront shape and on the pattern of the potential field is investigated by means of numerical simulations.

3. Numerical results

In this section we discuss the evolution of wavefronts, originated by an epicardial, intramural or endocardial point stimulus in the stratified model of the ventricular wall and we investigate the extracardiac potential field using respectively the zero order eikonal equation and the reduced potential model (12)(13). The values of the conductivity coefficients and of the constant ρ used in the numerical simulations are the following:

$$\sigma_l^e = 2.25\ 10^{-3}\ \Omega^{-1}\ \mathrm{cm}^{-1} \qquad \sigma_t^e = 1.68\ 10^{-3}\ \Omega^{-1}\ \mathrm{cm}^{-1}$$

$$\sigma_l^i = 2.75\ 10^{-3}\ \Omega^{-1}\ \mathrm{cm}^{-1} \qquad \sigma_t^i = 3.21\ 10^{-4}\ \Omega^{-1}\ \mathrm{cm}^{-1}$$

$$\rho^2 = 290.91\ \Omega\mathrm{cm}\ \mathrm{mm}^2\ \mathrm{msec}^{-2}$$

The calibration of the eikonal parameters requires also the knowledge of the function $\gamma(x_3)$ which qualitatively models the fibre rotation counterclockwise from the epicardium to the endocardium in the ventricular wall. Based on anatomical measurements (see e.g. [21]) the fibre rotation can be described qualitatively by the following analytical expression

$$\gamma(x_3) = -\frac{5}{18}\pi \sin \left[4\pi \left(x_{3/d}\right)^3 \right] - \frac{5}{36}\pi, \quad x_3 \in \left[-\frac{d}{2}, \frac{d}{2}\right]$$

and we shall assume a thickness d = 1 cm for the ventricular wall.

The numerical integration of the differential system (15) was carried out with a time step Δt ranging from 0.01 to 0.04 msec for a set of 2268 (1153) initial vectors $\underline{p}_0$ in the case of closed (open) wavefronts.

Figure 1 shows the orthogonal projections on horizontal, vertical and side planes of the computed surface S_t, 12.5 msec after an epicardial stimulus. The combined effects of anisotropic conduction and fibre rotation are revealed by the shape of the wavefront; more particularly, facing the propagation front, a characteristic clockwise torsion is observed which is correlated to the fibre rotation.

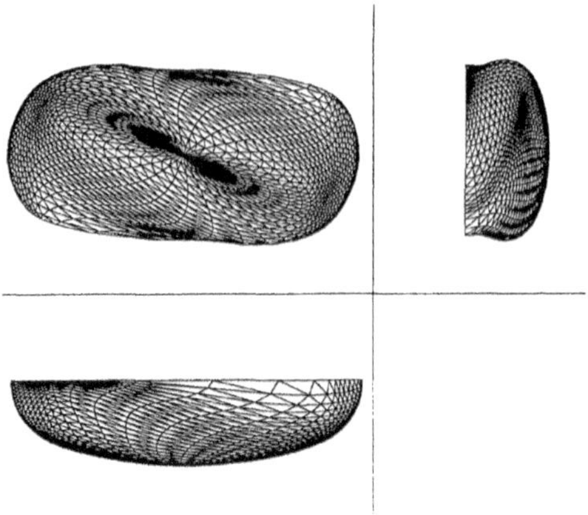

Figure 1. Orthogonal projections of the approximated wavefront at 12.5 msec after an epicardial stimulus.

It is expected that the wavefront equation should fail somewhere since it is derived from an approximate theory. In fact in the case of the fibre rotation, when the direction of highest velocity is not constant a folding of the wavefront onto the epicardial plane is observed corresponding to excitation pathways which penetrate into the interior wall and subsequently turn back toward the epicardial plane reaching it ahead of the epicardial isochrone, i.e. the basis of the wavefront spreading on the epicardial plane.

The appearance of the previous phenomenon implies that some time after the stimulus the front collides with the boundary from which the stimulation was started and also front-front collisions happen. As a consequence the validity of the zero order eikonal equation breaks down and the corresponding wavefront motion is described by this equation only in a time interval ranging between 10 and 20 msec depending on the conductivity coefficients and on the rotation function (in our case the time interval is of 12.5 msec). We remark that the wavefront folding is due to the combined effect of the anisotropy and fibre rotation. Consequently when considering the motion of a wavefront approximated by the zero order eikonal equation the corresponding potential field can be investigated only in the time interval stated above.

We now investigate the potential maps generated, in a bounded isotropic conducting medium, by the oblique dipole layers located on wavefronts spreading in a fibre system modelled as a packing of rotating planes of parallel fibres. We have simulated the potential on polyhedral surfaces lying at 0.5 cm and 1 cm from the **epicardial** surface, and having nodes corresponding to an array of electrodes surrounding the heart surface like the one used in the experimental set-up described in [3-4], related to an isolated dog heart placed in a cylindrical tank filled with a physiological solution. We focus our attention on the part of the surface facing the stimulating needle which delivers the epi-, endo- or intra- cardiac starting stimulus. In these simulations the reference zero potential was that of a node lying on the bottom of the tank.

Figure 2 displays the simulated potential maps at 9 and 12.5 msec after an

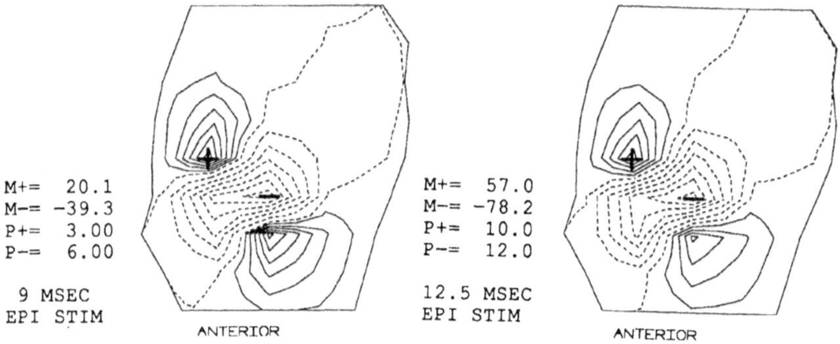

Figure 2. Simulated potential maps at 9 and 12.5 msec after an epicardial stimulus, plotted on a polyhedral surface lying at 0.5 cm from the epicardial surface.

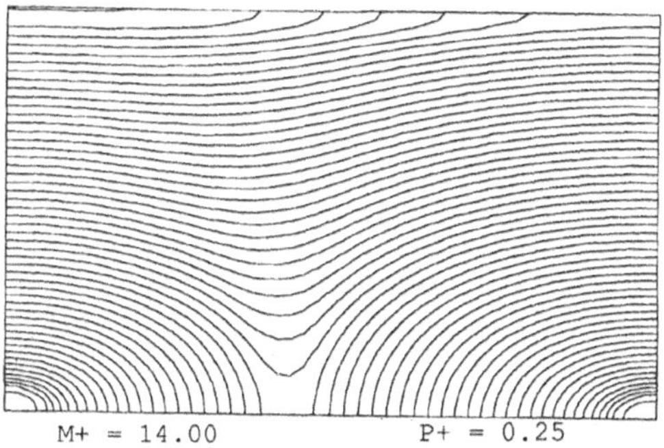

Figure 3. Isochrones of the activation time, plotted with a time step of 0.25 msec, generated by two non-simultaneous current stimuli delivered at the bottom vertices with a delay of 1.5 msec obtained by solving the reaction-diffusion system (16).

epicardial point stimulus and obtained by using $\sigma_l^i/\sigma_t^i = 10$ as an estimate of the intracellular anisotropic ratio. The pattern evolution displayed by the potential contour maps reveals that the line joining the two potential maxima undergoes a counterclockwise rotation indicating that the wavefront and the oblique dipole layer on it travel through rotating planes of parallel fibres. This behaviour was actually observed in experiments with *in situ* canine hearts (see [14],[20]). The failure of the zero order eikonal equation due to collision phenomena requires a higher order equation to cope with front-front and front-boundary collisions, like that given in (9), which incorporates the influence of the wavefront shape on the propagation velocity.

To assess the accuracy of the first order eikonal equation in (9) for anisotropic excitable media, we carried out numerical simulations of the propagation in a 2-D anisotropic insulated rectangular sheet H of parallel fibres. The differential system given below was integrated numerically, using a predictor- corrector scheme of second order in time treating implicitly the diffusion terms:

$$\begin{cases} c_m V_t + I(V) - \text{div } M_i \nabla V = \text{div } M_i \nabla U_e + i_{app} & \text{in } H \\ \text{div } (M_i + M_e)\nabla U_e = \text{div } M_i \nabla V & \text{in } H \\ \underline{n}^T (m_i + M_e)\nabla U_e = 0, \ \underline{n}^T M_i \nabla v = 0 & \text{on } \partial H \\ V(\underline{x},0) = 0 & \text{in } H \end{cases} \tag{16}$$

where $M_{i,e} = \text{diag } (\sigma_t^{i,e}, \sigma_l^{i,e})$, $H = [0,l_x] \times [0,l_y]$, with $l_x = 5$ mm, $l_y = 3$ mm and $I(V) = gV(1 - V/V_{th})(1 - V/V_p)$. In the numerical simulations we used the following parameter values:

$$C_m = 0.8 \ \mu F \ cm^{-2} \qquad \sigma_l^e = 2.22 \ 10^{-3} \ \Omega^{-1} \ cm^{-1}$$

$$G = 9.6 \ m\Omega^{-1} \ cm^{-2} \qquad \sigma_t^e = 1.05 \ 10^{-3} \ \Omega^{-1} \ cm^{-1}$$

$$V_p = 100 \ mV, \ V_{th} = 33.3 \ mV \qquad \sigma_l^i = 2.77 \ 10^{-3} \ \Omega^{-1} \ cm^{-1}$$

$$\chi^{-1} = 8 \ 10^{-4} \ cm \qquad \sigma_t^i = 6.15 \ 10^{-3} \ \Omega^{-1} \ cm^{-1},$$

a time step $\Delta t = 2\ 10^{-2}$ msec and space discretization steps $\Delta x = 0.1$ mm, $\Delta y = 3/50$ mm. We present the results obtained by applying non-simultaneous current stimuli at two vertices of the rectangular sheet, respectively at time 0 in one vertex and 1.5 msec in the other, for a duration of 0.5 msec. Actually in the numerical simulation the current stimulus is applied at 4 mesh points near each vertex.

Figure 3 displays the excitation wavefronts obtained as equilevel lines of the activation time $t = \psi(\underline{x})$ computed by linear interpolation. After a transient phase, two wavefronts start propagating in the interior of the domain then collide and merge.

The following eikonal problem is solved numerically by using a line relaxation technique:

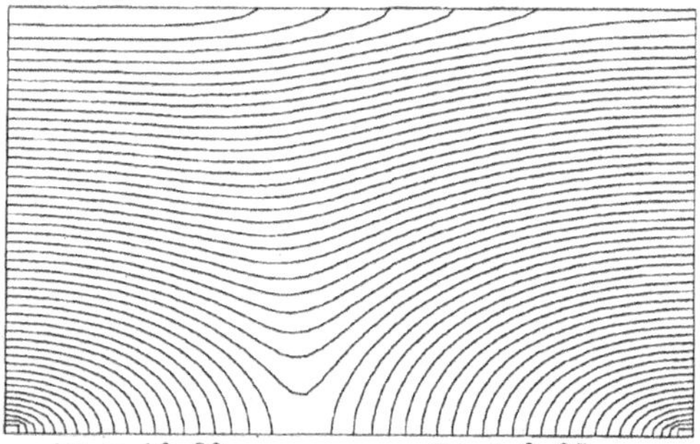

Figure 4. Isochrones of the activation time generated by two non-simultaneous current stimuli delivered at the bottom corners with a delay of 1.5 msec obtained by solving the first order eikonal equation.

$$\begin{cases} \text{Find } t = \psi(\underline{x}) : \\ -\frac{1}{C_m} \operatorname{div} \underline{q}(\underline{p}) + \frac{K\sqrt{g}}{C_m} \sqrt{\underline{q}(\underline{p})^T \underline{p}} = 1 \text{ in } H \\ \underline{n}^T \underline{p} = 0 \text{ on } \Gamma_N , \quad \psi = \psi_0 \qquad \text{on } \Gamma_D \end{cases}$$

where actually the stimulated corners are excluded. We assign the value 0 msec or 1.5 msec to ψ_0 on Γ_D, which represents the inner boundary defined by the two stimulated areas and Γ_N is the remaining part of the boundary of the rectangular domain. Moreover, for the cubic function I(V) the constant K, related to the eigenvalue problem (6), is given by $K = S\sqrt{1+S}$, $S = (V_p/2V_{th}) - 1$. The isochrones of the computed solution are displayed in Fig. 4. The comparison of Fig. 3 with Fig. 4 shows an excellent pattern match with a time discrepancy of less than 1 msec, which is just of the order of the rising time of the transmembrane potential.

This result shows that the first order eikonal equation correctly reproduces the main features of the wavefront propagation, such as the front-front and front-boundary collisions, the location of the collision points and the convergence of the wavefronts around the extinction point of the excitation.

Acknowledgements

This work was supported by: M.P.I. 40% Grants, C.N.R. under contract N. 88.00326.01, C.N.R Special Project on Informatic Systems and Parallel Computation, I.A.N. of C.N.R. (Pavia).

References

[1] Brebbia, C.A., Telles, J. & Wrobel, L. (1984). *Boundary Element Techniques - Theory and Applications in Engineering.* Springer-Verlag: Berlin, New York.

[2] Casten, R., Cohen, H. & Lagerstrom, P. (1975). Perturbation analysis of approximation to Hodgkin-Huxley theory. *Quart. Appl. Math.* **32**, 365-402.

[3] Colli Franzone, P., Guerri, L., Viganotti, C., Macchi, E., Baruffi, S., Spaggiari, S. & Taccardi, B. (1982). Potential fields generated by oblique dipole layer modelling excitation wavefronts in the anisotropic myocardium. Comparison with potential fields elicited by paced dog hearts in a volume conductor. *Circ. Res.* **51**, 330-346.

[4] Colli Franzone, P., Guerri, L., Taccardi, B., Tentoni, S. & Viganotti, C. (1982). Cardiac fibers orientation and potential fields: model studies. *Japanese Heart J.* **23(1)**, 279-281.

[5] Colli Franzone, P., Guerri, L. & Rovida, S. (1988). Macroscopic cardiac sources model and propagation wave in excitable media. In *Biomathematics and Related Computational Problems,* pp. 615-627,Ricciardi, L.M. (ed.). Kluwer Academic Publishers Group: Dordrecht.

[6] Colli Franzone, P., Guerri, L. & Rovida, S. (1988). Wavefront propagation in an activation model of the anisotropic cardiac tissue: asymptotic analysis and numerical simulations. Preprint of I.A.N.-C.N.R. **625**, 1-69 (to appear J. Math. Biol., 1989).
[7] Colli Franzone, P., Guerri, L. & Tentoni, S. (1990). Mathematical modelling of the excitation process in the myocardial tissue: influence of the fiber rotation on the wavefront propagation and the potential field. Submitted to *Math. Biosci.*[8] Courant, R. & Hilbert, D. (1966). *Methods of Mathematical Physics, Vol. II.* Interscience Publishers: New York, London.
[9] Di Francesco, D. & Noble, D. (1985). A model of cardiac electrical activity incorporating ionic pumps and concentration changes. *Phil. Trans. R. Soc. Lond.* **B 307**, 353-398.
[10] Ebihara, L. & Johnson, E.A. (1980). Fast sodium current in cardiac muscle. *Biophys. J.* **32**, 779-790.
[11] Fife, P.C. (1976). Singular perturbation and wavefront techniques in reaction-diffusion problems. *Proceedings SIAM-AMS vol. 10, Symposium on Asymptotic Methods and Singular Perturbations*, 23-49, New York.
[12] Fife, P.C. (1979). Mathematical aspects of reacting and diffusing systems. In *Lecture Notes in Biomathematics 28.* Springer Verlag.
[13] Fozzard, H.A. (1979). Conduction of the action potential. In *Handbook of Physiology, Vol. 1, The Heart, Sect. 2, The Cardiovascular System*,pp. 335-356, Berne, R.M., Sperelakis, N., & Geiger, S.R. (eds.). Am. Physiol. Soc., Bethesda, MD.
[14] Frazier, D.W., Krassowska, W., Chen, P.S., Wolf, P.D., Danieley, N.D., Smith, W.M. & Ideker, R.E. (1988). Transmural activations and stimulus potentials in three- dimensional anisotropic canine myocardium. *Circ. Res.* **63(1)**, 135-146.
[15] Gomatam, J. & Grindrod, P. (1987). Three-dimensional waves in excitable reaction-diffusion systems. *J. Math. Biol.* **25**, 611-622.
[16] Hodgkin, A.L. & Huxley, A.F. (1952). A quantitative description of membrane current and its application to conduction and excitation in nerve. *J. Physiol. (Lond.)* **117**, 500-544.
[17] Jack, J.J.B., Noble, D. & Tsien, R.W. (1983). *Electric Current Flow in Excitable Cells.* Clarendon Press: Oxford.
[18] Keener, J.P. (1980). Waves in excitable media. *S.I.A.M. J. Appl. Math.* **39(3)**, 528-548.
[19] Keener, J.P. (1986). A geometrical theory for spiral waves in excitable media. *S.I.A.M. J. Appl. Math.* **46(6)**, 1039-1056.
[20] Spaggiari, S.,Baruffi, S., Arisi, G., Macchi, E. & Taccardi, B. (1987). Effect of intramural fiber direction on epicardial isochrone and potential maps. (Abstract) *Circulation* **76** II-961.
[21] Streeter, D. (1979). Gross Morphology and Fiber Geometry of the Heart. In *Handbook of Physiology, Section 2, the Cardiovascular System*, Geiger, S.R. (ed.).
[22] Zykov, V.S. & Petrov, A. (1977). Role of the inhomogeneity of an excitable medium in the mechanism of self-sustained activity. *Biofizika* **22(2)**, 300-306. (English translation *Biophysics* **22**, 307-314.)
[23] Zykov, V.S. (1980). Analytical evaluation of the dependence of the speed of an excitation wave in a two-dimensional excitable medium on the curvature of the its front. *Biofizika* **25(5)**, 888-892. (English translation *Biophysics* **25**, 906-911.)
[24] Zykov, V.S. (1987). *Simulation of Wave Processes in Excitable Media.* Manchester University Press: Manchester and New York.

30. HYSTERESIS PHENOMENA IN TRANSITIONAL ZONES BETWEEN VARIOUS KINDS OF PERIODIC AND QUIESCENT BEHAVIOUR IN ELECTROPHYSIOLOGICAL MATHEMATICAL MODELS OF CARDIAC CELLS

M. Landau

CNRS at INRIA, Domaine de Voluceau - BP 105
78153 Rocquencourt, France

This study deals with a mathematical model of a system composed of coupled pacemakers (PM) and non-pacemaker (NPM) cardiac cells, when electrical coupling resistance and pacemaker cell size are allowed to vary. Broad outlines of classification were provided by direct numerical simulation for discrete values of the parameters. For a large PM cell size, the NPM cell is always activated, showing either a sustained rhythmic activity or subthreshold oscillations. When PM cell size is reduced, we observe, in addition to these NPM cell responses, a complete inhibition of the system at low values of coupling resistance. Then we turn to the systematic study of these phenomena with the continuation techniques of stationary and periodic solutions and detection of Hopf bifurcation points. In this new qualitative behaviour appears for the system. In two zones of values of the coupling resistance coexist stationary and periodic stable states, or two stable periodic states of large and low amplitude.

1. Introduction

The present study aims to analyse the effect of cell size and coupling resistance on the behaviour of a system constituted of coupled pacemaker (PM) and non-pacemaker (NPM) cardiac cells. In fact the zones of connection between the conducting system and the myocardium are critical in the heart. To simulate the behaviour of a single cell we used the two state model of Van Capelle & Durrer [1] which among more sophisticated descriptions [2,3] of cardiac action potential generation appeared suitable to our purpose. Changing characteristic constants in the model enables us to simulate for isolated cells, the spontaneous firing of a PM cell and the action potential following a stimulus in a NPM cell.

2. First elements of classification by direct numerical simulation

Subsequently these PM and NPM cells are coupled through an ohmic connection with electrical resistance R. To obtain a first classification of the possible behaviours of the system, we have simulated the PM and NPM action potentials for two diferent sizes of the PM cell and in each case for three values of the coupling resistance: R = 0.5, R = 4, R = 100. In all cases the NPM cell size remains constant.

2.1. *The case of a "large" PM cell*

(i) R = 0.5. For a tight coupling the PM cell induces a rhythmic behaviour of the NPM cell, with lengthening of the period as compared to the intrinsic period of the PM cell (compare Fig. 1 and Fig. 2a).
(ii) R = 4. When increasing the coupling resistance, the same kind of behaviour occurs. The period of the system increases as compared to the intrinsic period of the PM cell, but less than in case (i).
(iii) R = 100. For a large coupling resistance the PM cell beats near its intrinsic behaviour and the NPM cell presents only subliminal oscillations.

In summary, for a large PM cell size, we observe two kinds of behaviour for the NPM cell: efficient rhythmic activity of large amplitude or subliminal oscillations unable to stimulate the phenomena which induce the cardiac muscle contraction.

2.2. *The case of a PM cell of reduced size*

(i) R = 0.5. With tight coupling the system no longer beats rhythmically, the NPM cell has a loading effect on the PM cell, and even after an initial depolarization of +50 mV, the whole system comes back to a silent state (see Fig. 3a).
(ii) R = 4. Releasing the coupling between the cells, the system recovers its rhythmic activity (compare Figs. 3b, 2b and 1). The period increases as compared to the intrinsic period of the PM cell, and more than in the case of a large PM cell.
(iii) R = 100. At very large coupling resistance, the cells behave much as in the case of a large PM cell.

In summary, for a reduced PM cell size, we observe three kinds of behaviour complete inhibition of rhythmic activity, efficient rhythmic activity of high amplitude, subliminal oscillations.

Nevertheless, if simulation of the system for discrete values of the coupling resistance and PM cell size, gave us some idea of the possible behaviours of the system, this is far away from a complete classification.

3. **Classification of the behaviours of the system by continuation-bifurcation techniques**

The conclusion of the last paragraph leads us to undertake a more systematic study of the branches of stationary and periodic solutions when varying the coupling resistance. We did this by continuation and Hopf bifurcation techniques in the reduced PM cell size case with R as continuation parameter.

In Fig. 4, we represented a bifurcation diagram with in the abscissa the continuation parameter R and in the ordinate the maximum value, relative to time, of the NPM cell action potential. For small values of R the system has only one stable stationary solution represented by a continuous line. We have to reach R = 2.54 (point H on the diagram) to meet a Hopf bifurcation point indicating the existence of a branch of periodic solutions, while the

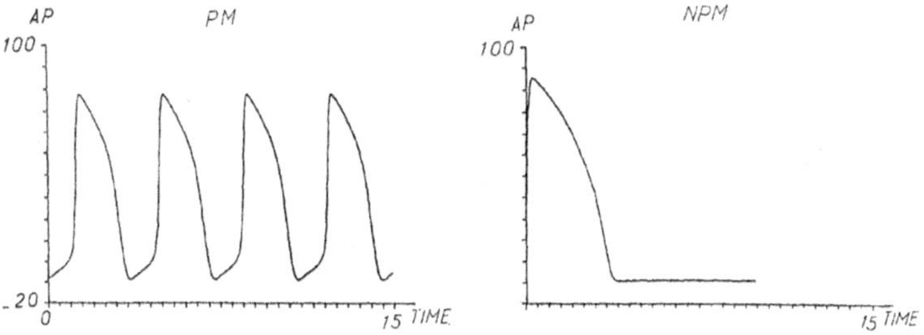

Figure 1. Action potential of isolated PM and NPM cells.

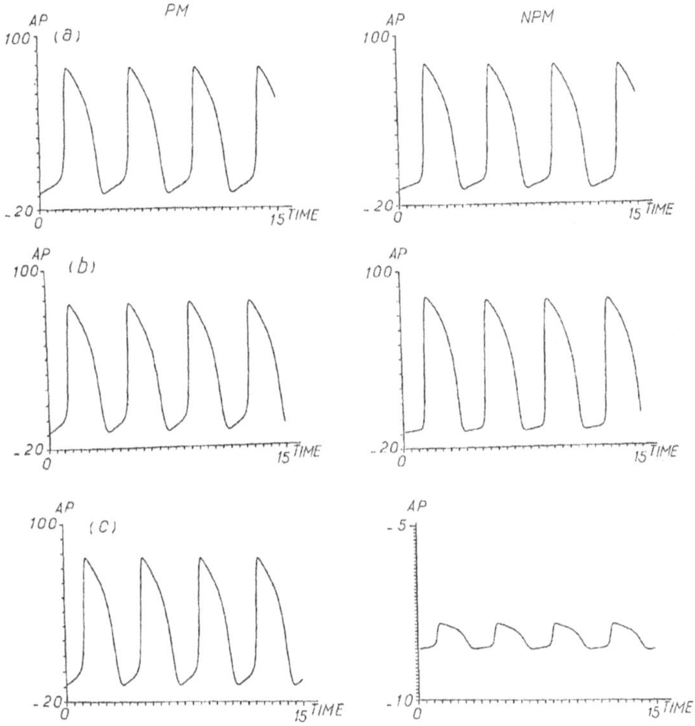

Figure 2. Behaviour of the coupled PM and NPM cells, for large PM size, at various coupling resistances. (a) R = 0.5. (b) R = 4. (c) R = 100.

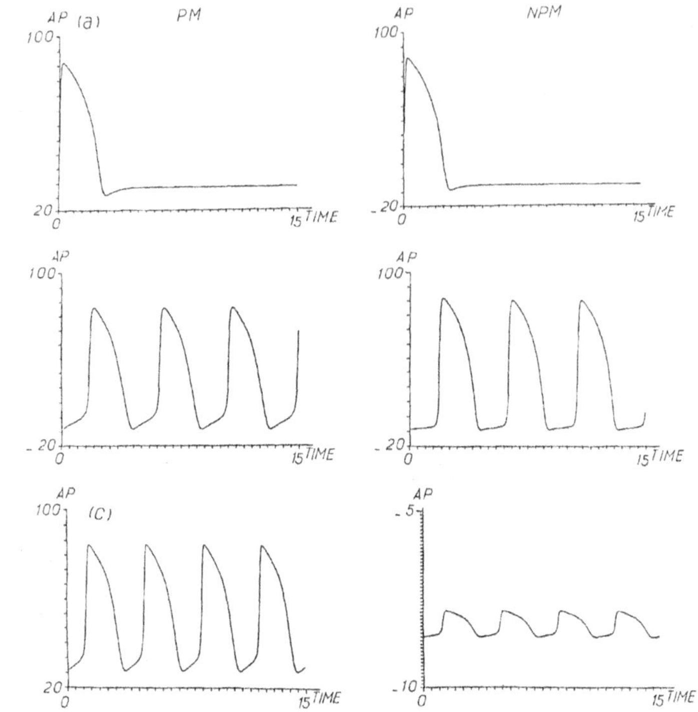

Figure 3. Behaviour of the coupled PM and NPM cells for reduced PM size at various coupling resistances. (a) R = 0.5. (b) R = 4. (c) R = 100.

stationary branch becomes unstable (dashed line in Fig. 4). From the subcritical Hopf bifurcation point emerges, for decreasing value of R, a branch of unstable periodic orbits (represented by hollow circles). At R = 2.33, we reach a periodic limit point L_1. From this value R increases again,

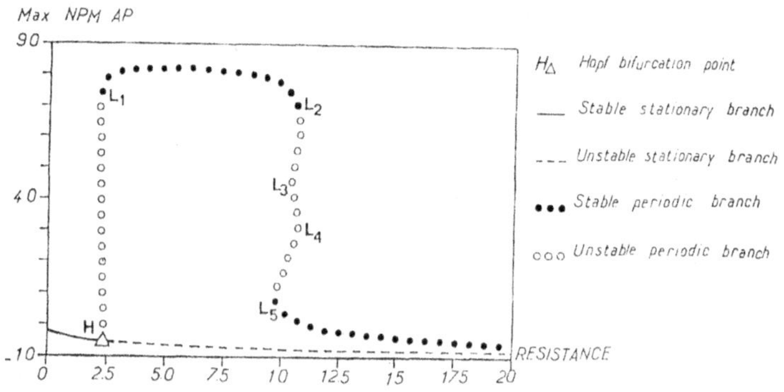

Figure 4. Bifurcation diagram showing stationary and periodic branches on the maximum NPM action potential-resistance plane.

and the periodic orbits become stable (represented by full circles). The continuation process goes on until R = 10.85 where the system presents a new limit point L_2 with change of stability of the periodic orbits. This piece of periodic branch after two turning points (L_3,L_4) without change of stability and a sharp fall of the amplitude of the NPM cell potential, reaches a last turning point L_5 at R = 9.74. From this value the whole remaining periodic branch is stable.

From this bifurcation diagram, the **stable behaviour of the system can be classified in five zones** of values for the coupling resistance. It is worthwhile to notice that direct numerical simulation of the system for a few discrete intuitive values of R led us only to three zones of behaviour. It is the advantage of the continuation to explore systematically a parametric field and so to have the opportunity to exhibit qualitatively different behaviours of the system, in relatively narrow zones of variations of the parameter: zones 2 and 4 of Table 1.

Table 1

Coupling reistance	0 2.33	2.54	9.74	10.85	
Zone	1	2	3	4	5
Behaviour	S	S & HAP	HAP	SP & HAP	SP

S = stationary solution, HAP = high amplitude periodic solution,
SP = subthreshold periodic solution

3.1. *Behaviour in Zone 2*

In this case the situation is better described in the blow-up of Fig. 4, in Fig. 5. For this zone of values of the coupling resistance, coexist two stable solutions: one stable stationary solution corresponding to the loading of the PM cell by the NPM one, and one large amplitude periodic solution corresponding to an efficient rhythmic activity of the system.

3.2. *Behaviour in Zone 4*

In this parametric area coexist two periodic solutions, one of large amplitude and one of low amplitude corresponding to subthreshold oscillations of the NPM cell.

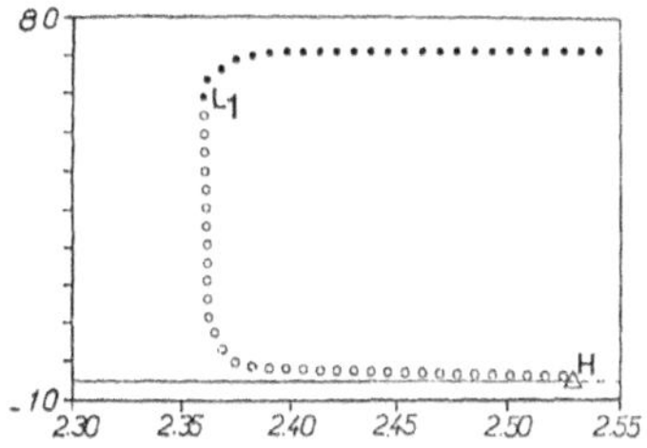

Figure 5. Magnification of the branches around the Hopf bifurcation point.

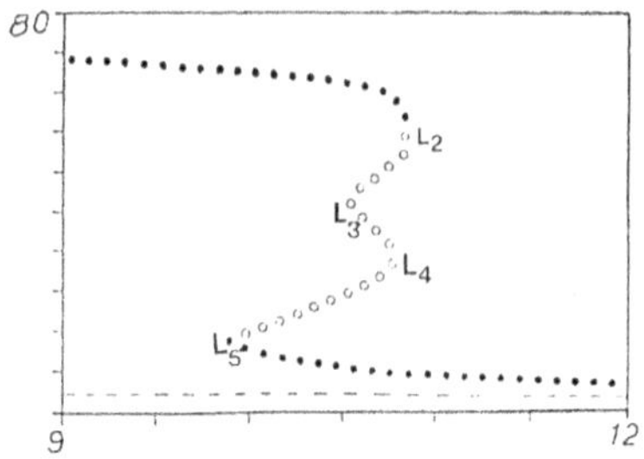

Figure 6. Magnification of the branches around the L_2 - L_5 limit points region.

3.3. *Two parameters continuation of the Hopf bifurcation point*

To get some insight into the influence of the size of the PM cell we followed the coordinates of the Hopf bifurcation point H with respect to the size of the PM cell (S_1) and to the coupling resistance R. Figure 7 sums up the results, also for the negative values of R, even if they have only mathematical significance. When the size of the PM cell decreases, the resistance for which we obtained a Hopf bifurcation point increases. Indeed the smaller the PM cell is, the more autonomy will be necessary to escape from the inhibition of the NPM cell. We observe also that beyond some value of S_1, there no longer exists any Hopf bifurcation point for positive values of R. In this case the PM cell always induces oscillations of the system. This is compatible with the observation we did by direct numerical solution.

4. Discussion

We pointed out in this study how a continuation-bifurcation study can advantageously follow a direct simulation done for intuitive values of the

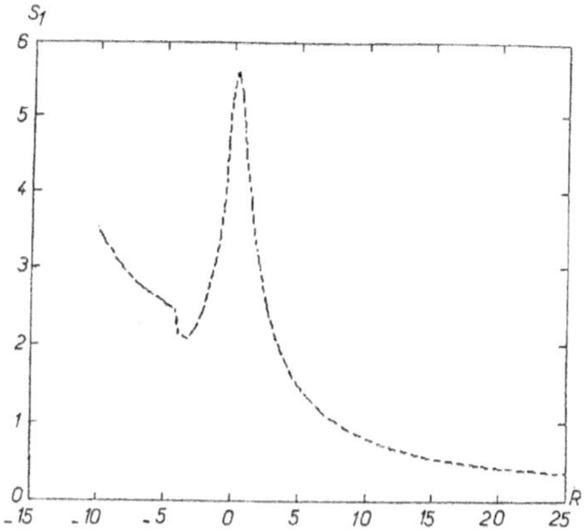

Figure 7. Locus of Hopf bifurcation point H in S_1 - R parameter plane.

parameters of a model. We were able by this methodology to give precise values to the limits of the parameteric area which define the various possible behaviours already detected by direct numerical simulation and even to discover new qualitative behaviours in relatively narrow zones of the parameters. In these zones coexist various kinds of stable solutions, and the associated hysteresis loops have been simulated.

It might be suggested that the coexistence of two stable solutions for the same values of coupling resistance would correspond with some varying activity of a parasystolic pacemaker, coming from complete absence to ectopic activity state and vice versa. This activity would depend on the degree of an "exit block" surrounding the parasystolic pacemaker. Biological and mathematical models of parasystole have been thoroughly studied by Jalife & Moe [4,5]. The present computational findings suggest that the apparent behaviour of a parasystolic pacemaker might also be modulated by slight variations of intrinsic local electrical properties of myocardial tissue.

Acknowledgement

I am very grateful to Dr. P. Lorente and Dr. J. Henry for valuable remarks during the completion of this study.

References

[1] Van Capelle, F.J.L. & Durer, D. (1980). Computer simulation of arrhythmias in a network of coupled excitable elements. *Circ. Res.* **47**, 454-466.

[2] Beeler, G.W. & Reuter, H. (1977). Reconstruction of the action potential of ventricular myocardial fibres. *J. Physiol. (Lond.)* **268**, 177-210.

[3] McAllister, R.E., Noble, D. & Tsien, R.W. (1985). Reconstruction of the electrical activity of cardiac Purkinje fibres. *J. Physiol. (Lond.)* **251**, 1-59.

[4] Moe, G.K., Jalife, J., Mueller, W.J. & Moe, B. (1977). A mathematical model of parasystole and its application to clinical arrhythmias. *Circulation* **56**, 968-979.

[5] Jalife, J. & Moe, G.K. (1979). A biologic model of parasystole. *Am. J. Cardiol.* **43**, 761-772.

31. DIMENSIONAL ANALYSIS OF THE VENTRICULAR FIBRILLATION ECG

Flavia Ravelli* and Renzo Antolini**

*Istituto per la Ricerca Scientifica e Tecnologica
Trento, Italy
**Dipartimento di Fisica, Università degli
Studi di Trento, Italy

From electrocardiographic data the correlation dimension of human ventricular fibrillation has been calculated. Since the algorithm does not converge towards a definitive dimension this arrhythmia cannot be associated with a low-dimensional chaotic dynamical system.

1. Introduction

Complex response patterns have been recently observed in a variety of cardiac preparations driven with external stimuli. Indeed, phase locking, period-doubling bifurcations, irregular activity have been shown both in self-oscillating preparation, like aggregates of embryonic heart cells [7] and in excitable non-pacemaker cardiac preparation, like Purkinje fibres and isolated heart [4,16,20]. Interestingly, complex rhythms obtained in experimental systems are similar to those observed in clinical electrocardiography during some cardiac arrhythmias [8]. In both cases, cardiac dynamics display some typical behaviours described and widely studied in other non-linear deterministic systems. Therefore lately there has been considerable interest in the application of non-linear dynamics theory to disturbances of cardiac rhythm.

The most important fatal arrhythmia is ventricular fibrillation (VF), characterized by low amplitude, rapid and irregular oscillations in the electrocardiogram (ECG) (Fig. 1). Theoretical models have suggested that these oscillations reflect abnormal organization of electrical activity due to the wanderings of multiple wavelets in the cardiac tissue [12,15]. Elegant experimental evidence of such complex behaviour has been obtained during atrial fibrillation by directly mapping the electrical activity of the isolated dog atrium [2]. The analogy between the results of electrophysiologic mapping and the computer simulation of Moe [15] is suggestive. More recent simulations on spatially distributed systems with variable refractory times have shown that a burst of rapid periodic stimulation will lead to auto-sustained excitation waves resembling fibrillation, and sometimes, passing through a period-doubling bifurcation [17]. This finding, associated with the observed beat-by-beat alternation of ECG waveforms in hearts susceptible to fibrillation [1], suggested the hypothesis that fibrillation might follow the subharmonic bifurcation route to chaos [17]. However, on the basis of spectral analysis, this hypothesis has been questioned [9]. In order

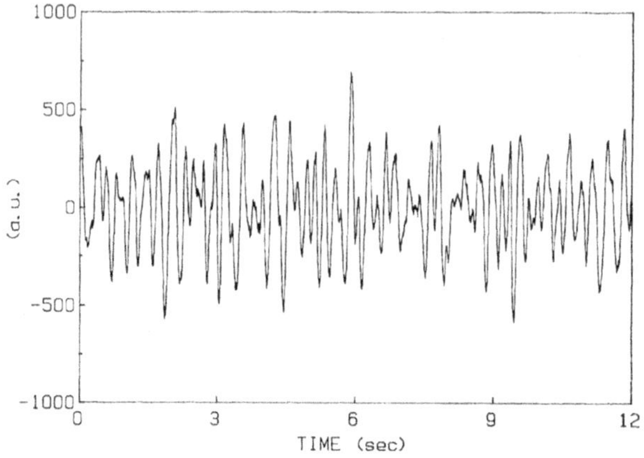

Figure 1. ECG of human ventricular fibrillation.

further to characterize cardiac fibrillation this contribution describes a dimensional analysis of VF-ECG.

2. Dimensional analysis

In recent years new techniques for the analysis of time series showing complex and irregular behaviour have been developed [14]. They are based on non-linear dynamic systems theory and greatly contributed to the transition from the analysis of deterministic chaos in a purely theoretical context to the quantitative determination of chaotic effects in experimental data. The methods consist in constructing an attractor in a suitable embedding space and calculating its dimension. They can be applied to time series obtained from measured data and have been used to investigate many experimental systems including biological ones [14]. Following these methods, it is now possible to evaluate the complexity of the dynamical system underlying fibrillation just by analysing ECG, a body surface recording of the electrical activity of the cardiac tissues.

Our studies were performed on four different episodes of VF extracted from the American Heart Association Database. The ECG signal was sampled at 250 Hz with 12 bits of precision. The digitized ECG constitutes the basic time series $\{Vi \; i = 1,\ldots,N\}$. The first step in the analysis is to reinterpret ECG signals as multidimensional geometrical objects. An n-dimensional vector $\{Vi^{(n)}\}$ is reconstructed from the one-dimensional time series by introducing a time lag T between the scalar data Vi [18]:

$$\{Vi^{(n)}\} = \{V_i, V_{i+T}, \ldots, V_{i+(n-1)T}\}$$

As time evolves, $\{Vi^{(n)}\}$ describes, in the phase space trajectories, the phase portrait of the system dynamics, Fig. 2 shows a two-dimensional phase portrait of the VF ECG constructed using 2000 data points with T equal to 64 msec, that correspond to the first zero crossing of the autocorrelation function. It displays complex trajectories with no obvious structure. From the phase portrait the dimension of the attractor, which is a measure of the dynamical system complexity, may be calculated. Calculation follows the method proposed

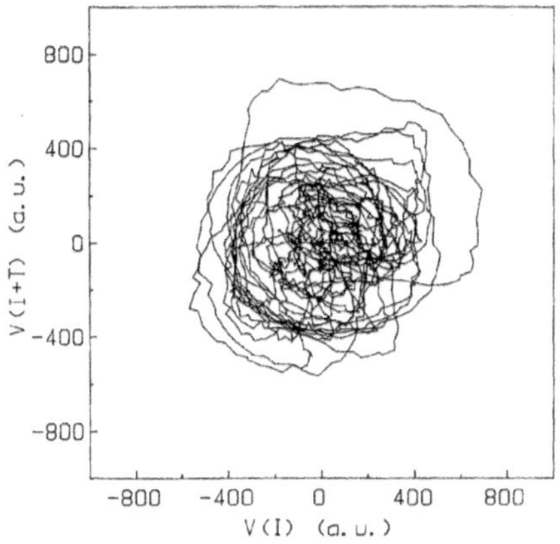

Figure 2. Two-dimensional phase-portrait of the VF episode displayed in Fig. 1. The portrait is generated by 2000 data points with T=64 msec corresponding to the time of the first zero of the autocorrelation function.

by Grassberger & Procaccia that has gained wide acceptance and has been tested experimentally in a number of cases [10,11]. It involves the computation of the correlation integral

$$Cn(R) = N^{-2} \sum_{\substack{i,j=1 \\ i \neq j}}^{N} \Theta(R - \| V_i^{(n)} - V_j^{(n)} \|) \qquad (1)$$

where Θ is the Heaviside function.

The correlation integral should scale as $Cn(R) \sim R^{D2(n)}$, where D2(n) is the correlation dimension for a time series of finite length. It is estimated as:

$$D2(n) = \lim_{R \to 0} \text{Log } Cn(R)/\text{Log}(R)$$

If D2(n), considering successively higher values of the embedding dimension n, reaches a saturation value the system represented by the time series should possess an attractor. The saturation value D2 is the dimension of the attractor. In practice, it is possible to approximate Cn(R) by averaging in (1) on a limited number M of origins Vi instead of N [13]. For our calculation we used 200 equally spaced reference points, the total number of the data being N=15000. A test of the independence of Cn(R) from M has been performed.

Figure 3a shows the logarithmic plot of correlation integrals for increasing n calculated with the data of one episode of ventricular fibrillation. To test whether a scaling region does exist the local slope

$$(\text{LogCn}(R_{i+1}) - \text{LogCn}(R_{i-1}))/(\text{LogR}_{i+1} - \text{LogR}_{i-1})$$

was calculated and plotted against LogCn(R). The plot of the slope for embedding dimension equal to 14, is shown in Fig. 3b. The inspection of these

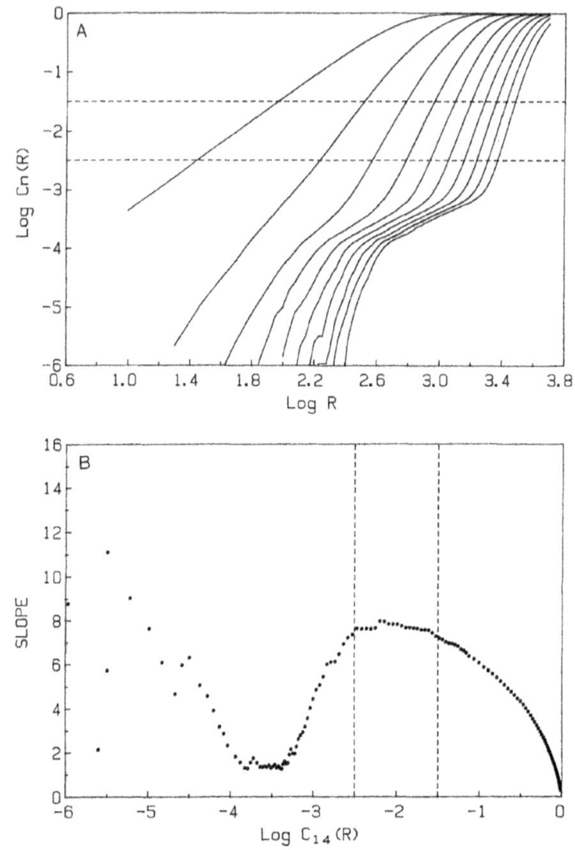

Figure 3. (a) Log-Log plot of the correlation integrals (n=2,4,6,...,20) for the VF episode used in Fig. 1 and 2. (b) The corresponding plot of the slope as a function of Log Cn(R) is shown for n=14. The two dotted lines define the range of linearity of the correlation integrals. This calculation is based on 15000 data points sampled at 250 Hz.

figures evidences three main features. At small R there is a great deal of variability in the slope. This effect is consistent with both the external noise and the poor statistics. At larger distances there is a distortion in the correlation curves and the local slope is ~ 1. This artifact is due to an excessive digitalization rate of 250 Hz for the VF ECG. In fact, Fig. 4 shows that by decreasing the sampling rate at 50 Hz the "shoulder" disappears. This spurious plateau disappears first at lower embedding dimension and the shoulder, if not correctly identified, can lead to an apparent convergence of the slope with the embedding dimension. To avoid this spurious dimension beside the undersampling a slight modification of the correlation integral has been proposed by Theiler [19]. Such an algorithm, by eliminating the shoulder, extends the range of linearity. The linear region of the curves, delimited in Figures 3a and 3d by two dotted lines, is at intermediate values of R. The dimension D2(n) was obtained by fitting a straight line to the logarithmic

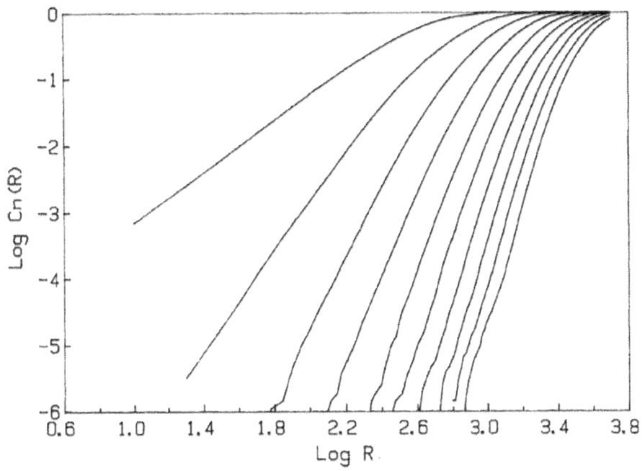

Figure 4. Log-Log plot of the correlation integrals. Same data as in Fig. 3a, except for a sampling rate of 50 Hz.

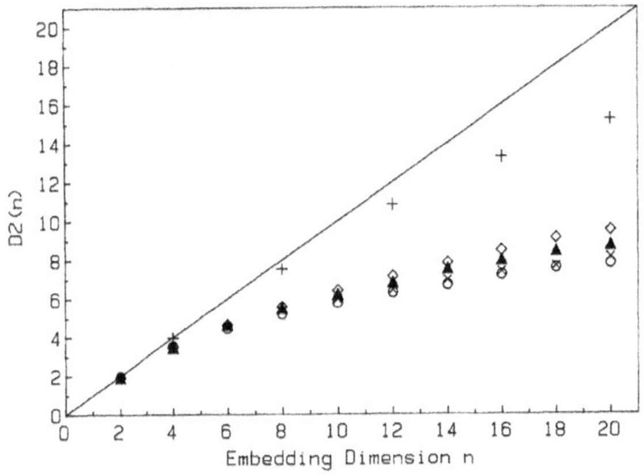

Figure 5. Correlation dimension versus embedding dimension n for the four analysed episodes of FV (▲, O,◊,x) and for a signal generated by a gaussian random process (+).

curve in that region.

Figure 5 summarizes the results by displaying D2(n) as a function of the embedding dimension n. The curves estimated from the four studied VF episodes look quite similar giving us confidence in their representativeness. They stay largely below those calculated for a pure random process, however with increasing embedding dimension they do not display a complete saturation. At n=20 a small increase is still observed, nevertheless D2(20) values are comparable, ranging from 7.7 to 9.5. Due to the lack of saturation the results do not support the existence of a low dimensional attractor underlying VF.

Dimension estimates based on data from real word complex systems require a careful check of the influence of the parameters considered in the computation algorithm on the results. First we consider the influence of the time delay T used in the phase space reconstruction. Theoretically, for an infinite time series the value of T may be chosen arbitrarily [18], on the

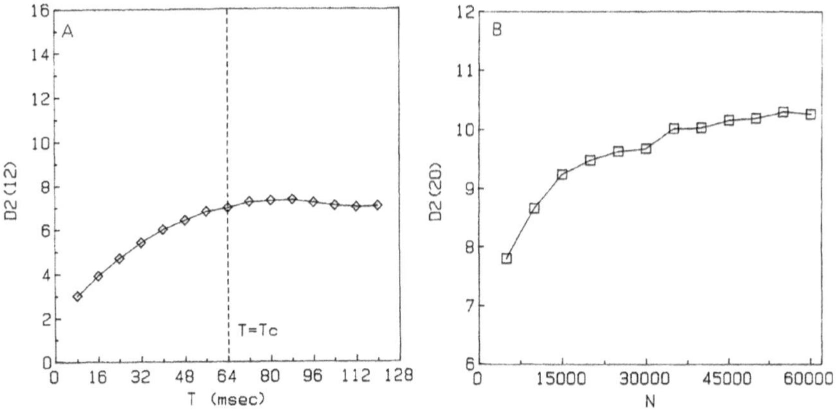

Figure 6. (a) Dependence of the correlation dimension upon the time-delay T and (b) upon the number of data points N for the VF episode described in the previous figures. The dotted line in panel (a) has been drawn at the time Tc corresponding to the first zero of the autocorrelation function.

best value of T [5], repeating the dimension calculation for different delay by selecting T in the range of stationarity may be a preferable procedure. The dependence of the correlation dimension at fixed n=12 as a function of the parameter T is shown in Fig. 6a for one VF episode. It can be seen that the dimension at first increases with T remaining approximately constant for $T \geq T_c$, where T_c is the time where the autocorrelation function has the first zero crossing. This behaviour has been observed in all four analysed cases. In our calculation we have used Tc as time-delay for phase portrait reconstruction and dimensional analysis. Actual values range from 52 msec to 100 msec.

Secondly we evaluate the number of data points N to be used to obtain consistent results. Figure 6b displays D2(20) as a function of the number of data points of the VF time series. The graph shows that for small data sets (N<15000) there is a substantial underestimation of the dimension but even 60000 data points are not sufficient to describe the full complexity of the system. The non-stationarity of the process of fibrillation for long time scales would probably contribute to the small increase observed with very large data sets (>15000 data points). Therefore to calculate the correlation dimension for VF ECG neither too small a data set giving very low dimension nor a data set including non-stationary phenomena can be used.

3. Conclusions

The results of this investigation provide no evidence that fibrillation reflects a dynamical system with a low dimensional attractor. Assuming that the characteristics of VF-ECG essentially reflect the intrinsic properties of the fibrillation dynamics, we can suppose the existence of a high dimensional chaotic dynamical system underlying VF. Since the number of data points required for convergence increases exponentially with the fractal dimension of the system [6], very long stationary time series would be required to obtain a

required for convergence increases exponentially with the fractal dimension of the system [6], very long stationary time series would be required to obtain a definitive value for the attractor dimension. The dynamics of VF, as seen in our data, seem more complex than the chaotic dynamics already described in experimental cardiac preparations [7,4,16]. This may be due to the coexistence of both temporal and spatial irregularities. From this aspect the analysis of VF data shares a problem common to other high dimensional systems such as full turbulence in fluid mechanics [14]. For these systems the data analysis is still an open problem.

Even if VF is not characterized by a dimension in a "strict" sense, the computed parameter D2(20) may be considered in a comparative sense. In fact, recent studies have shown that the normal ECG stems from deterministic chaotic dynamics with correlation dimension ranging from 3.2 to 5.2 [3]. The comparison between "dimensions" obtained during VF (our data range from 7.7 to 9.5) and sinus rhythm suggests a potential clinical relevance of this new descriptive parameter to be verified with further work.

Acknowledgements

We gratefully acknowledge financial support from Istituto Trentino di Cultura, CNR-GNCB, MPI 40% Biofisica. F.R. is a recipient of a fellowship from the Dottorato di Ricerca in Fisiologia (Torino).

References

[1] Adam, D.R., Smith, J.M., Akselrod, S., Nyberg, S., Powell, A.O. & Cohen, R.J. (1984). Fluctuations in T-wave morphology and susceptibility to ventricular fibrillation. *J. Electrocardiol.* **17**, 209-218.

[2] Allessie, M.A., Lammers, W.J.E.P., Bonke, F.I.M. & Hollen, J. (1985). Experimental evaluation of Moe's multiple wavelet hypothesis of atrial fibrillation. In *Cardiac Electrophysiology and Arrhythmias*, pp. 265-275. Zipes, D.P. & Jalife, J. (eds.). Grune and Stratton: Orlando.

[3] Babloyantz, A. & Destexhe, A. (1988). Is the normal heart a periodic oscillator? *Biol. Cybern.* **58**, 203-211.

[4] Chialvo, D.R. & Jalife, J. (1987). Non-linear dynamics of cardiac excitation and impulse propagation. *Nature* **330**, 749-752.

[5] Fraser, A.M. & Swinney, H.L. (1986). Using mutual information to find independent coordinates for strange attractors. *Phys. Rev.* **A33**, 1134-1140.

[6] Froehling, H., Crutchfield, J.P., Farmer, J.D., Packard, N.H. & Shaw, R. (1981). On determining the dimension of chaotic flows. *Physica* **3D**, 605-617.

[7] Glass, L., Guevara, M.R., Shrier, A.R. & Perez, R. (1983). Bifurcation and chaos in a periodically stimulated cardiac oscillator. *Physica* **7D**, 89-101.

[8] Glass, L., Goldberger, A.L., Courtemanche, M. & Shrier, A. (1987). Nonlinear dynamics, chaos and complex cardiac arrhythmias. In *Dynamical Chaos*, pp. 9-26, Berry, M.V., Percival, I.C. & Weiss, N.O. (eds.). Princeton Univ. Press: Princeton.

[9] Goldberger, A.L., Bhargava, V., West, B.J. & Mandell, A.J. (1986). Some observations on the question: is ventricular fibrillation "chaos"? *Physica* **19D**, 282-289.

[10] Grassberger, P. & Procaccia, I. (1983). Characterization of strange attractors. *Phys. Rev. Lett.* **50**, 346-349.

[11] Grassberger, P. & Procaccia, I. (1983). Measuring the strangeness of

strange attractors. *Physica* **9D**, 189-208.
[12] Krinskii, V.I. (1968). Fibrillation in excitable media. *Systems Theory Research (Prob. Kyb.)* **20**, 46-65.
[13] Layne, S.P., Mayer-Kress, G. & Holzfuss, J. (1986). Problems associated with dimensional analysis of electroencephalogram data. In *Dimensions and Entropies in Chaotic Systems. Quantification of Complex Behavior*, pp. 246-256, Mayer-Kress, G. (ed.). Springer-Verlag: Berlin.
[14] Mayer-Kress, G. (1986). *Dimensions and Entropies in Chaotic Systems. Quantification of Complex Behavior.* Springer-Verlag: Berlin.
[15] Moe, G.K., Rheinboldt, W.C. & Abildskov, J.A. (1964). A computer model of atrial fibrillation. *Am. Heart J.* **67**, 200-220.
[16] Savino, G.V., Romanelli, L., Gonzales, D.L., Piro, O & Valentinuzzi, M.E. (1989). Evidence for chaotic behavior in driven ventricles. *Biophys. J.* **56**, 273-280.
[17] Smith, J.M. & Cohen, R.J. (1984). Simple finite-element model accounts for wide range of cardiac dysrhythmias. *Proc. Natl. Acad. Sci. USA* **81**, 233-237.
[18] Takens, F. (1981). Detecting strange attractors in turbulence. In *Lecture Notes in Mathematics*, pp. 366-381, Rand, D.A. & Young, L.S. (eds.), Vol. 898. Springer-Verlag: Berlin.
[19] Theiler, J. (1986). Spurious dimension from correlation algorithms applied to limited time-series data. *Phys. Rev.* **34A**, 2427-2432.
[20] Zbilut, J.P., Mayer-Kress, G., Sobotka, P.A., O'Toole, M. & Thomas Jr., J.X. (1989). Bifurcations and intrinsic chaotic and 1/f dynamics in an isolated perfused rat heart. *Biol. Cybern.* **61**, 371-378.

32. FRONT PROPAGATION INTO UNSTABLE MEDIA: A COMPUTATIONAL TOOL

A. Babloyantz and J.A. Sepulchre*

Université Libre de Bruxelles, CP 231 - Campus Plaine
Boulevard du Triomphe, B-1050, Belgium
*Fellow of the Institut pour l'Encouragement de la
Recherche Scientifique dans l'Industrie et l'Agriculture

1. Introduction

Travelling waves in chemical and electrical active media have been observed experimentally as well as in model systems [1]. More recent advances in this field are reported in various contributions to this volume. However the majority of numerical and theoretical approaches in this field have been devoted to the excitable or bistable media [2].

In this paper we report numerical simulations which show the propagation of fronts in unstable media and the simultaneous propagation of phase waves behind these fronts [3]. In two- dimensional systems this gives rise to target patterns similar to those observed in active media. However it is seen that depending on the kinetic equations describing local oscillatory media and parameter values, the velocity of the front and the phase waves is either in the same direction or they point to opposite directions. Moreover the waves have the property of turning around the obstacles, and pouring into cavities. Now if, instead of a two-dimensional continuous medium, a network of coupled oscillators is considered, similar phenomena are expected. The numerical simulations indeed confirm this expectation.

The phase waves which propagate behind the front and in the opposite direction to the front velocity in a network of coupled oscillators may be used to perform "intelligent" tasks. To this network a physical space could be associated in which walls partition the space in complex configurations. The navigation problems in this space could be solved by the network of oscillators without prior teaching or expensive cost function evaluations [4].

In section 2 we describe the two-dimensional network of oscillators and discuss their properties. In section 3 we discuss how in continuous unstable media, phase waves propagate behind a front. Section 4 is devoted to the computational abilities of oscillatory networks.

2. Network of oscillators

Let us consider a two-dimensional square lattice of NxN elements. Each element

Nonlinear Wave Processes in Excitable Media
Edited by A. V. Holden *et al.*, Plenum Press, New York

is described by two variables X, Y and may exhibit oscillatory behaviour represented by a limit cycle.

The kinetic equations for a single unit may take the following form.

2.1. *The Brusselator scheme [5]*

$$\frac{dX}{dt} = A - (B+1)X + X^2Y$$
$$\frac{dY}{dt} = BX - X^2Y \tag{1}$$

Here A and B are assumed to be constant quantities. The homogeneous steady state of this system is X=A; Y=B/A. The standard linear stability analysis of this scheme shows that beyond a critical value $B_c = A^2 + 1$, the system switches into an oscillatory behaviour.

2.2 *The Maginu scheme [6]*

$$\frac{dX}{dt} = X - Y - X^3/3$$
$$\frac{dY}{dt} = (X - kY)/B \tag{2}$$

Here the reference state is X=0, Y=0, and the threshold for bifurcation is $B_c = k$. This model is a variant of the FitzHugh- Nagumo model. The latter is a simplified model describing the impulse propagation in nerve cells. Note also that when k=0, a simple change of variables leads to the van der Pol oscillator, with $\omega^2=1/B$:

$$\frac{d^2X}{dt^2} + (1 - X^2)\frac{dX}{dt} + \omega^2X = 0 \tag{3}$$

In the network, each oscillator is connected only to its first neighbours via a linear connectivity function. Thus the equations of the network are:

$$\begin{aligned}\frac{dX_{(i,j)}}{dt} &= f(X_{i,j}, Y_{i,j}) + \\ &+ C(X_{i+1,j} + X_{i,j+1} + X_{i-1,j} + X_{i,j-1} - 4X_{i,j}) \\ \frac{dY_{(i,j)}}{dt} &= g(X_{i,j}, Y_{i,j}) + \\ &+ C(Y_{i+1,j} + Y_{i,j+1} + Y_{i-1,j} + Y_{i,j-1} - 4Y_{i,j}) \\ & \qquad i,j=1,\ldots,N\end{aligned} \tag{4}$$

Functions f and g describe the kinetics of oscillatory units. Here C is the constant connectivity coefficient. The network is subject to zero flux boundary conditions.

An ensemble of interconnected oscillators described by eq. (4) is expected to exhibit collective behaviour similar to the ones observed in reaction diffusion systems. The connectivity coefficient C may be related to the Fick's diffusion coefficient $D = C(L/N)^2$ where L is the length of one side of the square. In the limit $N \to \infty$, the discrete operator used here tends to the Laplacian operator describing the diffusion phenomen [7]. In this limit, the network may also be considered as a continuous oscillating chemical media subject to diffusion processes. Under appropriate conditions concentric phase waves propagate in this medium. Therefore one may expect that even in networks of finite number of elements separated by a finite distance, target waves can appear. Numerical simulations show that for all three schemes considered in this paper, this is indeed the case.

The initial unperturbed state of the network described by eq. (4) is such that all the oscillators are at unstable non-oscillatory steady state. Therefore we speak of an unstable medium. A perturbation is introduced into the centre of the network. This can be achieved in two different ways. One may introduce a small local and transient fluctuation in the value of the variables describing a small number of cells. One can also modify slightly the parameters of several oscillators permanently. This case corresponds to a continuous signalling process in the network. In both cases, a front start from the perturbed region and propagates in the unstable medium. At the same time a wave propagates behind the front. According to the choice of functions f and g or the parameter values entering these functions, the direction of propagation of the waves is opposite or follows the direction of the propagation of the front.

Figure 1 shows the activity of a network of oscillators obeying eq. (1). For this scheme, as the front propagates into the unstable medium, the crest of the waves behind the front race toward the perturbation centre. This situation can be visualized in Fig. 2, where the radial profile of X is plotted against time.

One observes that the perturbation front invades the medium as it travels towards the right side of the system. However, whenever a new crest appears behind the front, it is pulled back to the left, towards the perturbation centre. This means that in Fig. 1, the rings behind the front are shrinking gradually.

This phenomenon of ingoing waves is not encountered in all networks of nonlinear oscillators. For instance with the van der Pol oscillators, the propagation of the front into the unstable system produces outgoing waves. Figure 3 shows the target waves obtained in this case. Here, as the front progresses, the rings are expanding.

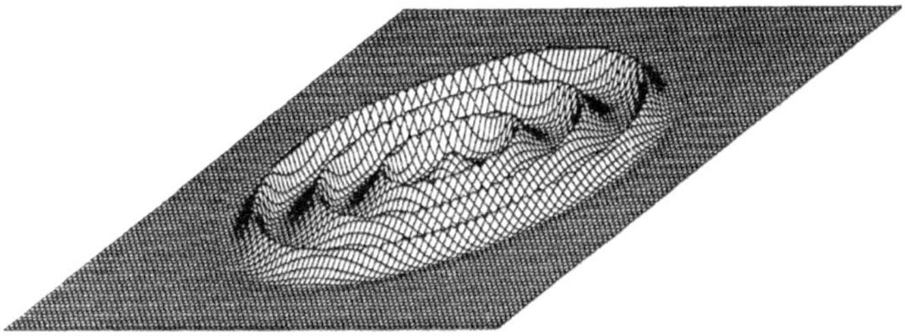

Figure 1. Spatial distribution of the X variable of the Brusselator (A=2; B=5.3; C=0.8). Behind the front which propagates into the unstable medium (X=2; Y=2.65), phase waves appear, and travel inward, towards the centre.

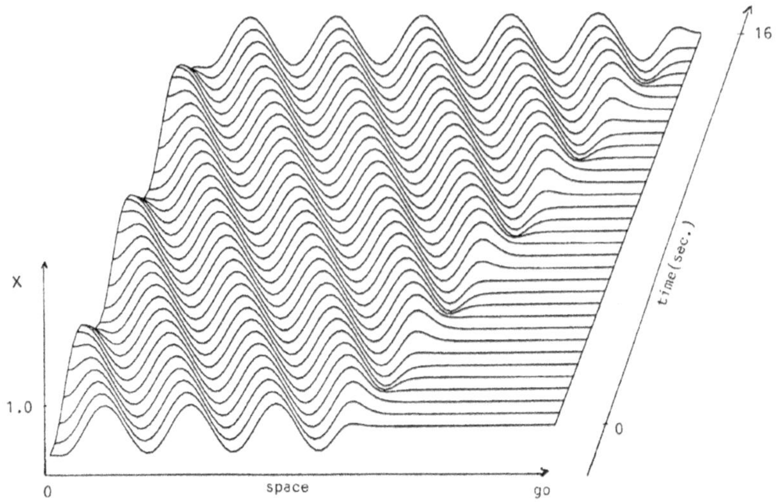

Figure 2. Evolution of a radial cross section of a system similar to the one in Fig. 1. Phase waves created in the unstable medium propagate in the opposite direction from the front.

The theoretical understanding of this phenomenon in two dimensions and in discrete media is not easy. However, the same behaviour is seen in one-dimensional continuous systems where an analytical approach is possible.

3. Front propagation in continuous media

Let us consider a linear array of NxN nonlinear oscillators coupled to their first neighbours. As was stated above, in the limit of $N \to \infty$, such a system represents a continuous reaction diffusion medium. In most cases the kinetics of such a system is well approximated by the following continuous Ginzburg-Landau equation for the complex amplitude W [8]:

$$\frac{\partial W}{\partial t} = (1 + i\omega_0)W - (1 + i\beta)|W|^2W + D\nabla^2W \qquad (5)$$

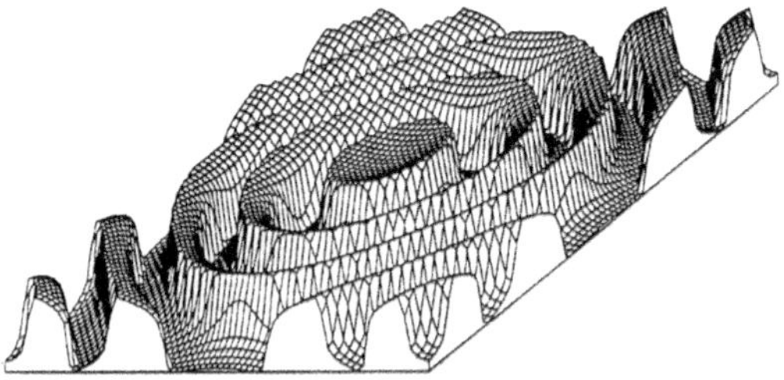

Figure 3. X profile of the van der Pol oscillator (ω=3; C=1). The phase waves observed in this system are expanding. The relaxational character of the temporal oscillations is reflected on the spatial form of the wave.

When D = 0, we obtain a limit cycle of frequency $\omega_0 - \beta|W|^2$. If the nonlinear dispersion coefficient β is positive, the period of the oscillations decreases when the amplitude increases. If $\beta < 0$, then the period decreases with the amplitude.

Starting from a state W=0, the system is perturbed at one end, and a front propagates from this point into the system. It has been shown that for sufficiently localized initial conditions, the velocity of this front approaches a value corresponding to marginal stability velocity $v^* = 2\,D^{1/2}$, which is the slowest stable propagating speed for such a front [9,10]. Behind the front, despite the fact that the critial mode has a zero wave- number, $k_c=0$, the numerical analysis shows the propagation of phase waves. We can determine analytically the velocity of these phase waves in the following manner.

Putting the plane wave $W = R\,e^{iq(x-ct)}$ in eq. (5) and writing this equation in a frame moving with velocity v*, we obtain:

$$q\,v^* = \beta(1-Dq^2) \tag{6}$$

Solving this equation for q, we deduce the selected wave number:

$$q = \sqrt{1/D}\,\frac{\sqrt{\beta^2+1}-1}{\beta} \tag{7}$$

The solutions of eq. (5) for which:

$$q > q_E = \sqrt{1/D}\left\{\frac{1}{3+2\beta^2}\right\}^{1/2} \tag{8}$$

are unstable with respect to the generalized Eckauss instability condition [11]. This is an instability associated with long wavelength perturbations. In terms of the parameters of the system, a simple calculation shows that condition (8) implies that β must be lower than β_c, where β_c is the positive real root of the polynomial $\beta^6 + 2\beta^4 - \beta^2 - 3$. ($\beta_c \approx 1.07$)

Equation (6) allows us to obtain the phase velocity of the phase waves emerging behind the front:

$$c = \frac{[\beta(1-Dq^2) - \omega_0]}{\beta(1-Dq^2)}\,v^* \tag{9}$$

We choose the parameters in eq. (5) such that $\omega_0 > \beta(1-Dq^2)$. In this case, the direction of propagation is thus determined by the sign of the nonlinear dispersion term β. Equation (3) implies that the phase velocity c has the opposite sign to v* whenever $\beta > 0$ (see Fig. 2).

If the kinetic equations of the oscillatory unit are such that in the corresponding Ginzburg-Landau equation $\beta=0$, then eq. (5) is not sufficient for the determination of the direction of the plane wave velocity. In such cases, in the amplitude equation, higher order contributions must be considered. For example, if the individual oscillators are described by the van der Pol model, it is easily found that in this case $\beta=0$. We have seen that computer simulation of such a network shows front propagation followed by phase waves.

The direction of the velocity of the waves is the same as the direction of the front propagation. If contributions of up to the fifth order are considered in the amplitude equation, the direction of velocity of the waves could be assessed with the help of a similar discussion as in the case of eq. (5).

4. Computation with coupled oscillators

The phase waves which propagate behind the front and in the opposite direction to the front velocity in a network of coupled oscillators may be used to perform "intelligent" tasks as will be shown in the next section.

Let us consider the following task. An object P of a certain size, for example a mobile robot, has to travel across three rooms and must reach a target point S called the goal. The objective is to drive the object along the shortest path between P and S without prior teaching and expensive cost function evaluation. Figure 4 depicts such a physical space.

To the physical space, we associate a network of oscillators (see Fig. 4) such that the corresponding Ginzburg-Landau equation shows the propagation of phase waves which travel in the opposite direction to the front. To the instantaneous state of point P and S of the physical space, we associate one or few oscillators in the network space. The internal walls of the physical space are represented in the network by deleting the corresponding oscillators in the network. The neighbouring cells to these walls obey the same boundary conditions as the one imposed on external walls, which are zero-flux boundary conditions.

In this navigational problem the target S is the signalling centre. Therefore, in the unstable network it constitutes the pacemaker centre. Thus the corresponding oscillators in S' must be perturbed slightly with respect to their unstable steady state values X=0, Y=0. When the pacemaker cells are switched on in the unstable medium a wave front is initiated at S' and propagates through the network. The units behind the front leave the unstable state and start to oscillate. Due to the existence of a phase gradient a travelling wave is selected and propagates behind the front. The crest of these target waves has the tendency to race towards the signalling centre S' Moreover these target waves have the property to go around the obstacles or pour into the cavities.

Figure 5 shows the progression of phase waves in time, in the network space. Eventually the front reaches the location in the network which corresponds to the object P. With the help of a simple external procedure, the project P at every instant measures the value of X variable in the network space in its immediate neighbouring oscillators (eight units) and moves towards the highest value of X. Thus in a sense, when the object P "sees" the

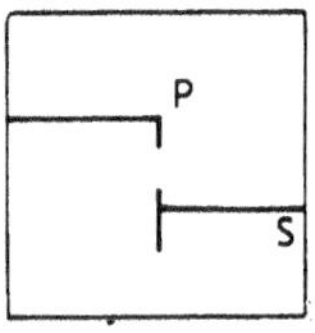

Figure 4. Representation of the free space for a mobile object P. This could be the schematic design of three rooms separated by walls. The object P has to reach the position S.

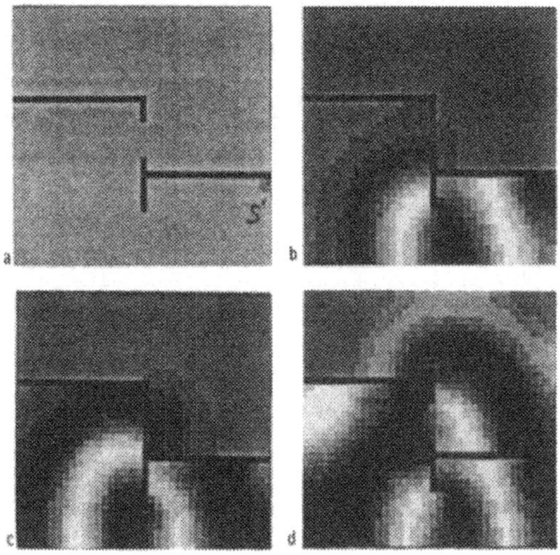

Figure 5. (a) Representation of a 38*38 network of oscillators, on which the physical space is mapped (see Fig. 4). The oscillators obey a Ginzburg-Landau equation with ω=2, β=1, D=1. the initial conditions are X=0, Y=0 for all the oscillators except for the oscillators in S' corresponding to the location S in the physical space, which are perturbed in order to generate the propagation of ingoing phase waves in the network. (b) and (c) As the front progresses, the crests of the waves (in dark) are moving back towards the perturbation centre located in S'. (d) The first crest has reached the position of the object P and has directed the latter towards the goal S.

first crest of the wave, it "sits" on it. As the crest shrinks, the object moves towards the pacemaker centre (see Fig. 2).

When the crest reaches the oscillator S', the object P is drawn to point S in the physical space following the shortest route. In our example, the navigational problem of a single object is illustrated. However if several objects were scattered in the three compartments of Fig. 4, they all would have reached the target centre S.

5. Conclusions

We have shown that in unstable media a front is generated from pacemaker centres and propagates in the system. In the same time phase waves propagate behind the front. We discussed the conditions under which the velocity of the phase waves is in the opposite direction to the velocity of the front. We have shown that the same phenomenon may be seen in discrete systems comprising a network of oscillators. Such a network may be used for performing "inteligent" tasks without prior teaching. The problem we solved here is just one example of the power of these networks. The properties of such networks may be combined with other properties of dynamical networks such as attractive fixed points, to perform still more complex tasks [12]. They may also serve as another layer in the conventional neural network systems [13,14]. Oscillatory networks have other interesting properties suitable for computational purposes. The research in this direction is in progress. We see this paradigm as an important alternative to the exploitation of minima of energy-like functions which is now the dominating use of artificial neural networks [15,16].

Acknowledgement

We are indebted to G. Dewel for helpful discussions. We also wish to thank A. Destexhe for his graphic software.

References

[1] For a review see: Field, R.J. & Burgers, M. (eds.) (1985). *Oscillations and Traveling Waves in Chemical Systems.* Wiley: New York.
[2] Tyson, J.J. & Keener, J.P. (1988). *Physica* **D32**, 327.
[3] Sepulchre, J.A., Dewel, G. & Babloyantz, A. Submitted to *Phys. Lett. A.*
[4] Babloyantz, A., Sepulchre, J.A. & Steels, L. Submitted to *Biol. Cyber.*
[5] Nicolis, G. & Prigogine, I. (1977). *Self-Organisation in Nonequilibrium Systems.*
[6] Maginu, K. (1979). *J. Diff. Eqs.* **31**, 130.
[7] Kubicek, M. & Marek, M. (1983). *Computational Methods in Bifurcation Theory and Dissipative Structures.* Springer- Verlag: New York.
[8] Kuramoto, Y. & Tsuzyuki, T. (1975). *Prog. Theor. Phys.* **54**, 687.
[9] van Saarloos, W. (1988). *Phys. Rev.* **A37**, 211.
[10] van Saarloos, W. (1989). *Phys. Rev.* **A39**, 6367.
[11] Stuart, J.T. & Di Prima, R.C. (1978). *Proc. R. Soc. Lond.* **A362**, 27.
[12] Haken, H. (1988). *Neural and Synergetic Computers.* Springer: Berlin.
[13] Hopfield, J.J. (1982). *Proc. Natl. Acad. Sci. USA* **79**, 2554.
[14] Rumelhart, D.E., Hinton, D.E. & Williams, R.J. (1986). In *Parallel Distributed Processing, Explorations in the Microstructure of Cognition, Vol. I.* MIT Press: Cambridge, Mass.
[15] Grossberg, S. (1976). *Biol. Cybernet.* **23**, 187.
[16] Kohonen, T. (1988). *Self-Organization and Associative Memory.* Springer: Berlin.

33. SPIRAL ARM WAVES IN GALAXIES

J.V. Feitzinger

Sternwarte Bochum, Castroper Str. 67
D-4630 Bochum, and
Astronomical Institut, Ruhr-University Bochum
P.O. Box 102148, D-4630 Bochum, FRG

1. Introduction

Spiral galaxies are often spectacular and pretty looking objects, and as such capture the imagination of many astronomers. The problems of the occurrence, formation and maintenance of spiral structure continue to be popular, and a vast literature exists on the subject. Athanassoula [1] gives a fine review on the spiral structure of galaxies with most emphasis on the density wave approach. Here we follow another concept.

Almost a decade ago the concept of the formation of galactic structures by stochastic self-propagating star formation (SSPSF) was developed [11,19]. These and almost all subsequent investigations of self-propagating star formation were done by computer simulations [4,10,12,25,26,27] or based on the specific assumptions underlying the computer models [24].

The first analytic approach to self-propagating star formation was made by Cowie & Rybicki [5] and by Balbus [2], but they assume a wavefront to exist and do not include any feedback mechanisms in their theory. Wavefronts were shown to exist under certain special circumstances in SSPSF-simulations by Freedman, Madore & Mehta [10].

Closely related to the theory of self-propagating star formation is the theory of the interstellar medium (ISM). Several investigations showed that in addition to the well-known static three phase state [18], time-dependent states may exist, as cyclic phase changes or burst-like runaway states [13,16,17]. Besides these pure ISM-models, where only the gas phases are considered, a few authors investigated the backreaction of the gas phase on star formation [3,28]. They found the same periodic or burst-like time structures as the investigators of the pure ISM-models.

These models did only consider the time development of the ISM, resp. stellar phase and did not take care of spatial couplings which might lead to propagating phenomena. The mathematical structure of these models was the same as that of models of chemical reactions with dissipative structures (e.g. [20]). In these chemical reaction models spatial coupling is introduced by adding a diffusion term to each equation. The resulting set of equations is called a system of reaction-diffusion equations (RD- equations). Self-

Nonlinear Wave Processes in Excitable Media
Edited by A. V. Holden *et al.*, Plenum Press, New York

organized spatial structuring is then possible. Therefore several authors introduced additional diffusion terms into their model equations [6,22,29]. Ferrini et al. [8] give a derivation of the diffusion term from a master equation.

Whereas Nozakura & Ikeuchi [23] investigated a pure ISM-model, Shore [29] and Feitzinger [6] developed models with diffusion of stellar matter. All authors find wave-like phenomena in their models, but only Shore [29] and Feitzinger [6] propose their models to represent analytical models of self-propagating star formation. Nozakura & Ikeuchi [23] mention the resemblance of what they call a "trigger wave", but they emphasize the difference between matter diffusion and the "diffusion" of star formation, which is characteristic of self-propagating star formation.

2. Basic considerations

As a starting point let us summarize the main features of the proposed RD-models [6,23,29]. The mass density of a galaxy is divided up into several distinct components. The mass density of the component i is denoted by ρ_i. In the above mentioned RD-models it is postulated that each component obeys a nonlinear partial differential equation:

$$\frac{\partial \rho_i}{\partial t} + \nabla \cdot (\vec{v}\rho_i) = F_i(\vec{\rho}) + D_i \Delta \rho_i \tag{1}$$

where D_i is a constant diffusion coefficient, Δ is the Laplace operator, $\vec{v}$ is the bulk velocity of the matter flow and $F_i(\vec{\rho})$ (in the simplest case) a nonlinear **function** of the density vector $\vec{\rho} = (\rho_i,\ldots,\rho_n)$. This set of equations describes the mass balance between the different components. The physical meaning of the terms in equation (1) is the following:

- $\nabla\cdot(\vec{v}\rho_i)$ describes the density change due to the bulk motion of matter which in spiral galaxies is differential rotation.
- $F_i(\vec{\rho})$ describes the creation and destruction of stars and the heating or cooling of gas.
- The $D_i\Delta\rho_i$-term was introduced to couple neighbourhood space points and to allow for wave-like phenomena which are well known to occur in similar sets of equations, e.g. in theoretical chemistry and biology [14,15,20]. A possible justification for this is given by Ferrini et al. [8].

To complete the discussion of the RD-models we state the further assumption usually made that the total density distribution is stationary and therefore the dynamics of the galaxy may be neglected. The same assumption is made in all SSPSF-simulations. Since the densities $\rho_i(\vec{r})$ should add up to give the total density $\rho(\vec{r})$, we may eliminate one equation from the system (1):

$$\rho(\vec{r}) = \sum_{i=1}^{n} \rho_i(\vec{r}) \tag{2}$$

If we impose the constraint of total mass conservation, $\rho(\vec{r})$ should obey the equation of continuity

$$\frac{\partial\rho}{\partial t} + \nabla\cdot(\vec{v}\rho) = \nabla\cdot(\vec{v}\rho) = 0 \tag{3}$$

where the first equality sign is due to the assumption of stationarity. Equation (3) is fulfilled identically, if ρ is taken to be axisymmetric and $\vec{v}$ is purely rotational motion about the axis of symmetry. This is an appropriate assumption for spiral galaxies. If we now sum up the system (1), use (2) and (3), we get:

$$\sum_{i=1}^{n} (F_i(\vec{\rho}) + D_i\Delta\rho_i) = 0 \tag{4}$$

Since the F_i describe the creation or destruction of different sorts of matter and since the total mass should be conserved and furthermore the galaxy is assumed to be in a state of **dynamical equilibrium**, we should have

$$\sum_{i=1}^{n} F_i(\vec{\rho}) = 0 \tag{5}$$

which means that there is no net creation or destruction of matter. Equation (5) immediately leads to

$$\sum_{i=1}^{n} D_i\Delta\rho_i = 0 \tag{6}$$

Nozakura & Ikeuchi [23] have obtained a nonlinear rigidly rotating spiral wave solution for a reaction-diffusion equation which describes the evolution of the interstellar medium on a differentially rotating galactic disk. A rigidly rotating spiral contour is a consequence of balance between winding effect of differential rotation and straightening effect of diffusion propagation. Therefore, differential rotations manifest themselves as tightening or loosening of spiralling patterns according to the direction of the rotation. The solution exist with asymptotic radial wavenumber limited to a certain range for a specified set of parameters.

3. The new approach

Neukirch & Feitzinger [21] developed a new approach concerning the handling of the self-propagating star formation.

We mentioned above that the basic assumption of self-propagating star formation is that the formation of stars at a point $\vec{r}$ at the time t is triggered by stars which were at the point $\vec{r} - \delta\vec{r}$ at the time $t - \delta t$. The distance $|\delta\vec{r}|$ and the delay time δt are not independent, but give the average spreading velocity $\vec{v}_T$ of the trigger mechanism:

$$\vec{v}_T = \frac{\delta\vec{r}}{\delta t} \tag{7}$$

As a starting point of the mathematical discussion, we again take the system of equations (1), but we now drop the diffusion term $D_i\Delta\rho_i$, which led to difficulties. We then get:

$$\frac{\partial \rho_i}{\partial t} + \nabla\cdot(\vec{v}\rho_i) = F_i[\vec{\rho}] \tag{8}$$

where $F_i[\vec{\rho}]$ means that F_i is not necessarily a **function** of its argument, but may have a more complicated dependence on $\vec{\rho}$. It is convenient to concentrate on the equation which will describe the propagation of star formation. Let the densities of the participating components be ρ_s for the stellar density and ρ_g for the gas density, where gas density means the density of the gas component in which stars are formed. We will not specify this any further. The following discussion may be easily extended to as many components as suitable. It is merely a matter of convenience to take only two components for the following discussion. The corresponding F will be called F_s.

First we split F_s up into a part F_{sc} which describes the creation of stars and into a part F_{sd} which describes the destruction of stars:

$$F_s = F_{sc} - F_{sd} \tag{9}$$

where F_{sc} and F_{sd} are positive definite. The interesting part for star formation is F_{sc}, let us therefore discuss this part. In the RD-models the asumption was made that the F_i were functions of the densities $\rho_i(\vec{r},t)$:

$$F_i = F_i(\vec{\rho}(\vec{r},t)) \tag{10}$$

As stated above the basic idea of self-propagating star formation is the triggering of star formation at location $\vec{r}$ at time t by stars which resided at $\vec{r} - \delta\vec{r}$ at time $t - \delta t$. Therefore F_s will depend on $\rho_s(\vec{r} - \delta\vec{r},\ t - \delta t)$. As we are dealing with a mean-field theory, we have to add up all influences of those stars over all space and all former times. Obviously, the more distant in space and time the less influence a star or a stellar association will have on the star formation process at location $\vec{r}$. To take this into account we introduce a weight function $\Phi(\delta\vec{r},\delta t)$ which goes to zero very fast for growing $|\delta\vec{r}|$ and δt. This is in analogy to the SSPSF-models, where it is assumed that only nearest neighbour cells have an influence on the star formation process in a specific cell and that this influence lasts only for one time step. The following ansatz is then made for F_{sc}:

$$F_{sc}(\vec{r},t) = \int_{-\infty}^{\infty} d\delta\vec{r} \int_0^{\infty} d\delta t\ \Psi(\rho_s(\vec{r}-\delta\vec{r},t-\delta t),\rho_g(\vec{r},t))\cdot\Phi(\delta\vec{r},\delta t) \tag{11}$$

where Ψ is a **function** of its arguments. This is the point where stochasticity enters. Φ may be considered as the probability density per unit time of a unit amount of stellar matter at place $\vec{r} - \delta\vec{r}$ at the time $t - \delta t$ to trigger star formation at the place $\vec{r}$ at time t. Φ is therefore taken to be positive definite and normalized to unity:

$$\int_{-\infty}^{\infty} d\delta\vec{r} \int_0^{\infty} d\delta t\ \Phi(\delta\vec{r},t) = 1 \tag{12}$$

Ψ specifies how much stars are formed from the gas density $\rho_g(\vec{r},t)$ due to a certain stellar density ρ_s at place $\vec{r} - \delta\vec{r}$ at time $t - \delta t$.

In this ansatz the nonlocal character of self-propagating star formation is explicit. We get an integro-differential equation for the development of the stellar density. We assume that the trigger mechanism (e.g. shock waves) expands spherically symmetrical in the absence of interstellar matter.

Therefore Φ will only depend on the propagation distance $|\delta\vec{r}|$. The direction of propagation of star formation will be determined by the dependence of Ψ on the gas density ρ_g.

Since in normal galaxies the maximum propagation length scale will be small compared with a galactic radius, we may expand $\Psi(\rho_s(\vec{r} - \delta\vec{r}, t - \delta t), \rho_g(\vec{r},t))$ into a Taylor series in $\delta\vec{r}$ and δt, keeping only the first order in δt and the second order in $\delta\vec{r}$. Introducing the abbreviations:

$$T = \int_{-\infty}^{\infty} d\delta\vec{r} \int_0^{\infty} d\delta t \ \delta t \ \Phi \tag{13}$$

$$D = \int_{-\infty}^{\infty} d\delta\vec{r} \int_0^{\infty} d\delta t (\delta\vec{r})^2 \ \Phi \tag{14}$$

and using (12) we get for F_{sc}:

$$\begin{aligned} F_{sc}&(\rho_s(\vec{r},t), \nabla\rho_s(\vec{r},t), \frac{\partial\rho_s}{\partial t}(\vec{r},t), \Delta\rho_s(\vec{r},t), \rho_g(\vec{r},t)) = \\ &\Psi(\rho_s(\vec{r},t),\rho_g(\vec{r},t)) - T\frac{\partial\Psi}{\partial\rho_s}\frac{\partial\rho_s}{\partial t} + D\frac{\partial^2\Psi}{\partial\rho_s^2}(\nabla\rho_s)^2 + D\frac{\partial\Psi}{\partial\rho_s}\Delta\rho_s \end{aligned} \tag{15}$$

Since we do not know Φ explicitly, we cannot calculate the numerical values of T and D from (13) and (14). T and D have a simple physical interpretation. T is the mean value of δt as (13) shows. Since δt represents the time necessary for the propagation of star formation for one event, T is a typical value of this time and therefore is the timescale of self-propagating star formation. Compared to SSPSF-computer models T should be of the order of one timestep of these models.

D is the mean value of $(\delta\vec{r})^2$. Since $|\delta\vec{r}|$ is the range of a single star formation event, $D^{1/2}$ may be interpreted as the average propagation length scale, which in comparison to the SSPSF-computer models would be of the order of one cell size of these models.

Given the average length scale and the average time scale the average propagation velocity may be defined as:

$$v_{prop} = \frac{D^{1/2}}{T} \tag{16}$$

The details of the mode analysis can be found in the paper by Neukirch & Feitzinger [21]. We gave one of our results. Several normal modes are shown in

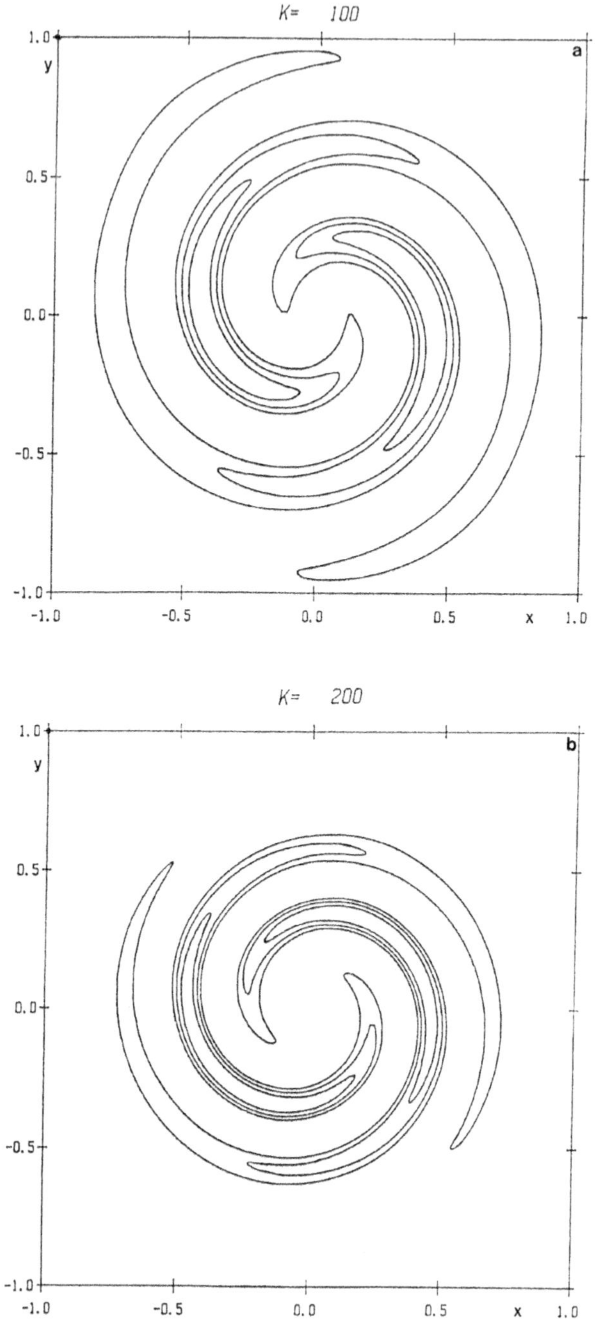

Figure 1 a and b.

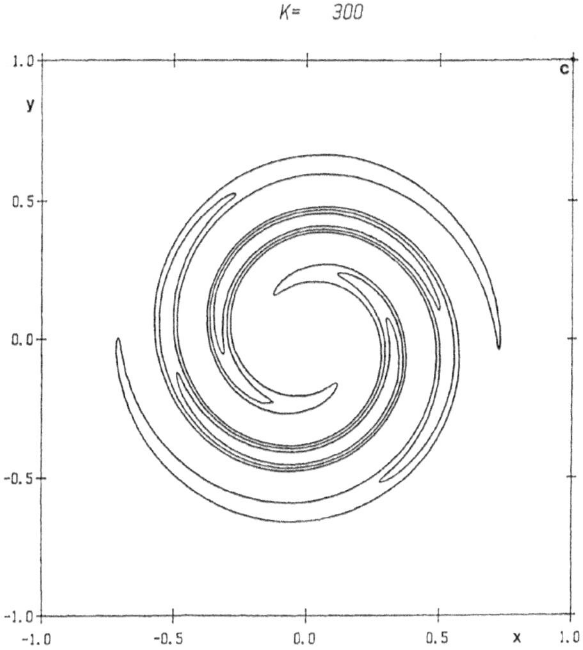

Figure 1. Density contours of the real part of normal modes for m = 2 are shown for the following values of K : 1a: K = 100, 1b: K = 200, 1c: K = 300. K depends on the maximum rotation velocity and the diffusion of the star forming processes.

Fig. 1 for m = 2 and different values of the parameter

$$K = \frac{v_0 R}{D \frac{\partial \Psi}{\partial \sigma}(0)_s} \tag{17}$$

so that the pictures show a sequence of increasing rotational velocity v_0. The resulting modes have a clear spiral shape. The equidensity contours wind up and are compressed with increasing velocity. We remind the reader that we chose a constant density distribution as the starting point of our analysis, so the shape of the contours will be altered if a more realistic total density distribution is chosen.

It has to be mentioned that in this constant density model, we will never have an isolated m = 2 mode instability, because if an m = 2 mode is unstable, there is also at least one unstable m = 1 mode and one unstable m = 0 mode. So we will possibly have some difficulty in getting symmetric two-armed structures in a constant density model. On the other hand, there remain a lot of further possibilities which may lead to symmetric structures in more elaborate models or in the nonlinear regime, but that will be a matter for future work.

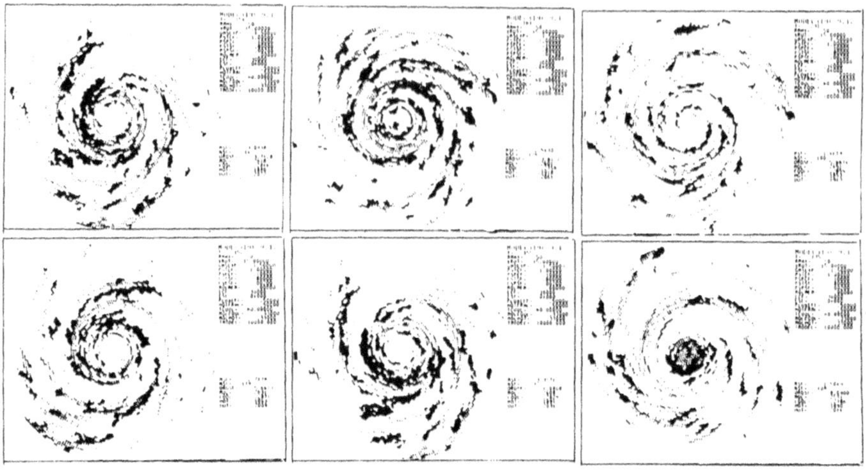

Figure 2. Evolution sequence of a model galaxy between time steps 152, 154, 155, 161, 175, 231 10^7 years. A milky way model is simulated. The size of the symbols is inversely proportional to the age of the star forming regions; the smallest points represent an age of 10^8 years.

4. Cell by cell simulations of self-propagating star formation and spiral structure

The model of self-propagating star formation uses local processes (200 pc cell size) in the interstellar medium to simulate the large scale cooperative behaviour of spiral structure in galaxies. The dynamic of the model galaxies is taken into account via the mass distribution and the resulting rotation curve: flat rotation curves are used. The interstellar medium is treated as a multiphase medium with appropriate cooling times and density history. The phases are: molecular gas, cool HI gas, warm intercloud and HII gas and hot coronal fountain gas. A detailed gas reshuffling between the star forming cells in the plane and outside the galactic plane controls the cell content. Two processes working stochastically are incorporated: the building and the decay of molecular clouds and the star forming events in the molecular clouds.

These models [7] are improving the former ones developed by Seiden, Gerola & Feitzinger [26]. The numerical receipt is the same. Figure 2 shows the simulation of spiral structure for a time series between 152.10^7 to 231.10^7 years. Figure 3 is the unsharp picture of one of our artificial computer galaxies. The luminosity distribution is scaled to resemble the luminosity of a real galaxy picture. The unsharpening is needed to overcome the cell structure.

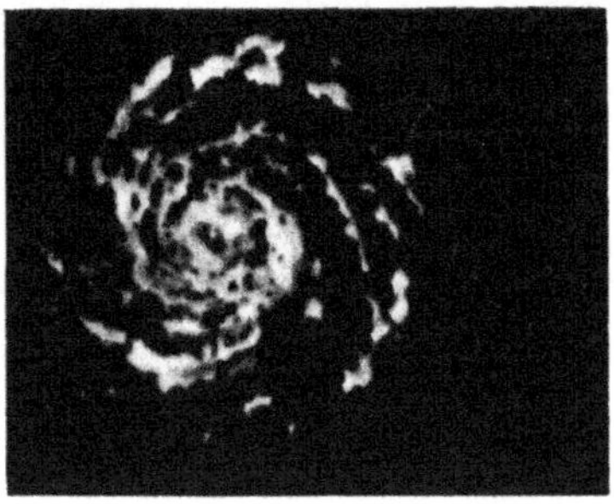

Figure 3. Example of an artificial galaxy: milky way model. Diameter 20 kpc, 7400 cells, maxium rotation velocity 250 km/s at 11 kpc. To suppress the cell structure of the computer simulated picture the reproductions were unsharpened. The luminosity distribution is scaled to match a real galaxy of SAb type, luminosity class II.

References

[1] Athanassoula, E. (1984). *Physics Reports* **114**, 319.
[2] Balbus, S.A. (1984). *Astrophys. J.* **277**, 550.
[3] Bodifée, G. & de Loore, C. (1985). *Astron. Astrophys.* **142**, 297.
[4] Comins, N.F. (1981). *Mon. Not. Roy. Astron. Soc.* **194**, 169.
[5] Cowie, L.L. & Rybicki, G.B. (1982). *Astrophys. J.* **260** 504.
[6] Feitzinger, J.V. (1985). *The Milky Way Galaxy*, p. 559. Reidel: Dordrecht.
[7] Feitzinger, J.V., Harfst, E., Perschke, M. (1990). In preparation.
[8] Ferrini, F., Marchesoni, F. & Shore, S.N. (1985). *Adv. Chem. Phys.* **62**, 493.
[9] Freedman, W.L. & Madore, B.F. (1982). *Astrophys. J.* **265**, 140.
[10] Freedman, W.L., Madore, B.F. & Mehta, S. (1984). *Astrophys. J.* **282**, 412.
[11] Gerola, H. & Seiden, P.E. (1978). *Astrophys. J.* **223**, 129.
[12] Gerola, H., Seiden, P.E. & Schulman, L.S. (1980). *Astrophys. J.* **242**, 517.
[13] Habe, A., Ikeuchi, S. & Tanaka, Y.D. (1981). *Publ. Astron. Soc. Japan* **33**, 23.
[14] Haken, H. (1977). *Synergetics*. Springer: Berlin.
[15] Haken, H. (1983). *Advanced Synergetics*. Springer: Berlin.
[16] Ikeuchi, S. & Tomita, H. (1983). *Publ. Astron. Soc. Japan* **35**, 56.
[17] Ikeuchi, S., Habe, A. & Tanaka, Y.D. (1984). *Mon. Not. Roy. Astron. Soc.* **207**, 909.
[18] McKee, C.F. & Ostriker, J.P. (1977). *Astrophys. J.* **218**, 148.
[19] Mueller, M.W. & Arnett, W.D. (1976). *Astrophys. J.* **210**, 670.
[20] Nicolis, G. & Prigogine, I. (1977). *Self-Organization in Nonequilibrium Systems*. Wiley: New York.
[21] Neukirch, Th., Feitzinger, J.V. (1988). *Mon. Not. Roy. Astron. Soc.* **235**, 1343.
[22] Nozakura, T. & Ikeuchi, S. (1984). *Astrophys. J.* **270**, 40.
[23] Nozakura, T., Ikeuchi, S. (1988). *Astrophys. J.* **333**, 68.
[24] Schulman, L.S. & Seiden, P.E. (1982). *J. Stat. Phys.* **27**, 83.
[25] Seiden, P.E. (1983). *Astrophys. J.* **266**, 555.

[26] Seiden, P.E., Schulman, L.S. & Feitzinger, J.V. (1982). *Astrophys. J.* **253**, 91.
[27] Seiden, P.E., Schulman, L.S. & Gerola, H. (1979). *Astrophys. J.* **232**, 702.
[28] Shore, S.N. (1981). *Astrophys. J.* **249**, 93.
[29] Shore, S.N. (1983). *Astrophys. J.* **265**, 202.

34. THREE DIMENSIONAL VORTICES IN ACTIVE MEDIA

A.V.Panfilov

Institute of Biological Physics
Pushchino, Moscow Region, 142292 USSR

Abstract

A short survey of the results of numerical investigation of the dynamics of 3-D vortices in excitable media on the FitzHugh-Nagumo type model is presented. The computational problems of 3-D numerical experiments are discussed. The dynamics of the scroll ring with an ideal circular and arbitrary untwisted filament, the properties of the twisted scroll wave and twisted scroll ring are considered. In the final section, comparison between the theory and the experiment is done.

Introduction

Vortices in excitable media provide impressive examples of self-organization phenomena in systems far from thermal equilibrium [13,18]. It is now possible to speak about two kinds of such vortices: two dimensional ones (in excitable media with a thickness much smaller than the vortex wavelength) and three dimensional vortices (scroll waves) arising in sufficiently thick media.

Scroll waves were discovered by A.T.Winfree when he studied the BZ reaction [32]. There is evidence for their occurrence in heart tissue [14]. No doubt, such scroll waves must occur in other types of excitable media. The first systematic investigation of scroll wave properties was performed by Winfree & Strogatz [36-39]. They described possible types of such vortices and determined the topologically allowed ones.

The next natural step in studying scroll waves is theoretical and experimental investigation of their dynamics. Because of considerable experimental difficulties, on the one hand, and the sharp rise of computer facilities, on the other, theoretical and numerical methods have played a significant role in investigating scroll wave dynamics. As a result, the dynamics of several classes of 3-D vortices has been studied in reaction-diffusion models. This dynamics has turned out to be very exciting.

In investigations of the scroll wave dynamics, the FitzHugh-Nagumo type models have played a significant role. Their simplicity and convenience make them the most appropriate for theoretical and numerical studies.

This paper is devoted to the investigation of the dynamics of 3-D vortices in such two-component models of the FitzHugh-Nagumo type. The structure of the paper is as follows. Section 1 describes some elementary topological properties of 3-D vortices; section 2 is devoted to the computational problems of 3-D numerical experiments, in section 3 the properties of a scroll ring with an ideal circular filament are described, in section 4 the drift of an arbitrary untwisted filament is considered, sections 5 and 6 are devoted to the twisted scroll wave and twisted scroll ring, and in the final section 7 a comparison between theory and experiment is made.

1. Types of 3-D vortices

We describe the types of 3-D vortices following [36-39]. The simplest type of a three-dimensional scroll wave (the so-called simple scroll (Fig.1a)) can be constructed in imagination by displacing a two dimensional spiral wave perpendicular to its plane of rotation. The locus of displacement of the rotation center is called the scroll wave's singular filament.

If a simple scroll is twisted along its filament, a different type of vortex occurs - a twisted scroll wave (Fig 1b).

When the filament of a simple scroll is bent into a closed ring, it creates a simple scroll ring (Fig 2a). If the filament of a twisted scroll is bent into a ring, it creates a twisted scroll ring (Fig 2b).It is obvious that for such a procedure the total twist of the scroll must be 360 degrees. In addition, knotted scroll waves and combinations of all the mentioned types of vortices are possible.

The topological classification of possible 3-D vortices was proposed by Winfree & Strogatz. They studied the anatomy of 3-D vortices and found some topological restrictions on plausible 3-D structures (exclusion principle). For example, they showed that the existence of a solitary twisted scroll ring is impossible. (Threading such a vortex, there must be another singular filament.)

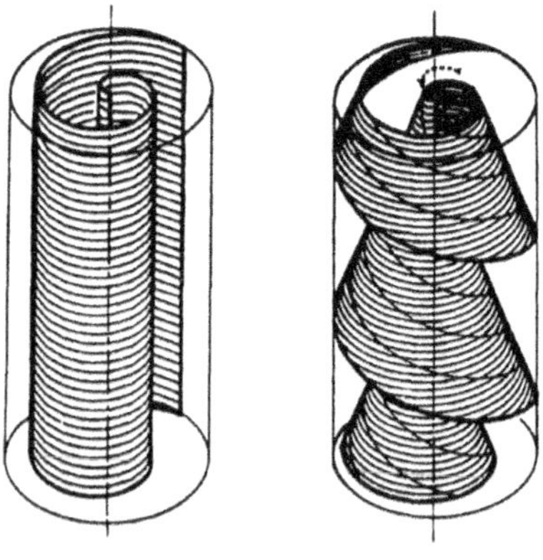

Figure 1. Simple (a) and twisted (b) scroll waves in three-dimensional active medium.

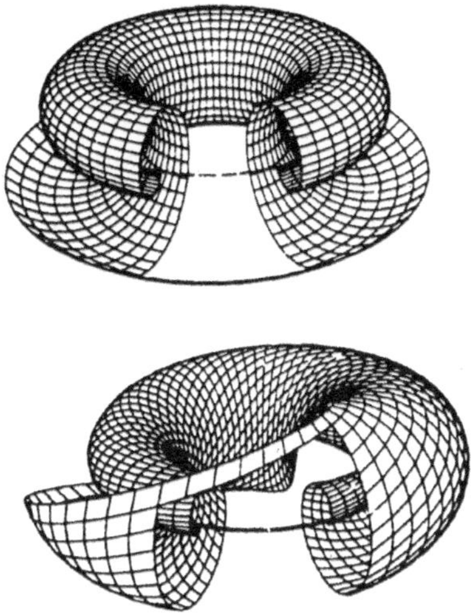

Figure 2. Untwisted (a) and twisted (b) scroll rings (A.T.Winfree [40]).

$$\tau\partial E/\partial t=D_E\Delta E+f(E)-g \qquad (1)$$

$$\tau\partial g/\partial t=D_g\Delta g+\varepsilon\varphi(E,g)$$

2. **Model, Method of computation**

In numerical experiments the following two-component FitzHugh-Nagumo model has been widely used:
where Δ is the three-dimensional Laplacian $\Delta=\partial^2/\partial x^2+\partial^2/\partial y^2+\partial^2/\partial z^2$; f and φ are nonlinear functions, D_E, D_g, τ and ε are parameters, ε being $\ll 1$.

This system can describe the dynamics of a chemically reacting system where E and g denote the concentrations of the reacting species, or it can describe the dynamics of a neurophysical medium for which E and g indicate the electric potential and outward current of repolarization, respectively. For the neurophysical model, D_g is zero.

The main difficulty of numerical experiments in active media is connected with the fact that usually an impulse has a steep front and a comparatively long plateau. For modelling such an impulse in one dimension a grid of 100-300 space points is necessary, and hence we need about 10^6-10^8 points for a three-dimensional grid. This leads to a tremendous waste of CPU time.

But it is possible to decrease the processor time if a specific kind of model (1) is used [19]. It is necessary for stable impulse propagation that at the impulse front the inhibitor concentration (g) should be small, when the activator concentration (E) is increasing, i.e. condition $\varepsilon \ll 1$ must be fulfilled. On the other hand when ε is small the impulse has a long plateau. The situation can be changed if we take a dependence of ε on the variable E such that ε is about 1 on the plateau and about 0.01 at the impulse front

[19]. In this case, we shall still have stable propagation of the impulse, but the impulse plateau will be much shorter. Using this approach, the speed of calculations in 3-D media can be increased from 10^3- to 10^5-fold. It makes possible direct 3-D modelling.

For a scroll ring with an ideal circular filament (Fig 2) the speed of computation can be increased, if the symmetry of the problem will be taken into account [19]. If we re-write for this case the Laplacian in (1) in the cylindrical coordinate system (ρ,φ,z) with the OZ- axis being directed along the torus axis of symmetry, then, considering that $\partial/\partial\varphi\equiv 0$, we obtain:

$$\Delta=\partial/\partial z \ +\partial \ /\partial\rho \ +1/\rho\partial/\partial\rho \tag{2}$$

The problem (1) with the Laplacian in the form of (2) is that of the rotation of a 2-D vortex in the (ρ,z)- plane with a modified Laplacian. Its calculation needs remarkably less computer time. But this approach is valid only for an ideal circular filament which is stable to disturbances of the symmetry.

At the end of the section, we give the form of the functions in model (1) which were used in refs. [16,19-25]:

$$f(E)=4E \text{ when } E<E1;$$

$$f(E)=-g_f(E-0.1) \text{ when } E1<E<E2; \tag{3}$$

$$f(E)=15(E-1) \text{ when } E>E2;$$

where E1 and E2 can be found using continuity conditions of the function f(E).

$$\varphi(E)=g_sE-g \tag{4}$$

$$\varepsilon(E)=1.82 \text{ if } E<0.01$$

$$\varepsilon(E)=0.06 \text{ if } 0.01<E<0.95 \tag{5}$$

$$\varepsilon(E)=2.0 \text{ if } E>0.95$$

3. **The dynamic properties of scroll rings**

This section is devoted to the dynamic properties of the scroll rings with an ideal circular filament. In the first part, the case of arbitrary diffusion coefficients D_E and D_g is considered. The second part examines the equal diffusion case ($D_E=D_g$), which is important for the experiments in the B-Z reagent.

3.1 *Case of arbitrary diffusion coefficients* $D_E\neq D_g$

The first type of 3D vortex obtained in numerical experiment in reaction-diffusion equations was the scroll ring (Fig 3a) [19]. The model (1) of 3-D active media in the one-component diffusion case ($D_g=0$) on a 30*30*30 grid was studied. The scroll ring obtained had surprising properties [19]. Its rotation was nonstationary: the position of its filament was not stabilized in

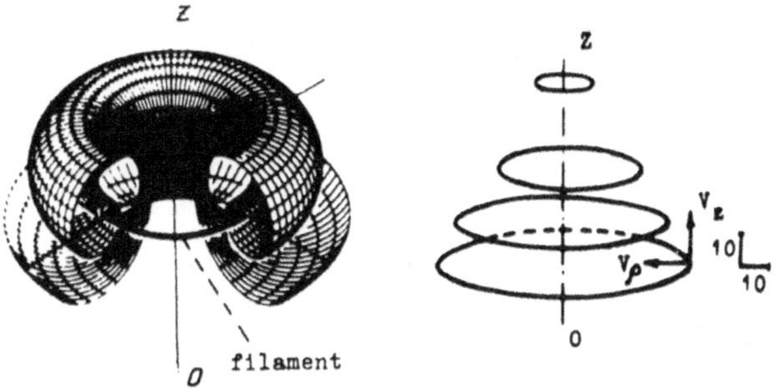

Figure 3. Scroll ring in a numerical experiment in model (1). a - general view, b - evolution of the scroll filament in time intervals ΔT=1000. Bottom, the initial location of the filament. (g_f=1., D_E=1., D_g=0.)

space, and it drifted in two directions (Fig 3b). The filament drifted gradually as a whole upward along the torus axis in the direction determined by the vector $\vec{n}=[\vec{r}\vec{\omega}]$, where $\vec{r}$ is the radius-vector of the distance from the rotation axis to any arbitrarily chosen point of the filament, $\vec{\omega}$ is the vector of the angular velocity of scroll rotation (ie., the direction of this drift was dependent on the direction of rotation of the scroll). At the same time, contraction of the filament was observed. When the length of the filament reached a certain critical value, collapse of the scroll occurred. The values of these velocities were very small, their maximum (before collapse) being two orders of magnitude less than the velocity of wave propagation. The velocities of the filament drift were found to be directly proportional to the curvature of the filament (Fig 4).

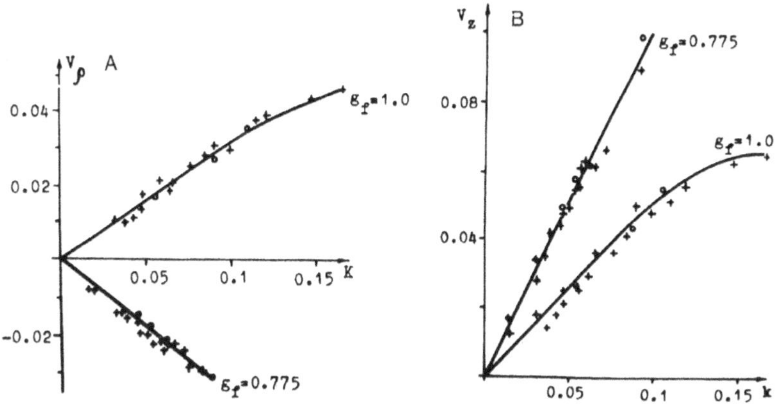

Figure 4. Velocities of the vertical (v_z) and radial (v_ρ) drift vs. the filament curvature (k) at different g_f. +, calculations in the cylindrical coordinate system; o, three dimensional calculations.

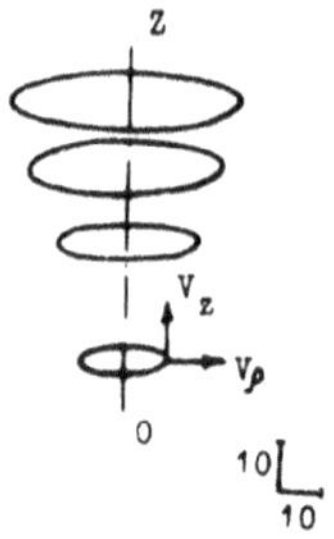

Figure 5. The extension regime of the scroll ring drift. Evolution of the scroll filament in time intervals $\Delta T=300$. Bottom, the initial location of the filament. ($g_f=0.775, D_E=1., D_g=0.$)

It is easy to estimate the life time of a scroll ring using this fact. If we take for the scroll radius the expression:

$$dr/dt \sim D_E/r, \tag{6}$$

then for the life time (T_l) of a scroll ring with a sufficiently large initial radius R_o ($R_o \gg R_C$, where R_C is the radius of the scroll ring collapse) we have:

$$T_l \sim R_o^2/2D_E - R_C^2/2D_E \sim R_o^2/2D_E \tag{7}$$

It is interesting to estimate the value of the life-time for different active media. In the BZ reaction, where D=0.12mm/min, the life-time for a scroll ring with a radius of 1 cm is about 400 min. For the heart tissue this estimation gives about 5-10 sec, a duration that is sufficiently long, because it takes about 20-40 rotations of the scroll.

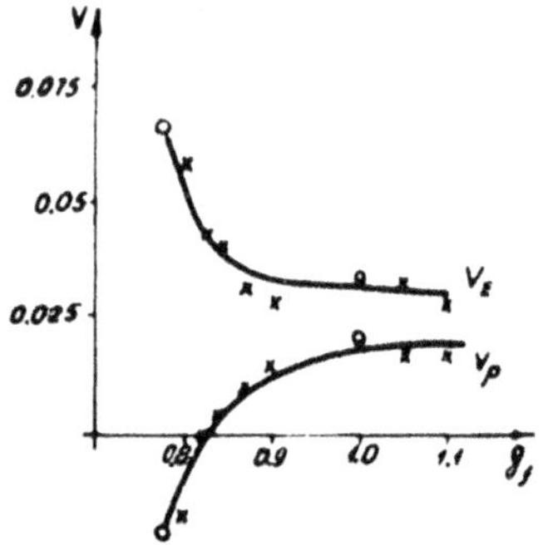

Figure 6. Velocities of the radial (v_ρ) and vertical drift (v_z) vs. g_f. The filament curvature k=0.063. +, calculations in the cylindrical coordinate system; o, three dimensional calculations.

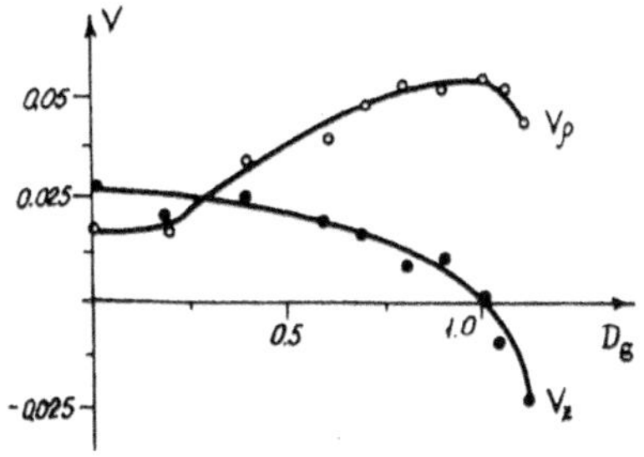

Figure 7. Dependence of the velocities of the radial (v_ρ) and vertical drift (v_z) on D_g. The filament curvature k=0.056. (D_E=1., g_f=1.)

Further investigations showed, however, that this regime of the scroll ring drift is not the only one. Under variation of the g_f parameter in model (1) a different regime of the scroll drift was revealed [25]. It consists of gradual extension and upward-directed drifting of the filament (Fig 5). In this case no collapse of the vortex occurs. But in a real active medium, the filament while drifting, comes to the boundary of the medium and disappears here. It was shown that this regime arises due to the decrease of the g_f parameter in model (1) (Fig 6), i.e. in media with low excitability. In the extension regime the drift velocities are also directly proportional to the filament curvature (Fig 4).

These two regimes were found in model (1) in the one component diffusion case (D_g=0). The two-component diffusion case was considered in [24]. It was shown that when the diffusion coefficient of the inhibitor D_g increased from 0 to D_E, a gradual increase in the radial velocity and a decrease of the vertical (drift) velocity were observed (Fig 7). At the point where D_E=D_g , the velocity of the radial motion reached a maximum and the vertical velocity became zero. When D_E<D_g, the vertical drift velocity changed its sign and the new regime with a different direction of the vertical drift of the scroll ring occurred (Fig 8).

To summarize, we can say that in numerical experiments three different regimes were obtained. The natural question arises as to when one or another of the regimes occurs. Let us try to give an answer.

The simplest case is to determine the vertical direction of drift. From the theory (see section 3.2), it follows that the velocity of the vertical drift is zero when the diffusion coefficients for all species are equal. So we can say that a change of the vertical velocity occurs when the diffusion coefficient of the inhibitor becomes greater than that of the activator. It is more difficult to find out when the change from contraction to extension occurs. Here we can give some heuristic reasons only: contraction of the scroll occurs when the ratio of the vortex period (T_v) to the minimal stable

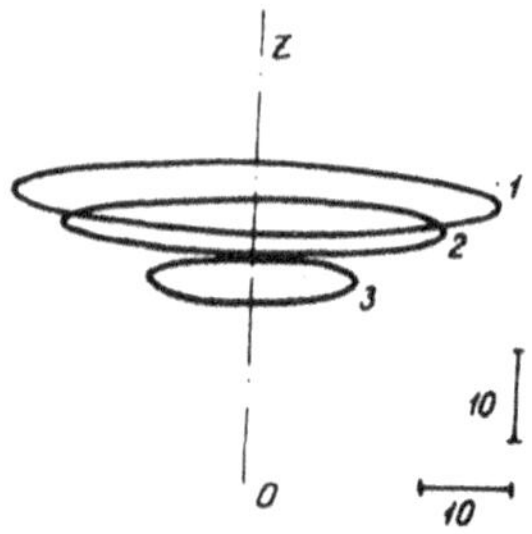

Figure 8. Drift of a scroll ring at $D_E<D_g$. (D_E=1.,D_g=1.1). Numbers are successive filament positions in time intervals ΔT=200.

time period in this medium (T_{min} due to refractoriness) is about 1, and when this ratio T_v/T_{min} is much more than 1, extension occurs. This is in conformity with numerical experiments and some qualitative considerations. The contraction of the filament is due to the fact that the velocity of the scroll wave in the direction of the torus axis (OZ) is more than in the other direction [6,9,19,24], because the concave front moves in the torus axis direction and the convex front moves in the other direction (here we consider a curvature component in the filament plane). But in the case of $T_v/T_{min} \gg 1$ another process, characterized by a sharp dependence of the vortex core radius on the excitability of the medium, should be considered [29]. (When the excitability increases, the core radius decreases). If we assume that a wave moving to the torus axis OZ has a greater effective excitability than when moving in the other direction, it must show a tighter turn in the first case and overall extension of the scroll occurs.

In this section, three drift regimes were described. Theoretically possible is the fourth one - the regime of extension and downward-directed drift of the filament. This regime could be observed when $D_g>D_E$ and $T_v \gg T_{min}$ It could be obtained, for example, by decreasing the excitability of an active medium when $D_g > D_E$. But in the case of model (1) this was impossible, because stable propagation of the wave disappeared when the g_f parameter was decreased.

Beside these regimes, some intermediate regimes exist. A regime between contraction and expansion was obtained in [5,17]. In this regime, the scroll ring ceases to contract, its radius stabilizes and only the vertical drift remains.

The stable regime of scroll ring rotation was revealed in [17], where a stable, steady rotating (non-drifting and non-expanding/shrinking) scroll ring near a wall was observed.

3.2 *Equal diffusion case*

The equal diffusion case is interesting from many points of view, in particular, because the Belousov-Zhabotinsky reaction [7] belongs to it, and because a quantitative analytical theory has been developed for this case, which is in good agreement with experiment. We describe this analytical approach, following papers [24,26].

Consider a three-dimensional active medium described by a reaction-diffusion equation:

$$\partial\vec{u}/\partial t=D\Delta\vec{u}+\vec{f}(\vec{u}) \qquad (8)$$

Here $\vec{u}$ is an n-dimensional vector, $\Delta = \partial^2/\partial x^2+\partial^2/\partial y^2+\partial^2/\partial z^2$ is a three-dimensional Laplacian, $\vec{f}$ is a nonlinear vector function accounting for the medium excitability, and D is a scalar.

We shall find a solution of (8) for problems with a cylindrical symmetry, in particular, for a scroll ring having a circular filament. Let us pass to the cylindrical coordinate system (r,φ,z) with the OZ axis coincident with the symmetry axis of a torus. Considering that $\partial\vec{u}/\partial\varphi\equiv0$ in view of the problem's cylindrical symmetry we obtain:

$$\partial\vec{u}/\partial t=D\Delta\vec{u}+D/r*\partial\vec{u}/\partial r+\vec{f}(\vec{u}) \qquad (9)$$

where $\Delta=\partial^2/\partial r^2+\partial^2/\partial z^2$.

The problem (9) describes rotation of a vortex in the (r,z) plane with a modified Laplacian (addition of $D/r*\partial\vec{u}/\partial r$). It is known for vortices in active media that their properties are determined by a small region in the vicinity of the vortex core, i.e. only perturbations within this region are significant. The value added to the Laplacian near the vortex core is $D/R*\partial\vec{u}/\partial r$ where R is the radius of the scroll ring filament. Therefore, to a first approximation, the term D/r in (9) can be replaced by a constant D/R. (For details, see [24].) When D/R is constant the addition can be removed by passing to a moving coordinate system with a velocity D/R.

$$\rho=r+D/R*t \qquad (10)$$

i.e. the filament contracts with a velocity of V_ρ without displacement along the OZ axis:

$$V_\rho=D/R=D*k \qquad (11)$$

$$V_z=0$$

where k=1/R is a filament curvature.

Thus, the basic conclusion is that in the equal diffusion case there takes place, independent of the shape of the vector-function f, pure contraction of the scroll ring with a velocity that is directly proportional to the curvature of the filament and to the diffusion coefficient. The vertical drift in this case is absent. Note that the analysis (and therefore eq.(11)) holds not only for stationary but also for nonstationary rotating vortices as well as for vortices with any number of arms.

Formula (11) obtained in [24] for the drift of the filament with an ideally circular shape has recently been confirmed and developed in the general analytical theory of Keener [10], in the kinematic theory of scroll rings drift [5] and in the λ-ω system [43].

In this section we have considered scroll rings with an ideal circular filament. The next section is devoted to the dynamics of an arbitrary untwisted scroll filament.

4 Dynamics of an arbitrary untwisted filament

In this section two methods for solving the problem of the drift of an arbitrary filament are considered: the method of differential equations for the filament drift and the method of solving the problem using some integral characteristics of the whole filament, in particular, the recently developed method of finding integral invariants for a planar filament.

4.1 *Equations for the filament drift*

The behaviour of a three-dimensional scroll wave with a filament of arbitrary shape was considered for the first time by Winfree & Guilford [41] who assumed that the drift of a filament is defined by its local geometry only. For example, in the equal diffusion case, it might turn out that each small arc of filament moves in the plane of its curvature at a rate proportional to the local curvature, regardless of anything else. In the general case, they supposed that the filament motion is determined by torsion, twist and curvature and knowing the dependence of the drift velocities on these characteristics (for example, from numerical simulations) one can calculate the evolution of the filament, solving the equations of differential geometry without solving three-dimensional reaction-diffusion equations [41]. In [41], an excellent project was proposed for solving this problem with the use of super computers.

A general analytical theory for the drift of a filament of arbitrary shape was developed by Keener [10]. Using ideas from the perturbation theory (multiscale techniques), Keener has shown that the filament evolution is determined only by its local geometry and has found equations for the filament drift. He has shown, in particular, that in the equal diffusion case eq. (11) holds not only for a circular but also for an arbitrary planar filament. In [9-12] a set of solutions of the equations of the filament drift were found (for elongated sources, for helical filament, for invariant knots, etc.).

4.2 *An integral invariant for the scroll filament*

In the equal diffusion case, an integral invariant for anarbitrary untwisted planar filament was revealed [26]. It has been found that the speed of decrease of an area, S, bounded by a planar scroll filament is given by $dS/dt=-2\pi D$ (where D is the diffusion coefficient) and is independent of the kind of model used to describe an active medium, and also of time and the filament shape. The following integral invariant exists:

$$\oint v\,ds=-2\pi D \qquad (12)$$

where ds is the differential of the filament length and v is the local contraction velocity. (Tracing around the filament contour, leaving its inside on the right.)

The invariant (12) exists not only for a closed filament but also for the

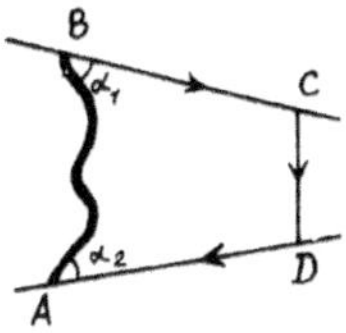

Figure 9. An unclosed planar filament AB of arbitrary shape. At the filament ends, the boundary conditions α_1=const, α_2=const are fulfilled.

case of a planar filament located between two interfaces at which the condition of constancy of the angle is fulfilled (Fig. 9). Any closed contour containing a filament can be considered. The speed of decrease of its area (similarly to (12)) is

$$dS/dt=D\Delta\alpha \tag{13}$$

where $\Delta\alpha$ is a change in the angle of the tangent vector during its motion along the filament.

In [26] this invariant was used to solve some problems on the drift of a planar filament. It was found that the life time (T_1) of a scroll ring with an arbitrary planar filament is:

$$T_1=So/2\pi D \tag{14}$$

where S_0 is the area bounded by the filament at the initial moment of time.

The problem of the motion of a Γ-shaped filament was solved. Consider an infinite Γ-shaped filament which is at right angle to the boundary of the medium and goes to infinity along a line parallel with the boundary (Fig.10). Assume that the filament takes a steady-state shape rapidly enough and with the course of time it shifts, as a whole, relative to infinity at a velocity v. Let us find this velocity. Consider a contour OCD, where the point C is sufficiently far from point O, so that the filament can, with a good accuracy, be considered parallel to AB. According to the integral invariant, the decrease in the area inside this contour is $dS/dt=-(\pi/2)D$, because the change in the angle along the filament is $-\pi/2$. Since the filament shape does not change with time, i.e. the filament simply slides along the boundary with a velocity v, the rate of change of the area equals $dS/dt=-vh$. By comparing these two equations it is easy to find the velocity of the filament:

$$v=\pi D/2h \tag{15}$$

Eq.(15) is in agreement with the formula obtained by Winfree and Jahnke [8,42] for this case.

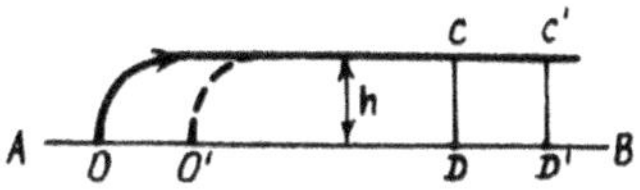

Figure 10. Drift of a Γ-shaped filament at a certain initial instant of time (solid line) and after time Δt (dashed line).

We determined the steady state position of a filament between two parallel interfaces. Consider a planar filament AB of arbitrary shape between two plane interfaces, for which the condition of orthogonality is fulfilled (Fig.11). The steady shape of the filament A'B' is a straight line orthogonal to the interfaces [9]. Let us determine this position using the integral invariant. Consider a contour ABCD. Since the change of the angle in moving along the filament is zero, it follows that the area of this contour remains unaltered, and therefore the area of contour ABCD is equal to that of contour A'B'CD. Hence

$$S1+S3=S2+S4 \tag{16}$$

Therefore, for the steady state position of the filament to be determined, it is necessary to draw a straight line orthogonal to the interfaces so that the sums of the areas on the left and on the right of the line are equal.

For a filament having initially the form of a straight line inclined to the boundaries, the problem of finding the characteristic time of reaching a steady-state position was solved in [9].

Using the integral invariant, some other problems of the filament drift in three-dimensions can be solved.

4.3 *Estimate of the filament length*

The length of a filament is one of the fundamental characteristics which is applicable not only to planar but also to arbitrary three-dimensional filaments. The problem of determining the rate of change of the filament length was considered in [4], where, on the assumption of a small twist and torsion, the expression for the rate of the change was found:

$$dL/dt=-b\oint_0^L k^2 ds \tag{17}$$

where b is the coefficient of proportionality between the curvature of the filament (k) and its velocity in the normal direction, i.e. the rate depends only on the normal component of the filament motion [4].

A study of the integral characteristics of the filament motion was performed in [28]. Using (17), the estimate of the lower limit for the rate of change of the length of an arbitrary closed filament was found:

$$|dL/dt| \geq > 4\pi^2 b/L \tag{18}$$

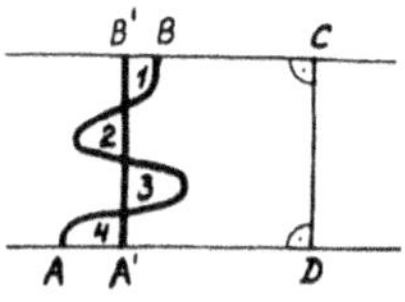

Figure 11. A filament between two plane boundaries. AB: initial position, A'B': steady-state position of the filament.

or

$$L(t) \leq < \sqrt{L_0^2 - 8\pi^2 bt}$$

where L_0 is the initial filament length.

In the equal diffusion case for a planar filament the expression (18) can be fulfilled by the estimate:

$$L(t) \geq > \sqrt{4\pi S_0 - 8\pi^2 Dt} \qquad (19)$$

where S_0 is the initial area bounded by the filament.

Using (18) and (19), in [28] the interval of change of L(t) is found:

$$\sqrt{4\pi S_0 - 8\pi^2 Dt} \leq < L(t) \leq < \sqrt{L_0^2 - 8\pi^2 Dt} \qquad (20)$$

The estimate (20) is more precise the closer the initial shape of a filament to a circumference, i.e. the closer L_0^2 to $4\pi S_0$.

5. The dynamics of twisted scrolls

In this section the dynamics of twisted scroll waves with an unclosed straight filament is considered.

5.1 *Stationary rotation*

Numerical studies [20,21] have shown that in a homogeneous medium, twisted scrolls with an open filament are unstable. Whatever the initial conditions, they always degrade to simple (untwisted) scrolls during their rotation (Fig 12).

It was shown that twisted scrolls naturally arise in inhomogeneous active

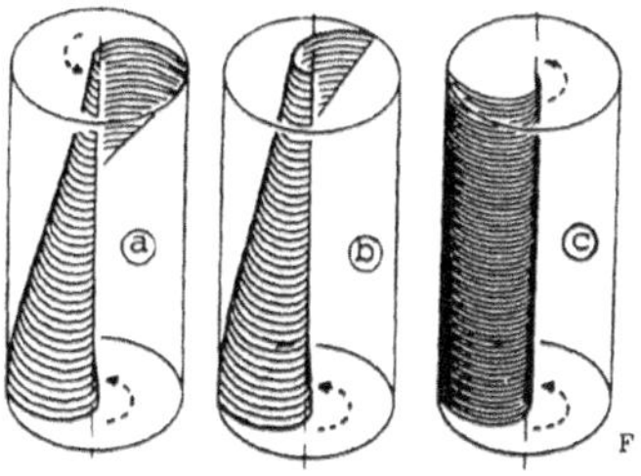

Figure 12. Degeneration of a twisted scroll to a simple one in a homogeneous medium (g_f=0.9).

a - the initial shape of the scroll (t=0), the phase shift in the upper and lower cross-sections being equal to π. The arrow indicates the direction of rotation.
b-t = 29.4, the phase shift reduced to $\pi/2$.
c-t = 59.3, no phase shift is observed.

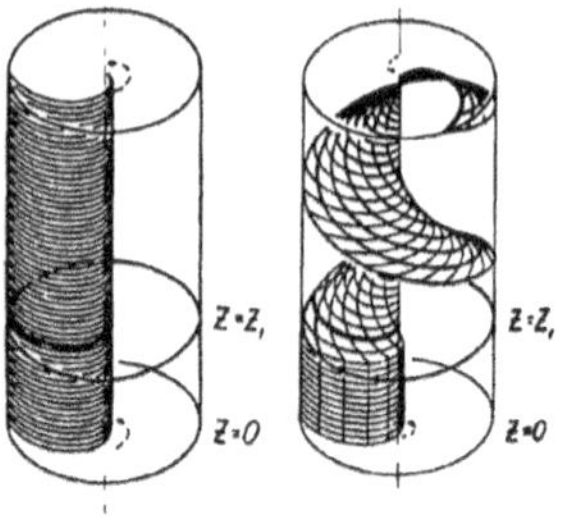

Figure 13. Generation of a twisted scroll.
a - initial position (simple scroll). At $z=z_1$, parameter τ is increased stepwise from $\tau=1$ to $\tau=2$.
b - the established shape of the twisted scroll.

media, with the inhomogeneity being directed along the scroll filament (Fig 13). In this figure, the stepwise heterogeneity in the parameter τ of model (1) was preset. The period of vortex rotation at Z < Z1 was half as large as at Z > Z1. The initial condition was in the form of a simple scroll. The situation was as follows. Near the boundary Z=Z1, a phase shift occurred because of the different velocities of scroll rotation in the upper and lower part of the medium. The wave of this twist, further propagated to the upper part only (i.e. to the region with a greater period). In the lower part, the simple untwisted scroll survived. Numerical experiments have shown that when the degree of twisting of the scroll increased, its period gradually decreased and when it became equal to the period in the lower part of the medium, stationary rotation occurred. The value of the rotation period was equal to the value of the vortex period in the lower part.

Figure 14 (curve 1) plots the phase shift vs. z for this case. It is seen that sufficiently far away from the boundary of inhomogeneity the phase shift gradient becomes constant (φ'=const). The same value $\varphi'=0.36$ is attained when the parameter τ is altered smoothly rather than stepwise (Fig. 14a).

The twist of the scroll depends not only on the value of inhomogeneity, but also on the specific inhomogeneity produced (Fig.14b). In [21] the established shape of the twisted scroll with gradients in g_f, g_s, and τ yielding the same period gradient $T=30+30*\theta(z-z_1)$ was studied. In this case the established value of twist with the parameter τ was about three times greater than with the parameters g_f and g_s.

So, numerical experiments show that if an inhomogeneous medium consists of a few layers of homogeneous media with different values of the periods, then in the whole medium a twisted scroll occurs and its period will be equal to the smallest of the period values in these layers.

5.2 *Unstationary regime*

In all the cases considered in the previous subsection only a stationary regime took place. Rotation of spirals in all the cross-sections z=const was along a circumference and no meandering was observed. Further numerical experiments showed [1,27] that this stable regime of rotation exists when the heterogeneity of a medium is not high. It was shown that with increasing

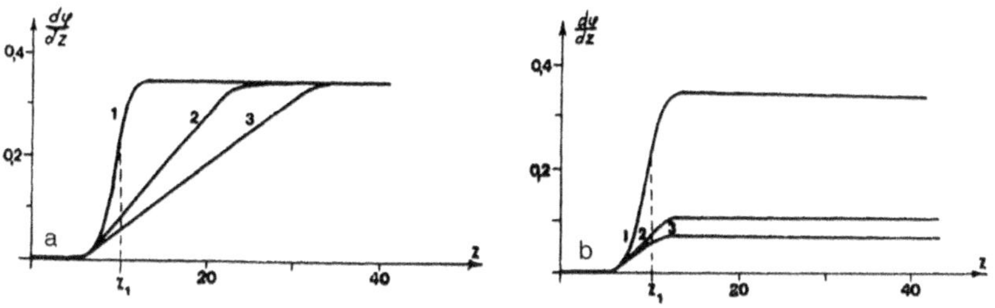

Figure 14. The degree of the scroll twistedness as a function of z.
a - different gradients in parameter τ. 1 - abrupt increase in τ, $\tau=1.+\theta(z-z_1)$, where $\theta(z)$ is the Heaviside function. 2,3 - smooth increase in τ, from $\tau=1$ to $\tau=2$ with different gradients: $\alpha=d\tau/dt=0.1$ (curve 2), $\alpha=0.05$ (curve 3).
b - set up of different inhomogeneities (gradients in τ, g_f, g_s).
Curve 1, $\tau=1.+\theta(z-z_1), g_f=0.9, g_s=1.0$; Curve 2, $g_f=0.9 -0.12\theta(z-z_1)$, $g_s=1.0, \tau=1$; Curve 3, $g_s=1. +0.5\theta(z-z_1), g_f=0.9, \tau=1.0$.

degree of heterogeneity, nonstationary regimes of rotation can be observed. These are the regime of radiating boundary and the meandering regime.

The regime of radiating boundary [27] arises in model (1) when heterogeneity in the parameter τ increases more than 3 times. In this case the following process takes place. A wave radiated periodically from the boundary of the longer-period region propagates to the lower part of the medium and perturbs the vortex rotation here. Figure 15 shows rotation of a vortex (bottom cross-section): 1 is a wave radiated by the boundary, 2 is a vortex

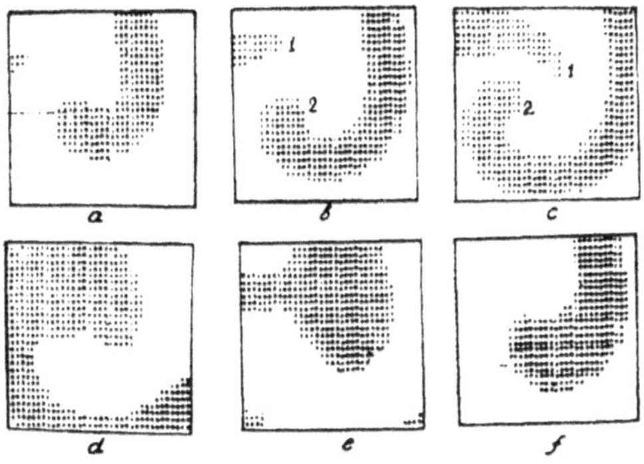

Figure 15. The regime of radiating boundary. The pattern of spiral rotation in the bottom part of the active medium is shown (the cross-section z=6 at $z<z_1$, see Fig 13). At $z=z_1$, parameter τ is increased stepwise from $\tau=1$ to $\tau=4$. The subsequent phases of the vortex rotation in time intervals $\Delta T=5$ are shown.

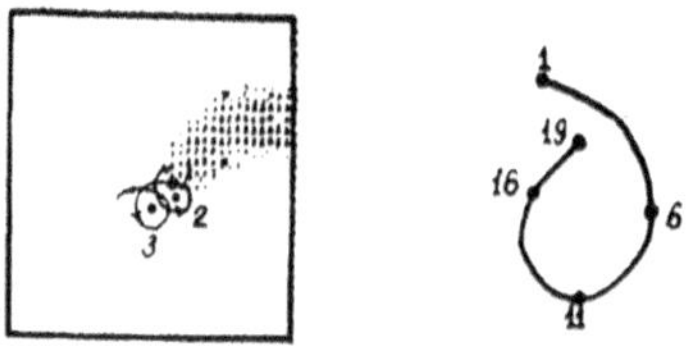

Figure 16. The meandering regime of the twisted scroll wave rotation.
a - the trajectory of rotation of a spiral wave in the top cross-section.
b - subsequent positions of the tip of the spiral wave in 1,6,11,16 and 19 rotations. At $z=z_1$, parameter τ is increased stepwise from $\tau=1$ to $\tau=2.5$.

wave. It is seen that the first wave perturbs rotation of the vortex.

In spite of the complexity of the regime, the vortex rotation in the upper part is periodic. However, the value of the period (t=62) is different from the minimal value (t=30, lower part) and from the maximal value (t=120, upper part).

This regime is likely to arise when the refractory period (R) of the part of the medium in which a twisted scroll occurs (the case of Fig 13, upper part) is more than the period of vortex rotation (T_v) in the part in which an untwisted scroll remains (the case of Fig. 13, lower part). Really, in the case of stationary rotation of a twisted scroll its period is equal to T_v. But on the other hand, the periods of any vortices in active media must exceed its refractory period R. So in the case $T_v<R$ we have a discrepancy between these two conditions. Since the period of a twisted scroll cannot be less than R, the break of the filament at the boundary and radiation of waves occur.

The regime of meandering [1] arises when heterogeneity is somewhat lower than in the regime of a radiating boundary. In this regime a minimal value of the period is established, but the rotation of a vortex is not circular. The tip of the vortex meanders along a complex trajectory (Fig. 16) which is similar to the unstationary regime of two dimensional vortex rotation [44] and to meandering of spiral waves in the BZ reagent [1,35].

To conclude this section, we note that the regimes of unstationary rotation of twisted scrolls has not yet been investigated well enough, and their systematic studies in different models and at different parameter values are necessary.

6. Dynamics of twisted scroll rings

The twisted scroll ring (Fig. 2b) is an object having nontrivial topological properties. Winfree & Strogatz [36-39] showed that the existence of a solitary twisted scroll ring is impossible. (Inside such a vortex must be another filament of singularity.)

First computations of twisted scroll rings in a reaction-diffusion model were done in [22,23]. The following properties of this type of vortex were observed:

1. Spiral rotation of a twisted scroll ring has a shorter period than that of a non-twisted ring of the same size.
2. The ring formed by a singular filament gradually contracts.
3. The twisted scroll ring's filament drifts along the OZ-axis in the direction of the cross product:

$$\vec{n}=[\vec{r}\vec{\omega}] \tag{21}$$

where $\vec{r}$ is the vector from the center of a curvature and $\vec{\omega}$ is the angular velocity of vortex rotation.

In refs. [22,23] vanishing of a twisted scroll ring was observed. After the ring disappeared, its central vertical vortex remained and continued to rotate in a stable manner.

A detailed study of the properties of twisted scroll rings was performed in [16], where expansion and contraction of curved twisted filaments was observed. It was shown that the vertical drift of a filament ring in the direction given by formula (21) is valid only for rings with a large radius. When the radius decreases, a sliding regime occurs, i.e. the ring filament slides along the linking filament, which can have a complicated form.

In [16] the process of disappearance of twisted scroll rings was investigated in details. It was shown that the vortex either dissipates to the unique spatially uniform stable steady state or breaks into open scrolls which untwist during the course of rotation but continue to rotate stably. In this paper, the effect of fusion of two linked twisted scroll filaments was also observed. As a result, two untwisted uncurved filaments or one untwisted scroll ring occurred, depending on the parameters. So, the process of the twisted scroll ring drift is closely connected with the problem of interaction of filaments.

7. **Dynamics of three dimensional vortices in experiment**

In conclusion, a few words about the comparison of theory and experiment in the BZ reaction.

First observations of **scroll rings** in the BZ reaction were reported by Winfree in 1973, 1974 [32,33,34] and by Welsh et al. in 1983 [31]. In these papers shrinkage and collapse of scroll rings were observed. It was shown that the period of their rotation is approximately equal to the rotation period of 2-D spiral waves.

The first quantitative comparison of the theory and experiment was done by Keener & Tyson [9]. They derived a formula for the dynamics of the elongated sources using the theoretical formula for scroll ring contraction in the equal diffusion case (see (11) section 3.2), and compared the results obtained with experimental data for the BZ reaction. A good quantitative agreement between experiment and theory was observed. For the diffusion constant the value $D=0.12mm^2/min$ was obtained.

Direct experimental verification of formula (11) was done by Agladze et al. [3] in liquid BZ reaction and by Pertsov & Aliev [30] in agarose gel.

Both papers have demonstrated a high accuracy of formula (11). Both in liquid and in agarose gel a value $D=0.12mm^2/min$ was found for the diffusion coefficient. These papers showed that shrinkage of a scroll ring proceeds without displacement of the filament in a perpendicular direction, i.e. $v_z=0$, which is also in conformity with formula (11).

In the recent papers of Jahnke et.al. [8] and Winfree & Jahnke [42] the motion of a scroll ring in agarose gel was studied. In [8,42] the direct proportionality between the velocity of shrinkage on filament curvature was demonstrated, but for the diffusion coefficient a value of $D=0.048mm^2/min$ was obtained, which is 2.5 fold less than in agarose experiment [30]. In [8,42], in contrast to [3,30] the vertical drift of a scroll ring was observed. The discrepancy is probably connected with different experimental (temperature) conditions in [30] and [8], and a special consideration is needed to clarify this question.

Motion of a filament of complex shape. In [26] in agarose gel the drift of a filament with a shape far from being circular and having both convex and concave segments is studied. As follows from the theory (section 4 (12)), the speed of decrease of an area bounded by the filament is constant and equals $2\pi D$. The experiment confirmed this fact with good accuracy. The bounded area decreased linearly with time and for the diffusion constant a value of $D=0.12mm^2/min$ was found.

The drift of a filament of a spiral shape was observed in [2,8]. From the theoretical point of view, this problem is equivalent to the problem of the motion of the Γ-shaped filament considered in section 4.2, i.e. the filament must move with a constant speed given by formula (15). Experimental data showed that the speed was constant but its value was greater than predicted by theory. In [8] the observed value was approximately 3.1 times greater, and in [2] 2.7 times greater than the theoretical value.

The difference observed can be connected with various factors: the failure of the orthogonality conditions at the bottom boundary, the inapplicability of formula (11) to the dependence of the shrinkage velocity on the curvature at large curvatures, or inaccuracy of the determination of the distance between the filament and the boundary in [2], because of impossibility of its direct measurements.

Twisted scrolls have not been observed directly as yet. But they served to explain in some cases the phenomenon of meandering of spiral waves in the BZ reaction [35]. It was shown [1] that nonstationarity in some cases is a result of 3-D effects of the nonuniformity of the reaction throughout the solution depth due to the nonuniform saturation with oxygen of the solution from the surface to the bottom. And meandering is an unstationary rotation of the twisted scrolls formed in such inhomogeneous conditions. In [1] it was shown both experimentally and theoretically that the meandering of a spiral wave has the same trajectory as the unstationary rotation of a twisted scroll wave (Fig.16) and that this meandering disappears in the absence of inhomogeneity.

Extensive investigations of the role of the meandering problems, especially in two dimensions, were performed in [15,42].

Acknowledgements

The author is grateful to Arun V. Holden, A.T. Winfree and R.I. Sel'kova for great help in the preparation of the manuscript.

References

[1] Agladze, K.I., Panfilov, A.V. & Rudenko, A.N. (1988). Nonstationary rotation of spiral waves: three-dimensional effect. *Physica* **29D**, 409-415.

[2] Agladze, K.I., Kuhnert, L., Linde, H., Krinsky, V.I.& Panfilov, A.V. Three-dimensional vortex with a spiral filament in a chemical active medium. Physica D to appear.

[3] Agladze, K.I., Kocharjan, R.A. & Krinsky, V.I. The direct observation of the collapse of the scroll rings in chemical active medium. Physica D, submitted.

[4] Biktashev, V.N. (1989). Interaction of a 3-D autowave vortex and external waves. Preprint NCBI, Pushchino, 24p.

[5] Braznik, P.K., Davydov, V.A. Zykov, V.S. & Mikhailov, A.S. (1987). Vortex rings in distributed excitable media. *ZETP* **93**, 1725-1736.

[6] Ding da-fu (1988). A plausible mechanism for the motion of untwisted scroll rings in excitable media. *Physica* **32D**, 471-487.

[7] Field, R.J., Noyes, R.M. (1974). Oscillations in Chemical Systems, Part 4. Limit Cycle Behavior in a model of a real Chemical reaction. *J.Chem. Phys.* **60**, 1877-1884.

[8] Jahnke, W., Henze, Ch., Winfree, A.T. (1988). Chemical vortex dynamics in three-dimensional excitable media. *Nature* **366**, 662-665.

[9] Keener, J.P. & Tyson, J.J. (1988). The motion of untwisted untorted scroll waves in Belousov-Zhabotinsky reagent. *Science* **239**, 1284-1286.

[10] Keener, J.P. (1988). The dynamics of three-dimensional scroll waves in excitable media. *Physica* **31D**, 269-276.

[11] Keener, J.P. Knotted Scroll Wave Filaments in Excitable Media. Physica D, in press.

[12] Keener, J.P. Knotted Vortex filament. In press.

[13] Krinsky, V.I. (1984). Autowaves: results, problems, outlooks. In: *Self-Organization.Autowaves and Structures Far from Equilibrium,* V.I.Krinsky (ed.). Springer-Verlag Heidelberg, 9-19.

[14] Medvinsky, A.B., A.V.Panfilov,A.V & Pertsov, A.M. (1984). Properties of rotating waves in three dimensions. Scroll rings in myocard. In *Self-Organization.Autowaves and Structures Far from Equilibrium,* V.I.Krinsky (ed.). Springer-Verlag Heidelberg, 195-199.

[15] Muller, S.C. & Plesser, Th. Dynamics of spiral centers in the ferroin catalyzed Belousov-Zhabotinsky reaction. paper in this volume.

[16] Nandapurkar, P.J. & Winfree, A.T. (1987). A computational study of twisted linked scroll waves in excitable media. *Physica* **29D**, 69-83.

[17] Nandapurkar, P.J. & Winfree, A.T. (1989). Dynamical stability of untwisted scroll rings in excitable media. *Physica* **35D**, 277-288.

[18] Nicolis, G. & Prigogine, I. (1977). *Self-Organization in Non-Equilibrium Systems.* Wiley: New York.

[19] Panfilov, A.V. & Pertsov, A.M. (1984). Vortex ring in three-dimensional active medium in reaction-diffusion system. *Dokl. AN. SSSR* **274**, 1500-1503.

[20] Panfilov, A.V., Rudenko, A.N. & Pertsov, A.M. (1984). Twisted scroll waves in three-dimensional active media. *Dokl. AN. SSSR* **279**, 1000-1002.

[21] Panfilov, A.V., Rudenko, A.N. & Pertsov, A.M. (1984). Twisted scroll waves in three-dimensional active media. In *Self-Organization. Autowaves and Structures Far from Equilibrium,* V.I.Krinsky (ed.). Springer-Verlag: Heidelberg, 103-105.
[22] Panfilov, A.V. & Winfree, A.T. (1985). Dynamical simulation of twisted scroll rings in three-dimensional excitable media. *Physica* **17D**, 323-330.
[23] Panfilov, A.V., Rudenko, A.N. & Winfree, A.T. (1985). Twisted scroll rings in three-dimensional active media. *Biofizica* **30**, 464-466.
[24] Panfilov, A.V., Rudenko, A.N. & Krinsky, V.I. (1986). Scroll rings in three dimensional active medium with two component diffusion. *Biofizica* **31**, 80-854.
[25] Panfilov, A.V. & Rudenko, A.N. (1987). Two regimes of the scroll ring drift in the three-dimensional active media. *Physica* **28D**, 215-218.
[26] Panfilov, A.V., Aliev, R.R. & Mushinsky, A.V. (1989). An integral invariant for scroll rings in a reaction-diffusion system. *Physica* **36D**, 181-188.
[27] Panfilov, A.V. & Rudenko, A.N. Unpublished result.
[28] Panfilov, A.V. & Aliev, R.R. Integral properties of the filament length. *Physica D* in preparation.
[29] Pertsov, A.M. & Panfilov, A.V. (1981).Spiral waves in active media. Reverberator in the FitzHugh-Nagumo model. In: Autowave processes in Systems with diffusion. *Acad. Sci. USSR, Gorky*, 77-84.
[30] Pertsov, A.M. & Aliev, R.R. 3D vortices in the gel BZ reaction. *Physica D*, in preparation.
[31] Welsh, B.J., Gomatam, J. & Burgess, A.E. (1983). Three-dimensional chemical waves in the Belousov-Zhabotinsky reaction. *Nature* **304**, 611-614.
[32] Winfree, A.T. (1973). Scroll-shaped Waves of Chemical Activity in Three Dimensions. *Science* **181**, 937-939.
[33] Winfree, A.T. (1974). Rotating chemical reactions. *Sci. Amer.* **230(6)**, 82-95.
[34] Winfree, A.T. (1975). Two kinds of waves in an oscillating chemical solution. *Faraday Symposia of the Chemical Society No. 9*, 38-46.
[35] Winfree, A.T. (1978). Stably rotating patterns of reaction and diffusion. In *Periodicities in Chemistry and Biology, Progress in Theoretical Chemistry,* Eyring, H., and Henderson, D., (eds.). Academic Press: New York, 4, 1-51.
[36] Winfree, A.T. & Strogatz, S.H. (1983). Singular Filaments Organize Chemical Waves in Three Dimensions: 1. Geometrically Simple Waves. *Physica* **8D**, 35-49.
[37] Winfree, A.T. & Strogatz, S.H. (1983). Singular Filaments Organize Chemical Waves in Three Dimensions: 2. Twisted Waves. *Physica* **9D**, 65-80.
[38] Winfree, A.T. & Strogatz, S.H. (1983). Singular Filaments Organize Chemical Waves in Three Dimensions: 3. Knotted Waves. *Physica* **9D**, 333-345.
[39] Winfree, A.T. & Strogatz, S.H. (1984). Singular Filaments Organize Chemical Waves in Three Dimensions: 4. Wave Taxonomy. *Physica* **13D**, 221-233.
[40] Winfree, A.T. & Strogatz, S.H. (1984). Organizing centers for three-dimensional chemical waves. *Nature* **311**, 611-615.
[41] Winfree, A.T. & Guilford, W. (1988). The dynamics of organizing centers: Numerical experiments in differential geometry.In *Biomathematics and Related Computational Problems,* Ricciardi (ed.). Kluwer Academic Publ.in Biomathematics and Related Computational Problems, 697-716.
[42] Winfree, A.T. & Jahnke, W.J. (1989). Three-dimensional scroll ring dynamics in the Belousov-Zhabotinsky reagent and the 2-variable Oregonator model. *J. Phys.Chem.* **93**, 2823-2932.

[43] Yakushevich, L.V. (1984). Vortex filament Elasticity in Active Medium. *Studia Biophysica* **100**, 195-200.
[44] Zykov, V.S. (1984). Modelling of the Wave Processes in Excitable Medium. Nauka: Moscow.

35. STRUCTURE AND DYNAMICS OF SPIRAL WAVES AND OF DEFECTS IN TRAVELLING WAVES

E. Bodenschatz*, A. Weber and L. Kramer

Physikalisches Institut der Universität Bayreuth
D-8580 Bayreuth, FRG
*now Department of Physics, University of California
Santa Barbara, CA 93106, USA

1. The envelope-equation

Systems with a bifurcation to an oscillatory homogeneous state (type 1) or to an oscillatory and spatially periodic state (type 2) appear in biology, chemistry and physics. The most prominent examples for type 1 are in biology the aggregation of social amoebae [2] and in chemistry the Belousov-Zhabotinskii (BZ) reaction [17]. Examples for type 2 are found in physical systems and are electro-convection of liquid crystals [19] and thermo-convection of binary-fluid mixtures [1]. These systems exhibit a Hopf bifurcation which may be super- or subcritical. Although in the case of the BZ reaction most experiments are performed in a parameter range, where the system behaves in an excitable manner, the description of the system as an oscillatory one captures many qualitative features.

Near a supercritical Hopf bifurcation, the physical quantities in a quasi-2-dimensional layer in the x-y-plane are essentially of the form

$$u = \varepsilon^{1/2}\,[A(X,Y,T)\,e^{i(\vec{q}_c\vec{x}+\omega_c t)} + B(X,Y,T)\,e^{i(\vec{q}_c\vec{x}-\omega_c t)} + c.c.\,]\,f(z) + \mathcal{O}(\varepsilon). \quad (1)$$

The wavevector $\vec{q}_c$ and the frequency ω_c describe the travelling-wave pattern at threshold of a bifurcation to type 2. For a bifurcation to purely oscillatory state (type 1) the $\vec{q}_c = 0$. The function f(z) captures the z dependence. In the case of type 2 the time- and space-varying envelopes A and B describe left- and right-going travelling waves, respectively. For type 1 either B = 0 or A = 0. Here the slow variables $X = \varepsilon^{1/2}x$ $Y = \varepsilon^{1/2}y$ and $T = \varepsilon t$ are introduced, which capture essentially the scaling of the coordinates on the control parameter ε, where ε measures the distance of the threshold of bifurcation.

For anisotropic convective systems with one preferred roll direction the so called complex Ginzburg-Landau equations for the envelopes A and B read

$$T_0\partial_T A + \varepsilon^{-1/2} v\partial_X A = [b_1\partial_X^2 + b_2\partial_Y^2 + 1 - c|A|^2 - d|B|^2]\ A \tag{2}$$

$$T_0\partial_T B - \varepsilon^{-1/2} v\partial_X B = [b_1\partial_X^2 + b_2\partial_Y^2 + 1 - c|B|^2 - d|A|^2]\ B. \tag{3}$$

For B=0 or A=0 the equation was first derived for plain Poiseuille flow [21]. The complex parameters b_1, b_2, c and d depend on the system. For binary fluid convection they have been calculated completely [20].

From now on we want to study mainly type 2 situations with only a left- or a right-going travelling wave. Then one of the amplitudes, say $|B|=0$ and going into a moving system the advection term can be transformed out of eqs. (2),(3). Thus in this case type 1 and type 2 are described by the same equation. Furthermore we consider the case $\arg(b_1) = \arg(b_2)$. Scaling length, time and amplitude appropriately we are left with one equation for the complex envelope A

$$\partial_T A = [(1+ib)(\partial_X^2 + \partial_Y^2) + 1 - (1+ic)|A|^2]\ A. \tag{4}$$

where b and c are now real. This equation is equivalent to the (λ,ω)-system derived by Hagan [9] for reaction diffusion systems.

2. Scaling properties of the envelope-equation

Let us consider solutions of eq. (4), whose time behaviour is described by a frequency Ω, thus

$$A(X,Y,T) = B(X,Y)\ e^{-i\Omega T}. \tag{5}$$

Then eq. (4) shows an interesting scaling property from one set of coefficients (b,c) to a one-parameter family of coefficients (b',c') as was pointed out before by [9,12,23]. Inserting eq. (5) into eq. (4) leads to

$$z\ \nabla^2 B + (1+i\Omega)/(1+ic)B - |B|^2 B\ 0, \tag{6}$$

where $z = |z|\ e^{i\theta} = \frac{1+ib}{1+ic}$. The scaling is possible if the phase θ of z is unchanged. This implies $z = \alpha^2\ z'$ with α real, which leads to

$$(b-c)/(1+bc) = (b'-c')/(1+b'c') \tag{7}$$

and

$$\alpha^2 = \mathrm{Re}(z/z') = \frac{(1+bb')(1+cc') + (b-b')(c-c')}{(1+c^2)(1+b^2)}. \tag{8}$$

Now by introducing the scaling

$$r = \beta\ r'$$
$$|B|^2 = \gamma^2\ |B'|^2 \tag{9}$$

eq. (6) leads to

$$(\frac{\alpha}{\beta})^2\ z'\nabla'^2 B' + \frac{1+i\Omega}{1+ic}\ B\quad ' - \gamma^2\ |B'|^2 B' = 0. \tag{10}$$

Then comparison of eqs. (6) and (10) leads to

$$\left(\frac{\alpha}{\beta\gamma}\right)^2 = 1 \tag{11}$$

and

$$\gamma^2(1+i\Omega') = \frac{1+ic'}{1+ic}(1+i\Omega). \tag{12}$$

From eqs. (11) and (12) Ω' and γ can be determined as

$$\gamma^2 = ((c-c')\Omega + 1 + cc')/(1+c^2) \tag{13}$$

$$\Omega' = (\Omega+c')/\gamma^2 - c$$

By using eqs. (9), (11) and (13) it is possible to scale an oscillating solution of eq. (4) of the form (5) from parameters (b,c) to another parameter set (b',c').

2. Periodic Solutions

It is well known that eq. (4) has periodic solutions of the form

$$A(\vec{X},T) = F\, e^{i(\vec{Q}\vec{X}-\Omega T)} \tag{14}$$

with $F^2 = 1 - Q^2$ and $\Omega = c + (b-c)Q^2$. For type 1 eq. (14) corresponds to phase waves and for type 2 to travelling roll-solutions with a displaced wavenumber. Obviously the range of existence of the periodic solutions is given by $F^2 \geq 0$ and thus $Q^2 \leq 1$.

For 1 + bc ≥ 0 ("Newell criterion" [18]) the periodic solutions are linearly stable within a narrower band. For 1 + bc → 0 the band shrinks to zero and for 1 + bc < 0 there do not appear to exist simple periodic solutions (Benjamin-Feir turbulence). The linear stability limit for 1+bc < 0 is well known [11,22,9,13,16]. The onset of instability with respect to a vanishing wavenumber is given by the generalized Eckhaus-criterion

$$Q^2 < Q^2_E = (1+bc)/(3+bc+2c^2). \tag{15}$$

In Fig. 1 a plot of eq. (15) is given in the case b = -c. For b = c one has the simple Eckhaus-criterion $Q^2_E = 1/3$. The periodic solution with $\vec{Q}=0$ is the last one to lose stability.

In the unstable range for 1+bc<0 it is known from 2-dimensional simulations of eq. (4) that the system evolves into a turbulent state with permanent nucleation and annihilation of dislocations [5] with well-defined average density of dislocations. If the system is initially in such a state and the parameters are changed somewhat into the stable parameter-range the system still behaves turbulently, so one has hysteresis [6]. Clearly this behaviour cannot be understood by means of the linear stability analysis of the simple periodic solutions.

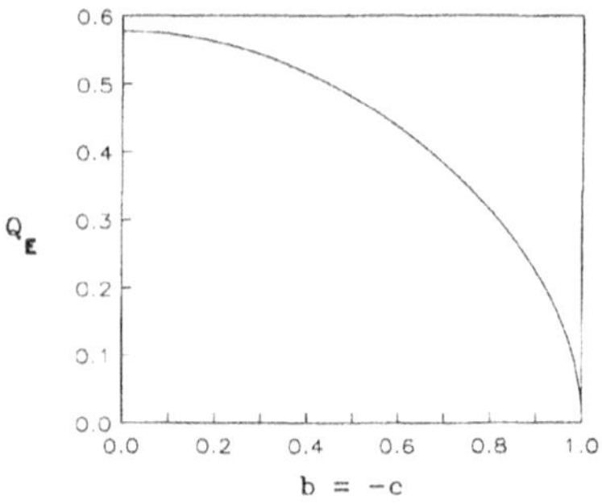

Figure 1. The generalized Eckhaus stability condition (eq. 15) in the case b = -c.

4. Defect Solutions - Spiral Waves

The envelope equation has defect solutions. The topology of a defect is such that the closed line integral of the gradient of the phase of A around the defect core is $\oint d\vec{s}\vec{\nabla}\phi = \pm 2\pi$ or, in other words, the defect core is descibed by a simple zero of the complex envelope A (A remains analytical near the core). The defect solutions of eq. (4) are approximately Archimedian spirals [9], which are often referred to as spiral waves. Now we want to review some of the properties of spiral waves. The defect solutions have the form in polar coordinates r, ϕ [9]

$$A(r,\phi,T) = D(r)\, e^{i(\pm\phi-\Omega T)} \tag{16}$$

with $\Omega = c + (b-c)Q^2$ and the complex field

$$D(r) = R(r)\, e^{i\theta(r)} \tag{17}$$

with the boundary conditions

$$R(r) \sim \alpha r\ ,\ \partial_r\theta(r) = 0 \text{ for } r \rightarrow 0, \tag{18}$$

$$\partial_r R = 0 \text{ and } \partial_r^2\theta = 0 \text{ for } r \rightarrow \infty.$$

The sign of ϕ gives the polarity and thus the orientation of rotation of the spiral core. By inserting the ansatz (17) into eq. (4) one obtains

$$(1+ib)\,[\partial_r^2 + 1/r\partial_r - 1/r^2] + (1+ic)(1-|D|^2)]D = -i(b-c)Q^2D. \tag{19}$$

This equation together with the boundary conditions (18) is a nonlinaer eigenvalue problem for Q. It was already explored by Hagan [9]. By a matched asymptotic expansion and by numeric analysis he showed, that for given parameters b,c there is only a single eigenvalue Q for spiral solutions. Far from the spiral core the solution looks like an Archimedian spiral

$$A(r,\phi,T) = D(r)\, e^{i(\pm\phi-\Omega T+Qr)} \quad \text{for} \quad r \rightarrow \infty. \tag{20}$$

Thus for given parameters (b,c) the defect selects far from its core a spiral wave with a wavenumber Q. Hagan [9] calculated the case b=0, but by using the

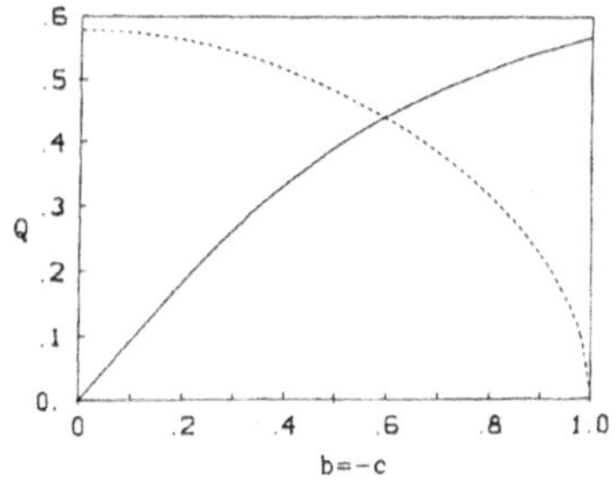

Figure 2. The asymptotic wavenumber Q of the 1-d spiral solution plotted as a function of b = -c (solid curve). The dashed curve shows the Eckhaus stability limit (eq. (15)) for periodic solutions. For b = -c = 0.594 both curves cross each other. Thus the 1-d spiral selects for b > 0.594 an asymptotic wavenumber, which lies in the unstable range.

scaling law (9) and (13) his results can be scaled to any parameters b and c.

Hagan's [9] analysis also shows that there exists an analytic one-dimensional defect solution, which however is unstable. In our analysis it turns out, that most of the interaction properties of the isolated spiral are qualitatively shown by the one-dimensional analogon (from now on referred to as 1-d spiral).

4.1. *One-dimensional case*

The 1-d spiral solution has the form [3]

$$A = R(X)\, e^{i(\theta \pm \Omega T)} \tag{21}$$

with $R(X) = (1-Q^2)^{1/2}\tanh(\alpha X)$,

$$\Omega = c + (b-c)Q^2, \quad \partial_X\theta = Q \tanh(\alpha X).$$

For α and Q we get

$$\alpha^2 = \{3b[8(c+b)^2+9(1-bc)^2+4bc]^{1/2} + b(5+9bc) + 4c\}/[4(2c+9b^3+7b)]$$

$$(b-c)Q^2 + 3\alpha Q + c - 2\alpha^2 b = 0, \tag{22}$$

where the positive square root of α has to be taken.

Equation (20) shows that a simple zero of the complex envelope A (spiral-core) emits travelling waves, where for given b and c the asymptotic wavenumber Q is determined by eq. (22) and the oscillation frequency of the solution is given by eq. (21). In Fig. 2 the asymptotic wavenumber Q is plotted as a function of b = -c (solid curve). The dashed curve shows the stability limit (15) for periodic solutions. For b = -c = 0.594 both curves cross each other. Thus the 1-d spiral selects in the case b = -c for b > 0.594 an asymptotic wavenumber which lies in the unstable range.

In the following we analyse numerically 1-d spiral solutions in an interval [0,L] in the parameter range 0.5 < b < 1 with b = -c (without loss of

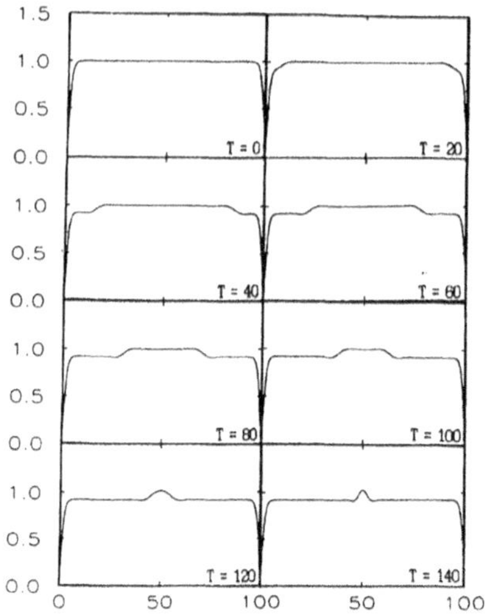

Figure 3. The time evolution of the modulus of A versus X at subsequent time intervals for b = -c = 0.5 and a system length of 100. In the collision region of the two counter-propagating waves the amplitude is locally increased (shock).

generality we always choose $b > 0$). The boundary conditions used in the simulations of the one-dimensional version of eq. (4) are

$$A = 0 \quad \text{at } x = 0, L . \tag{23}$$

In Fig. 3 the time evolution of the modulus of A is plotted versus X at subsequent time intervals for $b = -c = 0.5$. We use the initial conditions $A = \tanh(\alpha X) \tanh(\alpha(X-L))$ with α from (22) appropriate for $Q = 0$. The time evolution shows, that the boundaries begin to emit travelling waves with $Q \neq 0$. The local amplitude is lowered accordingly (see eq. (21)). After some time the counter-propagating travelling waves collide, which leads to a localized increase in the collision region, which is referred to as a shock. This solution with counter-propagating waves and the shock in the middle persist stationarily. Away from the shock the solutions appear to be equivalent to 1-d spiral solutions (21), where the selected wavenumber $|Q|$ and frequency Ω corresponds to the one given by eq. (22) [3]. Shocks were described and investigated by Howard and Kopell [10] for general (λ,ω)-systems and by Malomed [15] within the phase-diffusion approximation of eq. (4).

How does the system behave in the Eckhaus unstable regime? A typical time evolution of a system with $L = 100$ starting with the same initial conditions as before is plotted in Fig. 4 for $b = -c = 0.8$. At first the system behaves as in the stable case before. The amplitude is lowered at the boundaries, but then it rises again. Thus shock waves are emitted continuously from the boundaries. These shock waves propagate into the system and as they approach the middle the modulation of $|A|$ is enhanced. In some cases the enhancement leads to phase slips, where the modulus of A passes through zero. There exist boundary layers where A remains stationary. These layers become thicker in time and the modulation dies out completely and the solution stabilizes to two

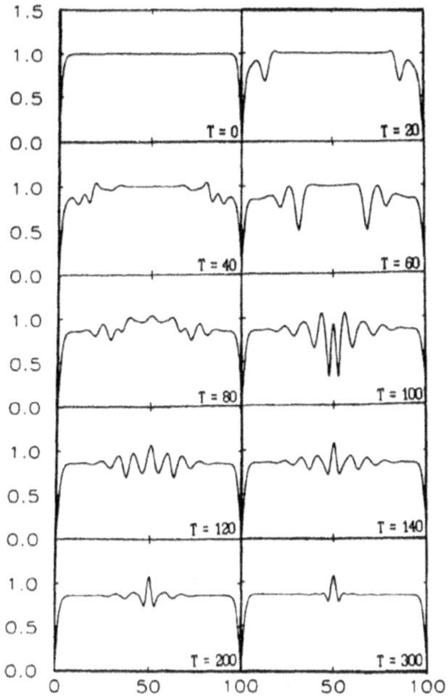

Figure 4. The time evolution of the modulus of A of two counter-propagating 1-d spiral solutions versus X for $b = -c = 0.8$ and a system length of 100. For these parameters the solution selects an Eckhaus unstable wavenumber, but after some time the solution restabilizes.

counter-propagating waves with a shock in the middle. Again the numerical solution away from the shock is equivalent to the 1-d spiral (21) and the selected wavenumber and frequency is to a good approximation the one given by eq. (22). Thus the system appears to stabilize in an Eckhaus unstable solution.

For b (= -c) 0.85 this stabilization depends on the length of the system. A plot of this length L_s versus $b = -c$ is given in Fig. 5 [3]. For $L > L_s$ the solution is turbulent whereas for $L < L_s$ the system relaxes into a stationary solution. L_s seems to diverge at $b = -c = 0.85$. It is also possible to stabilize solutions in the BF-unstable range. Then L_s approaches the asymptotic value of 26, which appears to be shorter than the 1-d spiral core. The divergence of L_s indicates that for $b = -c < .85$ the solutions can be stabilized even in a infinite system. We do not understand why an Eckhaus unstable state can apparently remain numerically stable.

4.2. *Spiral waves*

4.2.1. *Spirals in circular approximation.* The spiral in circular approximation has been discussed before (see eqs. (16)-(20)). Here we simulate

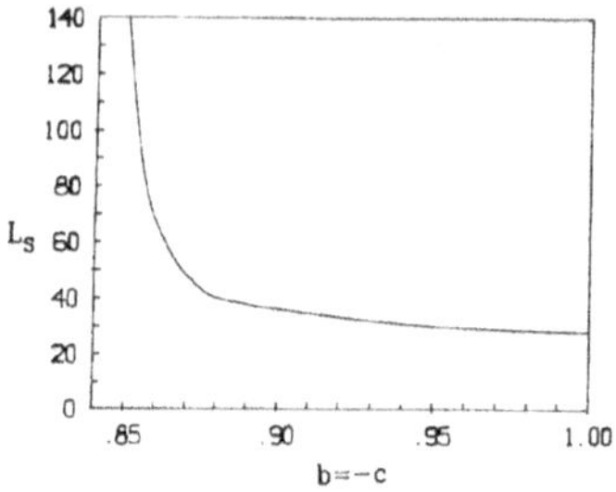

Figure 5. The critical length of the system L_s versus b = -c. For $L < L_s$ the solution restabilizes although an Eckhaus- unstable wavenumber is selected.

numerically the selection problem again and investigate the restabilizing properties of a spiral in a circular cell with radial symmetry. The procedure used is analogous to the one used for the 1-d spirals. In the simulation the radial eq. (19) is used, supplemented by the time derivative $\partial_T D$. In this case we choose the boundary conditions

$$X = 0: \qquad D = 0$$

$$X = L: \qquad \partial_r(D^* \partial_r D) - c.c. = 0, \; \partial_r(D^* D) = 0, \tag{24}$$

$$(\text{or: } \partial_r^2\theta = 0 \text{ and } \partial_r |A|^2 = 0).$$

It appears that most of the behaviour of the spiral in this circular approximation is qualitatively equivalent to that of the 1-d spiral. Thus the spiral in circular approximation shows in the range where the selected wavenumber is Eckhaus-unstable also restabilizing states. In Fig. 6 the selected wavenumber (solid line) and the Eckhaus stability boundary (eq. (15)) is plotted versus b = -c. The intercept of both curves is at b = -c = 0.63. Thus the selected asymptotic wavenumber is Eckhaus unstable for b = -c > 0.63. For b = -c ; 0.8 the system-length has to be decreased to restabilize the spiral in circular approximation. In Fig. 7 the length L_s needed for

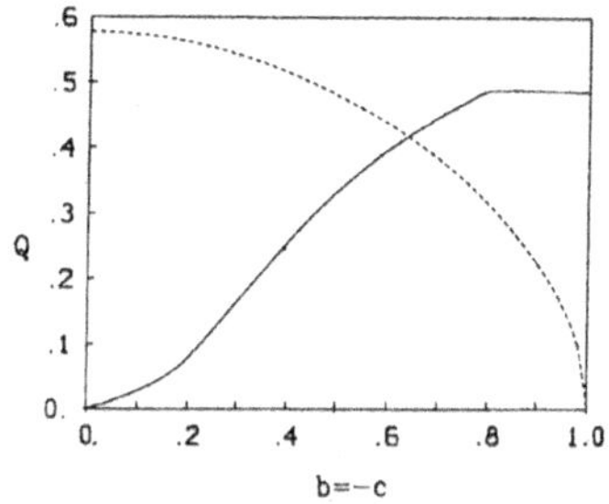

Figure 6. The asymptotic wavenumber selected by the spirals in circular approximation in dependence of b = -c (solid curve). The dashed curve shows the Eckhaus stability condition.

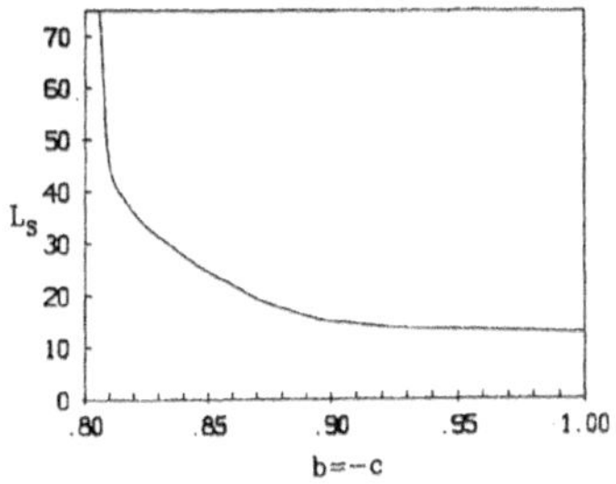

Figure 7. The critical length of the system L_s versus b = -c for the spiral in the circular approximation. For $L < L_s$ the solution restabilizes although an Eckhaus unstable wavenumber is selected.

stabilization is plotted versus b=-c. One sees that this length seems to diverge for b = -c ≈ 0.8, which coincides with the corner in the curve for the selected wavenumber in Fig. 6 [3]. By using the scaling law (chapter 2) comparison with the results by Hagan show good agreement up to b = -c = 0.8. Again we do not understand the apparent numerical stability of Eckhaus unstable states.

4.2.2. *Interacting spirals.* Here we are investigating in two-dimensional simulations of eq. (4) the properties of states with several spirals in the

Figure 8. The time evolution of a spiral pair in a 2-d simulation for b = -c = 0.5, a cell size $L_x = L_y = 100$ and an initial separation of $L_i = 24$. Plotted are the zeros of the imaginary and the real part of the solution. The last picture shows a stable final state, where the spiral cores build a kind of antiferromagnetic square lattice and one spiral is shrunk to the bare core.

Eckhaus-stable regime. We use a pseudo-spectral code [14,23] on a square cell of size L with periodic boundary conditions and initialize two defects in the system at $(0,L_i/2)$ and $(0,-L_i/2)$. (X = Y = 0 gives the center of the cell).

Figure 9 a. For legend see facing page.

The initial condition represents qualitatively a solution with 2 spirals with asymptotic wavenumber Q = 0 and frequency $\Omega = 0$ and therefore resemble the initial conditions used in the simulation of the spiral in circular approximation. In Fig. 8 the typical time evolution starting from the initial conditions with well separated spiral cores is shown in the region where the spirals select an Eckhaus stable wavenumber. The lines represent Re(A) = 0 and Im(A) = 0, respectively. The intersection of Im(A) = 0 and Re(A) = 0 gives the position of the spiral core. As was reported before [12] the spiral centers rotate and emit waves until the whole cell is filled. Where the waves collide

Figure 9 b. The phase of defect solutions in the two different types of systems. The picture (a) shows the defect solution in the case of a bifurcation to a purely oscillatory state (type 1), the picture (b) in the case of a bifurcation to an oscillatory and spatially periodic state. Points with the same phase have the same grey value.

shock structures appear, which are qualitatively similar to the shocks observed in the 1-d simulations.

If the distance L_s between spirals lies within some interval (for b = -c = 0.5: $14 < L_s < 30$) the spirals attract or repell each other until an equilibrium seperation (b = -c = 0.5: $L_E = 18$) is reached. As this spacing is approached the spirals begin to drift along the direction perpedicular to their connection line with a constant drift velocity. After some time the up/down symmetry breaks and one spiral stays behind at a fixed position (X_1, Y_1). The other one makes a fairly complicated motion and finally settles down near (X_1-L/2, Y_1-L/2). In this way a stationary "antiferromagnetic" square lattice with l = L/2 is formed. The spiral at (X_1,Y_1) now has become large in the sense that its emitted waves fill almost all space whereas the other spiral has been reduced to its bare core. This symmetry breaking can be accelerated by adding noise. A typical run is shown in Fig. 8. In the case of type 1 the experimentalist would directly observe the spirals. In the case of type 2 the periodic pattern is only modulated. In Fig. 9. grey shaded graphs of the spirals in type 1 and type 2 systems are given. In order to observe the spiral waves in convecting patterns (type 2) the asymptotic modulation wavenumber of the spirals has to be sufficiently large to be detectable.

By initializing the system directly in the anitferromagnetic arrangement we tested the stability of the arrangement within the given lattice for different spiral separations [12,23]. As a main result we obtained that the arrangement is stable up to a minimal separation of the spirals (or maximum spiral density) below which the spirals annihilate. The maximal spiral density goes to zero for $b = -c \to 0$ and increases for larger $b = -c$ approaching a constant value for parameters where the selected wavenumber becomes Eckhaus unstable.

From the 2-d simulations it is possible to measure the asymptotic wavenumbers of the spirals and the selected frequencies. It turns out that the selected asymptotic wavenumbers and frequencies are in good agreeement with the results from the simulation of the spiral in circular approximation. In the region, where the selected asymtotic wavenumber is Eckhaus unstable, restabilisation can now be achieved only for not too large system size. There the selected wavenumbers and frequencies agree up to $b = -c = 0.8$ with the spiral in circular approximation. For $b = -c > 0.8$ no restabilisation can be achieved by decreasing the system size. Either the system is is too small and all spirals annihilate or the system behaves in a turbulent manner with permanent nucleation and annihilation of spirals.

Thus for $b = -c$ 0.8 there appears to exist either the defect-free state or turbulent state. The turbulent state was already observed by Kuramoto [13] and was investigated in greater detail by Coullet et al. [7]. It would be interesting to study the possible differences of these two turbulent regimes.

In the Eckhaus unstable region the evolution of the sytem starting from the initial conditions described before is at the beginning analogous to the one of the spiral in circular approximation. Then new spirals are nucleated near to the shocks. In the case of the spiral in circular approximation the nucleation of the spirals corresponds to the occurrence of phase slips. The fact, that the spirals are nucleated near to the shocks is in qualitative

Figure 10. The time evolution of a 2-d spiral solution for $b = -c = 0.8$ and a cell size of $L_x = L_y = 60$. The solution restabilizes after some turbulent behaviour, although an Eckhaus unstable is selected.

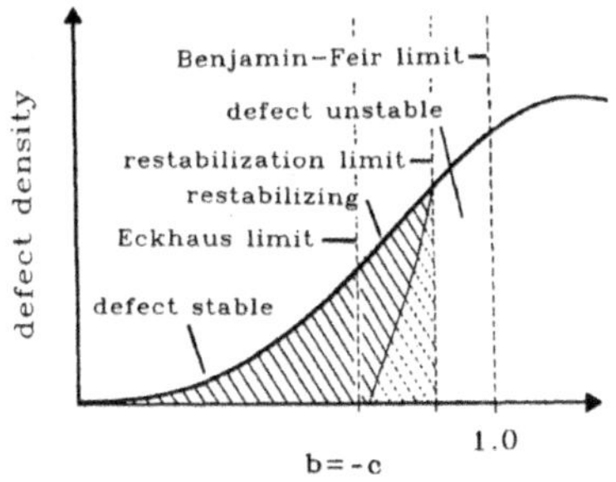

Figure 11. A schematic diagram of the defect density in dependence of b = -c. The numerical investigations lead to a division of the Benjamin- Feir stable range (1+bc<0). For further explanation see the text.

agreement with the location of the events of phase slips in the 1-d spiral simulations. The system then seems to be turbulent, but after a short time more and more spirals annihilate and for sufficiently small cells the system finally settles down into the antiferromagnetic lattice described before. A typical run for a restabilization in a system that is too large to achieve an antiferromagnetic lattice is shown in Fig. 10.

5. **Summary**

We have found that the Eckhaus instability plays a major role in the properties of the evolution of spirals in the Benjamin-Feir stable range 1+bc > 0. The Eckhaus instability, which is continuously triggered due to the spiral waves emitted from the cores, leads in some parameter range to a turbulent state. In another parameter range the quasi-turbulent state restabilizes to an antiferromagnetic square lattice of spirals. It has to be pointed out that all of the behaviour we described before is due to the fact of having initially spirals in the system. In the whole parameter range up to the Benjamin-Feir limit the homogeneous solution is stable. This behaviour could therefore be termed defect initialized turbulence.

The behaviour described above leads to the conclusion that the range up to the Benjamin Feir limit should be subdivided. In Fig. 11 a schematic diagram of the possible spiral densities versus b = -c is given for the two dimensional case. For b = -c < 0.63 the selected asymptotic wavenumber of the spirals is Eckhaus stable. In the simulation bounded states of spirals are found. This result seems to be consistent with recent analytic results on spiral-interaction [8]. If the density is high enough they order on anitferromagnetic square lattices. The spiral (defect) density seems to tend to zero when b = -c → 0. When the Eckhaus-limit is approached the spiral density goes to a finite value. This range where spiral-solutions build stationary stable patterns can be termed a spiral (or defect) stable range. If the parameters are in the Eckhaus unstable range (b = -c ; 0.63) we observed restabilized spiral patterns above a minimal density. The absolutely defect unstable range sets in for b = -c > 0.8. There the pattern is always turbulent with a continuous nucleation and annihilation of spiral pairs: We expect that the defect density behaves continuously when passing into the Benjamin Feir unstable range. In the one dimensional case there does not appear to exist a lower bound for the defect density in the restabilizing range and also the boundaries of the different regimes are slightly shifted.

This schematic diagram helps us to understand the hysteresis found by Coullet et. al.[6]. Our numerical results could be experimentally tested in type 1 sytems, like the BZ-reaction. Then to allow a description by the complex Ginzburg Landau equation it would be useful to drive the reaction in the Hopf bifurcation range. Another test could be possible for type 2 systems, like electro-convection of liquid crystals or thermo-convection of binary mixtures. The latter system is in 2d isotropic so our results can be applied only in 1d.

The main limitations of our results in 2d derive from the fact that a square cell was prescribed. A few simulations with a lattice made out of nine such cells give rise to the assumption that these arrangements of defects could be stable in a large lattice. But this point is not really clear and has to be investigated. Some discrepancies with simulations using a cellular automaton [4] may arise from this fact.

Acknowledgements

We would like to thank A. Newell, W. Pesch and W. Zimmermann for helpful discussions. This work was supported by the Deutsche Forschungsgemeinschaft SFB 213, Fonds der chemischen Industrie and Emil Warburg Stiftung Bayreuth.

References

[1] See the contributions of Ahlers, G., Cannell, D.S. & Heinrichs, R.S. (1987). In *Chaos 87, International Conference on the Physics of Chaos and Systems far from Equilibrium*, Duong-Van, M. (ed.). North Holland: Amsterdam; and (1987) *Nuclear Physics B* (Proc. Suppl.) **2**; Passner, A., Williams, H.L. & Surko, C.M., ibid.; Steinberg, V., Moses, E. & Fineberg, J., ibid..

[2] Alcantara, F. & Monk, M. (1974). Signal propagation during aggregation in the slime mould *Dictyostelium discoideum. J. Gen. Microbiol.* **85**, 321.

[3] Bodenschatz, E. (1989). *Pattern and Defects within the Weakly Nonlinear Analysis of Anisotropic Pattern-Forming Systems.* Ph.D. thesis, University of Bayreuth.

[4] Bohr, T., Pedersen, A.W. & Jensen,M.H. (1989). Transition to Turbulence in a Discrete Ginzburg-Landau Model, Preprint.

[5] Coullet,P. & Lega,J. (1988). Defect-Mediated Turbulence in Wave Patterns, *Europhys. Lett.* **7**, 511.

[6] Coullet, P., Gil, L. & Lega, J. (1988). Defect mediated Turbulence, preprint.

[7] Coullet, P., Gil, L. & Lega, J. A form of turbulence associated wirh defects, to appear in Physica D.

[8] Elphick, C. & Meron, E. (1989). Spiral Vortex Interactions, Preprint.

[9] Hagan, P.S. (1982). Spiral Waves in Reaction Diffusion Equations, *SIAM J. Appl. Math.* **42**, 762.

[10] Howard, L.N. & Kopell, N. (1977). Slowly Varying Waves and Shock Structures in Reaction-Diffusion Equations, *Stud. Appl. Math.* **56**, 95.

[11] Kopell, N. & Howard, L.N. (1973). Plane Wave Solutions to Reaction-Diffusion Equations, *Stud. Appl. Math.* **52**, 291.

[12] Kramer, L., Bodenschatz, E., Kaiser, M., Pesch, W., Weber, A. and Zimmermann, W. (1989). *Pattern and Defects in Liquid Crystals, in New Trends in Nonlinear Dynamics and Pattern Forming Phenomena: The Geometry of Nonequilibrium*, Coullet, P. & Huerre, P. (eds.). NATO ASI Series, Plenum Press.

[13] Kuramoto, Y. (1984). *Chemical Oscillations, Waves and Turbulence, Series in Synergetics.* Springer: Berlin.
[14] Lega, J. & Gil, L.: private communication.
[15] Malomed, B.A. (1983). Nonsteady Waves in distributed dynamical systems, *Physica* **8D**, 353.
[16] Malomed, B.A. (1984). Nonlinear Waves in Nonequilibrium Systems of the Oscillatory Type, Part I. Z. *Phys.* **55B**, 241.
[17] Müller, S.C., Plesser, T. & Hess, B. (1987). Two-dimensional Spectrophotometry of Spiral Wave Propagation in the Belousov-Zhabotinskii Reaction, Part I and II, *Physica* **24D**, 71; (1987) *Physica* **24D**, 87.
[18] Newell, A.C. (1974). Envelope Equations. *Lect. in Appl. Math.* **15**, 157.
[19] See e.g. Rehberg, I., Rasenat, S. & Steinberg, V. (1989). Travelling Waves and Defect-Initiated Turbulence in Electroconvecting Nematics, *Phys. Rev. Lett.* **62**, 756
[20] Schöpf, W. & Zimmermann, W. (1990). Results om wave patterns in binary fluid convection, to appear in *Phys. Rev.* **A41**, 2.
[21] Stewartson, K. & Stuart, J.R.(1971). A non-linear instability theory for a wave system in plane Poiseuille flow. *J. Fluid Mech.* **48**, 529. DiPrima, R.C., Eckhaus, W. & Segel, L.A.: Nonlinear wave-number interaction in near-critical two-dimensional flows. *J. Fluid Mech.* **49**, 705.
[22] Stuart, J.T. & DiPrima, R.C. (1978). The Eckhaus and Benjamin-Feir resonance mechanism, *Proc. R. Soc. London* **A362**, 27.
[23] Weber, A. (1989). *Two-dimensional Defect-Structures as Solutions of Ginzburg Landau Equations without Potential.* Diploma Thesis, University of Bayreuth.

36. IGNITION/EXTINCTION PHENOMENA AS DISSIPATIVE PHASE TRANSITIONS

D. Meinköhn

Fakultät für Physik, Universität Bielefeld

4800 Bielefeld 1, FRG

1. Representative physical problem

A particular class of reaction-diffusion systems may be defined by the representative example of a highly porous body (e.g. a catalyst pellet or a fuel agglomerate) residing in a stagnant ambient atmosphere which contains some or all of the gaseous reactant species. An exothermic chemical reaction is supposed to take place on the internal surface of the body, with an area much larger than its external surface area. Therefore, the reaction may be regarded as quasihomogeneous, i.e. as distributed over the entire volume of the system. The state of the system is given by the internal temperature T, the species concentrations Y_i and the boundary conditions which detail the ambient atmosphere and the transfer processes for heat, reactants and reaction products. The system is generally characterized by the absence of convection and in that its states result from an interaction of essentially different physical processes which comprise the activated process of internal heat release and the transport of mass and heat. It may be noted that the class of reaction-diffusion systems thus defined comprises, e.g. smokeless solid combustion, viscous deformation of solids (with distributed heat release due to internal friction), or electric conduction with distributed Ohmic losses.

Generally, the evolution of the system is specified by indicating which one of the interacting physical processes is of overwhelming influence. With regard to the internal temperature, two operating regimes may thus be distinguished: in the range of high temperatures, the system evolves under diffusion control, whereas for lower temperatures, control is by the chemical reaction.

2. Ignition/extinction concepts

In what follows, considerations are restricted to autocatalysed processes. This is equivalent to excluding ignition or extinction processes which are due to an external stimulus. For the remaining processes, two classes may be distinguished. There are

(a) "dynamic" ignition/extinction phenomena observed for systems evolving from arbitrary initial conditions which do not correspond to a stationary state of the system,

(b) "static" ignition/extinction phenomena for which the system is initially in a stationary state.

The present investigation intends to address static ignition/ extinction processes. According to concepts introduced by Semenov & Frank-Kamenetzkii, ignition or exctinction is conceived as loss of stability of the initial stationary state due to a slow continuous change in some control parameter. Generally, ignition is accompanied by an increase in temperature while extinction is characterized by a decrease in temperature. After ignition, the system may eventually regain a stationary state at a higher temperature, in which case ignition induced the system to transfer from a stationary state at low temperature to such a state at high temperatures which is equivalent to change from reaction control to diffusion control. But the model also allows for a different ignition process where the stationary system loses stability and commences an unstationary evolution without ever reattaining stationarity. By contrast, a "static" extinction process will always induce the system to change from an initial stationary state at higher temperatures to a final stationary state at lower temperatures, i.e. extinction is always associated with a change from diffusion control to reaction control.

A given regular stationary state of the system may be classified as to its stability. Therefore, the set of all stationary solutions decomposes into branches of stable stationary solutions and branches of unstable stationary ones. The singular stationary solutions which represent the stability boundary are termed "critical" solutions and may be classified as to whether they represent an incipient ignition or exctinction process or whether they represent a singular stationary solution of higher order.

Slightly subcritical stationary states represent an excitable system or a multi-stable system, depending on the order of the singular stationary state. Weak reaction waves (standing or travelling solitary waves) may be shown to connect the subcritical stationary states from a neighbourhood associated with the singular stationary state.

3. The mathematical model

The physical system is to be investigated with the help of a mathematical model resulting from simplifying assumptions due to which a single nonlinear differential equation is obtained which encompasses the characteristic features of ignition/extinction processes [1]:

$$\frac{\partial y}{\partial t} = \Delta y + \lambda(1-\xi y)^n \exp(y/(1+\beta y)) \quad \text{in } D$$

$$y = 0 \quad \text{on } \partial D \qquad (1)$$

Here, the variable y characterizes the internal state of the system (e.g. y = temperature), whereas λ, ξ, β, n designate control parameters. The system occupies a region D with the boundary ∂D, with t designating the time and Δ designating the Laplacian.

The stationary states are obtained as solutions from

$$\Delta y + \lambda(1-\xi y)^n \exp(y/(1+\beta y)) = 0 \quad \text{in } D$$

$$y = 0 \quad \text{on } \partial D \qquad (2)$$

Eq. (2) represents a quasilinear elliptic differential equation with Dirichlet boundary conditions.

The nonlinearity of Eqs. (1), (2) is entirely due to the function

$$w(y;\xi,\beta,n) = (1 - \xi y)^n \exp(y/(1+\beta y)) \tag{3}$$

Consequently, there are two types of control parameters: ξ, β, n which determine the shape of the nonlinear function $w(y;\xi,\beta,n)$, i.e. its convexity with respect to y, and λ which has no influence on the shape of w.

The class of reaction-diffusion systems represented by Eq. (1) is determined by the ranges of the control parameters, which are as follows

$$\lambda \geq 0, \quad \beta \geq 0, \quad n \lessgtr 0, \quad \xi \lessgtr 0 \tag{4}$$

Eq. (1) represents a nonisothermal reaction-diffusion system, whereby

$$w > 0 \tag{5}$$

On account of Eq. (5), a solution $y \equiv 0$ of Eq. (1) or Eq. (2) does not exist for arbitrary values of λ. Therefore, in order to discuss ignition and extinction processes, it is necessary to determine the set of stationary solutions from Eq. (2) which is a strongly nonlinear boundary value problem. Particular interest is in those singular solutions of Eq. (2) which form the stability boundary (i.e. the critical solutions).

A general solution approach may be based on an analogy between the operating regimes of the reaction-diffusion system and the phases of an equilibrium system. In order to establish this analogy, a state surface representing the stationary states of Eq. (1) will be derived with the help of analytic continuation. As Eq. (2) is in the form of a quasi-linear elliptic boundary value problem, the "derived" problem is a linear elliptic one and thus an integrability condition may be specified due to Fredholm's alternative whereby the solutions of Eq. (2) result in parametric representation with respect to a convenient parameter [2]. If the norm $\|y\|$ of y (e.g. $\|y\| = y_m$ with y_m designating the maximum of y in $D+\partial D$) is selected to be this parameter, then the solutions of Eq. (2) result in the following parametric representation:

$$\{y\|y\|;\xi,\beta n) \ , \ \lambda(\|y\|;\xi,\beta,n)\} \tag{6}$$

Consequently, the stationary states of Eq. (1) are represented by a state surface

$$\lambda(\|y\|;\xi,\beta,n) \tag{7}$$

which is defined in a finite dimensional state space spanned by $\|y\|$ and by the control parameters λ, ξ, β, n. The surface given by Eq. (7) resembles the state surface for equilibrium systems, with $\|y\|$ corresponding to a thermodynamic state variable of the "density"-type, whereas λ, ξ, β, n correspond to state variables of the "field"-type [3].

The "visible" contours of the state surface given by Eq. (7) define the singular stationary solutions, i.e. upon projecting the state surface into the subspace of control parameters λ, ξ, β, n (along the $\|y\|$-direction) an image of the contours is produced which represents the coexistence diagram (or bifurcation set). It comprises curves, surfaces, etc. consisting of those control parameter combinations for which the number of solutions of Eq. (2)

changes. Use of a state surface representation by which the control parameter λ is given in terms of $\|y\|$ and the other control parameters ξ, β, n turns out to be of advantage due to the fact that λ does not influence the convexity of the nonlinearity w (cf. Eq. (3)).

4. The stationary state surface in terms of cross sections

With the help of lower bounds of the state surface (Eq. (7)), necessary conditions for the various singularities of the coexistence diagram may be derived [4,5]. These lower bounds may then serve as initial data for a numerical derivation of the state surface by monotone iteration methods.

In order to illustrate the analogy between ignition/extinction phenomena and phase transitions, three types of critical stationary solutions of Eq. (1) will be discussed which have been derived with the help of lower bounds [5,6]. They have been shown to represent singular stationary states of first, second and third order.

4.1. *Singular solutions P_1, P_2 of first order*

Inspection of Eqs. (1), (2) reveals that the parameter λ controls the intensity of the energy release due to the chemical reaction, with the energy "driving" the system to stationary states away from thermodynamic equilibrium. For $\lambda = 0$, the chemical reaction is switched off and the regular solution $y \equiv 0$ of Eq. (2) is obtained, which represents thermodynamic equilibrium.

Increasing λ continuously from $\lambda = 0$ produces the "thermodynamic branch" [7] of stable stationary solutions which is bounded by the critical solution associated with P_1. Increasing λ beyond λ^* will induce the system to become unstationary whereupon, in the example of Fig. 1, it will be attracted by a corresponding stationary state on the second branch of stable stationary solutions which is bounded at lower temperatures by P_2. According to [7], this second branch comprises the "dissipative structures" which are separated from the thermodynamic branch by a singularity. The passage between the branches may be viewed as a "dissipative phase transition" [7]. As y represents the temperature, P_1 is interpreted as an ignition point due to the associated temperature increase upon passage from P_1 to the second branch of stable stationary solutions. Because of the difference in temperature, the thermodynamic branch represents the operating regime of reaction control, whereas for the dissipative structures, the states result from diffusion

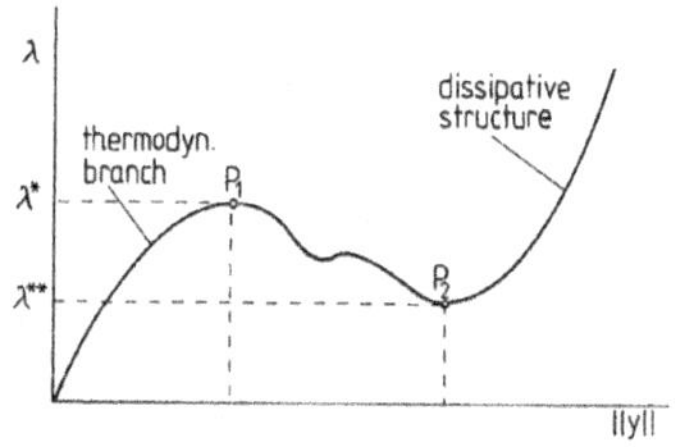

Figure 1. Cross section of the state surface with ignition point P_1, extinction point P_2.

control. P_2 thus turns out to be an extinction point. Projecting the curve of Fig. 1 along the $\|y\|$-direction onto the λ-axis furnishes a coexistence diagram consisting of the two points λ^* and λ^{**}.

4.2. *Singular solution P^+ of second order*

For certain ranges of the control parameters, a singular solution P^+ of second order appears. P^+ is equivalent to a critical point in phase diagrams of equilibrium systems of fluids, i.e. P^+ designates a phase transition of second order. The shaded region of Fig. 2 consists of unstable stationary states which cannot be physically attained. They are bounded by a curve C of critical states of Eq. (2). C consists of a branch C_1 of ignition points (e.g. P_1) and a branch C_2 of extinction points (e.g. P_2), with the two branches C_1,C_2 joining at P^+. Due to the existence of P^+, quasi-stationary processes exist in the region of stable stationary states which smoothly connect the states of the thermodynamic branch with the dissipative structures, i.e. without ignition or extinction jumps. Consequently, there is no essential difference between the operating regimes of reaction control and diffusion control because in the course of an appropriate quasi- stationary process, reaction control may gradually change over to diffusion control without any interruption by ignition or extinction. Projecting the state surface of Fig. 2 into the subspace of control parameters λ,ξ furnishes the coexistence diagram which is in the form of a curve consisting of two branches C* and C** which join in a cusp. The upper branch C* results as the image of the curve C_1 of ignition points whereas the lower branch C** represents the projection image of the curve C_2 which is composed of extinction points.

4.3. *Singular solution P^{++} of third order*

For an appropriate range of the control parameters, the state surface sketched in Fig. 3 is obtained.

A comparison of the local topology of Figs. 2 and 3 reveals that P_1^+ in Fig. 3 represents a critical point analogous to P^+ in Fig. 2. For very small and very large negative values of ξ, a cross section of the state surface of Fig. 3 results which is sketched in Fig. 4a. Here, the system ceases to be stationary due to ignition at $\lambda = \lambda^*$, but as there are no further stationary states for $\lambda > \lambda^*$, the system remains unstationary after ignition.

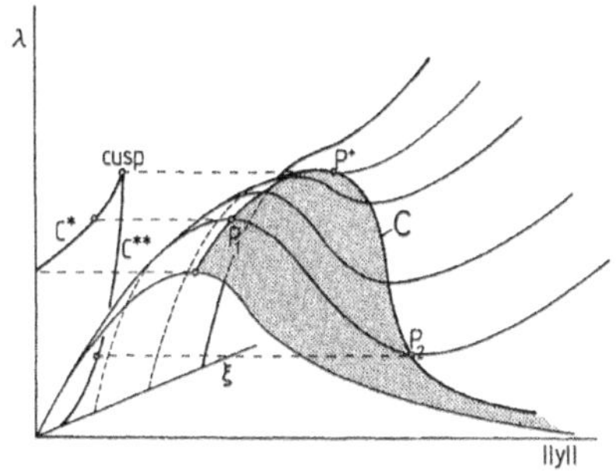

Figure 2. State surface and coexistence diagram for critical point P^+. The unstable stationary states are in the shaded region.

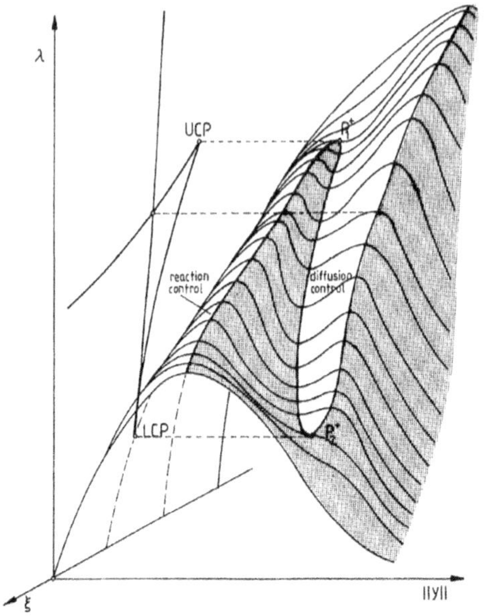

Figure 3. State surface and coexistence diagram for a pair of critical points P_1^+, P_2^+. The unstable stationary states are in the shaded region.

Consequently, the only stable structures are those of the thermodynamic branch, while ignition is equivalent to the system departing from the thermodynamic branch without eventually resettling in a branch of stable dissipative structures.

Increasing the parameter ξ from large negative values eventually leads to a cross section of the state surface depicted in Fig. 4c. Here, two branches of stable stationary solutions have appeared, with ignition points P_1,P_3 and an extinction point P_2. Therefore, a regime of diffusion control can be reached from the regime of reaction control on the thermodynamic branch via an ignition process at P_1 and, conversely, the operating regime may change from diffusion control to reaction control by extinction at P_2. A further increase of ξ leads by way of the intermediate cross section displayed in Fig. 4b to the cross section displayed in Fig. 4d. Here again two stable branches exist which are associated with reaction control and with diffusion control, but contrary to what was observed for Fig. 4c, the regime of diffusion control (at constant ξ) cannot be reached from reaction control. The only ignition point is at P_1 from where the system departs into permanently unstationary behaviour, while P_2,P_3 designate extinction points. Consequently, by keeping ξ at an appropriate value such that the cross section of Fig. 4d results, the regime of diffusion control can only be left but not reached.

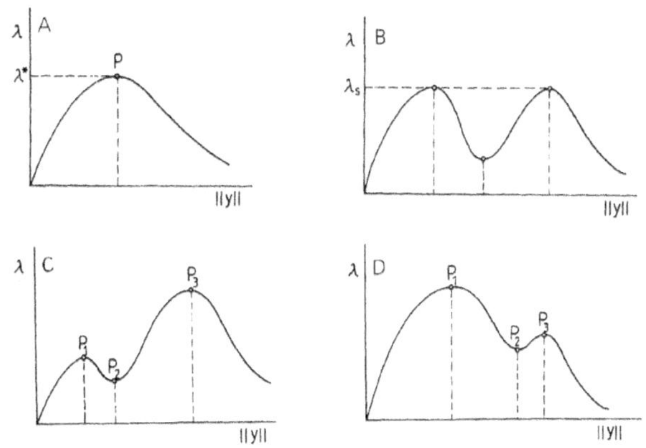

Figure 4. Various cross sections of the state surface of Fig. 3.

Both P_1^+ and P_2^+ of Fig. 3 depict critical points in the thermodynamic sense because

(i) they mark the appearance (or disappearance) of a second process regime (of diffusion control) beyond the regime of reaction control which is associated with the thermodynamic branch,

(ii) due to the topology of the state surface, there is no essential difference between both types of process regimes.

Projection of the state surface into the subspace of control parameters λ, ξ furnishes a coexistence diagram in the shape of a self-intersecting curve with two cusps which result as images of the critical points P_1^+ and P_2^+. The point of self-intersection is associated with the cross section displayed in Fig. 4b. Allowing for a variation of a third control parameter, e.g. β, leads to a 3-dimensional control parameter space spanned by λ, ξ, β with a coexistence diagram containing two lines of critical points ("critical lines" in thermodynamics [3]) passing through P_1^+ and P_2^+. These critical lines are found to intersect in a cusp marked P^{++} in Fig. 5. For Eq. (2), P^{++} represents a singularity of third order, a so-called "swallow-tail".

Conceptually, a critical point is equivalent to a continuous transition from one phase to another. Only for simple gas-liquid transitions in 1-component systems the critical points turn out to be isolated. The state manifold for complex thermodynamic systems may contain critical lines or, generally, critical manifolds of higher order which intersect [8]. Upon projection into the subspace of control parameters, a coexistence diagram (or bifurcation set) is obtained in which the intersection of critical manifolds give rise to characteristic singularities which may be classified as to their order (or multiplicity). In the spirit of Landau (e.g. [9], p. 493), Griffiths [10] introduced a definition of multiplicity for multi-component fluid systems based on coexistence of phases. The cross sections of the state surface of Fig. 3 which are displayed in Fig. 4 have been shown to allow for an ignition leading to a permanently unstationary evolution of the system. Therefore, the concept of coexisting phases cannot generally hold for stationary systems. But by basing the concept of multiplicity on the number of intersecting critical lines, the singularity P^{++} displayed in the bifurcation set of Fig. 5 is said

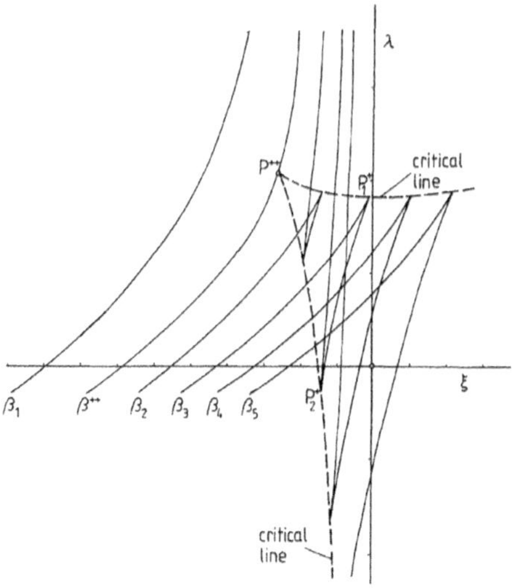

Figure 5. Contours of the state surface of Fig. 3 for varied values of parameter β.

to be of multiplicity 2 whereby P^{++} represents a bicritical point, as P^{++} marks the intersection of two critical lines.

5. Weak reaction waves in slightly subcritical systems

Slightly subcritical systems provide for stable and unstable stationary states in a close neighbourhood defined in terms of $||y||$ about the corresponding critical state. For an infinite reactive slab, Gray & Kordylewski [11,12] have found standing and travelling solitary waves which connect the stable and unstable stationary states of that neighbourhood.

For $\lambda < \lambda^*$ and small values of $|\lambda-\lambda^*|$, two stationary states - a stable and an unstable one - result for a singularity of first order which is depicted in Fig. 4a. In the limit of large activation energies, these two states are connected by an unstable standing wave of the pulse-type in the temperature distribution which represents an additional spatial structure.

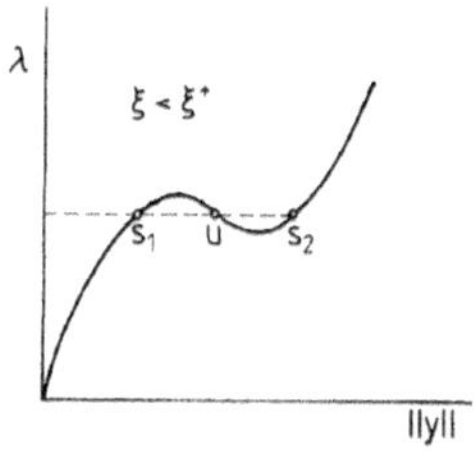

Figure 6. Cross section of the state surface of Fig. 2 for slightly subcritical ξ.

For $\xi < \xi^+$ and small values of $|\xi-\xi^+|$, a cross section of the state surface of Fig. 2 results which is displayed in Fig. 6.

Here, a neighbouring group of 3 stationary states occurs, with 2 states being stable (designated by s_1, s_2) and an unstable one designated by u. Under these conditions, the system turns out to be bistable.

According to [12], three solitary waves of the front-type exist which comprise a standing wave connecting the stable stationary states s_1 and s_2, and beyond that, two travelling waves connecting u with s_1 and u with s_2.

6. Conclusion

"Static" ignition and extinction phenomena are shown to be associated with critical stationary states which may be classified by employing concepts from the theory of phase transitions for complex fluid systems with no breaking of symmetry.

For slightly subcritical systems, additional spatio-temporal structures exist in the form of weak reaction waves which connect the stationary states in the neighbourhood of the critical stationary state.

References

[1] Aris, R. (1975). *The Mathematical Theory of Diffusion and Reaction in Permeable Catalysts.* Clarendon Press: Oxford.

[2] Meinköhn, D. (1981). Stability and Criticality of the Stationary Solutions of the Nonlinear Fourier Equation. *J. Chem. Phys.* **74**, 3603-3608.

[3] Griffiths, R.B. & Wheeler, J.C. (1970). Critical Points in Multicomponent Systems. *Phys. Rev.* **2A**, 1047-1064.

[4] Meinköhn, D. (1988). Disappearance of Criticality in Nonuniform Systems I. *SIAM J. Appl. Math* **48**, 536-548.

[5] Meinköhn, D. (1988). Disappearance of Criticality in Nonuniform Systems II. *SIAM J. Appl. Math.* **48**, 792-807.

[6] Meinköhn, D. Catastrophe Theory Concepts for Ignition/ Extinction Phenomena. In *Dissipative Structures in Transport Processes and Combustion,* Meinkohn, D. (ed.). Springer Ser. Synergetics, Springer, Berlin (in press).

[7] Nicolis, G. & Prigogine, I. (1977). *Self-Organization in Nonequilibrium Systems.* Wiley: New York.

[8] Knobler, C.M. & Scott, R.L. (1984). Multicritical Points in Fluid Mixtures - Experimental Studies. In *Phase Transitions and Critical Phenomena, Vol. 9*, Domb, C. & Lebowitz, J.L. (eds.). Academic Press: London.

[9] Landau, L.D. & Lifshitz, E.M. (1980). *Statistical Physics.* Pergamon Press: Oxford.

[10] Griffiths, R.B. (1975). Phase Diagrams and Higher-Order Critical Points. *Phys. Rev.* **12B**, 345-355.

[11] Gray, P. & Kordylewski, W. (1985). Standing Waves in Exothermic Systems. *Proc. R. Soc. Lond.* **A398**, 281-288.

[12] Gray, P. & Kordylewski, W. (1988). Travelling Waves in Exothermic Systems. *Proc. R. Soc. Lond.* **A416**, 103-113.

37. THE EFFECT OF CROSS MASS (SORET) TRANSPORT ON THE INITIATION, PROPAGATION AND STABILITY PROPERTIES OF A COMBUSTION WAVE

P.L. Garcia-Ybarra

Dept. Fisica Fundamental, UNED
Apdo. 60141, 28080-Madrid, Spain

1. The Soret (thermal diffusion) effect

Soret, or thermal diffusion, transport denotes the cross mass transport due to the local temperature gradient, in contrast to the more familiar Fick transport due to a concentration gradient. The phenomenon, first reported by Ludwig in 1856, takes its name from Ch. Soret, the Swiss physicist who during the 1870s carried out a series of experiments to measure the salt concentration differences appearing in some initially homogeneous liquid solutions subjected to a temperature gradient. The reciprocal phenomenon, a heat flux induced by a concentration gradient, was observed almost simultaneously by Dufour in 1872. (The references of these pioneering works can be found in [10].)

At about the same time, Tyndall [34] and later Aitken [2] reported the formation of a "dust free layer" around hot objects immersed in a dusty atmosphere. Although initially the phenomenon was not related to the Soret experiments in liquid phase, these were actually the first observations of this cross transport in gas phase.

The first successful theoretical treatment of the Soret effect was a monograph by De Groot [9] who later on published with Mazur the celebrated book [10] where this effect appears naturally, due to the Onsager reciprocal relations, as a byproduct of the general theory of non-equilibrium thermodynamics. On the other hand, the solution of the Boltzmann kinetic equation by Enskog [11] and Chapmann [4] provided a comprehensive theory on non-equilibrium gas mixtures that leads to analytical expressions allowing us to compute the transport coefficients.

In the next section, these expressions from the kinetic theory will be used to estimate the magnitude of the cross transports in gas mixtures relevant to combustion problems. In section 3, the influence of the thermal diffusion transport on premixed flames will be discussed: first on the plane steady propagation and then on the stability of this propagation. Section 4 is devoted to the analysis of the influence that this cross transport may exert

Nonlinear Wave Processes in Excitable Media
Edited by A. V. Holden *et al.*, Plenum Press, New York

in the ignition stage. Finally, some conclusions and general remarks are presented in section 5.

2. Mass and heat diffusion in a gas mixture

The Enskog-Chapmann expansion permits, in principle, the computation by systematic approximation of the coefficients controlling the transport mechanisms in multicomponent gas mixtures (see for instance the book by Ferziger & Kaper [12]). However, in general, the expressions obtained are too complicated to be useful and only under simplifying assumptions can they be analytically handled.

Neglecting barodiffusion and external forces, the heat and mass diffusive fluxes in a mixture of n+1 species can be written, respectively, as [12]

$$J_q = -\lambda \nabla T - p \sum_{i=1}^{n} (DT_i - DT_0) \nabla x_i \qquad (1)$$

$$J_i = -\rho_i \sum_{j=1}^{n} (D_{ij} - D_{i0}) \nabla x_j - \rho_i DT_i \nabla \ln T, \qquad (i=1,2,\ldots n) \qquad (2)$$

where T is the temperature, x_i and ρ_i $(i=0,1,\ldots,n)$ the molar fractions and partial densities, respectively, p the gas pressure and λ the thermal conductivity. D_{ij} and DT_i are the diffusion and thermal diffusion coefficients that depend on the molecules' characteristics and on the molar fractions and temperature. The precise definitions can be found in [12].

Let us assume that n (reactive) species are very diluted in an inert carrier gas. In such a case, the transport coefficients can be expressed as power series expansions in the molar fractions of the dilute species ($x_i \ll x_0 \approx 1$; $i = 1,2,\ldots,n$). To leading order [16,17]

$$D_{ii} - D_{i0} = \mathcal{D}_{i0} \left[\frac{1}{x_i} + 1 - \frac{m_i}{m_0} - \sum_{j=1\neq i}^{n} \frac{x_j}{x_i} \left(\frac{\mathcal{D}_{i0}}{\mathcal{D}_{ij}} - 1 \right) + O(x_i) \right] \qquad (3)$$

$$D_{ij} - D_{i0} = \mathcal{D}_{i0} \left[\frac{\mathcal{D}_{i0}}{\mathcal{D}_{ij}} - \frac{m_j}{m_0} \frac{\mathcal{D}_{j0}}{\mathcal{D}_{i0}} + O(x_i) \right], \qquad (i\neq j) \qquad (4)$$

$$DT_i = \mathcal{D}_{i0}\, \alpha_T^{i0} + O(x_i) \qquad (5)$$

where $m_i (i=0,1,\ldots,n)$ are the molecular mass, $\mathcal{D}_{ij}$ is the usual binary diffusion coefficient between species i and j, and α_T^{i0} $(i=1,2,\ldots,n)$ are the binary thermal diffusion factors between each species and the carrier.

In the limit case of infinite dilution ($x_i \rightarrow 0$) the expressions for the heat and mass fluxes reduce, respectively, to

$$J_q = -\lambda_0 \nabla T - p \sum_{i=1}^{n} \frac{m_0}{m_i} \alpha_T^{i0} \mathcal{D}_{i0} \nabla Y_i \tag{6}$$

FOURIER DUFOUR

$$J_i = -\rho_0 \mathcal{D}_{i0} (\nabla Y_i + \alpha_T^{i0} Y_i \nabla \ln T) \tag{7}$$

FICK SORET

here $Y_i (i=1,2,\ldots,n)$ are the reactant mass fractions and mass dilution has been assumed ($Y_i \ll Y_0 \approx 1$).

In both eqs. (6 & 7), direct and cross effects appear clearly distinguished and it is easy to get the relative order of magnitude

$$\frac{\text{DUFOUR}}{\text{FOURIER}} = O(\alpha_T^{i0} Y_i) \ll 1 \tag{8}$$

$$\frac{\text{SORET}}{\text{FICK}} = O(\alpha_T^{i0}) \tag{9}$$

Thus, as a consequence of the dilution, Dufour effect will be a small contribution to the heat flux and can be duly neglected in front of the Fourier term. However, the relative importance of the Soret effect is given by the thermal diffusion factor α_T^{i0}, which is often not negligible.

A rough estimation of the α_T^{i0} values can be obtained with the hard sphere model which yields [16,17]

$$\alpha_T^{i0} = \frac{\frac{2}{5}\frac{\gamma}{\gamma-1} Le_{i0} \frac{m_i}{m_0} \left(15 \frac{m_i}{m_0} - 7 \right) - 5 \left(\frac{m_i}{m_0} + 1 \right)}{13 + 16 \frac{m_i}{m_0} + 30 \left(\frac{m_i}{m_0} \right)^2} \tag{10}$$

where $\gamma=(c_p/c_v)_0$ is the specific heats ratio of the carrier species and $Le_{i0} \equiv (\lambda/\rho c_p)_0/\mathcal{D}_{i0}$ is the Lewis number associated to the species i and the carrier.

By recalling the dependence of $\mathcal{D}_{i0}$ on the molecular mass, eq. (10) provides the following values of α_T^{i0} in the limiting cases of light and heavy diluted species,

$$\text{light limit:} \quad \frac{m_i}{m_0} \to 0 \; ; \quad \alpha_T^{i0} \to -\frac{5}{13} \tag{11}$$

$$\text{heavy limit:} \quad \frac{m_i}{m_0} \to \infty \; ; \quad \alpha_T^{i0} \to \frac{\gamma}{5(\gamma-1)} Le_{i0} \tag{12}$$

The last result is very remarkable. It corresponds to a quasi- Lorentzian gas mixture [21]: the molecules of the diluted species are large and massive. Then, they diffuse slowly and the associated Lewis number takes on large values. The result of eq. (12) says that the thermal diffusion factor α_T^{i0} will be as large as Le_{i0} and reveals the Soret effect as the most effective mass transport mechanism (provided a temperature gradient exists) [30], overwhelming the Fick diffusion mechanism.

3. Soret transport and normal burning velocity of premixed flames

Premixed flames (deflagrations) are non-linear chemical waves driven by mass and heat diffusion which provide continuous feeding of reactant materials as well as removal of the products and released heat [35,31]. As the activation energy of the chemical reaction is typically very large in combustion, the reaction freezes suddenly in the unburnt gases zone, where the temperature falls down to the **ambient** temperature value. In this non-reactive, preheating zone, the reactive species concentration also rises to the fresh gas values. In these processes, convection and diffusion compete or coooperate to sustain the heat and mass fluxes required inside the reaction zone. Thus, it is clear that an accurate study of the combustion wave properties needs a detailed description of the prevailing diffusive mechanisms. Due to the strong temperature and species concentration gradients occurring in flames, the use of the Fourier and Fick laws for heat and mass diffusion require some reconsideration. The analysis made in the previous section supplies more complete, self-consistent expressions for the diffusive fluxes and has been applied successfully to combustion problems [16,17]. In fact, the hypothesis of diluted mixture is actually a good approximation in many practical situations (e.g. burning fuel in air, the nitrogen plays the role of inert gas carrier).

3.1. *The plane flame propagation*

The "scenario" displayed about a premixed flame at the beginning of this section was exploited by Zeldovich & Frank-Kamenetski [36] to get a first estimation of the normal burning velocity of a plane premixed flame. Their solution corresponds to the leading order of the asymptotic theory developed some years later [3] (the connection of this solution with the corresponding one for moderate values of the activation energy, studied by Kolmogorov, Petrowskii & Piskunov, is clarified in [37]). In the following we will sketch the asymptotic theory while emphasizing the effect induced by the Soret transport. More details on the general theory can be found in [8].

It is possible to study the fundamental aspects of the premixed flame propagation with the simple model of a one-step irreversible exothermic reaction, controlled by an Arrhenius law. This reaction brings the reactive species from a metastable state at room temperature to the state of chemical equilibrium at the adiabatic flame temperature. In that case the flame structure is governed by just two coupled reaction-diffusion (heat and mass) equations that written in suitable units are [17]

$$\frac{d\theta}{d\xi} - \frac{d^2\theta}{d\xi^2} = w \tag{13}$$

$$\frac{dy}{d\xi} - \frac{1}{Le}\frac{d^2y}{d\xi^2} = -w + \frac{\alpha_T}{Le}\frac{d}{d\xi}\left[\frac{y}{\theta+\sigma}\frac{d\theta}{d\xi}\right] \tag{14}$$

Here $\theta \equiv (T-T_{-\infty})/(T_b-T_{-\infty})$ is the dimensionless temperature and $y \equiv Y/Y_{-\infty}$ is the reduced mass fraction of the reactant. The subscripts "$-\infty$" and "b" refer to the fresh upstream and burned downstream conditions, respectively.

$\sigma \equiv T_{-\infty}/(T_b - T_{-\infty})$ is related to the gas thermal expansion and ξ is the dimensionless coordinate normal to the flame front, measured in flame thickness units, $\lambda/\rho c_p v_b$, v_b being the normal burning velocity of the flame. The chemical reaction effects are included in w. Le and α_T are the Lewis number and the thermal diffusion factor of the reactant in the carrier. The last term on the r.h.s. of eq. (14) accounts for the Soret transport.

The system of eqs. (13 & 14) must be solved subject to the following boundary conditions

$$\text{in the fresh side:} \quad \xi \longrightarrow -\infty, \; \theta = 0, \; y = 1 \tag{15}$$

$$\text{in the burned side:} \quad \xi > 0, \; \theta = 1, \; y = 0 \tag{16}$$

The method of high activation energy asymptotics [3] provides the jump relations for θ,y and their derivatives, through the reactive sheet located around $\xi = 0$. The inverse of the Zeldovich number, $Ze \equiv E_a/kT \longrightarrow \infty$, is a small parameter in which all the variables are expanded. Due to the Arrhenius term, the chemical reaction is exponentially small in the preheating zone, $\xi < 0$, and it is neglected to all algebraic orders in Ze^{-1}. Thus by taking $w = 0$, the profiles of temperature and mass fraction are readily obtained in this zone by a direct integration

$$\theta = e^{\xi} \tag{17}$$

$$y = \frac{Le \; e^{Le\xi}}{(\theta+\sigma)^{\alpha_T}} \int_{\xi}^{0} \frac{(\theta+\sigma)^{\alpha_T}}{e^{Le\xi'}} \, d\xi' \tag{18}$$

Eq. (18) contains the distortion induced by the Soret effect on the classical exponential profile at which it reduces for vanishing Soret transport, $y(\alpha_T=0) = 1-e^{Le\xi}$, as would be expected. These distortions have been detected from experimental results by Fristrom & Monchick [14] by giving a new interpretation to old experimental data on flame structure.

However, the leading order of the flame burning velocity v_b is determined by the first derivatives of θ and y at the origin. By integration of eq. (14) from $\xi \longrightarrow -\infty$ to the origin $\xi=0$, it is found that, due to the boundary conditions (15 & 16), there is no contribution from the Soret term

$$\left[\frac{y}{\theta+\sigma} \frac{d\theta}{d\xi} \right]_{-\infty}^{0} = 0 \tag{19}$$

The chemical reaction is assumed to be confined around $\xi = 0$ in a region of relative thickness Ze^{-1} compared to the preheating zone thickness. To fix the origin without ambiguity the best choice is to locate it at the position where the outer solution for the mass fraction vanishes to all orders [24]. Thus when the Soret term of eq. (14) is integrated through the reaction zone, then matching with the vanishing outer solution gives, once more, no contribution

$$\left[\frac{y}{\theta+\sigma} \frac{d\theta}{d\xi} \right]_{Ze\xi\to-\infty}^{Ze\xi\to\infty} = 0 \tag{20}$$

Then, we arrive at the surprising and interesting result that even though thermal diffusion may produce measurable distortions in the reactant concentrations profile through the flame, the value of the flame burning velocity remains unaltered by this effect at the leading order in the activation energy asymptotic theory [15].

The extension of this result to the more realistic case of a two-component reaction is straightforward and can be found in [16,17] where the first correction in dilution of the reactive species was also retained.

3.2. *Modification of the local burning velocity in non-planar flames*

When a plane flame front is distorted, tangential gradients of temperature and concentration appear on each surface element of the front. These fluxes modify the normal undisturbed removal and arrival of heat and matter and alter locally the chemical reaction rate [24,33,37]. Furthermore, the gas expansion due to the chemical heat release generates complex streamline patterns that for their part produce additional convective fluxes. This feed-back mechanism couples strongly the purely reaction-diffusion problem with the associated fluid mechanic problem and adds the inherent difficulties concerning the Navier-Stokes equations [5]. Up to date, the whole problem has only been solved for infinitesimal disturbances (linear stability analysis) in the long wavelength limit [6,13,18,28] although progress has been made in the non-linear problem [7] and for more complex flows [19,20,29].

The analysis shows that the effects of the diffusive mechanism inside the flame thickness can be described by only one dimensionless group, the Markstein number $\mathcal{M}$, that plays the role of an effective surface tension for the flame front surface. This number includes two different effects. Firstly, the pure diffusive relaxation of the distorted front that gives always a positive (stabilizing) contribution to $\mathcal{M}$. Secondly, the change produced in the chemical reaction velocity by the heat and mass fluxes that the diffusive relaxation originate, as mentioned earlier. The sign of this contribution depends on the prevailing diffusive coefficients. With the hypothesis $Le-1 = O(Ze^{-1})$ and $\alpha_T = O(Ze^{-1})$, the calculation gives the result [16,17]

$$\mathcal{M} = \left[\begin{array}{ll} (1+\sigma)\ln\dfrac{1+\sigma}{\sigma} \;\ldots\ldots\ldots\ldots\ldots\ldots & \text{Diffusive relaxation} \\ \left.\begin{array}{l} Ze(Le-1)\,\dfrac{\sigma}{2(1+\sigma)}\,\mathcal{D}(\sigma) \\ Ze\,\alpha_T \left[\dfrac{\mathcal{D}(\sigma)}{2(1+\sigma)} - \dfrac{\ln^2\frac{1+\sigma}{\sigma}}{4} \right] \end{array}\right\} & \ldots\text{Reaction rate change} \end{array} \right. \tag{21}$$

where

$$\mathcal{D}(\sigma) \equiv \int_0^{1/\sigma} x^{-1}\ln(1+x)\,dx \tag{22}$$

Generally, in gases the Lewis number takes on values close to unity, thus, eq. (21) shows that the Soret transport is as important as the direct transport in determining the value of the Markstein number and therefore, in the stability properties of a flame front. The Fourier (stabilizing) and Fick (destabilizing) fluxes always compete to modify the reaction rate and the net effect is proportional to Le-1. On the other hand, the effect of the Soret flux depends on the sign of α_T.

Note that for the quasi-Lorentzian mixture discussed in section 2 (e.g. lean flames of heavy fuels) where α_T and Le take on large and positive values, all the contributions in $\mathcal{M}$ have positive sign and eq. (21) predicts a strong stabilizing effect of the diffusive mechanisms. In the opposite case of a light fuel, Le-1 and α_T will take on negative values (see eq. (11)) giving small (or even negative) values of $\mathcal{M}$ and promoting unstable flame fronts. Then, in both limits the thermal diffusion phenomenon reinforces the net effect produced by the direct transports.

A corollary of this conclusion is that Soret effect can make attainable the oscillatory instability predicted for very large values of the Lewis number [24]. In that case the Soret effect increases the stability of the flame and the overstability condition would be reached easily. This was in fact the prediction in the constant density approximation [15] (no gas expansion) but the actual state of the theory that takes into account thermal expansion does not permit the study of oscillatory bifurcation.

4. Implications on the ignition stage

At room conditions (low enough temperature and pressure) the reactive species that can eventually lead to a propagating flame exhibit metastable states with typical characteristic times (given by the chemical time under such conditions) of the order of several hundreds of years [37]. To initiate a flame, this time has to be shortened by an external energy supply. In practice this is done in different ways: sparks, hot or catalytic bodies, adiabatic compression, etc.

The whole process of flame initiation is very complex [25] involving distinct stages where the Soret transport can play a significant role. However, the problem presents serious difficulties and a complete theory is still lacking.

In this section we will consider the ignition of a combustible gas flowing over a hot flat plate, held at temperature T_w. Due to the high activation energy, the chemical reaction is restricted to a very thin layer sticking to the plate where the large values of the temperature induce non-negligible reaction rates. In the remainder of the boundary layer the chemical reaction is frozen and the temperature and concentration profiles result from the balance between convection and diffusion [26]. Using the well known boundary layer approximation, all the variables are functions of only one similarity variable η. The flow configuration and the temperature distribution are described by the Blasius streamfunction $f(\eta)$ and by the Polhausen profile, respectively [32].

For vanishing Soret transport and negligible chemical consumption the reactive species concentration in this frozen zone is constant and equal to the value in the mainstream $Y(\eta) = Y_\infty$. When Soret transport is considered, because of the temperature gradient, the concentration is no longer homogeneous inside the thermal boundary layer and, depending on the sign of

α_T, either a fuel rich (α_T<0, light fuel) or a fuel lean (α_T>0, heavy fuel) gas layer will develop around the plate.

According to eq. (11), the contribution of thermal diffusion for light fuels remains always small ($|\alpha_T| \le 5/13$) and this case can be studied as a regular perturbation around the solution with vanishing thermal difusion. However, the case of fuels with large molecular mass deserves special attention because of eq. (12) implies that the Soret transport will be the dominant diffusive transport [22,23].

When the Soret contribution is included, the mass fraction of a diluted species in a fully developed laminar boundary layer obeys the equation

$$Sc^{-1}\Psi'' + [f + \alpha(\ln T)']\Psi' + \alpha(\ln T)''\Psi = 0 \tag{23}$$

with the boundary conditions (BC's)

$$\Psi = 1 \quad \text{at} \quad \eta \longrightarrow \infty \tag{24}$$

$$Sc^{-1}\Psi' + \alpha\Psi\,(\ln T)' = 0 \quad \text{at} \quad \eta = 0 \tag{25}$$

Here Ψ is the species mass fraction normalized with the value in the oncoming stream and the prime denotes the derivative with respect to the similarity variable η. Furthermore, $Sc=\nu/\mathcal{D}$ (≡ratio of momentum diffusivity to mass diffusivity) is the Schmidt number and the parameter $\alpha \equiv \alpha_T/Sc$ measures the thermal diffusion strength. By using eq. (12) and Eucken's relation [12] we get

$$\alpha \equiv \frac{\alpha_T}{Sc} = \frac{9\gamma - 5}{20(\gamma - 1)} \tag{26}$$

in the quasi-Lorentzian limit. For a diatomic gas carrier, $\gamma=7/5$, this result gives $\alpha=19/20$, which provides a superior limit value for this parameter.

When $Sc \gg 1$, but $\alpha=O(1)$, the effect of the second derivative term in eq. (23), that takes into account Fick diffusion, becomes negligible almost everywhere and the solution with BC's (24 & 25) becomes a singular perturbation problem. Note that the coefficient of Ψ' vanishes at a well defined value of the similarity variable. For very large $\eta(T' \longrightarrow 0,\ f \longrightarrow \infty)$ the coefficient is positive whereas for small $\eta(T' < 0,\ f \longrightarrow 0)$ it is negative, thus it vanishes at some intermediate value $\eta = \eta_c$. As a consequence of η close to η_c, the diffusion term cannot be neglected and an inner sublayer develops around this critical value η_c where diffusion becomes comparable to both convection and thermal diffusion. The location of this critical line is implicitly given by the relation

$$[f + \alpha(\ln T)']_{\eta=\eta_c} = 0 \tag{27}$$

that can be numerically solved for η_c.

Away from the inner diffusive sublayer ($\eta \neq \eta_c$), an asymptotic expansion in the limit $Sc^{-1} \longrightarrow 0$ for the solution of eq. (23) can be written in the form of two series, corresponding to algebraic and exponential orders respectively

[27]. To leading order the solution is

$$\Psi^{\pm}_{out} = C^{\pm}_1 \exp\left(-\int_{\eta^{\pm}}^{\eta} \frac{H(\tilde{\eta})}{G(\tilde{\eta})} d\tilde{\eta}\right) + C^{\pm}_2 G(\eta)^{-1} \exp\left(\int_{\eta^{\pm}}^{\eta} \frac{H(\tilde{\eta})}{G(\tilde{\eta})} d\tilde{\eta}\right)$$

$$\exp\left(-Sc\int_{\eta^{\pm}}^{\eta} G(\tilde{\eta})d\tilde{\eta}\right) \tag{28}$$

where the superscripts (+) and (-) stand for the upper ($\eta>\eta_c$) and lower ($\eta<\eta_c$) zones, respectively and we have introduced the functions

$$G(\eta) \equiv f + \alpha(\ln T)' \tag{29}$$

$$H(\eta) \equiv \alpha(\ln T)'' \tag{30}$$

By choosing $\eta^+ > \eta_c$ and $\eta^- = 0$ as the integration limits in eq. (28), the boundary conditions (24 & 25) impose

$$C^-_1 = C^+_2 = 0 \qquad C^+_1 = 1 \tag{31}$$

and the remaining unknown is determined by the matching condition with the solution in the inner zone.

To study this inner zone, $\eta\approx\eta_c$, the appropriate stretching variable that introduces a non negligible contribution from the diffusion term is

$$\Lambda = (Sc\ G_c')^{1/2}(\eta-\eta_c) \tag{32}$$

where the subscript (c) refers to the value of the function at $\eta=\eta_c$. To leading order, the inner form of eq. (23) is

$$\frac{d^2\Psi_{inn}}{d\Lambda^2} + \Lambda \frac{d\Psi_{inn}}{dL} - \delta\Psi_{inn} = 0 \tag{32}$$

with

$$\delta = -H_c/G_c' \tag{34}$$

Eq. (33) can be easily transformed into the parabolic cylinder equation [1] and the general solution written as the following linear combination involving the Whittaker function U

$$\Psi_{inn} = e^{-\Lambda^2/4} \left[E_1\ U\left(\frac{1}{2} + \delta,\ \Lambda\right) + E_2\ U\left(\frac{1}{2} + \delta,\ -\Lambda\right)\right] \tag{35}$$

where the constants E_1 and E_2 are determined by matching with the outer solutions.

Matching with the solution in the upper region provides the value of E_2,

$$E_2 = \left[\frac{\Gamma(1+\delta)}{(2\pi)^{1/2}} \frac{(1/2+\delta)^{1/2+\delta}}{(G'_c\ Sc)^{\delta}} \right]^{1/2} (\eta^+ - \eta_c)^{-\delta}$$

$$\exp\left[\int_{\eta^+}^{\infty} \frac{H(\eta)}{G(\eta)}\, d\eta + \int_{\eta_c}^{\eta^+} F(\eta)\, d\eta \right] \qquad (36)$$

where we have used the usual gamma function Γ and introduced

$$F(\eta) \equiv H(\eta)/G(\eta) + \delta/(\eta - \eta_c) \qquad (37)$$

Matching with the lower region solution imposes

$$E_1 = 0 \qquad (38)$$

and

$$E_2 = -C_2^- \left[\frac{\Gamma(1+\delta)}{(2\pi)^{1/2}} \frac{(G'_c\ Sc)^{1+\delta}}{(1/2+\delta)^{1/2+\delta}} \right] \eta_c^{\delta}\, (G'_c)^{-1}$$

$$\exp\left\{ \int_0^{\eta_c} [F(\eta) - ScG(\eta)]\, d\eta \right\} \qquad (39)$$

By equating (39 & 36), the constant C_2^- is determined and we can finally write down the expression for the concentration profile in the region close to the wall

$$\Psi^-_{out}(\eta) = \left\{ \frac{(G'_c)^{1/2-\delta}}{(\eta^+ - \eta_c)^{\delta}} \left(\frac{1/2+\delta}{Sc} \right)^{1/2+\delta} \exp\left[\int_{\eta_c}^{\eta^+} F(\eta) d\eta + \int_{\eta^+}^{\infty} \frac{H(\eta)}{G(\eta)}\, d\eta \right] \right\}$$

$$\frac{\exp\left\{ \int_{\eta}^{\eta_c} [Sc\ G(\tilde{\eta}) - F(\tilde{\eta})] d\tilde{\eta} \right\}}{[-G(\eta)\cdot(\eta_c - \eta)^{\delta}]} \qquad (40)$$

Eq. (40) evaluated at $\eta = 0$ provides the mass fraction of the massive species, the fuel, on the wall. As the function $G(\eta)$ is negative in this region, our result shows that the relative fuel concentration on the wall is exponentially small in the Schmidt number.

In the frame of the purely thermal ignition theory (i.e. neglecting fuel consumption), the ignition distance is shown to be inversely proportional to the fuel mass fraction on the wall. Then, eq. (40) predicts an exponential increase in the ignition distance with respect to the value advanced by the classical theory. Due to this strong fuel concentration reduction, the importance of fuel consumption is enhanced and the validity of the purely thermal theory restricted.

5. Conclusions

The Soret (thermal diffusion) effect has been shown to have a significant influence on the properties of premixed flames. A surprising exception is the normal burning velocity of a plane flame because the particular dependence on temperature and fuel concentration of the Soret contribution gives a vanishing overall effect through the flame thickness. Although this result also holds for a two reactant flame, actual flames exhibit a large number of intermediate species (radicals) whose distribution should be also notably influenced by the Soret effect. These species are often decisive in the flame propagation, however, the theoretical understanding of multistep chemistry flames is yet a challenging problem and our conclusion cannot be generalized.

With respect to the stability properties of a flame front, the Soret effect modifies locally the reactant arrival rate and produces local changes in the chemical reaction velocity which are of the same order of magnitude than the changes induced by the direct transports. Thus, the Soret effect should be always retained for an accurate theoretical prediction of the flame front stability.

The problem of premixed flame ignition with the Soret effect is not completely solved up to date, but the known partial results show that the Soret effect induces exponential delays in the ignition of heavy fuels by hot bodies. Even more, the ignition may occur within the boundary layer instead of on the body surface, because when the Schmidt and Zeldovich numbers are of a comparable order of magnitude, the fuel concentration in a width zone over the body becomes exponentially small and comparable to the chemical consumption. In this case the fuel depletion would occur around the fuel stagnation line and the ignition is expected at this line.

Acknowledgements

This work has been supported by the Dirección General de Investigación Cientifica y Técnica (DGICYT) Spain, under contracts PB87-0588 and PB88-0159.

References

[1] Abramowitz, M. & Stegun, I. (1972). In *Handbook of Mathematical Functions*, Chapter 19. Dover.
[2] Aitken, J. (1884). Collected Works. *Trans. Roy. Soc. Edinb.* **32**, 84-113.
[3] Bush, W.B. & Fendell, F.E. (1970). Asymptotic Analysis of Laminar Flame Propagation for General Lewis Number. *Combust. Sci. Tech.* **1**, 421-428.
[4] Chapman, S. (1916). On the Law of the Distribution of Molecular Velocities, and on the Theory of Viscosity and Thermal Conduction, in a Non-Uniform Simple Monatomic Gas. *Phil. Trans. Roy. Soc. A* **216**, 279.

[5] Clavin, P. & Williams, F.A. (1982). Effects of Molecular Diffusion and of Thermal Expansion on the Structure and Dynamics of Premixed Flames in Turbulent Flows of Large Scale and Low Intensity. *J. Fluid. Mech.* **116**, 251-282.
[6] Clavin, P. & García-Ybarra, P. (1983). The Influence of the Temperature Dependence of Diffusivities on the Dynamics of Flame Fronts. *J. Theor. App. Mech.* **2**, 245-263.
[7] Clavin, P. & Joulin, G. (1983). Premixed Flames in Large Scale and High Intensity Turbulence Flow. *J. Phys. Lettres* **44**, L1-12.
[8] Clavin, P. (1985). Dynamical Behaviour of Premixed Flame Fronts in Laminar and Turbulent Flows. *Prog. Energy Combust. Sci.* **11**, 1-59.
[9] de Groot, S.R. (1945). *L'Effet Soret, Diffusion Thermique dans les Phases Condensées.* North Holland.
[10] de Groot, S.R. & Mazur, P. (1984). *Non-Equilibrium Thermodynamics.* Dover.
[11] Enskog, D. (1917). Kinetische Theorie der Vorgänge in mässig verdünten Gasen. *Inaug. Dissertation.* Uppsala.
[12] Ferziger, J.H & Kaper, H.G. (1972). *Mathematical Theory of Tansport Processes in Gases.* North Holland.
[13] Frankel, M.L. & Sivashinsky, G.I. (1982). The Effect of Viscosity on Hydrodynamic Stability of a Plane Flame Front. *Combust. Sci. Tech.* **29**, 207-224.
[14] Fristrom, R.M. & Monchick, L. (1988). Two Simple Approximations to the Thermal Diffusion Factor and their Applications to Flame Studies. *Combust. Flame* **71**, 89-99.
[15] García-Ybarra, P. & Clavin, P. (1981). Cross-Transport Effects in Nonadiabatic Premixed Flames. *Prog. Astro. Aero.* **74**, 463-482.
[16] García-Ybarra, P. (1982). Estudio de las Propiedades Difusivas de una Llama de Premezcla y de su Influencia sobre la Estabilidad de la Propagación Plana Estacionaria. *Thesis, U.N.E.D. Madrid.*
[17] García-Ybarra, P., Nicoli, C. & Clavin, P. (1984). Soret and Dilution Effects on Premixed Flames. *Combust. Sci. Tech.* **42**, 87-109.
[18] García-Ybarra, P. (1986). Flame Front Stability with General Intermolecular Interaction Potential. *Prog. Astro. Aero.* **95**, 115-128.
[19] García-Ybarra, P. & Borghi, R. (1986). Stability Study of an Oblique Flame Front Model. *Prog. Astro. Aero.* **105**, 296-319.
[20] García-Ybarra, P. & Borghi, R. (1986). Étude de la Stabilité des Flammes Prémélangées Obliques. *J. theor. App. Mech.* **Special issue**, 157-172.
[21] García-Ybarra, P. & Rosner, D.E. (1989). Thermophoretic Properties of Nonspherical Particles and Large Molecules. *AIChEJ* **35**, 139-147.
[22] García-Ybarra, P. & Castillo, J. Influence of Thermal Diffusion on Combustible Gas Ignition by a Hot Plate. *Proc. 3rd Int. Seminar on Flame Structure.* Nauka: Moscow, in press.
[23] García-Ybarra, P. & Castillo, J. (1989). Flat Plate Boundary Layer Ignition with Fuel Thermal Diffusion. *Prog. Astro. Aero.*, submitted.
[24] Joulin, G. & Clavin, P. (1979). Linear Stability Analysis of Non-Adiabatic Flames: Diffusional-Thermal Model. *Combust. Flame* **35**, 139-153.
[25] Joulin, G. (1985). On Point Source Initiation. In *Combustion and Nonlinear Phenomena*, Clavin, Larrouturou & Pelcé (eds.), pp. 29-50. Les Editions de Physique. Les Ullis.
[26] Méndez, F., Treviño, C. & Liñán, A. (1986). Premixed Combustion in Boundary Layers for Moderate Values of the Zeldovich Numbers. *Combust. Sci. Tech.* **48**, 129-149.
[27] Nayfeh, A. (1973). *Perturbation Methods.* John Wiley & Sons.

[28] Pelcé, P. & Clavin, P. (1982). Influence of Hydrodynamics and Diffusion upon the Stability Limits of Laminar Premixed Flames. *J. Fluid Mech.* **124**, 219-237.

[29] Pelcé, P. (1985). Effects of Gravity upon Propagation of Flames in Tubes. *J. Physique* **46**, 503-510.

[30] Rosner, D.E. (1980). Thermal (Soret) Diffusion Effects on Interfacial Mass Transport Rates. *PhysicoChemical Hydrodynamics* **1**, 159-185.

[31] Rosner, D.E. (1986). *Transport Processes in Chemically Reacting Flow Systems.* Butterworths.

[32] Schlichting, H. (1968). *Boundary Layer Theory.* McGraw Hill.

[33] Sivashinsky, G.I. (1977). Diffusional Thermal Theory of Cellular Flames. *Combust. Sci. Tech.* **15**, 137-145.

[34] Tyndall, J. (1870). *Proc. R. Instn. Gt. Br.* **6**, 3.

[35] Williams, F.A. (1985). *Combustion Theory.* Benjamin/Cummings.

[36] Zeldovich, Ya.B. & Frank-Kamenetski, D.A. (1938). A Theory of Thermal Propagation of Flame. *Acta Phys. USSR* **IX**, 341-350.

[37] Zeldovich, Ya.B., Barenblatt, G.I., Librovich, V.B. & Makhviladze, G.M. (1985). *The Mathematical Theory of Combustion and Explosion.* Consultants Bureau.

38. EXCYCLON DYNAMICS

E. Dulos, J. Boissonade and P. De Kepper

Centre de Recherche Paul Pascal/CNRS

Université de Bordeaux I, Avenue Schweitzer
33600 Pessac, France

We use an open two-dimensional spatial reactor, to create sustained crescent-shaped excitation waves (**"excyclons"**) propagating in the same azimuthal direction in a thin annular layer of BZ medium. They can produce formerly unknown two-dimensional wavetrain patterns. Excyclons spread evenly around the annulus, as is often the case for pulse trains on ring fibres. However, depending on their number, the resulting rotating wavetrain structures may present either the highest symmetry or topological defects that develop in the radial direction. These defects may be locked in relative position or permanently exchanged from one excyclon to another. A dispersion relation is established. Numerical simulations based on a two-variable Tyson-Fife model of the reaction exhibit good qualitative agreement with experimental observations.

1. Introduction

One of the most fascinating properties of chemical reactions evolving far from equilibrium is their capability to develop spatial structures. Experiments on spatial self-organization phenomena in chemical systems [1-3] have essentially yielded concentric ring and spiral wave patterns in excitable media [4-15]. Most of these studies were performed on different variants of the Belousov-Zhabotinskii (B.Z.) reaction, though similar patterns have also been observed in other single phase isothermal reactions [13,14].

Until recently, experiments were performed in closed reactors (Petri dishes, test tubes) [4-15] where the systems irreversibly and uncontrollably evolve towards their thermodynamic equilibrium. The observed structures are, of this sort, transient phenomena that can last but for a brief while. In contrast, theoretical studies of reaction-diffusion patterns essentially limit themselves to the discussion of asymptotic properties [1,2]. The exclusive use of closed reactors has hindered the development of other types of reaction-diffusion structures than the target and spiral patterns mentioned above. The so-called mosaic structures which have also been reported [16,17] result from convective motions in the fluid due to adverse density gradients of various origins [17,18] and thus do not qualify as genuine chemical dissipative structures [19].

To overcome the limitations of closed systems, a number of open reactors have been developed where different types of sustained chemical wave patterns, some of them yet unknown in closed reactors [20] have been produced. Presently, three types of open spatial reactors have been designed. They differ in their geometry and their dimensionality. They are: (i) The **Couette Flow Reactor** [21,22], a one-dimensional reaction-diffusion system which is discussed in detail by Boissonade et al. in these proceedings [23]. (ii) The **Disc Reactor** [24], a thin disc of gel uniformly fed from one side with an appropriate mixture of reactive solutions. It is a two-dimensional system able to sustain and control spiral wave structures. (iii) The **Ring Reactor** which is also a two-dimensional reactor but with radial feeds from the two rims of an annular gel strip. This reactor, first proposed by Noszticzius et al. [25], is used in the present experiments.

In the first section of the report, we briefly describe the latter piece of apparatus and the experimental conditions to produce sustained travelling waves. The experimental results are gathered in the second section. Wave patterns including or not including irregularities (defects) are described, a dispersion relation is obtained and the implications of these features are discussed. Finally, preliminary results on numerical simulations with a two-variable model of the reaction, are reported in a third section.

2. The annular reactor and the experimental conditions

2.1. *The reactor*

A scheme of the reactor is presented in Fig. 1. The core of the device is a flat, thin annular strip of polyacrylamide gel (1 mm thick, 10 mm wide) with a median diameter of 30 mm. This porous transparent annulus is squeezed (Fig. 1b) between a white teflon bottom and a transparent plexiglass cover which allows for observations and monitoring with a video camera.

The rims of the gel (Fig. 1a) are kept in contact with two stirred solutions A and B of well defined compositions. The chemicals diffuse radially into the gel where they meet and react. The gel is practically inert and avoids any parasitic convective phenomenon. The concentrations at the rims are kept constant by a permanent supply of solutions A and B (Fig. 1b). The steady temperature of the whole system is maintained by an external water jacket.

2.2. *Experimental conditions*

In all reported experiments, we deal with a BZ reaction with ferroin as catalyst. Ferroin, an iron-orthophenanthrolin complex ion, also acts as a redox colour indicator, blue when oxidized (Fe^{+3}), red when reduced (Fe^{+2}).

As shown in Table 1, neither solution A nor B contains all the components of the reaction. The diffusion of the reactants establishes radial concentration gradients of the major chemical species, except for ferroin which is present on both sides of the annulus, so that there is no radial gradient of the colour indicator. The typical settling time of the steady state gradients ($t \sim l^2/D$) is about 30 hours since small ionic species have a diffusion coefficient $D \simeq 10^{-5} cm^2.sec^{-1}$. The concentrations of the reactants in solutions A and B are chosen in such a way that, at steady state, a circular active zone of excitability develops in the middle of the gel.

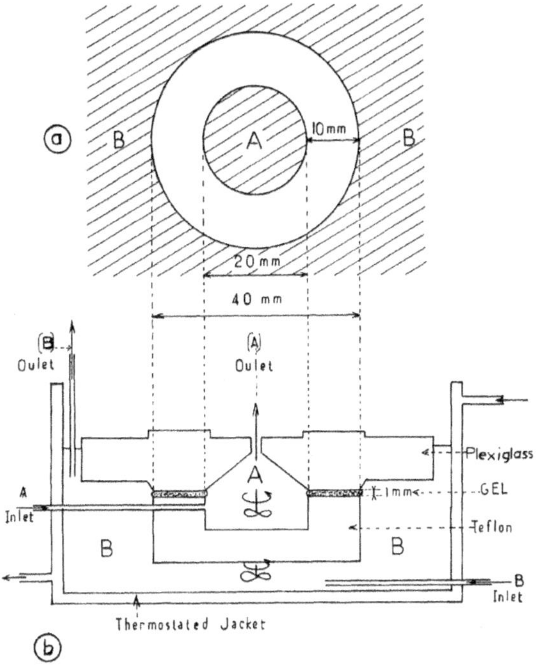

Figure 1. Schematic view of the annular reactor.
a) Top view of the annulus of gel: only the two rims of the gel are in contact with the two solutions A and B.
b) Cross-section of the whole piece of apparatus: the polished surface of the plexiglass cover above the annulus of gel allows for the observation from above of the patterns which develop in the gel. The permanent feed of solutions A and B and the stirring maintain a constant homogeneous composition at the boundaries.

Table 1. Composition of the solutions A and B. Concentrations are given in mole/l.

Solution A	Solution B
Ferroin 2.25 x 10^{-3}	Ferroin 2.25 x 10^{-3}
$NaBrO_3$ 0.3	H_2SO_4 0.2
	Malonic acid 2.25 x 10^{-3}

2.3. *Operating technique*

Triggered waves can be induced by different types of perturbations in this excitable medium. In the following experiments, excitations were produced with a focused light beam from a halogen lamp. The perturbation triggers a blue oxidized centre (Fig. 2a) which immediately grows and parts into two waves propagating along the annulus, in opposite directions (Fig. 2b) until they annihilate in head-on collision. In order to obtain a permanent wave, one of the twin waves must be eliminated by means of a strong light flash. We are then left with a single blue crescent-shaped excitation wave travelling in a circle (Fig. 2c) in a red bulk medium. We call **excyclon** such a self-sustained wave.

In a similar way, n light perturbations can result in n pairs of opposite waves. By eliminating those propagating in one direction, we are left with n permanent waves, all propagating in the other direction. Then, excyclons rearrange until they all rotate at the same speed and form an equally spaced wave pattern (Fig. 2d), the so-called **Pinwheel structure** [25].

All results reported in this paper were obtained under the same experimental conditions, i.e. composition as in Table 1, temperature 15°C, identical light intensity.

2.4. *Dispersion relation*

Sequences of n excyclons arranged in a chemical pinwheel constitute genuine sustained wavetrains. The propagation of such trains of chemical excitation waves can be compared to those observed in other excitable media.

From a dynamical point of view, excitation waves are schematically composed of a wavefront followed by a refractory zone where various recovery processes take place, bringing the medium back to the initial steady state (or "rest state"). The propagation velocity in a wavetrain is known to depend upon the distance (or time interval) between two successive waves. In other words, there is a **dispersion** of speed with the wavelength. Indeed, the instantaneous propagation velocity of a wave is essentially determined by the recovery level of the medium just ahead of the wavefront; the recovery level is a function of the time elapsed since the previous wavefront passed at the same location. The relationship between velocity and wavelength (or period), the **dispersion relation**, is a characteristic of the medium.

Theoretical dispersion curves have been calculated from numerical analysis of the Hodgkin-Huxley (HH) [26] and the FitzHugh-Nagumo (FHN) [27] models of the nerve impulse and of Oregonator [28-32] models of the BZ reaction.

If we now limit our interest to the field of chemistry, the experimental dispersion relations found in the literature were only obtained on transient properties of target and spiral patterns [8-12,30] since the reacting media were never fed with fresh reactants, so that chemical composition drifts were unavoidable during the experiments. The use of open spatial reactors allows a much more satisfactory approach to the asymptotic behaviour since the spatial structures obtained can be indefinitely sustained in permanently fed media.

In all the pinwheel structures presented in this paper, the n excyclons are equally spaced (Fig. 2d) on a circle of circumference $l = 94.25$ mm. Their wavelength is thus $\lambda = l/n$. Their period, the time to complete a cycle, has been measured for wavetrains with $1 \leq n \leq 12$. In our experimental conditions,

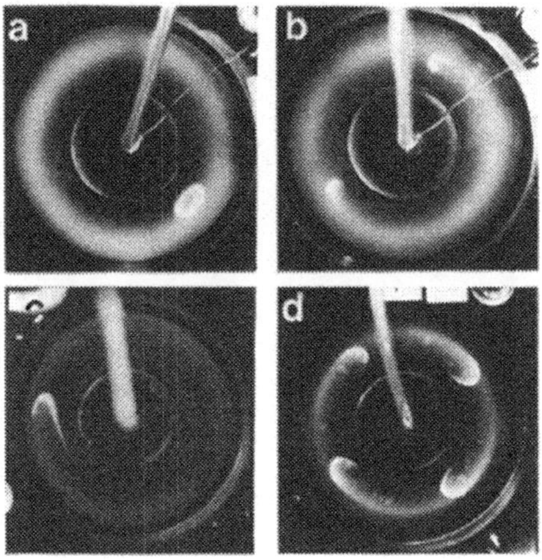

Figure 2. Creation of an excyclon and a symmetric pinwheel structure. In this and other pictures, the blue excitation centre and the waves appear bright and the red bulk medium much darker. In some cases, carbon dioxide bubbles (pale grey) resulting from the reaction can also be seen, trapped under the gel strip.
a) An excitation centre about 3 seconds after it has been created. Temperature 20°C.
b) A pair of crescent-shaped waves resulting from the partition of the excitation centre of Fig. 2a, propagate in opposite directions. Temperature 20°C.
c) A solitary excyclon rotating clockwise. It was obtained by eliminating one of the twin waves of Fig. 2b. Temperature 15°C.
d) A pinwheel structure of four equally spaced and identical excyclons. Temperature 15°C.

twelve is the maximum number of exyclons that we have been able to sustain over an extended period of time. The situation, of course, is identical to that of a ring fibre [26,33,34] that can only sustain a limited number of pulses since the refractory zone behind the wavefront requires that spacing is larger than a critical wavelength for the following wave to propagate. It can be seen in Table 2 that, as n is increased, T also increases. This result shows that, as generally found in theoretical studies and in experimental measurements in closed systems, waves with long wavelength travel faster than waves with shorter wavelength.

A dispersion relation has been determined from our experiments. The curve C (λ) of the velocity versus the wavelength, gathers the results of two sets of experiments (Fig. 3). This dispersion relation exhibits a monotonic dependence of velocity on wavelength.

Monotonic as well as nonmonotonic dispersion relations are found in a number of experimental and theoretical studies of excitable biological systems. Nonmonotonic dispersion curves are often associated with irregular pulse spacing; this problem is discussed in the next paragraph.

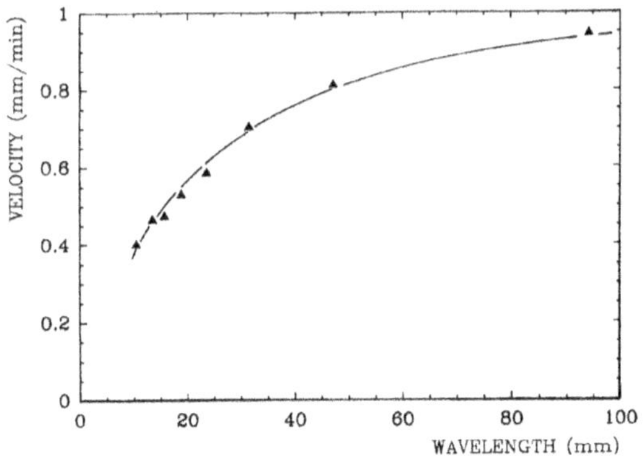

Figure 3. The monotonic dispersion relation obtained from two different sets of experiments, for the same experimental conditions.

Table 2. Period T and propagation velocity C as a function of the number n and wavelength λ of excyclons. Each value of T or C is averaged over several experiments (λ = l/n where the circumference l is 94.25 mm).

n	λ(mm)	C(mm/min)	T(min)
1	94.25	0.942	100
2	47.12	0.812	116
3	31.42	0.70	134
4	23.56	0.59	161
5	18.85	0.53	178
6	15.71	0.47	199
7	13.46	0.46	203
9	10.47	0.40	235
12	7.854	0.38	245

3. **Topology of wave patterns**

The problem of irregularities in the spacing within wavetrains has been experimentally studied in relation to the nerve impulse propagation [33,34]. The irregularities generally appear as pulse pairing or grouping phenomena which lead to bunches of waves of different wavelengths propagating at the same velocity. Theoretical studies of simple one-dimensional dynamical models clearly demonstrate that this pulse grouping is associated with a nonmonotonic dispersion relation [26,35,36] which itself is a direct consequence of a nonmonotonic return to the rest state after a solitary impulse. In the field of chemistry, calculations by Karfunkel & Kahlert [28] for the full Oregonator model (three variables), of the excitable BZ reaction in a one-dimensional ring geometry, leads to a nonmonotonic dispersion relation resulting from damped oscillations in the return to the homogeneous steady state. As a consequence of this nonmonotonic behaviour, they also show that, for some intermediate range of pulse frequency, there exist several stable pulse spacings leading to the same propagation velocity and that waves can organize into groups of two or more travelling pulses with different wavelengths.

As mentioned above, our experimental system leads to a monotonic dispersion relation which is consistent with excyclons regularly spaced in the azimuthal direction. However, we are dealing here with a two-dimensional system since, even though sharp concentration gradients of the major reactants are set across the gel strip, the excitability properties expand over a quite wide region in the radial direction. We may thus expect that instabilities develop in the radial direction. In fact, radial irregularities in the size of excyclons are readily observed in our experiments.

In the patterns made of a small number n of excyclons, all the waves have identical shape and size and are equally spaced (Fig. 2d); but, such fully periodic sequences seem to be limited to systems with $n \leq 4$. Indeed, for $n > 4$, the wave patterns obtained always include at least one excyclon which considerably differs from the others, both in shape and size but, contrary to pulse trains in one-dimensional systems, these sequences of excyclons do not exhibit noticeable irregularities in the spacing. An atypical excyclon can be seen in a train of six, in Fig. 4. This excyclon presents an outgrowth at the rear of the wavefront, along the inner rim of the gel. This expansion of atypical excyclons of the pinwheel pattern takes place in the radial direction of the gel. Such radial expansions of the waves are the only topological defects that we have yet been able to observe. Thus, it can be assumed that our two-dimensional system prefers to use the radial rather than the azimuthal dimension when an arrangement of highest symmetry becomes impossible.

A pinwheel structure of 12 excyclons can be seen in Fig. 5a. It is composed of 3 identical groups of 4 waves, every group made of 1 large and 3 regular excyclons. This ternary symmetry could be maintained when, by eliminating 1 (resp. 2) regular excyclon(s) in each group, a pinwheel structure of 9 (resp. 6) excyclons was produced. As shown in Figs. 5b and 5c, such systems are also made of 3 identical groups composed of 1 large and 2 (resp. 1) regular excyclons. Each of these ternary structures remained stable for as long as the same experimental conditions were maintained without perturbation of any kind, i.e. for more than 36 hours.

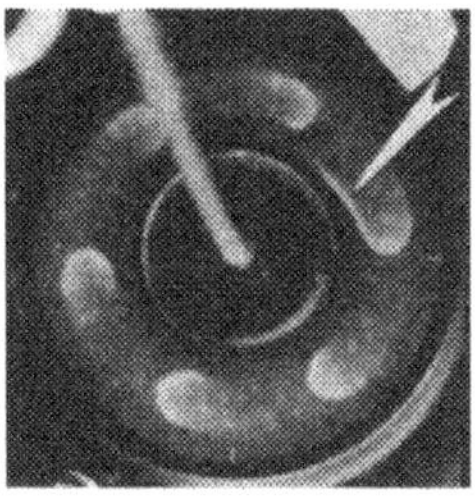

Figure 4. Pinwheel structure of six excyclons, with one atypical excyclon (arrow). The whole wave structure rotates as a solid body.

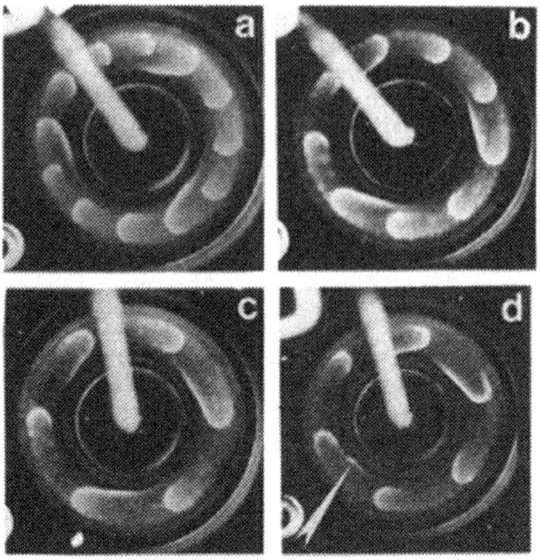

Figure 5. Complex pinwheel structures with several atypical excyclons: a,b,c) Structures with three equally distributed atypical excyclons locked in relative position, with respectively 12, 9 and 6 excyclons.
d) Snapshot of an unlocked pattern of six excyclons, more than 30 hours after an asymmetric perturbation of structure c. Excyclon outgrowths permanently develop and are expelled (arrow) at different positions.

In all previous structures, the atypical excyclons were locked in relative position and the wave pattern would rotate as a solid body after the initial settling period (~ 12 h). No excyclon ever changed in shape during the observation time. In other cases, on the contrary, excyclons permanently changed in size and as a result, the atypical excyclon seemed to migrate from one position to another. In such cases, no frozen pinwheel structure could be observed. A snapshot of such an unlocked structure is shown in Fig. 5d. This structure was obtained after an asymmetric perturbation of the structure of Fig. 5c.

Generally, the velocities measured for patterns corresponding to the same number of excyclons are quite reproducible from one set of experiments to another but the wave patterns can change considerably, as one can see by comparing the 6 excyclon patterns in Figs. 4, 5c and 5d. The two last patterns were obtained during the same set of experiments while the pattern in Fig. 4 was obtained in a separate experimental run. We still cannot decide if these differences originate in some uncontrolled experimental parameter or in the existence of a great number of stable solutions.

4. Numerical simulations

Preliminary numerical studies of a two-dimensional reaction- diffusion model with appropriate boundary conditions and an elementary representation of the BZ chemistry, show that the essential features of the experimental observations can be captured. The model is a two-variable reduced Oregonator extensively studied by Tyson & Fife [29] as a skeletal representation of the cerium catalyzed BZ reaction. The model equations in a rescaled form, where the bromate, the organic substrate, and the hydrogen ion concentrations remain explicit, are:

$$\varepsilon\frac{\partial u}{\partial \tau} = u(AH - u) + \frac{\mu AH - u}{\mu AH + u}qBv + D\Delta_r u$$

$$\frac{\partial v}{\partial \tau} = AHu - Bv + D\Delta_r v$$

where u and v are respectively the fast and slow variables of the model, and represent the main autocatalytic species (bromous acid) and the oxidized form of the catalyst (in our experiments the Fe^{+3}.[phenantroline]$_3$ blue complex). The quantities A, B, H and q are control parameters and represent respectively the bromate, the total organic substrate (essentially malonic + bromomalonic acid), the hydrogen ion concentrations, and an ad hoc stoichiometric factor which accounts for complex kinetic features. This latter parameter depends among other things on the bromate/organic substrate concentration ratio and, on the basis of stoichiometric arguments, cannot be greater than 4 [37,38]. The quantities ε and μ are the ratios of several rate constants and are independent of chemical concentrations. D is the diffusion coefficient taken identical for u and v and Δ_r is the laplacian operator. For fixed values of A, B, and H, the theoretical analysis [37,38] shows that the homogeneous system spontaneously oscillates for $1/2 < q \leq 1 + \sqrt{2}$. A steady reduced state is obtained for $q > 1 + \sqrt{2}$.

For the sake of simplicity, the annular gel strip is modelled as a rectangular space grid of 500 x 100 with periodic boundary conditions in the **Ox** direction (left-right in the figures); hence curvature is neglected. The period corresponds to the circumference l of the annular reactor. We use Dirichlet boundary conditions in the **Oy** direction (up-down in the figures). Concentrations of A, B, and H in the strip are supposed to be solely determined by diffusion and to be always locally in large enough amounts for their consumption to be negligible, so that they can be appropriately accounted for by permanent linear gradients from one edge to the other. To mimic the experiments, the fixed concentrations at the upper edge are set to A = 0, B = 0.02M, H = 0.4M and at the lower edge to A = 0.3M, B = 0, and H = 0. The changes in q (as a function of composition) are difficult to determine. To reduce to the essentials, we assume a linear dependence of q in the radial direction. For chemical reasons, q should be larger at large than at low values of the bromate/organic substrate concentration ratio. This type of approximation has already been used satisfactorily by Tam et al. [22] for modelling the dynamics of the BZ reaction when operated in a continuous Couette flow reactor with asymmetric boundary conditions. The values q = 4 and

Figure 6. Computed solitary excyclon. Shaded representation of variable v which corresponds to the blue wave (clear in the figures) of Fe^{+3} complex in the experiments. With $\varepsilon = 0.025$, $\mu = 4.10^{-4}$, and $D = 10^{-5} cm^2/s$.

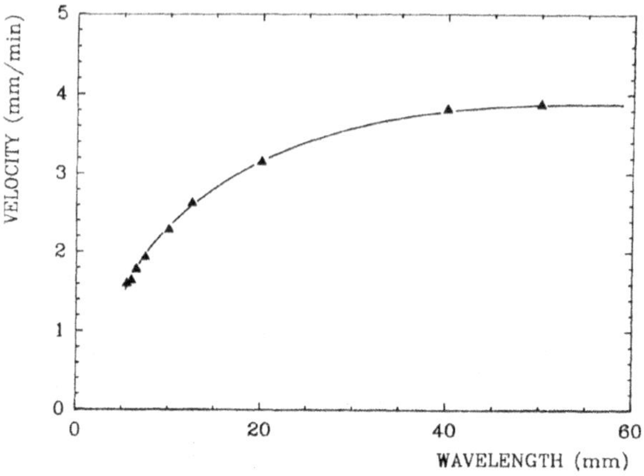

Figure 7. Computed dispersion curve velocity vs. wavelength.

q = 2 were respectively chosen at the lower and upper edges. In these conditions, the system exhibits a reduced steady state where activity waves can only be triggered by a localized superthreshold perturbation, as in experiments. The dynamical equations are integrated with the simple explicit Euler method which has been shown to be satisfactory in the study of travelling waves [39] and is sufficient for this preliminary work. The computed shape of a solitary excyclon, shown in Fig. 6, is in excellent agreement with the experimental observations (Fig. 2c). It exhibits a crescent shape with, at the rear, a longer extension along the bromate rich edge. A velocity dispersion curve was determined by simulating the propagation of such a solitary wave for different periods in the **Ox** direction. The results are gathered in Fig. 7. The general shape of the computed dispersion relation compares very well with experiments (Fig. 3) though the critical minimum wavelength and the asymptotic velocity are smaller than the experimental values by a factor of 2. Moreover, the computed dynamics of two excyclons in a 5 cm long strip showed excyclon rearrangements where the refractory zone of one excyclon leads to the separation of part of the tail of the following excyclon (Fig. 8), as often seen in experiments.

5. Conclusion

Our experimental work provides a striking example of the enhanced possibilities offered by the use of open spatial reactors in the study of reaction-diffusion structures. In this respect, the Ring Reactor proves to be particularly suited for the study of the dynamics of wavetrains with periodic boundary conditions. Controlled numbers of waves can be introduced in the ring. Well defined boundary feed conditions can be finely tuned and maintained for any length of time, which is crucial to test the stability of the observed wave patterns. We have thus shown that the waves can either organize in the highest possible circular symmetry compatible with the symmetry of the feed or develop topological defects. Different types of defects are readily observed: defects locked in relative position or "fluid" defects permanently changing relative positions. The dynamics of defects is of fundamental importance in the theoretical studies on the onset of turbulence. Thus, the dynamics of excyclons in pinwheel structure could become an experimental model for the

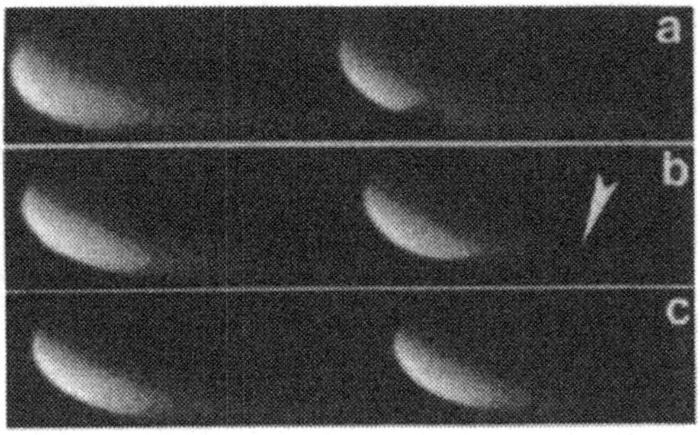

Figure 8. Computed interaction of two excyclons on a 5 cm periodic annular reactor. Shaded profile of v at three different times: a) $t = t_0$; b) $r = t_0 + 132$ s, arrow points to the separation of a piece of the wave; c) $t = t_0 + 490$ s. Same parameter values as in Fig. 6.

transition to turbulence in a restrained geometry system.

It is also noteworthy that the observed defects to the maximum symmetry unfold in the radial direction, that is, in the direction of the chemical gradients. In this respect, it should be interesting to determine the relation that may exist between the excyclon dynamics in the Ring Reactor and the gradient induced spatio-temporal structures observed in the Continuous Couette Flow Reactor described in these Proceedings [20].

Acknowledgement

This work has been supported by the BP Venture Research Unit

References

[1] Field, R.J. & Burger, M. (eds.) (1985). *Oscillating and travelling waves in chemical systems.* Wiley.
[2] Markus, M., Miller, S.C. & Nicolis, G. (eds.) (1987). *From chemical to biological organization.* Springer.
[3] Vidal, C. & Pacault, A. (1982). In *Evolution of order and chaos,* Haken, H. (ed.). Springer.
[4] Zaikin, A.N. & Zhabotinskii, A.M. (1970). *Nature* **225**, 535.
[5] Zhabotinskii, A.M. & Zaikin, A.N. (1973). *J. Theor. Biol.* **40**, 45.
[6] Hanusse, P., Vidal, C. & Pagola, A. In ref. [2], p. 99.
[7] Smoes, M.L. (1980). In *Dynamics of synergetic systems,* Haken, H. (ed.), p. 80. Springer.
[8] Pagola, A., Ross, J. & Vidal, C. (1988). *J. Phys. Chem.* **92**, 163.
[9] Agladze, K.I. & Krinski, V.I. (1982). *Nature* **296**, 424.
[10] Muller, S.C., Plesser, T. & Hess, B. (1987). *Physica* **D24**, 71.
[11] Foerster, P., Muller, S. & Hess, B. (1988). *Science* **241**, 685.
[12] Nagy-Ungvarai, Z., Muller, S.C., Tyson, J.J. & Hess, B. (1989). *J. Phys. Chem.* **93**, 2760.
[13] De Kepper, P., Epstein, I.R., Kustin, K. & Orban, M. (1982). *J. Phys. Chem.* **86**, 170.

[14] Nagy-Ungvarai, Z., Muller, S.C. & Hess, B. (1989). *Chem. Phys. Lett.* **156**, 433.
[15] Sevcikova, A. & Marek, M. (1983). *Physica* **D9**, 140.
[16] Avnir, D. & Kagan, M. (1984). *Nature* **307**, 717.
[17] Nagypal, I., Bazsa, G. & Epstein, I.R. (1986). *J.A.C.S.* **108**, 3635.
[18] Micheau, J.C., Gimenez, M., Borckmans, P. & Dewell, G. (1983). *Nature* **305**, 43.
[19] Nicolis, G. & Prigogine, I. (1977). *Self-organization in nonequilibrium chemical systems.* Wiley.
[20] Boissonade, J. (1988). In *Dynamic and stochastic processes. Theory and applications.* Lima, R., Streit, L. & Videla Mendes, R. (eds). Lisbonne Workshop. Lecture Notes in Physics, Springer.
[21] Ouyang, Q., Boissonade, J., Roux, J.C. & De Kepper, P. (1989). *Phys. Lett.* **134**, 282.
[22] Tam, W.Y., Vastano, J.A., Swinney, H.L. & Horsthemke, W. (1988). *Phys. Rev. Lett.* **61**, 2163.
[23] Boissonade, J., Ouyang, Q., Arneodo, A., Elezgaray, J., Roux, J.C. & De Kepper, P. In these Proceedings.
[24] Tam, W.Y., Horsthemke, W., Noszticzius, Z. & Swinney, H.L. (1988). *J. Chem. Phys.* **88**, 3395.
[25] Noszticzius, Z., Horsthemke, W., McCormick, W.D., Swinney, H.L. & Tam, W.Y. (1987). *Nature* **329**, 619.
[26] Miller, R.N. & Rinzel, J. (1981). *Biophys. J.* **34**, 227.
[27] Rinzel, J. & Keller, B. (1973). *Biophys. J.* **13**, 1313.
[28] Karfunkel, H.R. & Kahlert, C. (1977). *J. Math. Biol.* **4**, 183.
[29] Tyson, J.J. & Fife, P.C. (1980). *J. Chem. Phys.* **73** 2224.
[30] Keener, J.P. & Tyson, J.J. (1986). *Physica* **D21**, 307.
[31] Dockery, J.D., Keener, J.P. & Tyson, J.J. (1987). *Physica* **D30**, 177.
[32] Jahnke, W., Skaggs, W.E. & Winfree, A.T. (1989). *J. Chem. Phys.* **93**, 740.
[33] Arshavskii, Y.I., Berkinblit, M.B. & Dunin-Berkovskii, V.L. (1965). *Biophysics* **10**, 1160.
[34] Ramon, F., Vergara, J. & Moore, J.W. (1973). *Biophys. J.* **13**, 131 (Abstr.).
[35] Donati, F. & Kunov, H. (1979). *IEEE Trans. Biomed. Eng. BME* **23**, 23.
[36] George, S.A. (1977). *Biol. Cybernetics* **26**, 209.
[37] Field, R.J. In ref. [1], Ch. 2, 55.
[38] Tyson, J.J. In ref. [1], Ch. 3, 93.
[39] Nandapurkar, P.J. & Winfree, A.T. (1987). *Physica* **D29**, 69; Lugosi, E. & Winfree, A.T. (1988). *J. Comp. Chem.* **9**, 689.

39. INTERFACIAL INSTABILITIES, WAVES AND SOLITONS EXCITED BY THE MARANGONI EFFECT

M.G. Velarde and X.-L. Chu

U.N.E.D., Facultad de Ciencias
Apartado 60 141, Madrid 28 080, Spain

For **deep** liquid layers we provide an extension of the Stokes- Boussinesq theory to describe (space and time modulated limit cycle) nonlinear capillary-gravity waves excited and sustained by the Marangoni effect, i.e. by the variation of surface tension with temperature or concentration of surface active material. The latter may appear as the result of chemical activity in the liquid or other processes including adsorption, etc. at the surface, coupled to the hydrodynamics in the bulk of the layer. On the other hand, in parallel with the preceding approach, for **shallow** liquid layers we provide an extension of the (nonlinear and dispersive) Korteweg-de Vries theory to describe the soliton excitation triggered by the same Marangoni effect and dissipation.

1. Introduction

Generally, interfacial waves or interfacial oscillations are damped by viscosity [1-3]. However, if a nonequilibrium distribution of surfactant, temperature, chemical potential or an electric field,... is imposed in the liquid, with mass and/or energy transfer across the surface, **the Marangoni effect, i.e. stresses induced by the variation of the surface tension from point to point along the surface may transform the "thermal" or "chemical",... energy into convective motion, overtaking the viscous dissipation and thus sustaining the wave motions.**

Pearson [4], and Sternling & Scriven [5] were the first to provide threshold conditions for sustained spontaneous convection at the interface of two fluids. They clearly proved that the Marangoni stresses due to surfactant concentration or/and temperature gradients along the interface generate convective motions that are transmitted to the bulk in the form of steady cellular convection (Bénard cells) or oscillatory convection [6]. The latter case may correspond to waves that appear as either longitudinal or transverse waves or a combination of both [7-11].

Although Sternling & Scriven were aware of transverse motions in experiments, and referred to them as "localized stirring with rippling and twitching of the interface" they, however, treated the interface as flat in order to simplify the mathematical problem. This excluded the transverse waves in their analysis. Another approximation in S&S's model was the elimination of surface adsorption although they considered a surface-active solute. However, solute adsorption and eventual accumulation at the surface takes place in liquids with surface-active solutes and drastically affects stability.

The Pearson and Sternling & Scriven theories have had subsequent developments by several authors. Sanfeld & Linde and their collaborators [12-25] studied liquid-liquid interface models taking due account of deformable interfaces and solute adsorption at the interface. However, they only discussed the long wavelength longitudinal waves, just one of the various possible modes. It was also shown in their results that the deformability of the interface has negligible influence on the stabiity of these waves. An extension of these earlier theories for transverse and longitudinal waves as well as an extension of the KdV theory to include dissipation and the Marangoni effect has recently been provided by the present authors [24-36]. Thus this chapter is a succinct account of our findings with recipes for experimental tests, particularly, with liquids having a minimum in their surface tension variation with temperature or surfactant concentration [37-40].

2. The case of a liquid open to air and the role of surfactants or other thermal constraints

2.1. *Linear disturbance equations*

Let us, for simplicity, consider a liquid layer of infinite depth. This is Stokes' approach [1]. The motionless undisturbed surface is located at z=0. The linearized equations that disturbances upon the quiescent state obey are:

$$\nabla * v = 0 \tag{1}$$

$$\rho \frac{\partial v}{\partial \tau} = -\nabla p + \eta \nabla^2 v \tag{2}$$

$$\frac{\partial c}{\partial t} - \beta^c w = D\nabla^2 c \tag{3}$$

where $v = (u,w)$, p and c are velocity pressure and concentration of the surfactant, respectively. ρ is the density of the liquid, $\eta = \rho\nu$ the dynamic viscosity and ν the kinematic viscosity, D is the mass diffusivity and $\beta^c = (\delta c/\partial z)_0$ the gradient of surfactant at the quiescent state. Note that our analysis carries over almost *verbatim* to the heat transfer problem and we shall mark the difference only when strictly necessary. The disturbances obey the following (linearized) boundary conditions

$$\frac{\partial \zeta}{\partial t} = w \tag{4}$$

$$-T_0 \nabla_\Sigma^2 \zeta + g\rho\zeta - p - 2\eta \frac{\partial w}{\delta z} = 0 \tag{5}$$

$$\left(\frac{\partial T}{\partial \Gamma}\right)_0 \nabla_\Sigma \gamma - \eta \left(\nabla_\Sigma w + \frac{\partial u}{\partial z}\right) = 0 \tag{6}$$

$$\frac{\partial \gamma}{\partial t} + \Gamma_0 \nabla_\Sigma * u_\Sigma - D_\Sigma \nabla_\Sigma^2 \gamma + D\frac{\partial c}{\partial z} = 0 \tag{7}$$

$$\gamma = k^1 (c - \beta^c \zeta)_\Sigma \tag{8}$$

where ζ is the surface deviation from the z=0 level and Γ is the excess surfactant concentration at the surface. The subscript "0" indicates a value in a reference state, γ is the disturbance upon Γ_0, g is the gravity acceleration and T is the surface tension. Subscript "Σ" accounts for either a value taken on the surface or a derivative along the surface.

For universality in presentation, we introduce new units to rescale the quantities in the equations. The capillary length:

$$l = \sqrt{\frac{T_0}{g\rho}} \tag{9}$$

is chosen as our length scale; $\nu/l, l^2/\nu, \nu^2\rho/l^2, \beta^c l$ and Γ_0 are used as units for velocity, time, pressure, surfactant concentration and excess surface concentration, respectively. Thus, eqs. (1)-(8) become:

$$\nabla * \mathbf{v} = 0 \tag{10}$$

$$\frac{\partial \mathbf{v}}{\partial t} = - \nabla p + \nabla^2 \mathbf{v} \tag{11}$$

$$\frac{\partial c}{\partial t} - w = S^{-1}\nabla^2 c \tag{12}$$

with boundary conditions at z=ζ

$$\frac{\partial \zeta}{\partial t} = w \tag{13}$$

$$- \frac{1}{SC}\nabla_\Sigma^2\zeta + \frac{B_0}{SC}\zeta - p - 2\frac{\partial w}{\partial z} = 0 \tag{14}$$

$$\frac{HE}{SH_z}\nabla_\Sigma \nu + \left(\nabla_\Sigma \nu + \frac{\partial u}{\partial z}\right) = 0 \tag{15}$$

$$HS\left(\frac{\partial \gamma}{\partial t} + \nabla_\Sigma * u_\Sigma - S_\Sigma^{-1}\nabla_\Sigma^{-2}\gamma\right) + \frac{\partial c}{\partial z} = 0 \tag{16}$$

$$\gamma = \frac{H_z}{H}(c - \zeta)_\Sigma \tag{17}$$

Note that although we have used the same notation for v,p,c,γ and ζ in these equations as in the dimensional case, all the units have changed and they are dimensionless now. The following dimensionless parameters have also been used: $S = \nu/D$, Schmidt number; $S_\Sigma = \nu/D_\Sigma$, surface Schmidt number; $C = \eta D/lT_0$, capillary number; $B_0 = \rho g l^2/T_0$, Bond number (for simplicity is taken equal to unity); $E = -(\partial c/\delta\Gamma)_0 \, k^1\beta l^2/(\eta D)$, surfactant (elasticity) Marangoni number; $H = \Gamma/(\beta^c l^2)$, surface excess surfactant number; and $H_z = k^1/l$, Langmuir adsorption number.

2.2. *Transverse waves: threshold values*

Assuming for simplicity that surface adsorption and accumulation have negligible effects for transverse surface waves eqs. (15)-(17) reduce to

$$\frac{Ea^2}{S}(c - \zeta) + \left(\frac{\partial^2}{\partial z^2} + a^2\right) w = 0 \tag{18}$$

$$\frac{\partial c}{\partial z} = 0 \tag{19}$$

Equation (11) gives the dynamic evolution equation of the liquid layer. It is valid in the volume as well as at the surface - a part of the liquid. On the other hand, we have a kinematic relation (13) on the surface. Then a straightforward stability analysis yields [24,27] that when

$$4a^2 + \frac{Ea^3}{s\omega\sqrt{2S\omega}} > 0 \quad \left(\text{or } -E < \frac{4S\omega\sqrt{2S\omega}}{a}\right) \tag{20}$$

the oscillatory convection will be damped out. In contrast, when

$$4a^2 + \frac{Ea^3}{s\omega\sqrt{2S\omega}} < 0 \quad \left(\text{or } -E > \frac{4S\omega\sqrt{2S\omega}}{a}\right) \tag{21}$$

the oscillation grows exponentially. At the neutral state

$$-E = \frac{4S\omega\sqrt{2S\omega}}{a} \tag{22}$$

The kinetic energy dissipated by viscosity and that produced by surface tension work just compensate each other, giving a sustained oscillator. Thus the oscillation frequency is given by the dispersion relation

$$\frac{B_0 + a^2}{SC} a - \omega^2 = 0 \tag{23}$$

By taking dE(a,w(a))/da = 0, the necessary condition for minimum yields the neutral curve, i.e. the threshold values for sustained transverse waves:

$$E_c^T = 4\sqrt{10}\left(\frac{6S}{5\sqrt{5}\,C}\right)^{3/4} \approx -7.931\left(\frac{S}{C}\right)^{3/4} \tag{24}$$

$$\omega_c^T = \sqrt{\frac{6}{5^{3/2}}}\frac{1}{\sqrt{SC}} = \frac{0.7326}{\sqrt{SC}} \tag{25}$$

$$a_c^T = \frac{1}{\sqrt{5}} = 0.4472 \tag{26}$$

Table 1 accounts for numerical estimates at two different values of the gravitational acceleration, g and 10^{-4} g [24,27].

Table 1. Interfacial oscillations: Quantitative predictions for the case of an open surface of a liquid layer heated from the air side. β is negative for liquids with no minimum (or maximum) in their surface tension. Note the relevance to spacebound low/microgravity experiments.

	Water		Mercury		Tin		$NaNO_3$	
β (K/cm)	$1.9 \cdot 10^3$	5.85	441	1.39	55	1.8	$5.2 \cdot 10^3$	16.54
Period (sec)	0.14	143	0.12	119.50	0.17	146	0.14	137.8
Penetration depth (cm)	$1.5 \cdot 10^{-2}$	0.48	$0,5 \cdot 10^{-2}$	0.15	$6 \cdot 10^{-3}$	0.18	$1.8 \cdot 10^{-2}$	0.56
Gravitational acceleration	g	10^{-4}g	g	10^{-4}g	g	10^{-4}g	g	10^{-4}g

2.3. *Longitudinal waves: threshold values*

To avoid some tedious and rather irrelevant complications we first simplify the problem. We suppose that the deformability of the surface has negligible influence on the longitudinal wave motion in agreement with Lucassen's finding (8). Thus we set the capillary number, C, to zero. Also, it is known that surfactant accumulation on the surface affects mainly high frequency oscillatory convection and the frequency of longitudinal waves normally is small, so that the surfactant accumulation number H can be neglected. With these assumptions eqs. (13)-(14) reduce to

$$w = 0 \tag{27}$$

$$\frac{Ea^2}{S} c + \frac{\partial^2 w}{\partial z^2} = 0 \tag{28}$$

$$SH_z \frac{\partial c}{\partial t} = - \frac{\partial c}{\partial z} \tag{29}$$

Then a straightforward stability analysis [24-26,28] yields that when the energy dissipation and Marangoni work compensate each other

$$1 + \frac{Ea^2}{S^{3/2}\omega^2} = 0 \tag{30}$$

Thus one expects that sustained longitudinal oscillations occur at the surface. Eq. (54) was the pioneering result obtained by Lucassen [9]. The stability analysis for the oscillator gives

$$-E = \begin{cases} > \dfrac{S^{3/2}\omega^2}{a^2} & \text{explosion} \\[2ex] < \dfrac{S^{3/2}\omega^2}{a^2} & \text{damped motion} \end{cases}$$

In the neutral case we also have

$$SH_z\,\omega^2 + \frac{Ea^3}{S\omega} = 0 \tag{31}$$

Thus eqs. (20) and (31) characterize the longitudinal oscillation of the air-liquid interface.

Thus one expects that sustained longitudinal oscillations occur at the surface with dispersion relation

$$a_c = N_z\,\omega\,S^{1/2} \tag{32}$$

which generalizes earlier findings [7].

Comparison of (24) and (31) shows that both thresholds depend on the Schmidt (Prandtl) number. However, transverse waves are directly related to the interfacial deformation (measured by C) whereas longitudinal waves rather depend on Langmuir's adsorption (measured by H_z). Cross-over from one to another mode of instability is given by the condition: when S is greater (smaller) than $C^3/(7.931\,H_z^2)^4$ we have longitudinal (transverse) waves first.

3. The liquid-liquid interface

3.1. *Capillary-gravity waves: threshold values*

Now let us consider the deformable interface separating two liquids of different densities, viscosities and mass diffusivities and assume again the problem to be two-dimensional, i.e. with horizontal and vertical coordinates x and z, respectively. If the liquid below is labelled "one" the evolution of infinitesimal disturbances upon the motionless state on either side of the liquid-liquid interface is given by a straightforward extension of the equations described in Section 2. The analysis leads to the following dispersion relation

$$(1 + N_\rho)\omega^2 = \frac{B + a^2}{CS}a \tag{33}$$

which is the natural extension of (38). Using the expression for $N_\rho = \rho_2/\rho_1$, B (also denoted Bo), C, S, a=lk and $\omega = \Omega l^2 \nu_1$ (33) becomes

$$(\rho_1 + \rho_2)\Omega^2 = (\rho_1 + \rho_2)gk + \sigma_0 k^3 \tag{34}$$

which is the standard dispersion relation for gravity-capillary waves at a liquid-liquid interface. For the air-liquid interface $\rho_2=0$, thus $\rho_1 + \rho_2 \approx \rho_1 - \rho_2 \approx \rho_1 \approx \rho$. Note that Ω and k have, respectively, the units of s^{-1} and cm^{-1}.

The threshold for instability corresponds to

$$Ea^2(1 - N_D^{1/2}/N_\eta) > 0 \tag{35}$$

i.e.

$$\mathrm{sign}(E) = \mathrm{sign}(1 - N_D^{1/2}/N_\eta) \tag{36}$$

with $N_D = D_2/D_1$ and $N_\eta = \eta_2/\eta_1$.

Generally, the sign of the Marangoni number is given by the sign of the volume gradient of the surfactant β_i, say. Then if β_i is positive, i.e. the mass flux is from liquid "one" to liquid "two" (β_2 has always the same sign as β_1) eq. (36) demands

$$\eta_2/\eta_1 - \sqrt{D_2/D_1} > 0 \quad (\beta_1, \beta_2 > 0) \tag{37}$$

If, however, the gradient is negative we rather have

$$\eta_2/\eta_1 - \sqrt{D_2/D_1} < 0 \quad (\beta_1, \beta_2 < 0) \tag{38}$$

Then denoting by "f" and "t" the transport direction from phase f to phase t we see that (37) and (38) are just the same condition

$$D_f/D_t > (\eta_f/\eta_t)^2 \tag{39}$$

irrespective of the sign of the gradients. Condition (39) is the necessary condition to have sustained oscillations at the liquid- liquid interface. This is achieved when the Marangoni number reaches the threshold value

$$E_c = - \frac{4SN_\eta (1 + N_z^{-1} N_D^{-1/2}) B^{1/2}}{C(1 + N_\rho)(N_D^{1/2} - N_\eta)} \tag{40}$$

for a frequency

$$\omega_c = \frac{\sqrt{2}\, B^{3/4}}{\sqrt{SC(1 + N_\rho)}} \tag{41}$$

and a wavenumber

$$a_c = \sqrt{B} \tag{42}$$

Note that when B=1, i.e. when we set the capillary length to one, eqs. (41) and (42) reduce to

$$\omega_c = \frac{\sqrt{2}}{\sqrt{SC(1 + N_\rho)}} \quad \text{and} \quad a_c = 1$$

respectively. These are the dimensionless values of the parameters that correspond to the onset of overstability at a deformable liquid-liquid interface. Again, as in Section 2.2, we also have here clear-cut predictions for an experimental test.

3.2 *Longitudinal waves: threshold values*

We have indeed the natural extension of the problem posed in Sections 2.2 and 3.1 and thus we shall limit ourselves to a discussion of the results found. To have oscillatory behaviour we must satisfy [28,30]

$$(D_f/D_t) > (\nu_f/\nu_t) \tag{43}$$

where "f" and "t" stand for "from" and "to", a way of indicating how the surfactant is being transported from and to the volume. Condition (43) is a condition for overstability in Marangoni convection obtained some time ago by Sanfeld and collaborators [7]. However, contrary to their findings and due here to the role of surfactant adsorption at the interface this condition is a necessary, albeit not sufficient, condition for overstability. Further constraints must be satisfied and they are specified below.

Specifically the threshold value found is

$$E_c = -\left[\frac{(N_D^{1/2} + N_\mu)(N_\rho^{1/2} N_\mu^{1/2} - 1)}{S\Gamma\Pi_1\Pi_2}\right] \left[\frac{(1 + N^{-1} N_D^{-1/2}) S^2}{\Pi_1}(1 + N_\rho^{1/2} N_\mu^{1/2})\right] \tag{44}$$

with the dispersion relation

$$\omega_c = a_c \frac{(N_D^{1/2} + N_\mu)\ (N_\rho^{1/2} N_\mu^{-1/2} - 1)}{S^{1/2}\Gamma\Pi_1\Pi_2} \tag{45}$$

where $\Pi_1 = (N_\rho N_D / N_\eta)^{1/2} - 1$ and $\Pi_2 = N_D^{1/2} - 1$. The value E_c is the minimal value of the elasticity Marangoni number needed to sustain longitudinal interfacial convective oscillations of frequency ω_c. Using the fact that both ω and a must be positive numbers, we get from (43) that the following relationship must be satisfied [28,30]

$$\nu_f/\nu_t < 1 \tag{46}$$

in order to have oscillations. Our results generalize Lucassen's earlier finding as he only considered damped motions. Here we see that with strong enough dissipation, i.e. for Marangoni numbers larger than E_c these longitudinal oscillations can be sustained along the interface even if it is not deformed. On the other hand our results delineate a more restricted domain of (surfactant- induced) interfacial oscillations than the domain reported by earlier authors [7].

4. **Nonlinear transverse waves**

4.1. *Nonlinear disturbance equations*

If we now consider the nonlinear extension of the problem posed in Section 2.1 thus limiting ourselves to transverse oscillations, we have in dimensionless form

$$\nabla * \underline{v} = 0 \tag{47}$$

$$\frac{\partial w}{\partial t} + u\frac{\partial w}{\partial x} + w\frac{\partial w}{\partial z} = - \frac{\partial p}{\partial z} + \nabla^2 w \tag{48a}$$

$$\frac{\partial u}{\partial t} + u\frac{\partial u}{\partial x} + w\frac{\partial u}{\partial z} = - \frac{\partial p}{\partial x} + \nabla^2 u \tag{48b}$$

and

$$\frac{\partial \theta}{\partial t} + u\frac{\partial \theta}{\partial x} + w\frac{\partial \theta}{\partial z} = w + P^{-1}\nabla^2\theta \tag{49}$$

where we have explicitly indicated in a two-dimensional problem the full nonlinear disturbance system. Besides, in order to relate our analysis to the standard Bénard problem we take here β as temperature gradient rather than surfactant gradient. β is positive when heating the layer from below. Thus eq. (49) is Fourier's heat equation. P is the Prandtl number, $P=\nu/\kappa$, with κ the thermometric conductivity of the liquid. Its counterpart in mass transfer is the Schmidt number, S.

The b.c (4)-(8) at the open surface need to be extended to the nonlinear case. Thus we now have

$$\frac{\partial \xi}{\partial t} = w - u\frac{\partial \xi}{\partial x} \tag{50}$$

$$p - \frac{B_0}{CP}\xi = + \frac{1}{N^3}\left[\frac{1}{CP} - \frac{M}{P}(\theta - \xi) \right] \frac{\partial^2 \xi}{\partial x^2}$$

$$= \frac{2}{N^2}\left[\frac{\partial w}{\partial z} - \left(\frac{\partial u}{\partial z} + \frac{\partial w}{\partial x} \right)\frac{\partial \xi}{\partial x} + \frac{\partial u}{\partial x}\left(\frac{\partial \xi}{\partial x} \right)^2 \right] \tag{51}$$

$$- \frac{M}{P}\left[\frac{\partial(\theta - \xi)}{\partial x} + \frac{\partial \theta}{\partial z}\frac{\partial \xi}{\partial x} \right]$$

$$= \frac{1}{N}\left\{ \left(\frac{\partial u}{\partial z} + \frac{\partial w}{\partial x} \right)\left[1 - \left(\frac{\partial \xi}{\partial x} \right)^2 \right] + 4\frac{\partial w}{\partial z}\frac{\partial \xi}{\partial x} \right\} \tag{52}$$

and

$$\frac{\partial \theta}{dz} = 0 \tag{53}$$

We see nonlinear contributions like the second term in the r.h.s. of the kinematic b.c. (50). Equation (53) prescribes the heat flux at the open surface. $N = (1 + |\partial\xi/\partial x|^2)^{1/2}$. Again the air is assumed to be passive and weightless with respect to the liquid.

4.2. *Limit cycle oscillations*

The simplest approach to the above posed nonlinear problem is the single-mode analysis which is expected to be a useful description in a small enough neighbourhood of the onset of overstability. Moreover, the more we move into low gravity the larger the capillary length becomes, thus providing greater relevance to the single-mode approximation. On the other hand, transverse interfacial disturbances are expected to penetrate little in the liquid; the penetration depth depends indeed on the wavelength and frequency excited and on the viscosity of the liquid. The latter assumption gives relevance to the "potential" flow approximation to the time-dependent convection or in other terms to the limitation of the study to the high-frequency motions only. Thus for an arbitrary disturbance $f(x,z,t)$ we set $f(x,z,t) \approx f(z,t)\exp(iax)$ and $\omega_0^2 = (B_0 + a^2)a/CP$. The latter is Laplace's law (potential flow). Using them, eq. (50) becomes

$$\frac{\partial\xi}{\partial t} = w + \xi\frac{\partial w}{\partial z} \tag{54}$$

On the other hand eq. (47) at $z=\xi$ is

$$\frac{\partial w}{\partial t} = -\frac{B_0 + a^2/N^3}{PC}a\xi + \frac{M}{PN^3}(\theta - \xi)a^3\xi - 2a\frac{1 + a^2\xi^2}{N^2}\frac{\partial w}{\partial z} - a^2\left(1 + \frac{2a\xi}{N^2}\right)w + \left(1 - \frac{2a\xi}{N^2}\right)\frac{\partial^2 w}{\partial z^2} \tag{55}$$

The zeroth-order (linear) disturbances are ξ_0,

$$w_0 = \frac{\partial\xi_0}{\partial t}e^{az} \tag{56}$$

and

$$\theta_0 = \xi_0 e^{az} + \frac{a}{\sqrt{2P\omega_0}}\left(\frac{1}{\omega_0}\frac{\partial\zeta_0}{\partial t} - \xi_0\right)\exp\left(\sqrt{2P\omega_0}\, z\right) \tag{57}$$

where ω_0 denotes the harmonic frequency (41).

Consideration of the nonlinear terms in eqs. (54) and (55) up to cubic terms leads to

$$\frac{d^2\xi}{dt^2} + \delta\frac{d\xi}{dt} + [\omega_0^2 - \omega_0(\delta-4a^2)]\xi$$

$$= -\omega_0^2 a\xi^2 + a\left(\frac{d\xi}{dt}\right)^2$$

$$+ (\delta - 4a^2)\left(\frac{d\xi}{dt} - \omega_0\xi\right) 2a\xi - 8a^3\xi\frac{d\xi}{dt}$$

$$-a^2\xi\left(\frac{d\xi}{dt}\right)^2 + \frac{11}{2}a^2\xi^2(\delta-4a^2)\left(\frac{d\xi}{dt} - \omega_0\xi\right) + 16a^4\xi^2\frac{d\xi}{dt} + \frac{3a^5}{2CP}\xi^3 \qquad (57)$$

where

$$\delta = \frac{Ma^3}{\sqrt{2}\,(P\omega_0)^{3/2}} + 4a^2 \qquad (58)$$

At $\delta=0$ we recover the results of Section 2.2. Positive (respectively, negative) values of δ account for subcritical (respectively, supercritical) motions. We have checked that indeed eq. (57) possesses a limit cycle solution. This has been done both with the computer and using the time-derivative expansion procedure [31].

Note that for an effective gravitational acceleration of, say, 10^{-4} g, with g the standard value on earth, the predicted period of oscillation at onset is of the order of one to two minutes for most liquids. Threshold temperature gradients at overstability are of the order of a few K/cm for mercury and other liquids, including water. Thus, as already said in Section 2, aboard a spacecraft, in a low/microgravity environment the crucial test of our predictions can be obtained by making an experiment with a water-alcohol solution or some other liquids with a minimum of surface tension versus temperature [37-40]. The suggested experiment is Bénard convection. Before the minimum is reached one expects steady polygonal cells (Bénard cells), whereas past the minimum, i.e. in the region where the surface tension of the liquid increases with increasing temperature, oscillations are predicted. For soliton excitation in Section 5, we come back to the same experiment albeit with different initial conditions.

4.4. *Space modulated oscillations: extension of Stokes' theory*

When due consideration is given to the full space dependence, the **transverse** Kelvin-Laplace wave motion is described by the following set of nonlinear **partial** differential equations

$$\frac{\partial\zeta}{\partial t} = w - \frac{1}{a_c} w \frac{\partial^2\zeta}{\partial x^2} \qquad (59)$$

and

$$\frac{\partial w}{\partial t} = -\frac{a_c}{SC}\left\{Bo-\frac{\partial^2}{\partial x^2}\right\}\zeta - \frac{3a_c}{2SC}\left(\frac{\partial^2\zeta}{\partial x^2}\right)\left(\frac{\partial\zeta}{\partial z}\right)^2 - 2\,a_c^2 w + 2\,\frac{\partial^2 w}{\partial x^2}$$

$$+ 4\,a_c^2 w\left(\frac{\partial\zeta}{\partial x}\right)^2 + 4\,a_c(1-2a_c\zeta)\,\frac{\partial}{\partial x}\left\{w\,\frac{\partial\xi}{\partial x}\right\} + \frac{E_0 a_0}{S\,\omega_0\sqrt{2S\omega_0}}$$

$$\left\{-a_c\frac{\partial^2\zeta}{\partial x^2} + \frac{3}{2}\left[\frac{\partial}{\partial x}\left(\frac{\partial\zeta}{\partial x}\right)^2\right]\frac{\partial}{\partial x} + \left[1-2a_c\zeta+\frac{3}{2}\left(\frac{\partial\xi}{\partial x}\right)^2\right]\frac{\partial^2}{\partial x^2}\right\}\left(\frac{\partial\zeta}{\partial t}-\omega_c\zeta\right) \tag{60}$$

which constitute our extension of Stokes-Boussinesq water wave theory to a dissipative liquid layer subjected to Marangoni stresses. Work is at present underway in order to obtain solutions of these nonlinear partial differential equations.

5. Solitons excited by the Marangoni effect

5.1. *Introduction*

In the preceding sections we have discussed the onset and eventual nonlinear sustainment of some interfacial oscillations. These oscillations were either transverse or longitudinal waves. The former may be capillary-gravity waves that in a closed container are expected to develop as standing transverse periodic motions at the open surface of a liquid or at the interface between two liquids when the Marangoni effect is operating. In all the cases considered with the small viscous penetration depth or the high frequency limit the approximations used permitted assuming the "infinite" depth of the layer. Now we turn to purely travelling motions in the form of solitary waves with restriction to **shallow** layers. Work is now in progress toward a derivation of results for arbitrary depth (nonlinear Schroedinger theory).

5.2. *Dissipative KdV equation and threshold values*

We consider a shallow horizontal liquid layer of thickness "h" initially at rest and, as in the preceding sections, subjected to a transverse thermal gradient. Disturbances upon the quiescent state obey the continuity and Navier-Stokes equations to which we now add, to a first approximation, either Fourier's heat equation or Fick's mass diffusion equation. These equations are supplemented with the corresponding nonlinear boundary conditions at the bottom and at the open surface. Once more, for simplicity we shall restrict consideration to a two-dimensional geometry with x and z denoting the horizontal and vertical coordinates, respectively. β is the thermal gradient that again corresponds either to heat or mass diffusion (always positive when the heating is from the liquid side). Again as before θ denotes either temperature disturbance or surfactant concentration disturbance, so that D is heat or mass diffusivity. η and ν still account for dynamic and kinematic viscosity, respectively; $\eta=\rho\nu$, with ρ the liquid density. All other quantities have already the meaning assigned in the preceding sections. ∇^2 represents the laplacian that with subscript Σ restricts its action to the open surface. The "shallow layer approximation" demands that h<<1 (depth much smaller than wavelength of motions) and thus we can disregard buoyancy effects in the Navier-Stokes equations. Here "1" is the "wavelength" or maximum horizontal extent of the interfacial deformation ζ. For self-consistency in this section we recall the equations to be used. They are (46)-(48)

$$\nabla * v = 0 \tag{61}$$

$$\frac{\partial}{\partial t} v + v * \nabla v = -\frac{1}{\rho}\nabla p + \nu \nabla^2 v \tag{62}$$

$$\frac{\partial}{\partial t}\theta = \beta w + k\nabla^2\theta \tag{63}$$

with b.c. at the rigid bottom z=0

$$w = u = 0 \tag{64}$$

and

$$\theta = 0 \tag{65}$$

and b.c. at the surface $z=h+\zeta$

$$\frac{\partial}{\partial t}\zeta = 2 - u\frac{\partial}{\partial x}\zeta \tag{66}$$

$$p = \rho g + 2\mu\frac{\partial}{\partial z}w - \sigma_0 \nabla_\Sigma^2 \zeta \tag{67}$$

$$\left[\frac{\partial}{\partial T}\sigma\right] \nabla_\Sigma(\theta - \beta\zeta) - \eta\left[\frac{\partial}{\partial z}u + \frac{\partial}{\partial x}w\right] = 0 \tag{68}$$

and

$$\frac{\partial}{\partial z}\theta = 0 \tag{69}$$

Note that we considered the heat equation as a linear disturbance upon the nonlinear Navier-Stokes problem. Then a straightforward extension of the derivation presented by Lamb [41, Section 6.1] for the standard KdV equation yields here [35,36]

$$\xi_t + \xi_y + \frac{5}{2}\varepsilon\xi\xi_y - \frac{30 - B_0}{60B_0}\delta^2\xi_{yyy} = \frac{\gamma}{2}\xi_{yy} \tag{70}$$

where $\delta = 2(6)^{1/2}(4 + M)/3$ Re, with $Re = Col/\nu$ and $Co^2 = gh$. As before B_0 is the Bond number. $\varepsilon = a/h$, $\xi = \zeta/a$, with "a" the maximum value attainable by ζ. We see that setting $\gamma = 0$, eq. (70) provides the Korteweg-de Vries equation [41-43]. $\gamma = 0$ demands either $\nu = 0$, i.e. no viscosity, which is an irrelevant case in our analysis, or $M = -4$. The latter result indicates that for such a negative value of the Marangoni number the open surface of the liquid layer heated from the air side is excitable in the form of a KdV soliton. Whether or not the soliton is stable can only be decided by studying the role of the nonlinear part in eq. (63) left out here. However, what we can safely say is that, due to the Marangoni effect, $M = -4$ defines the threshold for soliton excitation in a quiescent shallow liquid layer subjected to a transverse thermal gradient. If, however, M is positive we are in the case of Bénard convection and we know that at M = 80 there is the onset of steady polygonal (mostly hexagonal) cell patterns. Thus here again an experiment with a liquid layer heated from below could provide a clear-cut test of our prediction. Once more it suffices to operate with a liquid having a minimum in the surface tension versus temperature curve. Before the minimum we expect steady patterned convection and past the minimum here the prediction is the solitary travelling wave. Note that as the KdV equation corresponds to a genuine nonlinear excitation the experimenter must strongly excite the liquid with, say, a sudden jump in the temperature gradient or once the soliton is mechanically excited, say, then heat and/or mass transfer with the Marangoni effect is expected to help in sustaining this excitation.

Acknowledgements

This chapter is based upon research sponsored by CICYT (Spain) Grant PB 86-651 and by an EEC Grant. Both authors acknowledge fruitful discussions with Prof. Ph. Drazin, J.K. Koster, H. Linde, A. Sanfeld and R. Sani and Dr. M. Hennenberg. The first author wishes to acknowledge the hospitality of the CNLS at Los Alamos Laboratory and the Center for Low-Gravity Fluid Mechanics, University of Colorado at Boulder, Colorado. The second author wishes to acknowledge the hospitality at the Service de Chimie Physique, Université libre de Bruxelles.

References

[1] Lamb, H. (1932). *Hydrodynamics.* Dover: New York.
[2] Landau, L.D. & Lifshitz, E.M. (1959). *Fluid Mechanics.* Pergamon: Oxford.
[3] Levich, B.G. (1962). *Physicochemical Hydrodynamics.* Prentice-Hall, Inc.: Englewood Cliffs, N.J.
[4] Pearson, J.R. (1958). *J. Fluid Mech.* **4**, 489.
[5] Sternling, C.V. & Scriven, L.E. (1959). *AIChE Journal* **5**, 514. See also Scriven, L.E. & Sternling, C.V. (1964). *J. Fluid Mech.* **19**, 321.
[6] Velarde, M.G. & Normand, C. (1980). *Sci. American* 243, 78.
[7] Legros, J.C., Sanfeld, A. & Velarde, M.G. (1987). In *Fluid Sciences and Materials Science in Space,* Walter, H.U. (ed.), pp. 83-140. Springer-Verlag: New York.
[8] Velarde, M.G. (ed.) (1988). *Physicochemical Hydrodynamics. Interfacial Phenomena.* Plenum Press: New York.
[9] Lucassen, J. (1968). *J. Trans. Faraday Soc.* **64**, 2221, 2231.
[10] Lucassen-Reynders, E.H. & Lucassen, J. (1969). *Adv. Colloid Interface Sci.* **2**, 347.
[11] Van den Tempel, M. & Lucassen-Reynders, E.H. (1983). *Adv. Colloid Interface Sci.* **18**, 281.
[12] Hennenberg, M., Sørensen, T.S. & Sanfeld, A. (1977). *J. Chem. Soc. Trans. Faraday II* **73**, 48.
[13] Hennenberg, M., Bisch, P.M., Vignes-Adler, M. & Sanfeld, A. (1978). In *Dynamics and Instability of fluid Interfaces,* Sørensen, T.S. (ed.), pp. 227-259. Springer-Verlag: Berlin.
[14] Sanfeld, A., Steinchen, A., Hennenberg, M., Bisch, P.M., Van Lamsweerde-Gallez, D. & Dale-Vedove, W. (1978). In *Dynamics and Instability of Fluid Interfaces,* Sørensen, T.S. (ed.), pp. 168-204. Springer-Verlag: Berlin.
[15] Sørensen, T.S., Hansen, F.Y, Nielsen, J. & Hennenberg, M. (1977). *J. Chem. Soc. Trans. Faraday II* **73**, 1589.
[16] Sørensen, T.S., Hennenberg, M. & Hansen, F.Y. (1978). *J. Chem. Soc. Trans. Faraday II* **74**, 1005.

[17] Sørensen, T.S. (1978). In *Dynamics and Instability of Fluids Interfaces,* Sørensen, T.S. (ed.), pp.1-74. Springer-Verlag: Berlin.
[18] Linde, H. & Leschcke, K. (1966). *Chem. Ing. Tech. (GDR)* **39**, 65.
[19] Linde, H. & Kunkel, E. (1969). *Z. Warme-Stoffubertr.* **2**, 60.
[20] Linde, H. & Schwartz, P. (1974). *Chem. Tech. (GDR)* **26**, 455.
[21] Linde. H. (1978). In *Dynamics and Instability of Fluid Interfaces,* Sørensen, T.S. (ed.), pp. 75-119. Springer- Verlag: Berlin.
[22] Reinchenbach, J. & Linde, H. (1981). *J. Colloid Interf. Sci.* **84**, 433.
[23] Linde, H. (1982). In *Convective Transport and Instability Phenomena,* Zierep, J. & Oertel, H. (eds.), pp. 256-296. Bruan-Verlag: Karlsruhe.
[24] Garcia Ybarra, P. & Velarde, M.G. (1987). *Phys. Fluids* **30**, 1649.
[25] Velarde, M.G., Garcia Ybarra, P. & Castillo, J. (1987). *Physiochem. Hydrodyn.* **9**, 387.
[26] Velarde, M.G. & Chu, X.-L. (1989). *Phys. Scripta* **T25**, 231.
[27] Chu, X.-L. & Velarde, M.G. (1988). *Physicochem. Hydrodyn.* **10**, 727.
[28] Chu, X.-L. & Velarde, M.G. (1989). *J. Colloid Interf. Sci.* **131**, 471.
[29] Velarde, M.G. & Chu, X.-L. (1988). *Phys. Lett.* **A131**, 403.
[30] Velarde, M.G. & Chu, X.-L. (1989). *Il Nouvo Cimento* **D 11**, 707.
[31] Chu, X.-L. & Velarde, M.G. (1989). *Il Nuovo Cimento* **D 11**, 1615.
[32] Chu, X.-L. & Velarde, M.G. (1989). *Il Nuovo Cimento* **D 11**, 1631.
[33] Chu, X.-L. & Velarde, M.G. (1989). *Phys. Lett.* **A136**, 126.
[34] Velarde, M.G. & Chu, X.-L. (1989). In *Phase Transitions in Soft Condensed Matter,* Riste, T. & Sherrington, D. (eds.), pp. 139-143. Plenum Press: New York.
[35] Chu, X.-L. & Velarde, M.G. (1989). Submitted to *Europhys. Lett.*
[36] Velarde, M.G. & Chu, X.-L. *Interfacial Instabilities.* World Scientific: London, in preparation.
[37] Vochten, R. & Petré, G. (1973). *J. Colloid Interf. Sci.* **42**, 320.
[38] Motomura, K., Iwanage, S.-I., Hayami, Y., Uryu, S. & Matuura, K. (1981). *J. Colloid Interface Sci.* **80**, 32.
[39] Desré, P.J. & Joud, J.C. (1981). *Acta Astronaut.* **8**, 407.
[40] Legros, J.C., Limbourg-Fontaine, M.C. & Petré, G. (1984). *Acta Astronaut.* **11**, 143.
[41] Lamb, G.L. (1980). *Elements of Soliton Theory.* John Wiley: New York.
[42] Whitham, G.B. (1974). *Linear and Non-linear Waves.* John Wiley: New York.
[43] Drazin, P.G. & Johnson, R.S. (1989). *Solitons: An Introduction.* Cambridge Univ. Press: Cambridge.

40. THE SIMULATION OF CHEMICAL WAVES IN EXCITABLE REACTION-DIFFUSION-CONVECTION SYSTEMS BY FINITE DIFFERENCE AND FINITE ELEMENT METHODS

Th. Plesser*, R.D. Kingdon and K.H. Winters

Theoretical Physics Division
Harwell Laboratory, Didcot, Oxon OX11 ORA, and

*Max-Planck-Institut für Ernährungsphysiologie
Rheinlanddamm 201, D-4600 Dortmund
West Germany

1. Introduction

Waves formed by the propagation of gradients in a supporting medium are an ubiquitous phenomenon in the gaseous, liquid and solid phases of matter. The wave motion and shape are maintained by a supply of energy either internal or external to the system. Familiar examples are water waves and acoustic waves coupled to mechanical energy sources. In recent years a new wave phenomenon has become the focus of scientific interest, namely the propagation of concentration gradients in chemical solutions. Here the medium is active rather than passive, with the internal energy source being provided by the participating chemical reactions. The most thoroughly investigated example of this phenomenon occurs during the oxidative bromination of an organic substrate in the presence of a catalyst. This process, the so-called Belousov-Zhabotinskii (BZ) reaction, was discovered in the fifties by the Russian Belousov who observed in a well-stirred reaction mixture the periodic variation of the concentration of the oxidized form of the cerium redox couple Ce^{3+}/Ce^{4+}. Later, Zhabotinskii investigated the underlying reaction mechanism and published results of the first experiments on moving waves in a thin layer of the mixture [1].

The BZ reaction in well-stirred batch solutions and in continuously-fed stirred tank reactors (CSTR) shows a broad spectrum of dynamic behaviour, such as multiple steady states, limit cycles, and even chaos. In an unstirred thin layer (~ 1 mm) the coupling between the reaction dynamics and diffusion creates a rich variety of moving patterns due to changes in local concentration, e.g. plane waves, solitary circular waves, repeated circular waves (the so-called target patterns), and spirals. In visual observation of such waves one sees moving redox fronts formed by a fast transition from the reduced (red) to the oxidized (blue) state at the wave front, followed by a slow relaxation from the oxidized to the reduced state at the back of the wave if the redox couple ferroin/ferriin is used as the catalyst. Typically, the spatial width of the waves travelling through the solution is of the order of a millimeter. Recent work has revealed other reactions which show spatial and temporal behaviour similar to the BZ reaction. A review of the subject is presented in [2]; the latest summary of experimental results can be found in [3].

The types of patterns observed in chemical solutions which manifest the BZ reaction are not a peculiarity, and similar behaviour is found in other media, such as biological membranes and tissues, and in colonies of social amoeba during their aggregation. The BZ reaction has emerged as a prototype for studying pattern formation since the mixture can easily be prepared from cheap and stable chemicals. The most important of many chemical mechanisms of the BZ reaction was formulated in a mathematical model, the Oregonator, containing three time-dependent variables in five irreversible reactions, one of which forms an autocatalytic step [4]. The Oregonator has been analysed in great detail and it is able to account for many of the peculiar features exhibited by the BZ reaction [4,5]. In analytical and numerical work on spatial pattern formation the mechanism was simplified later by reasonable assumptions to a set of two parabolic differential equations, and it was demonstrated that these equations have wave solutions of the observed kind [6-8]. Subsequently the importance of curvature and dispersion was realised and through the application of singular perturbation methods significant progress was made in describing the geometry and the kinematics of waves, leading to quantitative comparison with experiments [9-12].

These theoretical achievements have their counterparts on the experimental side, in which the development of techniques for the precise measurement of concentration in one [13] and two dimensions [14,15] has advanced significantly our detailed knowledge of the kinematics and the shape of the waves [16-18] as well as of transient effects during their formation [15]. Recent studies of fluid flow induced by chemical waves paved the way for in depth analysis of the complex coupling between convection, diffusion, and reaction [19,20]. With the information available from these recent theoretical and experimental studies, it is now timely to consider a numerical approach aimed at solving the underlying set of differential equations, using appropriate models and boundary conditions that simulate the experimental conditions as closely as possible.

The paper is arranged as follows. The next section gives a short overview of the chemistry of the BZ reaction and the reduction of the chemical scheme to a set of equations manageable by analysis. Section 3 contains a description of the software packages applied, namely FACSIMILE/CHEKMAT [21] for finite-difference and ENTWIFE [22] for finite-element calculations and bifurcation studies, and discusses their relative merits. Section 4 presents results of calculations in one and two dimensions of the transients during wave formation, the profiles, and the speed of waves. An approach that includes the effects of convection is formulated in Section 5.

2. The chemistry of the Belousov-Zhabotinskii reaction

The mixture usually prepared for studies of the BZ reaction contains malonic acid $CH_2(COOH)_2$ as the organic substrate, bromous compounds as the electron acceptors in the oxidation process, a metal ion as catalyst undergoing periodic oxidation/reduction cycles, and sulphuric acid providing an acidic medium in the range of pH=10.5. The reaction converts the organic substrate into CO_2 and brominated organic compounds. Most of the work leading to the elucidation of the reaction mechanism and to the determination of the rate constants has been done with Ce^{3+}/Ce^{4+} as catalyst, and its use has also formed the basis of experiments on temporal patterns in the BZ reaction. Spatial patterns have been investigated with the catalyst ferroin/ferriin $Fe(phen)^{2+}/Fe(phen)^{3+}$ due to its easily visible red to blue transition. However, with recent improvements of the observation techniques waves in cerium-catalyzed reactions are now measurable in the UV range [23,17,18].

The fundamental mechanism responsible for the dynamical behaviour of the system is the so-called Field-Körös-Noyes (FKN) mechanism. For most theoretical work and numerical simulations the following scheme, the Oregonator model has been applied [4].

$$
\begin{aligned}
&A + W \xrightarrow{k_1} U + P \\
&U + W \xrightarrow{k_2} 2P \\
&A + U \xrightarrow{k_3} 2U + 2V \\
&2U \xrightarrow{k_4} A + P \\
&B + V \xrightarrow{k_5} hW
\end{aligned}
\tag{1}
$$

where $A = BrO_3^-$, B = bromomalonic acid + malonic acid, P = HOBr, $U = HBrO_2$, $V = Ce^{4+}$, $W = Br^-$. k_1 to k_5 are rate constants and h is a stoichiometric factor. The factor h counts the number of moles of Br^- liberated during the reduction of one mole of Ce^{4+}. The last equation approximates a reaction mechanism not yet fully understood. Mass action kinetics leads to a set of three differential equations if A and B are considered to be constant so as to simulate the concentrations of inexhaustible pools of chemical free energy. The equations can be scaled in such a way that the following system of dimensionless quantities appears [10].

$$
\begin{aligned}
&\varepsilon du/dt = qw - uw + u - u^2 \\
&dv/dt = u - v \\
&\varepsilon' dw/dt = -qw - uw + \hat{f}v \\
&\varepsilon = 10^{-2},\ \varepsilon' = 10^{-4},\ q = 2\times10^{-4},\ \hat{f} = 3
\end{aligned}
\tag{2}
$$

Since $\varepsilon' < \varepsilon < 1$, w can be considered as a fast variable. From the approximation $\varepsilon' = 0$ we obtain $w = \hat{f}\, v/(u + q)$, and taking diffusion into account the reduced set of partial differential equations describing the patterns in an unstirred solution layer reads

$$
\begin{aligned}
&\varepsilon \partial u/\partial t = \varepsilon^2 \nabla^2 u + u - u^2 - \hat{f}\, v(u-q)/(u+q) \\
&\partial v/\partial t = \varepsilon \nabla^2 v + u - v
\end{aligned}
\tag{3}
$$

The same dimensionless diffusion constant ε is assumed for both species u and v. These equations are the starting point used by Keener & Tyson [10] for their analysis by singular perturbation methods. In the present paper the equations are solved numerically in one and two dimensions using finite-difference and finite- element codes.

3. Two software packages for solving partial differential equations

Two software packages have been considered for the numerical solution of the reaction-diffusion equations, FACSIMILE/CHEKMAT [21] and ENTWIFE [22], both developed at the Harwell Laboratory by different groups for different purposes. FACSIMILE/CHEKMAT has been developed for the solution of large systems of stiff ordinary differential equations. The algorithm, an improved version of Gear's implicit technique [24], is embedded in a powerful command language. The structure of the language shows two distinct influences: programming in FORTRAN, and the formulation of chemical mechanisms in terms of chemical equations. The second influence came from the first applications for which FACSIMILE was originally designed, the simulation of biochemical pathways. The solution of partial differential equations with FACSIMILE became possible when the package was extended by a powerful set of array and matrix commands. These commands allow an easy and flexible implementation of finite difference schemes in one, two, and three dimensions. It is left totally to the user to formulate his own discretization scheme, but the most simple scheme with no-flux boundary conditions is implemented as a default. An example of the code for the reaction-diffusion equations (3) is given in Table 1. In addition, the package is supplemented with a nonlinear fitting algorithm for the estimation of unknown parameters, for example rate constants, from sets of experimental data. Printed output facilities are very flexible and FORTRAN-like statements can be invoked if necessary. Graphical output is fairly poor and limited to line printer plots. The strength of FACSIMILE lies in its robust integration scheme using sparse matrix techniques so that large systems can be solved with manageable memory requirements.

The ENTWIFE code was developed for the solution of partial differential equations in one, two or three spatial dimensions. Its original applications were to the solution of problems in fluid flow and heat transfer governed by the Navier Stokes and energy transport equations. However, the ease with which sets of equations can be programmed in the code has led to other applications in such diverse areas as semiconductor device modelling, atomic structure calculations and pitting corrosion. Thus, its extension to reaction-diffusion systems such as (3) is straightforward. In the code a standard Galerkin formulation of the finite-element method is implemented for the solution of the equations in the steady state or at a particular time step. The recent introduction of orthogonal mapping techniques makes the solution of free-boundary problems possible [25]. The variables in the equations being solved can be approximated by quadratic, linear or piecewise-constant functions. The distribution of elements that discretize the spatial domain is controlled by the user and regions of local refinement can be specified where improved resolution is required. The time dependence of the solution is obtained by Gear's method using a finite difference discretization of the time coordinate. A Crank-Nicholson scheme is also available and the package is supplemented with a powerful set of algorithms for the localization and analysis of bifurcation points. The operation of the ENTWIFE package is controlled by a free-format command language in hierarchical structure with the choice of keywords designed to give natural, English-like constructions. The graphical output facilities are extensive and include plots of contours and vectors and linegraphs showing the variation of a variable in space or time.

4. Results

4.1. *Wave excitation in 1-D and 2-D: Thresholds and transients*

The spatio-temporal development of a perturbation applied to the stationary state of a reaction-diffusion system governed by the partial differential equations (3) is expected to follow a complicated transient phase. As a first approach to its analysis, some information can be deduced by considering the temporal behaviour of the reaction kinetics without diffusion, as described by the following ordinary differential equations:

$$\varepsilon\dot{u} = u - u^2 - \hat{f}\, v\, (u - q)/(u + q)$$

$$\dot{v} = u - v \qquad (4)$$

where the dot indicates differentiation with respect to time. The nullclines $\dot{u} = 0$ and $\dot{v} = 0$ are shown in Fig. 1 superimposed with the vectors of the flow and with eight trajectories for the parameter values $q = 2.0\times10^{-4}$ and $\hat{f} = 3$.

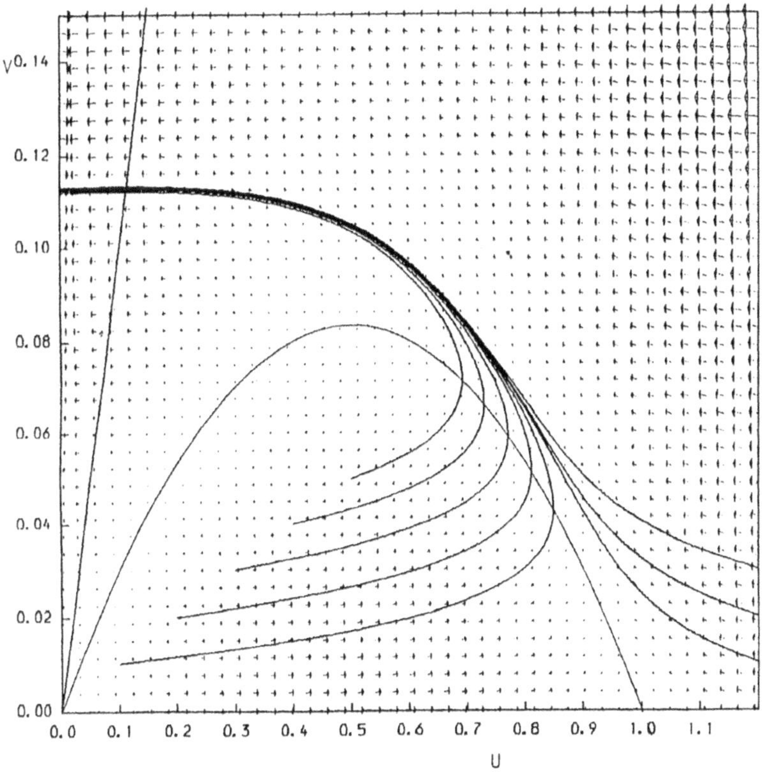

Figure 1. Phase plane diagram with the nullclines and the flow of the vector field for the dynamics of (2). As examples trajectories with various starting points are superimposed.

The stationary point $u_o = v_o$ is stable if it lies on the left side of the minimum as shown. The flow vectors indicate that the system undergoes a large excursion if the initial values are in the bell-shaped area where the flow vectors have positive u components. Systems with this characteristic dynamical behaviour are called excitable. When the trajectory has crossed the nullcline $\dot{u} = 0$ the sign for the u component changes and the v component tends to increase. At high v values a fast transition to low u values is followed by a slow decrease of v at almost constant u. It is noteworthy that the trajectories originating within a large area of low v values form a small band for v greater than the maximum of the bell-shaped part of the u nullcline. In general, the changes in u are faster by a factor of $1/\varepsilon$ than those in v due to the term $\varepsilon\,\dot{u}$ in the first equation of (4).

In a spatially extended medium without mixing, each volume element is able to respond to a proper perturbation in the same manner as displayed in Fig. 1. The local perturbation generates gradients which affect neighbouring elements by diffusion. Due to the different time scales the perturbation may evolve into a wave and move through the medium as a solitary undamped wave.

As a prerequisite of the numerical analysis of dispersion relations [26] and curvature effects [9-12] in excitable media we study first the transient behaviour of a perturbation applied locally to a domain with extension r in a one dimensional system of length L (Fig. 2). In the following calculations the rectangular perturbation is set at the left boundary starting at the coordinate x=0 and extending to x=r, the spatial parameter of the calculations. The "intensity" of the perturbation in the range 0 to r is constant and given by the values $u(t=0)=u_p$ and $v(t=0)=v_p$ of the dynamic variables. All elements for x>r are at the stationary state value $u(t=0)=u_o$ and $v(t=0)=v_o$. Figure 2 shows the evolution of the perturbation after a short time. The initial state of the system at t=0 is indicated by dashed lines. The evolution in time of a specific spatial element reflects the excitability of

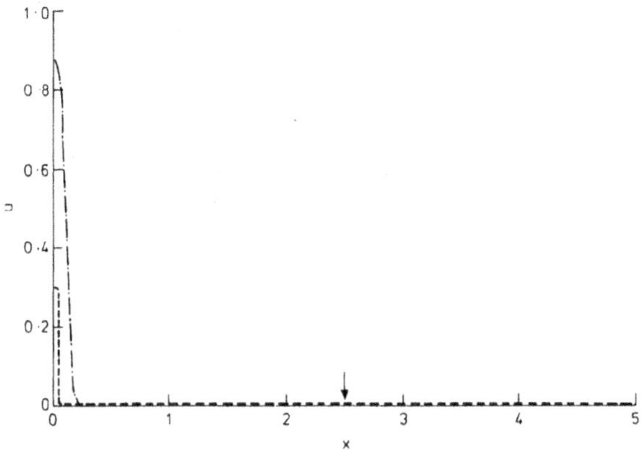

Figure 2. Local perturbation of the variable u to test for the formation of a wave, (---) initial perturbation $u_p=0.3$ and $v_p=v_o$ at t=0; (-.-.) first maximum of excitation at t=0.06. The arrow indicates the element at x = L/2 for which the trajectory has been plotted in Fig. 3.

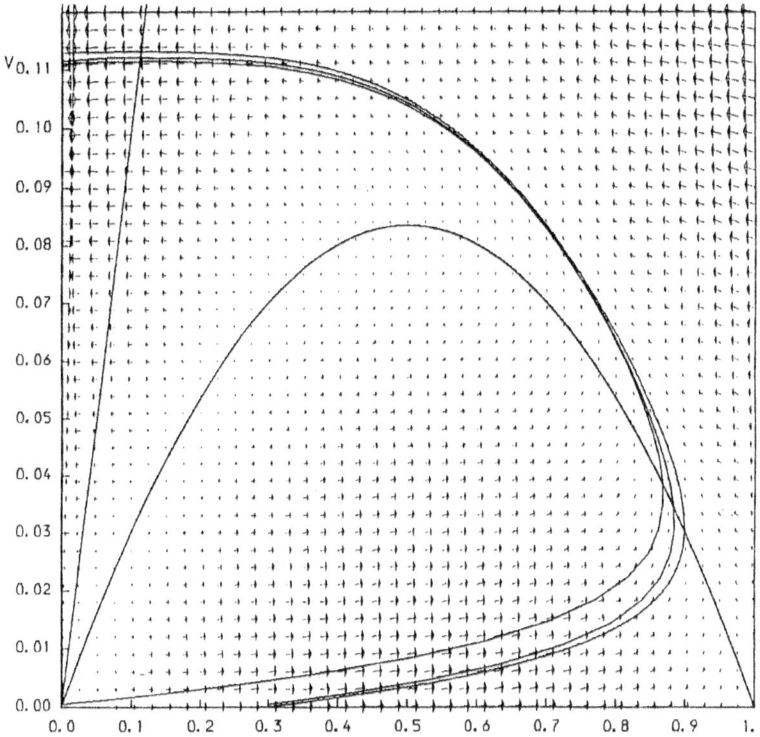

Figure 3. Phase plane representation of wave formation stimulated by a rectangular perturbation $u_p = 0.3$ and $v_p = v_o$ of two elements in a 1-D reaction-diffusion calculation. (a) trajectory of the system without diffusion; (b) element at x=L/2; (c) left-most element at x=0.

the medium as shown in Fig. 3 for two elements, at the left boundary x=0 and at x=L/2.

The transient behaviour becomes more complex if the parameter is in a range where the diffusion due to the gradient at x=r affects the excursion at the origin. Typical examples are presented in Fig. 4 for u_p=0.5 and v_p=v_o. In the region of u_p=0.5 the vector field has its highest positive u component, which drives the system to still higher u values and overcomes the smoothing effect of the diffusion term. The balance between diffusion and excitability becomes more critical if the perturbation is placed close to the nullcline $\dot{u}$=0.

One may distinguish between three different situations:

(i) $u=u_p$ in $0 \le x \le r$ $\quad$ $u=u_o$ in $r < x \le L$

$v=v_o$ in $0 \le x \le L$

(ii) $u=u_p$ in $0 \le x \le r$ $u=u_o$ in $r < x \le L$

$v=v_p$ in $0 \le x \le r$ $v=v_o$ in $r < x \le L$

(iii) $u=u_p$ in $0 \le x \le r$ $u=u_o$ in $r < x \le L$

$v=v_p$ in $0 \le x \le L$

Case (i) with $u_p=0.3$ and $r=0.043$ is illustrated in Figs. 2 and 3 and was already mentioned in the previous chapter. The perturbation develops as expected for an excitable medium: after a transient phase in which the concentration u has a bell-shaped distribution, u grows and spreads into neighbouring cells, forming a wave. No critical radius r_c at which diffusion is able to counteract the excitability and to suppress the generation of a wave could be found within the numerical limits. In case (ii), where a perturbation is set for both variables, $u=0.015$ and $v=10v_o=3.98 \times 10^{-3}$, a critical length of $r_c = 2.6 \times 10^{-3}$ was detected. A perturbation below the critical extension damps out and the system relaxes back to the stationary state $u=u_o$ and $v=v_o$.

Figure 5 shows the dynamics of the leftmost element for various values of r in calculations with initial conditions $u_p=0.015$ in r and $v_p=10v_0$ in all cells, case (iii). Here the initial state is close to the nullcline $\dot{u}=0$ in the

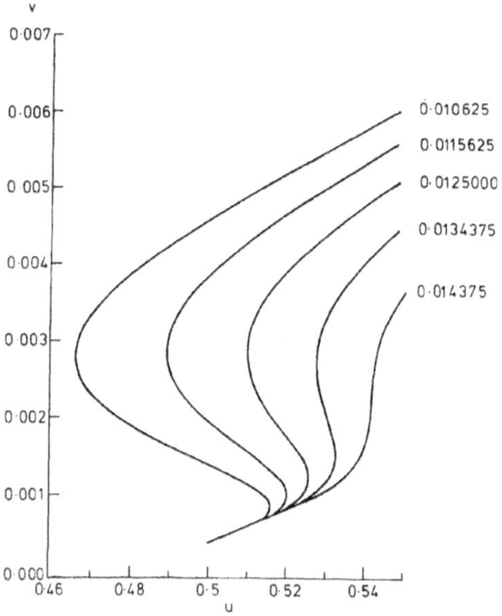

Figure 4. Trajectories describing the dynamics of the leftmost element at x=0 in a series of 1-D reaction-diffusion calculations for various extensions r of the perturbation $u_p = 0.5$, $v_p = v_o = 3.98 \times 10^{-4}$. Only the area close to the starting point is shown.

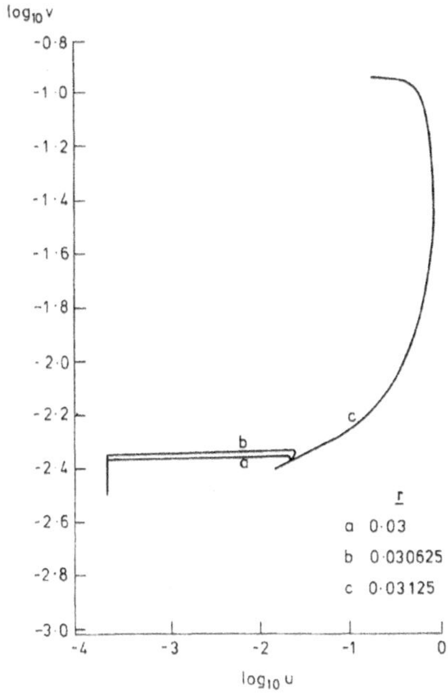

Figure 5. Trajectories describing the dynamics of the leftmost element in 1-D reaction-diffusion calculations for an increasing spatial extension r of the perturbation $v_p = 10\ v_o = 3.98\times10^{-3}$, u_p = 0.015. (a) $r = 3\times10^{-2}$ no wave; (b) $r = 3.06\times10^{-2}$ no wave; (c) $r = 3.13\times10^{-2}$ formation of a wave.

region $\dot{u}>0$ so that diffusion can move the system easily to a region with $\dot{u}<0$. The calculated critical radius below which no wave evolution was observed is $r_c \sim 3\times10^{-2}$ spatial units or 51 μm. A similar behaviour has been observed in the case of the degenerate FitzHugh-Nagumo equations [27]. It is clear from Fig. 5 that the spatial parameter r plays a role as bifurcation parameter in the problem; as the extent of the perturbation increases through a critical value r_c an abrupt change in behaviour occurs.

Similar calculations have been performed for wave formation in the 2-D case for $v_p = 3.98\times10^{-3}$ and u_p = 0.015 on a grid with 40 by 40 elements (Fig. 7). Due to the coarseness of the grid the critical radius can only be estimated within an interval of 0.03 to 0.05 spatial units. This estimate is just above the critical extension of the perturbation in the 1-D case.

4.2. *The speed of travelling waves*

A characteristic property of waves in excitable media is the velocity of motion, a parameter very sensitive to the discretization of the problem in space and time. As a first test, we examined the precision of the

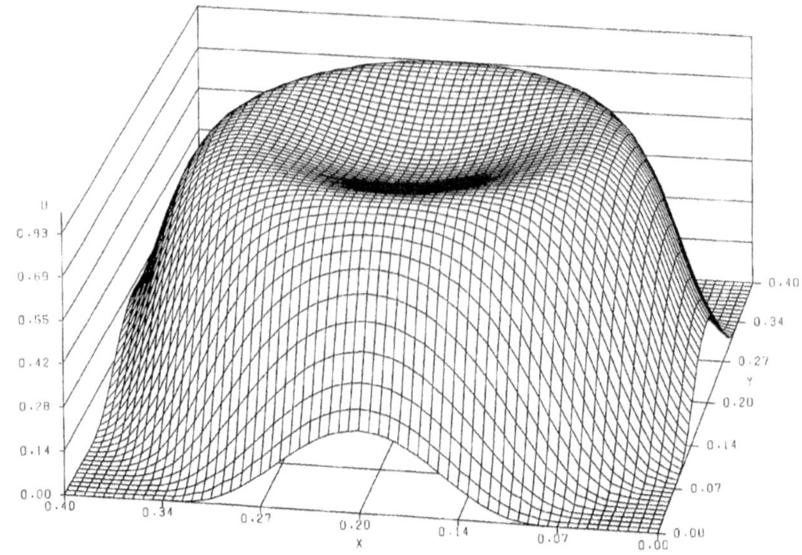

Figure 6. Two-dimensional circular wave evolving from a perturbation u_p = 0.015, $v_p = 3.98\times10^{-3}$.

FACSIMILE/CHEKMAT calculations using an example for which the speed is exactly known, namely the degenerate FitzHugh-Nagumo equation with the kinetics $f(u) = u(1-u)(u-q)$, $0 < q < 1/2$ [27,28]. In a reaction-diffusion system with such kinetics the speed of a wave is given by $c = \sqrt{2}\ (1/2-q)$. Our computed speeds for various values of q are slightly lower than the theoretical figures but the deviations do not exceed 3%. Similar agreement has been obtained by a comparison with the results of Murray [6] and of Manoranjan & Mitchell [29].

In the test cases mentioned the speed of a travelling wavefront is calculated; in contrast the solutions of (3) represent the motion of an asymmetric pulse with a leading front and a relaxing back, which together with the steepness of the leading front may account for the sensitivity of (3) to spatial discretization as shown in Fig. 7. The speed on a linear 1D array (denoted by O,+,□) increases sharply if the length per cell exceeds about 0.015 spatial units. This holds for FACSIMILE/CHEKMAT with its sophisticated time stepping algorithm (+,O) as well as for the simple Euler integration scheme (□). The symbol (+) marks data from runs with total length L=5 and the points (O) are for L=10 spatial units. The triangles (Δ) represent wave speeds calculated with the Euler method on a 1D array with periodic boundary conditions, i.e. on a ring. Obviously this curve exhibits a much larger range where the speed is independent of the length per cell. Similar calculations could not be done by FACSIMILE/CHEKMAT because the periodic boundary conditions lead to an intolerably high number of convergence failures. The reasons for this have yet to be determined.

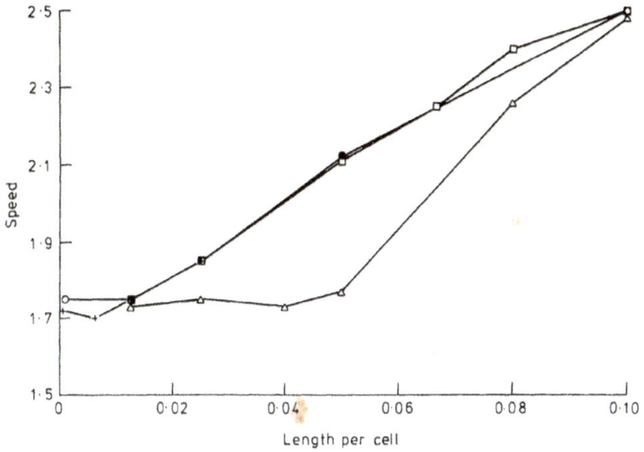

Figure 7. Wave speeds in an excitable medium calculated for system (3) as a function of the discretization parameter "length per cell". (+) FACSIMILE L=5; (O) FACSIMILE L=10; (□) Euler 5≤L≤20 and number of cells between 200 and 800; (Δ) Euler with periodic boundary condition, 10≤L≤20 and number of cells between 200 and 800.

4.3. *Dispersion relations*

The velocity of a travelling wave depends on the diffusion coefficients of the chemicals involved and on the kinetics of the reaction. The reaction kinetics may behave differently when a wave is moving into a medium which exists in its stationary state than when it moves in the wake of a wave generated earlier in time. It is explained in detail in [26] that the first wave has the maximum speed possible in the system and a following wave travels at lower velocity, so that the distance between the waves increases with time. After a very long time the distance becomes so large that the second wave is moving into a region which has by then relaxed back to the stationary state following the excitation due to the first wave. At such distances the second wave also attains the maximum speed.

The complex behaviour just described is quantified in the dispersion relation, that is the functional relation between wave velocity and distance or wavelength $c=c(\lambda)$ or between wave velocity and period T, $c=c(T)$. The dispersion relation has been calculated mathematically by singular perturbation methods [26] and numerically by the solution of a set of ordinary differential equations obtained from (3) by the introduction of moving frame coordinates. This transformation assumes a constant shape for the travelling wave. We tried to model in 1-D a pair of waves, initiating its propagation in one direction with the shortest possible displacement. After some straightforward tests we realized that in 1-D the generation of a pair of waves travelling in the same direction is a non-trivial task despite the fact that FACSIMILE/CHEKMAT has powerful facilities to resume the simulation after a break and a modification of parameters and/or initial conditions. Normally a perturbation imposed on the back of a wave dies out. Sometimes the perturbation generates a wave moving in the opposite direction to the first one. Finally we found that a fully developed wave has to be imposed on the back of the first wave as an initial condition for both variables u and v. The next problem was then to determine the shortest distance between the waves for which the second wave survives. Multiplication of the spatial coordinate by a

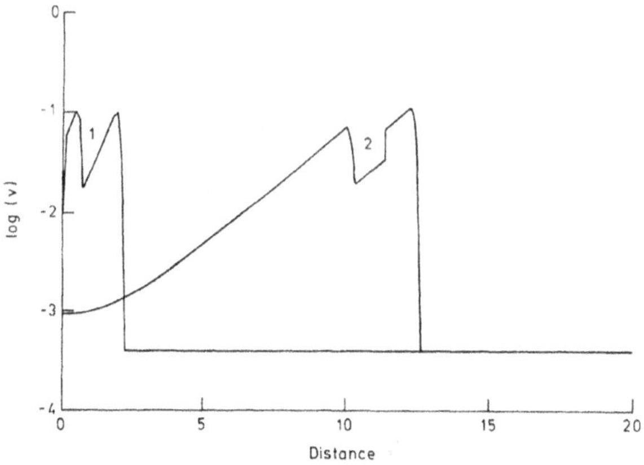

Figure 8. Two sets of concentration distribution for the initiation of a pair of waves travelling in the same direction.

factor lower than unity is one way to achieve that goal. It makes the wave profiles steeper and the wavelength shorter. Two examples of initial concentration distributions are shown in Fig. 8. In each case the value of the controller variable v has to be less than $2x10^{-2}$ at the minimum between the waves; any perturbation that would increase the value of v at a point between the trough and the peak of the leading wave will die out.

Figure 9 exhibits an overlay of a sequence of snapshots taken from a run initialized with the concentration distribution marked 1 in Fig. 8. The leading wave is taken as the frame of reference. Its profile, the peak at the far right, is therefore plotted only once. The second wave glides down the back of this wave with increasing time. Snapshots of this process are plotted for 6,8,10,20 and 30 time units. The profile of the second wave attains its final shape after about 10 time units.

The details discussed so far are quantified by the dispersion relation as shown in Fig. 10. The curves are the predicted outcome of different approaches. The solid lines are from the simulations of two waves moving in the same direction on a linear array. The upper horizontal solid lines show the speed of the leading wavefront. The other solid lines exhibit the dispersion of the speed of the second wave moving in the wake of the first one. Different initial conditions produce a significant difference with respect to the shortest distance and to the dynamics at the beginning.

The dashed line represents a result obtained from the solution of a set of ordinary differential equations derived from (3) by the introduction of a moving reference coordinate system as explained above. The curve is the upper branch of Fig. 7 in [26] after transformation from period to wavelength. A bias of 0.32 is added to its maximum speed, adjusting the curve to our value of 1.87. This was done only for a clearer presentation of the data.

The circles mark the dispersion relation for (3) calculated on a 1-D array with a periodic boundary condition. The speed was determined from the time taken for the wave to reappear at the same position after one cycle. The minimum array length, i.e. the abscissa labelled "distance" in the graph, was reached when the leading front of the wave moved into its own back at a

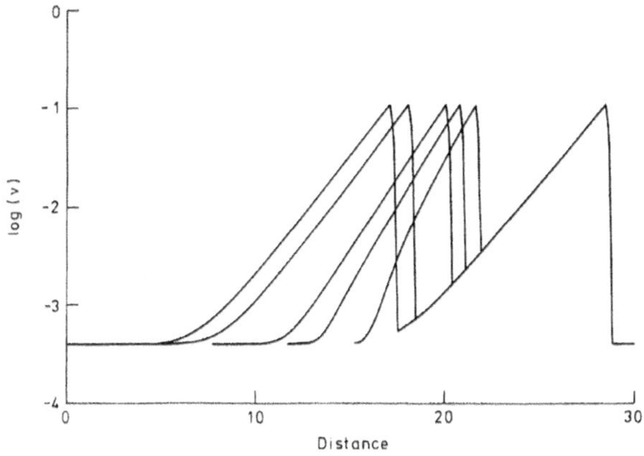

Figure 9. An overlay of a sequence of snapshots taken from a run simulating two waves which travel in the same direction. The leading wave, shown by the profile to the right, is taken as a frame of reference. The second wave slides down on the back of the first one. The snapshots are taken at 6,8,10,20 and 30 time units.

controller concentration level just below $v = 2x10^{-2}$, the same value as found in the two wave case.

5. Hydrodynamic flow in excitable reaction-diffusion systems

The results presented in the previous section related specifically to the reaction-diffusion equations (3) for the BZ reaction, whose solution describes the spatial patterns of concentration that develop in time from an initially homogeneous mixture of the reactants. Such models represent an idealization of the BZ reaction, not only because of their assumptions about the nature of the chemical mechanisms, but also because they ignore the hydrodynamic effects that will inevitably occur in the fluid layer in which the reaction takes place. Hydrodynamic flow will typically arise from differences in temperature or concentration within the bulk of the fluid, giving rise to buoyancy forces, or from similar differences at the free surface of the layer leading to gradients in the surface tension. An example of the way in which hydrodynamic effects can influence the pattern formation is provided by the experiments of Miike et al. [20], which revealed in particular the entrainment of fluid that can result as a BZ wave passes across the surface of an uncovered layer.

A proper description of hydrodynamic effects in excitable reaction-diffusion systems would couple necessarily the chemical equations for the concentration with the Navier-Stokes equations for the fluid flow. The accurate numerical solution of these coupled equations would present a considerable challenge, and would compound the difficulties that we described in the previous section in relation to the computation of wave-like solutions of the chemical equations (3). While our long-term aim is to carry out such direct simulations of the hydrodynamic and reaction- diffusion equations, we are first isolating and studying specific effects that contribute to the coupling, in order to build up our understanding of the coupled processes. As one example, we are developing a model of the entrainment of fluid by a BZ wave that was observed by Miike et al. [20]. In this model we represent the BZ wave by a region that moves over the surface of the fluid and which alters the

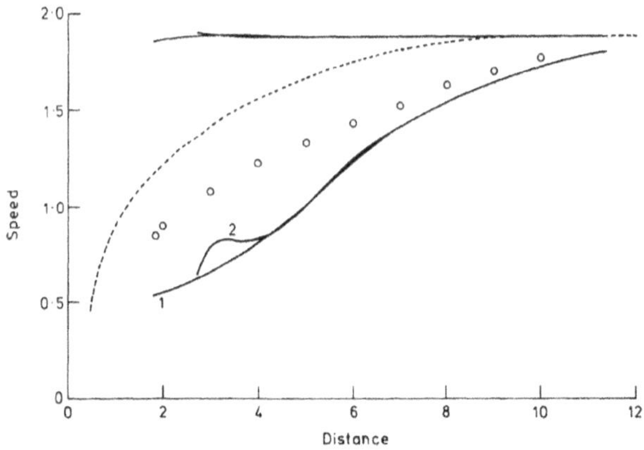

Figure 10. Dispersion relations for system (3) as a function of the distance of two waves travelling in the same direction. (---) numerical solution from [26] with a bias of + 0.32. (o) numerical solution of (3) with periodic boundary conditions; (-) horizontal lines: speed of the leading wave; (—) lower lines: (1) started with initial concentration distribution (1) of Fig. 8; (2) started with initial concentration distribution (2) of Fig. 8.

surface tension over that region, thereby establishing a convective flow that moves with the wave. By transforming to a reference frame moving with the wave we reduce the problem to the solution of the steady-state equations for Marangoni convection in which the wave speed appears as an additional parameter in the advective terms. We are also developing techniques for solving free-boundary problems that will represent explicitly the deflection of the surface that accompanies Marangoni convection. For these new developments the ENTWIFE package will be the most appropriate tool.

6. **Discussion**

Two software packages have been evaluated for the solution of parabolic partial differential equations of the reaction-diffusion-convection type. It was straightforward with FACSIMILE/ CHEKMAT to formulate the discretization by finite differences and to calculate various aspects of travelling waves in an excitable medium. The code was very reliable for the standard calculations of travelling waves in one and two dimensions, but some difficulties were encountered when trying to generate more than one wave train.

The investigation of the initial conditions for the generation of a moving wave by a perturbation in a two variable model of the BZ reaction shows that in 1-D and 2-D spatial extension of the perturbation is a critical parameter. This result questions the idea that the curvature is the only decisive parameter in the process of pattern formation in an excitable medium, at least for the generation of circular waves and target patterns [11].

A detailed analysis of the spatial discretization reveals that the parameter "length per cell" is the most critical one for the calculation of pulse shaped waves. It is interesting to note that the reasonable limit of this parameter increases by about a factor of three if a linear 1-D array is replaced by a ring, i.e. if periodic boundary conditions are applied. The

Table 1.

```
* FACSIMILE/CHEKMAT;
* BZ MODEL;
* J.P. KEENER AND J.J. TYSON;
* PHYSICA 21D, 307-324 (1987);
*;
* ;
* DECLARATION OF PARAMETERS;
PARAMETER Q 2.0E-4  EPS 1.0E-2  FAK 3.0  EPSINV;
PARAMETER LENGTH 5.0  H  HALF  LD  RD;
PARAMETER <600> WORK;
PARAMETER <60> DISTX;
* DECLARATION OF VARIABLES;
VARIABLE <60>  U  V;
* ;
COMPILE INITIAL;
H = LENGTH/FLOAT(60);
HALF = H/2;
DO 10 FOR #10 = 0 (1) 59;
DISTX<#10> = HALF + FLOAT(#10) * H;
LABEL 10;
EPSINV = 1.0/EPS;
* TRANSPORT COEFFICIENTS FROM THE RIGHT (RD) AND THE LEFT (LD);
LD = EPS/(H**2);
RD = EPS/(H**2);
ARRAY <60>  WORK;
U = 3.98E-4;
V = 3.98E-4;
ARRAY END;
* SET PERTURBATION;
U<0> = 0.3;
**;
COMPILE EQUATIONS;
ARRAY <60>  WORK;
'U = EPSINV * (U - U**2 - FAK*V*(U-Q)/(U+Q));
'V = U - V;
ARRAY END;
PATTERN IGNORE 5;
TRANSPORT <60> U  LD RD;
TRANSPORT <60> V  LD RD;
LABEL 5;
**;
COMPILE PRINT;
  DISTI = DISTX<0>;
  UI = U<0>;
  VI = V<0>;
  PSTREAM 2;
  DISTI = DISTX<19>;
  UI = U<19>;
  VI = V<19>;
  PSTREAM 3;
**;
PSTREAM 2 8 4;
DISTI  UI  VI  TIME;
**;
PSTREAM 3 9 4;
DISTI  UI  VI  TIME;
**;
WHEN
     TIME = 0 + .005 * 500 % CALL PRINT;
**;

EXEC OPEN 8 "TK8.OUT" NEW;
EXEC OPEN 9 "TK9.OUT" NEW;

BEGIN;
STOP;
```

calculation of the dispersion relation from a 1-D model of two waves following each other became very difficult, to our surprise, in that such a system is very sensitive to the initial arrangement of the waves, i.e. to their distance and shape. This observation raises questions about wave generation in experimental 1-D systems.

Similar calculations require a more careful approach using the finite element package ENTWIFE. The second-order approximation of u and v generates negative values which are amplified due to the nature of the model equations that describe the excitable medium. Also, it is unclear which of the time-stepping algorithms available in ENTWIFE is appropriate for this problem. Some improvements of the ENTWIFE code have to be made before its powerful bifurcation and path following features can be applied to time-dependent reaction-diffusion systems.

Acknowledgements

This work was supported by the Stiftung Volkswagenwerk, Hannover and by Grant GR/D/13573 from the Science and Engineering Research Council of the U.K. One of us (Th.P.) acknowledges gratefully the hospitality of Prof. J.D. Murray and the stimulating atmosphere at the Centre for Mathematical Biology in Oxford, England.

The computations were done on the CRAY-2 at the Harwell Laboratory and on a CONVEX 201 at the Max-Planck-Institut. We thank Mr. J. Heitkötter for the implementation of FACSIMILE on the CONVEX and its integration in a graphics environment.

References

[1] Zaikin, A.N. & Zhabotinskii, A.M. (1970). *Nature* **224**, 535-532.
[2] Field, R.J. & Burger, M. (eds.) (1985). *Oscillations and Traveling Waves in Chemical Systems*. Wiley: New York.
[3] Ross, J., Müller, S.C. & Vidal, C. (1988). *Science* **240**, 460-465.
[4] Field, R.J. & Noyes, R.M. (1974). *J. Chem. Phys.* **60**, 1877-1884.
[5] Tyson, J.J. (1985). In *Oscillations and Traveling Waves in Chemical Systems*, Field, R.J. & Burger, M. (eds.), pp. 93-144. Wiley: New York.
[6] Murray, J.D. (1976). *J. Theoret. Biol.* **56**, 329-353.
[7] Tyson, J.J. & Fife, P.C. (1980). *J. Chem. Phys.* **73**, 2224-2237.
[8] Rovinsky, A.B. (1986). *J. Phys. Chem.* **90**, 217-219.
[9] Zykov, V.S. & Petrov, A.A. (1977). *Biophysics* **22**, 307-314.
[10] Keener, J.P. & Tyson, J.J. (1986). *Physica* **21D**, 307-324.
[11] Tyson, J.J. & Keener, J.P. (1988). *Physica* **32D**, 327-361.
[12] Zykov, V.S. (1988). *Modelling of Wave Processes on Excitable Media*. Winfree, A.T. (transl. & ed.). Manchester University Press.
[13] Wood, P.M. & Ross, J. (1985). *J. Chem. Phys.* **82**, 1924-1936.
[14] Müller, S.C., Plesser, Th. & Hess, B. (1986). *Naturwissenschaften* **73**, 165-179.
[15] Pagola, A. & Vidal, C. (1987). *J. Phys. Chem.* **91**, 501-503.
[16] Müller, S.C., Plesser, Th. & Hess, B. (1987). *Physica* **24D**, 71-86; *Physica* **24D**, 87-96.
[17] Nagy-Ungvarai, Zs., Tyson, J.J. & Hess, B. (1989). *J. Phys. Chem.* **93**, 707-713.
[18] Nagy-Ungvarai, Zs., Müller, S.C., Tyson, J.J. & Hess, B. (1989). *J. Phys. Chem.* **93**, 2760-2764.

[19] Plesser, Th., Miike, H., Müller, S.C. & Winters, K.H. (1987). *Harwell Laboratory Report TP.1267.*
[20] Miike, H., Müller, S.C. & Hess, B. (1988). *Phys. Rev. Lett.* **61**, 2109-2112.
[21] Curtis, A.R. & Sweetenham, W.P. (1987). *FACSIMILE/CHEKMAT User's Manual, Harwell Laboratory, Oxfordshire.*
[22] Winters, K.H. (1985). *ENTWIFE User Manual, Harwell Laboratory Report R.11577.*
[23] Nagy-Unvarai, Zs., Müller, S.C., Plesser, Th. & Hess, B. (1988). *Naturwissenschaften* **75**, 87-89.
[24] Gear, C.W. (1971). *Numerical Initial Value Problems.* Prentice Hall Inc.: Englewood Cliffs, New Jersey.
[25] Cliffe, K.A., Wheeler, A.A. & Winters, K.H. (1989). *Harwell Laboratory Report M.3731.*
[26] Dockery, J.D., Keener, J.P. & Tyson, J.J. (1988). *Physica* **30D**, 177-191.
[27] Manoranjan, V.S. (1984). In *Mathematics in Medicine and Biomechanics,* pp. 73-80, Roach, R.F. (ed.). Shiva Publishing Ltd.: Nantwich.
[28] Rinzel, J. & Terman, D. (1982). *SIAM J. Appl. Math.* **42**, 1111-1137.
[29] Manoranjan, V.S. & Mitchell, A.R. (1983). *J. Math. Biol.* **16**, 251-260.

41. SPATIAL TEMPERATURE DISTRIBUTION DURING THE HETEROGENEOUSLY CATALYZED OXIDATION OF ETHANOL

N.I. Jaeger, M. Liauw and P.J. Plath

Institut für Angewandte und Physikalische Chemie
Universität Bremen, Bibliothekstrasse
Postfach 330 440, D-2800 Bremen 33

1. Introduction

Oscillations in heterogeneous catalytic reactions are normally observed on a macroscopic scale either as product oscillations in the outlet of the reactor or as oscillations of the average temperature of the catalyst bed. As a prerequisite for these observations the coupling of a large number of active catalyst centres across a supported metal catalyst is required.

Synchronization of all parts of the surface cannot necessarily be expected and a number of authors have demonstrated by *in situ* measurements the existence of nonuniformities of the surface temperature [1,2] and of concentrations of adsorbed species [3,4]. A mathematical model of the oscillatory catalytic oxidation of carbon monoxide has also led to the result that oscillations can be accompanied by spatio-temporal structures on the surface of the catalyst [5]. The observation of complicated global product or temperature oscillations can therefore be due to the superposition and coupling of a number of oscillating regions on the surface of the catalyst [6,7].

In the case of the oxidation of ethanol on a Pd/Al_2O_3 supported catalyst, localized areas can be synchronized by a supporting silver plate due to the high thermal conductivity of the silver. Based on these results a set of experiments was carried out where the coupling of spatially separated oscillating areas can be studied. Moreover, a variation of a bifurcation parameter related to the temperature can be established across the catalyst bed without changing the external constraints.

2. Experimental

The experiments were carried out in a continuous flow reactor system described in [8]. A special catalyst holder was constructed by arranging six silver plates (diameter 2.5 mm, thickness 0.13 mm) at the corners and in the centre of a pentagon as an insert on a Teflon plate of low thermal conductivity (Fig. 1). A NiCr-Ni thermocouple was soldered to each silver plate.

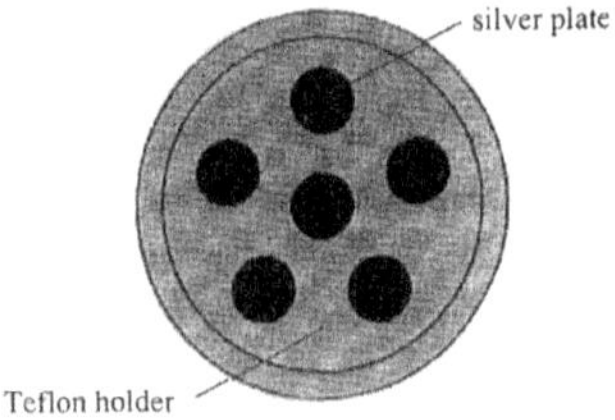

Figure 1. The catalyst holder. Six silver plates as an insert on a Teflon plate.

Care was taken that the thermocouples and the silver plates were as similar in their thermal properties as possible. The response time of the thermocouples to temperature changes was $t_{1/2} < 0.5$ s, which is small compared to the period of a typical temperature oscillation (30 s).

An amorphous 5 wt-% Pd/Al_2O_3 catalyst (Fluka) with a grain size of 32 - 45 µm was used (average size of the Pd crystals: 3 - 4 mm). In the experiments 46 mg of the catalyst was spread on the holder covering Teflon and silver plates. With a mean density of 0.83 g/cm^3, the thickness of a uniform catalyst layer was estimated to be 0.42 mm.

A flow of nitrogen (99.996%, Messer-Griesheim) was first passed through an ethanol-filled saturator at 28.0°C ± 0.1°C, then through a condenser kept at 22.0°C ± 0.1°C to ensure saturation. It was then mixed with an unloaded stream of nitrogen and a stream of oxygen (99.995%, Messer-Griesheim) controlled by thermal mass flow controllers (HI-TEC; precision ±1%). The gas stream was then preheated to the desired reaction temperature and led into the reactor.

3. Results and discussion

The ignition of the reaction could be induced by increasing the temperature of the reactor, or by changing the gas phase concentrations of the reactants. The ignition of the reaction did not occur simultaneously on all six silver plates, but rather with time lags of up to a minute.

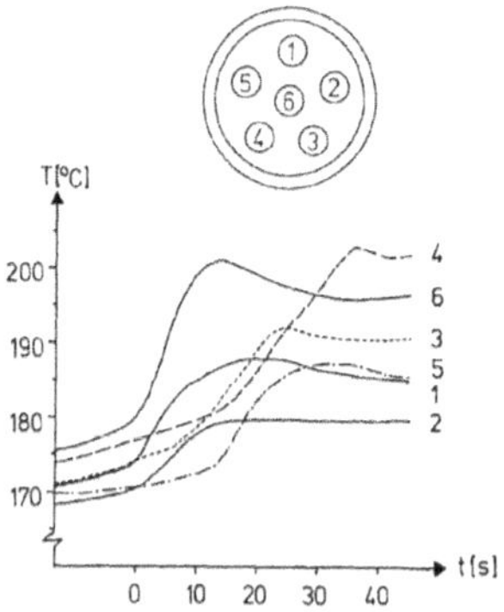

Figure 2. Ignition: Temperatures on all six silver plates.

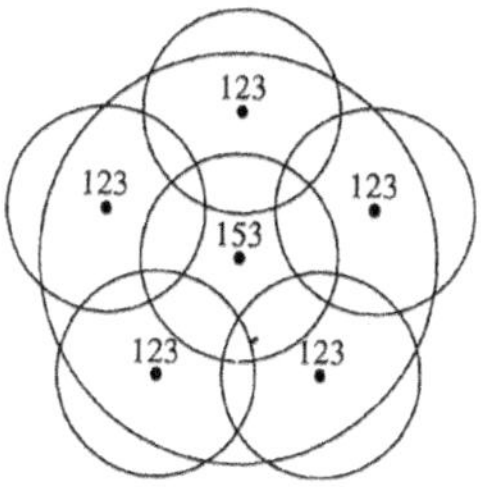

Figure 3. Temperature distribution in the model: temperatures in arbitrary units.

This behaviour is shown in Fig. 2. A catalyst layer was uniformly spread on the holder. At constant oxygen concentration (12.7 vol-%) and flow rate (47 mL/min), ignition was effected by stepping the ethanol concentration in the feed from 1.8 vol-% to 3.6 vol-%. The ignition occurred almost simultaneously on the silver plates no. 1, 2, and 6, and with a time lag of up to 17 s on the other plates.

The temperature on the silver plates no. 1 and 6 was observed while the ignition was induced six times. The shape of the temperature curves as well as their temporal relation were the same in all six experiments, indicating a high reproducibility with an unchanged catalyst layer.

The differences in the steady state temperature measured on equivalent silver plates reflect an inhomogeneity in the spatial distribution of conversion due to unavoidable inhomogeneities in the distribution of the catalyst powder and differences in the heat capacities of the plates.

The significantly higher temperature of the central plate cannot be accounted for by inhomogeneities and is explained by the following model:

> Let us assume a homogeneous heat production within all the elements of the catalyst volume. Depending upon the heat transfer and the cooling rate, there will be a given amount of heat which yields the measured temperatures of the silver plates in one time step. This special amount of heat corresponds to an effective volume of the catalyst. For simplicity this volume is modelled by a cylinder. In this way the effective volume is correlated to an effective radius. Those parts of the cylinder which project beyond the borderline of the holder do not contribute to the temperature of the corresponding silver plate. Figure 3 shows that the model leads to a straightforward explanation of the higher temperature measured on the central silver disk.

A proportionality between amount of catalyst and local conversion is supported by experiments in which the catalyst was distributed across the simple holder in a wedge shaped layer (see Fig. 4a).

A decreasing layer thickness corresponds to a decreasing average temperature of the silver plate in the ignited state in the same direction. The temperature gradient reflects a rather low thermal conductivity of the catalyst powder between the silver disks. The temperature distribution at the same time corresponds to a variation of a bifurcation parameter. A bifurcation of the reaction into the oscillatory state can be achieved by lowering the temperature of the reactor while the reaction is in the ignited state.

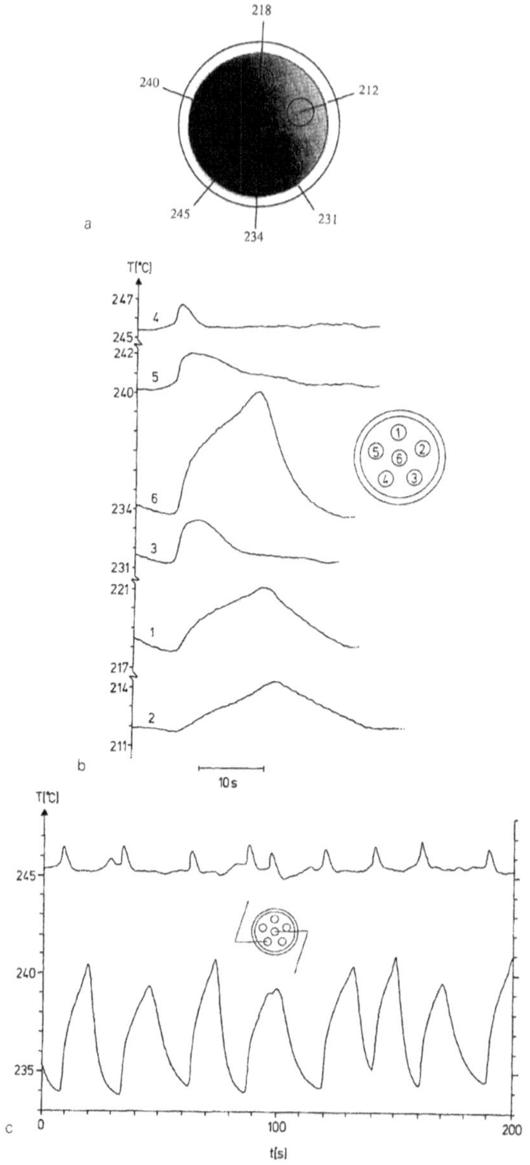

Figure 4. (See facing page.)
(a) Temperature distribution in a wedge shaped catalyst layer in the oscillatory state: the temperatures of the minima in °C. (b) Temperature oscillations in the wedge shaped catalyst layer, shown for two plates. (c) Shown for all six plates.

Figure 4b demonstrates that a local oscillator has been established on the central silver plate which drives the temperature oscillation on the surrounding plates.

While plates no. 3, 4, and 5 close to the bifurcation point can be driven, plates 1 and 2 far from a bifurcation point at lower temperatures merely reflect the oscillation of the central plate via heat conduction (Fig. 4c).

References

[1] Brown, J.R., D'Netto, G.A. & Schmitz, R.A. (1985). In *Temporal Order*, Rensing, L. & Jaeger, N.I. (eds.), p. 86. Springer Series in Synergetics, Vol. 29. Springer: Berlin.

[2] Onken, H.U. & Wolf, E.E. (1988). *Chem. Eng. Sci.* **43**, 2251.

[3] Kaul, D.J. & Wolf, E.E. (1985). *J. Catal.* **91**, 216.

[4] Kaul, D.J. & Wolf, E.E. (1985). *J. Catal.* **93**, 321.

[5] Gerhardt, M. & Schuster, H. (1989). *Physica* **D36**, 209.

[6] Tsai, P.K., Maple, M.B. & Herz, R.K. (1988). *J. Catal.* **113**, 453.

[7] Wicke, E. & Onken, H.U. (1988). *Chem. Eng. Sci.* **43**, 2289.

[8] Jaeger, N.I., Ottensmeyer, R. & Plath, P.J. (1986). *Ber. Bunsenges. Phys. Chem.* **90**, 1075.

42. A NEW PATTERN OF SPREADING EXCITATIONS IN CELLULAR AUTOMATA: MODELS FOR HETEROGENEOUS CATALYTIC REACTIONS

P.J. Plath

Institut für Angewandte und Physikalische Chemie

Universität Bremen, Bibliothekstrasse NW 2
D-2800 Bremen 33, FRG.

1. **Introduction**

Heterogeneous catalytic reactions are complex chemical systems. This is even the case for the oxidation of CO which looks simple with respect to the mechanistic overall formula but it is of high complexity. There are conditions under which this reaction occurs in an oscillatory state [1-8] or even in a chaotic regime [9-11]. The patterns which arise during CO-oxidation by means of palladium support catalysts possess self-similar time series of CO-conversion and CO_2-production.

The self-similar patterns can be modelled fairly well using one-dimensional binary cellular automata [12,13]. The simulations give rise to the question of the semantic interpretation of such one-dimensional automata, since a nearly two-dimensional layer of the catalyst was used for the chemical experiment.

It is the aim of this article to present some new ideas which may allow more realistic models for this complex reaction to develop.

2. **Experimental facts**

Using palladium crystals incorporated in a zeolite matrix [11,14,15] or supported by amorphous Al_2O_3 [10,14] as a catalyst (20 mg catalyst with 0.5 wt-% Pd), one can easily find constraints for running the reaction in a self-similarly oscillating state. Using low flow rates (9 l/h) of the reactants (0 - 3 vol-% CO (Messer-Griesheim, 15 vol-% CO in N_2-gas mixture for detection) in synthetic air (Messer-Griesheim, 21 vol-% O_2 in N_2; NO and HC < 0.1 vpm), one can achieve sudden breaks in the CO_2-production the total conversion of which is almost in the range 94-98%.

The pattern of the time series of conversion is of most interest. It is characterized by:

(a) very short breaks of conversion; the sizes of the breaks are very different;
(b) the sequence of the breaks, which are self-similar with respect to their sizes;
(c) the fact that after a while the complex patterns as a whole become periodic with a period length of at least 20 minutes.

Within 20 mg of the zeolite support catalyst 10^{17} palladium particles are distributed. These particles are single crystals [16,17], 1 - 10 nm in diameter. The above authors found both Pd and PdO signals in the x-ray spectra of the catalyst after reaction regardless of the starting material: Pd (totally reduced palladium support catalyst) or PdO (totally oxidized palladium support catalyst) [see also 14].

They therefore assumed a phase transition betwen Pd and PdO as the essential process causing the oscillations. Within the framework of these ideas, the palladium crystals act as stores, and there are critical values correlated to the filling of the palladium with oxygen which characterize the phase transitions: Pd/PdO and PdO/Pd.

It can be shown [14,18,19,24] that O_2 is not adsorbed at PdO below 218°C. However, CO can be adsorbed at PdO even at low temperatures (T > 76°C) reducing PdO to Pd without any excess oxygen [20]. At high temperatures the reduction process is faster than at lower temperatures.

On the other hand, oxygen can be chemisorbed at Pd at temperatures below 200°C [21], and forms a solid solution of oxygen in the Pd bulk above this temperature [21,22,23]. As we have shown crystallographically [10,14], the PdO phase can be achieved even during the catalytic oxidation of CO. Thus one can assume that there exist at least two kinetic phases:

- first, the production of CO_2: $CO + 1/2\ O_2 \rightarrow CO_2$, which takes place in the Pd-surface.
This exothermic reaction is accompanied by the formation of a solid solution of oxygen in the Pd-bulk. If one reaches the threshold for the phase transition of Pd → PdO, a second kinetic phase occurs.
- The second kinetic phase is characterized by the reduction process of the PdO-bulk: $PdO + CO \rightarrow Pd + CO_2$.
However this process is slow compared with the first one, which is a surface reaction. So, only a negligible amount of CO_2 is produced within a comparable time interval.

There is another experimental fact which cannot be neglected in modelling: there are about 10^{17} palladium particles distributed in the catalyst. With respect to the experimental time series (Fig. 1), all of them have to act in synchrony, especially if one regards the large breaks in the CO-conversion. So there must be coupling amongst the catalytically active palladium particles.

Furthermore, these palladium single crystals are incorporated in about 10^7 zeolite crystals. Therefore, there should be another mechanism responsible for the coupling of the reactions among all the zeolite crystals.

The catalyst was pressed and afterwards slightly ground [14]. As a result of this treatment, the catalyst is composed of several (10^2 - 10^3) agglomerates of the zeolite crystals. These macroscopic agglomerates (size:

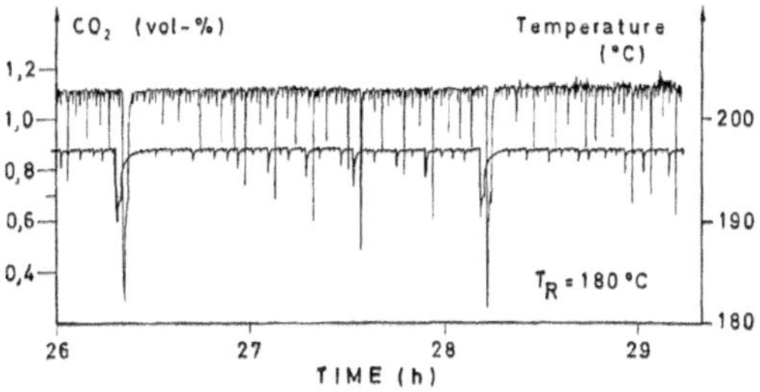

Figure 1. Self-similar oscillations of the heterogeneous CO- oxidation at a Pd-support catalyst (14,7 wt-% Pd incorporated in zeolite-X single crystals; Pd-single crystals of about 10 nm in diameter; 21,4 mg catalyst, size of the pellets: 0.1 - 0.28 mm in diameter; reactor temperature T_R = 180°C; flow rate of the reactants 9 l/h, CO concentration in the inlet: 1,10 vol-%). The upper curve shows the CO_2-concentration in the outlet, whereas the lower curve exhibits the corresponding temperature of the catalyst holder. Both curves are shifted against each other for experiments reasons, but there is no phase shift between both the oscillations. The period length of the self-similar regions is of about two hours.

10^{-1} - 10^{0} mm in diameter) might be regarded as the elementary reactors of the catalytic arrangement.

With respect to the macroscopic character of the catalytic reaction the time scales of all processes inside the elementary reactors are so small that we can neglect them, if we wish to describe the macroscopic behaviour of the system. Each of these elementary reactors thus forms one unit of a new kind in a "macroscopic chemical reaction".

3. The models

3.1. *The one-dimensional exor-rule automaton*

It is the aim of this work to describe only the pattern of those time series mentioned above. The time series is a macroscopic feature of the system. So it is not intended to return to the molecular level of description, or even to the level of the Pd or PdO single crystals.

For this purpose, let us divide the whole catalyst into a number of regions representing the elementary unit reactors. In the mathematical formalism of the model, the abstract expressions of these regions are the cells. The state of the cells may be zero (0) - if the unit reactor is productive, or one (1) - if the unit reactor is unproductive. For simplicity, the cells are arranged linearly. Each cell may have only one nearest neighbour - for instance the lower one.

A simple dynamics is easily introduced into this artificial reaction system by [17]:

$$z(i,t+1) = (z(i,t) + z(i-1,t)) \bmod 2 \qquad (1)$$

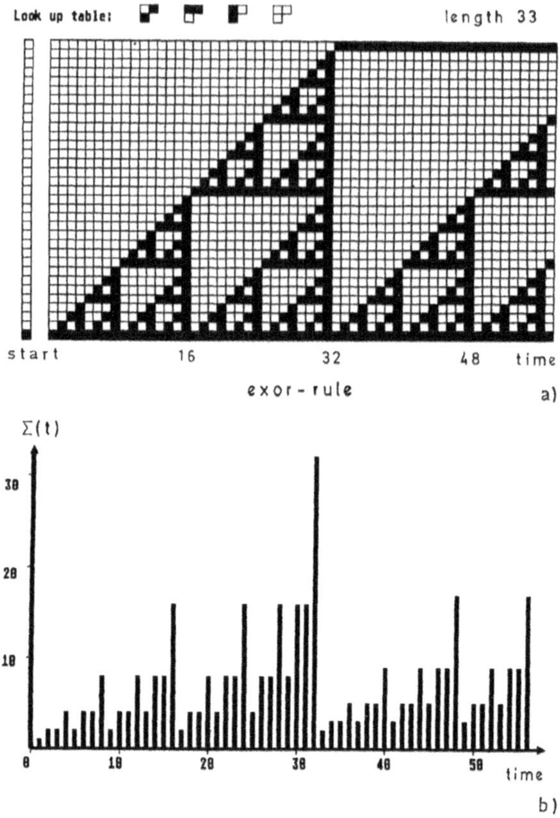

Figure 2. (a) Temporal development of a one-dimensional automaton the length of which is 33 cells. The transformation rule (1) is the exor-rule. The rule is shown in the look-up table. (b) Temporal development of the corresponding function $\Sigma = \Sigma(t)$ of the state sums of the automaton.

Figure 2a shows the temporal development of this automaton. The resulting fractal pattern in space-time is the discrete version of the Sierpinsky gasket. If we sum up all states $z(i,t)$ of the automaton at a given time t, then this sum $\Sigma = \Sigma_i z(i,t)$ is as a function time t: $\Sigma = \Sigma(t)$. The plot of the graph of this function (Fig. 2b) approximates qualitatively very well the experimental time series.

The essential features of the function $\Sigma = \Sigma(t)$ are:

(i) it exhibits very sharp and also very large peaks, which resemble the breaks in the conversion of the chemical system;

(ii) the temporal sequence of the sum Σ has a self-similar structure which is correlated to the fractal structure of the automaton in space-time;

(iii) because of the finite length of the automaton the function $\Sigma = \Sigma(t)$ becomes periodic in time.

There are many possible ways to increase greatly the similarities between the one-dimensional automata model and the time series of the chemical experiments [12,13,17]. For example, one can enhance the number of neighbourhood cells which will influence the future state z(i,t+1) of the i-th cell. In this way the next state plot of the automaton resembles well the Poincaré plot of the experimental time series [17].

3.2. *A two-dimensional automaton with an extended exor-rule dynamic*

However, this powerful but simple model does not reflect the two- dimensional arrangement of the catalyst in the experiment. For this reason, let us take as our starting point an isolated cell placed in the centre of a two-dimensional quadratic lattice. The requirement to reproduce the sharp breaks in conversion forces us to assume the essential ideas of the one-dimensional model:

firstly, only an exposed cell in a connected area of cells, with each of them in the state z(i,t)=1, creates at time (t+1) an isolated neighbouring cell z(i/j,t+1) the state of which at time t is z(i/j,t)=0;
secondly, all cells of the connected block of cells with z(i,t)=1 together reach the state z(i,t+1)=0 within only one time step;
and thirdly, an isolated cell in the state z(i,t)=1 is followed by several neighbouring cells z(i/j,t+1)=1.

The rules in Fig. 3 are based on these basic ideas. They can be formulated exactly, in the following manner:

$$z(i,t+1) = \begin{cases} 1 & \text{if } z(i,t)=0 \text{ and } \quad (z(i/j,t)) = 1 \\ 0 & \text{if } z(i,t)=0 \text{ and } \quad (z(i/j,t)) = 0 \end{cases}$$
$$z(i,t+1) = \begin{cases} 1 & \text{if } z(i,t)-1 \text{ and } \quad (z(i/j,t)) = 1 \\ 0 & \text{if } z(i,t)=1 \text{ and } \quad (z(i/j,t)) \geq 2 \end{cases} \qquad (2)$$

(z(i/j,t))=k means the number of all cells j neighbouring the cell i which are in the state k. The cell i is its own neighbour. Using rule (2), the automaton becomes covered with several clusters of cells with z(i,t)=j at one time. These blocks are followed by a set of isolated cells at time (t+1) which are in the state z(i,t)=1.

Again summing the states of all cells, we obtain a function $\Sigma = \Sigma(t)$ which also exhibits:

(a) sharp and very large peaks;
(b) a self-similar structure in time; and
(c) periodicity.

Figure 4 shows one state of the automaton (at time t = 90) and the correlated function $\Sigma = \Sigma(t)$. Σ is proportional to the degree of coverage of the automaton with cells in state one. Since the cells in this state z(i,t)=1

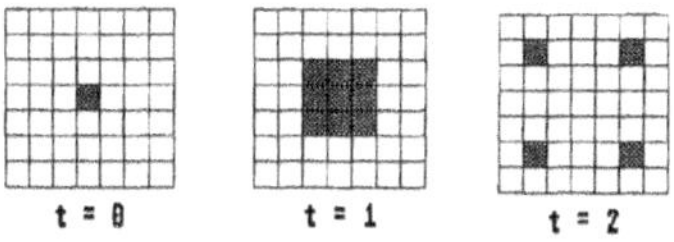

Figure 3. Graphical representation of the transformation rule (2) for the two-dimensional automaton.

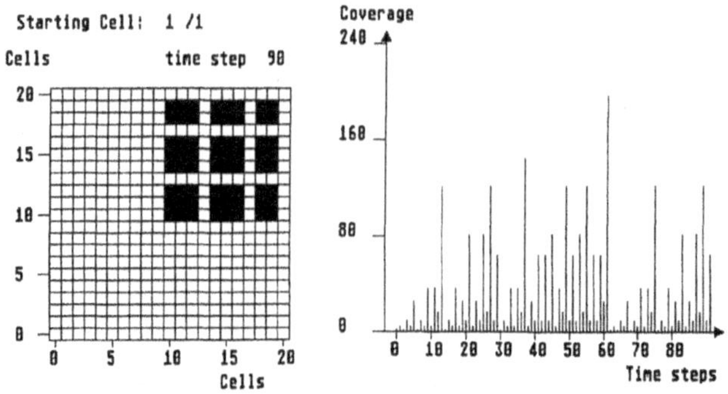

Figure 4. A two-dimensional automaton with the dynamic (2) of **"cooperative annihilation and exposed adsorption"**. The state of the automaton is shown at time t=90 and the development of its state sum $\Sigma = \Sigma(t)$ up to this time step. Starting with the cell (1,1) the period length of $\Sigma(t)$ becomes 62 time steps because of the spatial symmetry, whereas the period length of the automaton is 124.

represent the non-productive unit reactors, the function $\Sigma = \Sigma(t)$ resembles the sequence of breaks in the experimental time series. Depending upon the position of the starting cell, different patterns of the sum $\Sigma = \Sigma(t)$ can be achieved.

3.3. *The one-dimensional vector automaton model*

In both the one-dimensional and the two-dimensional automata discussed above, the state of a cell is given by the number zero or one. These numbers represent the productive or the non- productive kinetic phases of the elementary reactors. However, the filling state of the reactors, for example with oxygen, and their temperatures have so far been neglected. But describing the state of a cell with a vector instead of a scalar, filling and temperature can also be taken into account.

Now the dynamics consists of transformation rules for all three components of the vector state:

$$z(i,t) = \begin{pmatrix} p(i,t) \\ f(i,t) \\ T(i,t) \end{pmatrix} = \begin{pmatrix} \text{phase} \\ \text{filling} \\ \text{temperature} \end{pmatrix} \qquad (3)$$

The transformation rules of the phases p(t) depend on the filling f(t) of the cells and the temperature T. The rules are given in look-up tables like the following one:

filling		f(i,t+1)	0	1	2	3	4	5	6	
phase	if p(i,t)=0	p(i,t+1)	0	0	0	0	1	1	1	(4)
phase	if p(i,t)=1	p(i,t+1)	0	0	1	1	1	1	1	

The total rule consists of a set of 25 such tables for different temperatures. The whole set represents a discrete bifurcation diagram of the phase changes in the filling and temperature plane.

There is a hysteresis included in these rules (4). The transformation rules for the filling state are given by:

$$f(i,t+1) = f(i,t) + c_f$$
$$c_f = f(p(i,t),\ p(i-1,t),\ T(i,t), T(i,t)) \qquad (5)$$

and they depend upon the phase p(i,t) and the temperature T(i,t) of each cell i and its lower neighbour (i-1).

The transformation rules for the temperature T(i,t) of each cell i are now required:

$$T(i,t+1) = 1/2 * (T(i,t) + c_T - a_k * T(i,t) + T(i-1,t))$$
$$c_T = f(p(i,t)) \qquad (6)$$

where c_T is correlated to the heat production of the elementary reactor and a_k is its cooling rate. The factor 1/2 results from the formation of the arithmetic average. Figure 5 shows the spatial and temporal development of such a linear cellular vector automaton for a given set of parameters. In this way one can reproduce the fractal structure of the automaton in space-time without referring to the crude rules mentioned at the beginning.

This short example may give a vague idea on how to model this reaction by using one-dimensional cellular automata.

3.4. *An excitable automaton with a fractal framework*

One aspect has not yet been taken into account, namely the non- uiform distribution of the elementary reactors within the catalyst. There are two very different methods for modelling the non-uniform distribution of the reactor. One may assume an arbitary distribution of a low amount of reactors and hence a stochastic algorithm for selecting cells within a given framework. This might be quadratic or hexagonal [25]. For a large amount of reactors this distribution will lead to uniform but stochastic distribution. There is another possibility which is favoured by Gerhardt & Schuster [26,27] in their

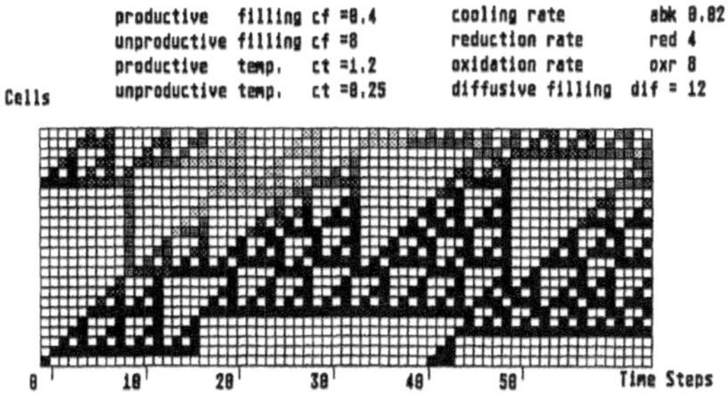

Figure 5. Development of a one-dimensional cellular vector automaton with a fractal structure in space-time. The dynamic of this automaton is given in the equations (4), (5) and (6). Only cells in the state z(i,t)=1 are coloured. The black cells are of low temperature, the grey cells are of high temperature.

"infection"-model. There, averaging of the neighbouring cells will lead to a quasi-uniform behaviour of the automaton. In this way the corners of the quadratic "circles" in the square lattice become well-rounded.

The other method starts from the assumption that the catalyst is self-similarly distributed on the holder. This means that there exist holes of all sizes, without any catalyst or catalytic area, which are equally distributed at all scales. In reality this will be true only for a small range of scales [27-34]. Nevertheless, one can form a very simple cellular automaton of finite size which allows the modelling of such a distribution [17]. In the simplest case the automaton works on the Sierpinsky lattice, where each vertex is a cell and its neighbourhood is defined by the adjacent vertices. If the lattice is octahedrally closed, the corresponding graph becomes four regular. This resembles the von Neumann neighbourhood in the square lattice. In the case of an infinite lattice, the growth rate dimension becomes D = 1,58 (Fig. 6).

In order to model the self-similar experimental time-series, one could assume that this automaton acts as an excitable medium and a travelling soliton wave of excited states will pass through the framework. After this, the cells of the automaton will return in the same state as before. Therefore an excitable automaton can be characterized by its ability to propagate signals undamped over long distances [35].

To construct such an excitable automaton, one can take either three different states $z_k(i,t)$; $k = 1,2,3$ of the i-th cell at time t, or one can use only two present states $z_l(i,t)$; $l = 1,2$ and the past state $z_m(i,t-1)$; $m = 3$ as the third state [17]. Since it is preferred to keep the two states model using the productive and the non-productive state of the elementary reactors, the second possibility is used.

$$z(i,t+1) = \begin{cases} 1 \text{ if } z(i,t) = 1 & \text{and } z(i,t-1) = 0 \\ 0 \text{ if } z(i,t) = 1 & \text{and } z(it-1) = 1 \text{ reactive desorption} \\ \text{and } & (z(i/j,t) = 1) \geq 2 \end{cases}$$

$$z(i,t+1) = \begin{cases} 1 \text{ if } z(i,t) = 0 & \text{and } z(i,t-1) = 0 \text{ adsorption} \\ 0 \text{ if } z(i,t) = 0 & \text{and } z(i,t-1) = 1 \\ \text{and } & (z(i/j,t) = 1) \geq 1 \end{cases} \tag{7}$$

The states z(i,t-1) of the past time (t-1) will retard the state change from z(i,t)=1 to z(i,t+1)=0. This "historic" rule reflects the hysteresis behaviour of the elementary reactors in the catalyst.

If one searches for circular patterns in this automaton, one has to take into consideration the specific geometry of the underlying Sierpinsky lattice. As elsewhere, a circle on this framework is defined by the set of all vertices or points which can be reached by a path of minimal length, which resembles the radius in the two-dimensional Euclidean space. Within this fractal framework a "circular" wave degenerates to two or three straight lines which are somehow orthogonal. These lines need not be connected at any point. If such a straight line reaches the exposed vertices of the lattice, it creates in general two new lines, which are orthogonal both to each other and to their mother line. If these lines touch the borderlines and meet each other in a

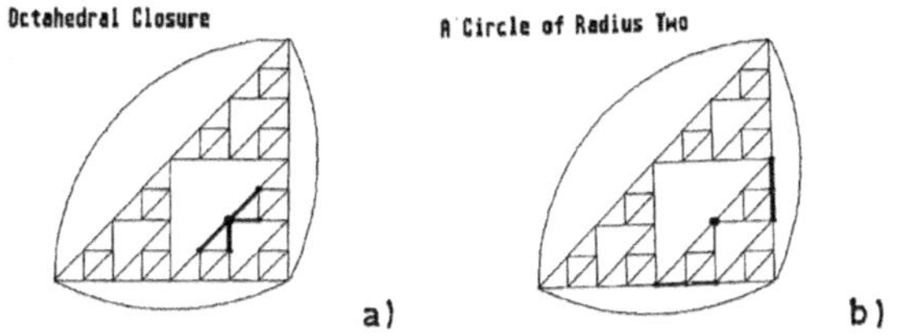

Figure 6. Finite Sierpinsky graph with octahedral closure.
(a) Each vertex of this graph is of degree four.
(b) A "circle" on this graph consists of unconnected straight lines. In this special case there are two lines belonging to the central point (6,2). The "radius" of the "circle" is of the length two.

common vertex, they will annihilate each other.

Figure 7 demonstrates the spreading of such a "circular" wave without dispersion similar to a soliton on a Sierpinsky lattice [17]. If one starts with one or only a few cells in the state z(i,0)=1, one obtains travelling "circles" intersecting each other. But in general all of them will vanish after a while by annihilating each other or by touching the borderlines. In some respects, the sequence of the sate sum resembles the experimental time series of the conversion. But if one compares these sums with the total amount of cells, these sums are too small to explain the large breaks in conversion.

Starting, however, with a large amount of cells in the state z(i,0)=1 which might be randomly distributed, the "coverage" of the automaton oscillates in time with strong periodicity [17]. However, these oscillations do not resemble the experimental time series, because they do not have a self-similar structure in their corresponding time series. So the aforementioned rules (7) represent the dynamics of the spreading of an

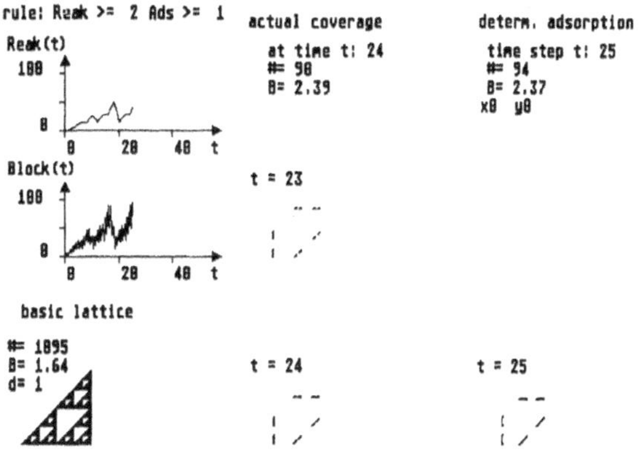

Figure 7. Spreading of a "circular" wave on a Sierpinsky lattice. The starting point of this wave was the point (0,0) which is in the left corner of the lattice. The "circle" consists of three "lines", which are not necessarily connected anywhere. These lines vanish at the time t = 34.

excitation in an excitable fractal automaton, but they do not reflect the dynamic of the chemical system under consideration.

3.5. *The mechanism of cooperative annihilation and exposed adsorption operating on an excitable fractal automaton*

The simple automata which I discussed at the beginning model the chemical observations better than the mechanism of a spreading "circle" on a fractal automaton. On the other hand, it is just the fractal lattice structure of the automaton which is the basic model for an ideal amorphous system. So one should combine the best of both models. I should therefore like to transfer the dynamics (7) of the "cooperative annihilation and exposed adsorption" which I used at the beginning onto the fractal framework. Let me first emphasize again what the main ideas of this rule are. In the one-dimensional as well as in the two- dimensional case one can observe the temporary formation of clusters of cells with $z(i,t)=1$. These clusters vanish by **cooperative annhilation** after very short times:

$$z(i,t)=1 \longrightarrow z(i,t+1)=0 \quad \text{if} \quad (z(i/j,t) = 1) \geq 2 \qquad (8)$$

Observing this process one can ascertain an incessant birth of temporarily growing clusters. **Adsorption** ($z(i,t)=0 \longrightarrow z(i,t+1)=1$) at the borderlines of these clusters occurs only at those cluster cells which are **exposed**. This means that the state change

$$z(i,t)=0 \longrightarrow z(i,t+1)=1 \quad \text{if} \quad (z(i/j,t)=1) = 1 \qquad (9)$$

only happens if just one cell j neighbouring the cell i is in state $z(j,t)=1$. During this adsorption process the temporary excitations move chaotically around the framework of the automaton.

This mechanism of **cooperative annihilation and exposed adsorption** is the essential dynamics for obtaining an automaton structure which is fractal in space-time. Figure 8 shows some clusters at time $t = 20$ occurring during the temporal development of the fractal automaton, whereby the dimension of its underlying framework is $D=1.66$. The self-similar time series of the state sums of the automaton is correlated to its fractal structure in space-time.

As one can see, there is a spreading of a "circular" front of two diffusive "lines", but behind this front new clusters are permanently being created, but survive only temporarily. But it can also be seen that almost the whole lattice of the automaton is covered with cells in state $z(i,t)=1$.

Therefore large state sums are possible and similarity between this model and the experimental time series increases.

4. Conclusion

Within this article I have presented several automata to model the heterogeneous oxidation of CO by a Pd-support catalyst. A new mechanism of **cooperative annihilation and exposed adsorption** has been developed, to describe the self-similar structure of the experimental time series by cellular automaton models.

This mechanism is studied by the use of different cellular automata of dimension between 1 and 2. It could be shown that this mechanism always

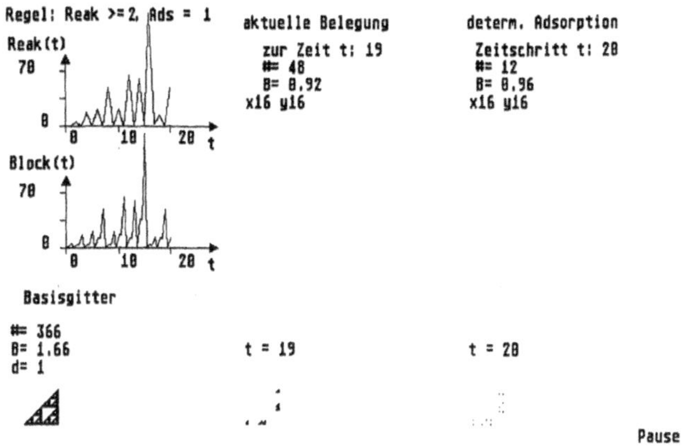

Figure 8. Snapshot of the spreading of a "circular" wavefront on a Sierpinsky lattice. The starting point was the point (16,16). The dynamic of the spreading front is the mechanism of the **"cooperative annihilation and exposed adsorption"**. The corresponding sequence of the state sums (compare the graph of the function Block = Block(t)) is self-similar within time. The function Reak(t) represents the number of cells which are changing their state z(t) = 1 to z(t+1) = 0. The values reak and ads mean the number of neighbouring cells in the state z(t)=1 and z(t)=0, which are needed for successful reactive desorption (z(t)=1 → z(t+1)=0) and deterministic adsorption (z(t)=0 → z(t+1)=1) respectively. V is the number of cells in the basic automaton or the number of cells in the state z(t)=1 in the instantaneous state of the automaton respectively. B is the growth rate dimension and d the distance between a cell and its nearest neighbours.

produces the interesting self-similar structures with respect to the temporal sequence of the sum of states of the cells. The propagating fronts of the fractal structure in space- time enables us to transport signals undamped over long distances. therefore one can put together all the automata which are ruled by the mechanism of the **"cooperative annihilation and exposed adsorption"** into the class of the excitable automata. But their dynamic works indeed quite different from that of a circularly spreading excitation.

The essential new feature of this dynamic is the fractal structure of the spreading excitation in space-time. That is neither the well-known circle nor the spiral, but a third, really new structure which has not previously been discussed comprehensivlely.

Acknowledgement

I am indebted to Dr. Karin Möller who makes her experimental work available to me.

References

[1] Beusch, H., Fieguth, P. & Wicke, E. (1972). *Chemie-Ing. Techn.* **44**, 445-451.

[2] Plichta, R.T. & Schmitz, R.A. (1979). *Chem. Eng. Commun.* **3**, 387-398.

[3] Rathausky, J. & Hlavacek, V. (1981). *J. Chem. Phys.* **75**(2), 749-756.

[4] Wicke, E. (1983). *Nova Acta Leopoldina* **55(257)**, 3-23.

[5] Ertl, G., Norton, P.R. & Rüstig, J. (:1982). *Phys. Rev. Lett.* **49**, 177-180.

[6] Varhhese, P., Carberry, J.J. & Wolf, E.E. (1978). *J. Catalysis* 55, 76-87.

[7] Imbihl, R., Cox, M.P. & Ertl, G. (1986). *J. Chem. Phys.* **84(6)**, 3519-3534.

[8] Slin'ko, M.M., Jaeger, N.I. & Svensson, P. (1989). *J. Catalysis* **118**, 349-359.

[9] Razon, L.F., Chang, S.-M. & Schmitz, R. (1986). *Chem. Eng. Sci.* **41(6)**, 1561-1576.

[10] Jaeger, N.I., Möller, K. & Plath, P.J. (1986). *J. Chem. Soc. Faraday Trans. I* **82**, 3315-3330.

[11] Plath, P.J., Möller, K. & Jaeger, N.I. (1988). *J. Chem. Soc. Faraday Trans. I.* **84**, 1715-1771.

[12] Plath, P.J. (1989). *Proceedings of the 4th IPSO-Konferenz in Rostock.*

[13] Plath, P.J., Meyer, N. Höth, Th. & Woydack, D. Submitted to *Z. phys. Chem.*

[14] Möller, K. (1984). Thesis: *Untersuchung des dynamischen Verhaltens der CO-Oxidation an Pd-Trägerkatalysatoren;* Fachbereich Biologie/Chemie der Universität Bremen.

[15] Svensson, P. (1988). Thesis: *Die Oxidation von Kohlenmonoxid an palladiumdotierten Zeolithen unter Einfluss einer pulsförmigen Eduktzufuhr variabler Zusammensetzung;* Fachbereich Biologie/Chemie der Universität Bremen.

[16] Kleine, A., Ryder, P.L., Jaeger, N.I. & Schulz-Ekloff, G. (1986). *J. Chem. Soc. Faraday Trans. I.* **82**, 205-212.

[17] Plath, P.J. (ed.) (1989). In *Optimal Structures in Heterogeneous Reaction Systems,* p. 1-25. Springer Series in Synergetics, Vol. 44. Springer-Verlag: Berlin, Heidelberg, New York.

[18] McKinney, P.V. (1933). *J. Am. Chem. Soc.* **55**, 3626.

[19] Imbihl, R. (1984). Thesis: *Nichtgleichgewichs- Phasenübergänge bei der katalytischen Oxidation von CO an Pt(100);* Phys. Chem. Inst. der Ludwigs-Maximilian Universität, München.

[20] McKinney, P.V. (1932). *J. Am. Chem. Soc.* **54**, 4498.

[21] Palazow, A., Chang, C.C. & Kokes, R.J. (1978). *J. Catal.* **36**, 338.

[22] Cambell, C.T., Foyt, D.C. & White, J.M. (1977). *J. Phys. Chem.* **81**, 491.

[23] Ertl, G. & Koch, J. (1972). In *Adsorption-Desorptio Phenomena,* Ricca, F. (ed.), p. 345. Academic Press: New York.

[24] Imbihl, R. (1989). In *Optimal Strucutres in Heterogeneous Reaction Systems,* Vol. 44,Plath, P.J. (ed.), p. 26. Springer-Verlag: Berlin, Heidelberg, New York.

[25] Markus, M. (1989). Lecture at the NATO Advanced Research Workshop, *Nonlinear Wave Processes in Excitable Media*, Leeds; see also this volume.

[26] Gerhardt, M. & Schuster, H. (1989). *Physica* **D36**, 209-221.

[27] Gerhardt, M. (1987). Thesis: *Mathematische Modellierung der Dynamik der heterogen katalysierten Oxidation von Kohlenmonoxid: Numerische Behandlung eines diskreten mathematischen Modells von über Diffusion miteinander gekopopelter chemischer Speicher;* Fachbereich Biologie/ Chemie der Universität Bremen.
[28] Pfeifer, P. & Avnir, D. (1983). *J. Chem. Phys.* **79(7)**, 3558-3565.
[29] Avnir, D., Farin, D. & Pfeifer, P. (1983). *J. Chem. Phys.* **79(7)**, 3566-3571.
[30] Farin, D. & Avnir, D. (1987). *J. Phys. Chem.* **91**, 5517-5521.
[31] Farin, D. & Avnir, D. (1988). *J. Am. Chem. Soc.* **110**, 2039-2045.
[32] Schmidt, P.W. (1989). In *The Fractal Approach to Heterogeneous Chemistry*, Avnir, D. (ed.), p. 67-79. John Wiley & Sons Ltd.: Chichester, New York, Brisbane, Toronto.
[33] Kaye, B.H. (1989). *A Random Walk Through Fractal Dimension.* Verlag Chemie: Weinheim.
[34] Spindler, H. & Becker, K. (1987). *Chem. Techn.* **39**, 139.
[35] Gerhardt, M., Schuster, H. & Tyson, J.J. (1990). Preprint; to appear in *Science*.

43. PROPAGATION OF CHEMICAL WAVES IN DISCRETE EXCITABLE MEDIA: ANISOTROPIC AND ISOTROPIC WAVE FRONTS

Christian Kurrer and Klaus Schulten

Beckman Institute and Department of Physics
University of Illinois at Urbana-Champaign
405 N. Mathews, Urbana, IL 61801, U.S.A.

1. Introduction

Cellular automaton theory has been recognized as a useful tool for the study and the simulation of wave processes in excitable media (see [2]) and presents an alternative to the integration of partial differential equations (pde), as used by Tyson et al. [10]. By incorporating only the most important characteristics of the dynamical system into the automaton rules, one can reproduce the dynamical properties of a large system of coupled nonlinear oscillators within a fraction of the computing time that is needed to integrate numerically corresponding partial differential equations.

For the purpose of a computer simulation one divides an excitable medium into a large array of elements. For ease of programming, these elements are arranged on a regular quadratic lattice. Since such a lattice breaks the isotropy of the medium, anisotropic and thus "unrealistic" wave geometries can result. In the framework of pde's this problem of "spatial stiffness" is overcome by dividing the excitable medium into a sufficiently fine mesh. Simulations typically use lattices with a lattice constant of the order of 1/50 of the wavelength of the resulting wave (see e.g. [10]).

The dynamics of excitable media are given by the local properties of the elements and by the coupling of elements. This coupling, i.e. the response of one element to the action or state of another element, is usually assumed to decrease with growing distance between the elements. Different directions with respect to the lattice axes will exhibit different nearest neighbour spacings and, as a result, one excitable element will elicit different responses along these axes.

The more sensitive the response of excitable elements is to small variation in the elements' stimulation, the more a resulting wave pattern will reflect the anisotropy of the lattice. Near their excitation threshold, excitable media show such sensitivity, and therefore the wave fronts resulting from computer simulations are prone to reflect the geometry of the underlying lattice. This is in sharp contrast to simulations of diffusion or heat flow, the dynamics of which is not sensitive to small fluctuations of any dynamical variable.

In the cellular automaton approach to the simulation of excitable media, the problem of artificial wave geometries is accentuated by the discrete dynamics of the excitable elements. For example, if the automaton rules are such that one automaton can excite a finite set of neighbouring automata and does not affect other automata, the polygonal outline of the set of neighbouring automata will be reflected by the excitation wave front.

In simulations based on the integration of pde's one can alleviate this problem by using finer lattices of excitable elements and shorter time steps. Similarly, in automaton simulations, one can go over to models with more states, with more elaborate rules and with interactions involving larger neighbourhoods. However, such procedure is in contrast to the original aim of the cellular automaton approach which is to reduce the complexity of the system and the computing time.

The question arises then, how simple modifications of the lattice automata models can lead to natural, i.e. isotropic, lattice-independent geometries of the excitation wave front.

2. **Random distributions of excitable elements**

Isotropic wave fronts arise in nature in the case of homogeneous isotropic excitable media, e.g. in the case of the well-known Belousov-Zhabotinsky reaction. However, the underlying homogeneous conditions cannot be reproduced on a digital computer. But there exist in nature also systems with a blatantly discrete character that give rise to isotropic wave fronts, an example being colonies of slime mold amoebae. A single amoeba can be considered a discrete excitable element. In an amoeba colony, isotropic wave fronts are not achieved because of an underlying homogeneous excitable medium, but are achieved rather because of a random distribution of amoebae. Such a system, however, can be readily reproduced in computer simulations by a lattice with a larger fraction of empty lattice sites. Figure 1 compares two extreme situations: the left side shows a regular lattice of excitable elements, that will give rise to polygonal wave fronts, whereas the random distribution on the right on a large spatial scale side will yield isotropic wave fronts.

The question is how to implement random distributions best on a computer. Mackay [4] distributed 1000 excitable elements randomly on a 2D surface to simulate successfully the cAMP wave propagation in slime mold colonies. He distributed the excitable elements without any regard to an underlying lattice, i.e. the coordinates of elements were chosen continuous insofar as this was possible on a computer. Another implementation was realized by Markus & Hess (see their contribution in this book [5]). In their approach, one cell

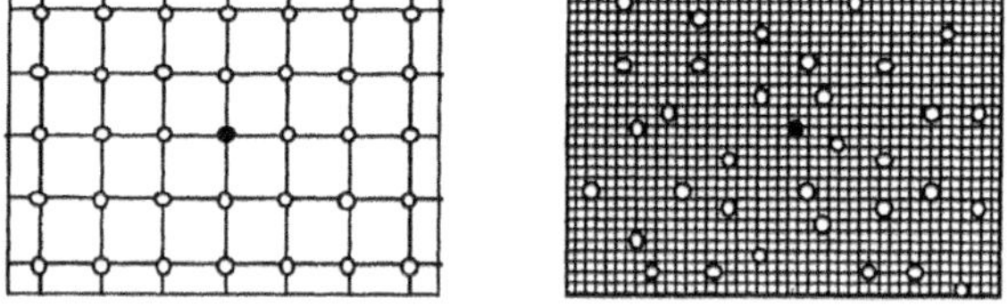

Figure 1. Fully occupied lattice of excitable elements (left side) and random distribution of excitable elements (right side). The situation on the right side will be interpreted as a partially occupied lattice.

was put randomly anywhere in each of the cells of a square lattice. Also in the approach of these authors, elements are essentially represented by a continuous position. In comparison to regular lattices, in these approaches the information about the neighbourhood of an excitable element is totally or partially lost. The distances between the excitable elements can have any values; thus the strength of the interaction has to be evaluated for every pair of excitable elements within a certain neighbourhood.

In this contribution we like to suggest a new approach which is reflected in the partially occupied lattice in Fig. 1. In our approach, automata are distributed on the lattice points of a fine-grained lattice with a low number of lattice points occupied. Each automaton is then represented by a discrete lattice position. In our approach, the simple neighbourhood relations typical of automaton simulations are preserved, with the only extension that some sites will not response to stimuli by neighbouring sites because they do not contain an excitable element. As the occupancy, i.e. the fraction of lattice sites occupied by excitable elements, is lowered, the distance over which automata interact has to be gradually increased such that an automaton remains to interact with approximately the same number of neighbouring automata.

Our approach not only benefits from having very simple neighbourhoods, but also allows the degree of randomness and isotropy to increase continuously by lowering the occupancy of a lattice. It would be ideal to use in simulations a lattice occupancy that is low enough to ensure isotropic wave propagation, and is high enough to limit the additional computational effort caused by the use of large, sparsely occupied lattices.

We have therefore investigated the properties of the transition from polygonal wave fronts of fully occupied lattices to isotropic wave fronts for sparsely occupied lattices. Our study has been based on the realization that the aforementioned transition in partially occupied lattices of excitable media is closely related to the phenomenon of percolation.

3. Coupling of excitable elements

Chemical media capable of generating spontaneous spatio-temporal order can be generated by chemical reactions with nonlinear kinetics involving activating and inhibiting substances [9,7,6]. The coupling between the local nonlinear kinetics is realized in most natural systems by diffusion of either the activating or the inhibiting substance, or of both substances. In fact, diffusion is itself a nonlinear process. This means that the diffusing substances reach nearby neighbours not only earlier than more distant ones, but the nearby neighbours are also reached with a higher velocity. More distant neighbours will either be reached with a very slow velocity or not be reached at all.

For our investigations, we described the propagation of the excitation from one element to the neighbouring elements by a law that is typical for diffusive processes, namely a $\sqrt{t}$-law. In this case the time needed for the excitation of one element to reach another element at a distance d is described by the power-law

$$t \sim d^2. \tag{1}$$

This law might not be realistic for all excitable media, but it yields a good model to investigate the effect of lowering the occupancy of a lattice of excitable elements. Equation (1) ensures that only the nearest neighbours of an excitable element are reached with maximal velocity. Furthermore, this property implies that besides the nearest neighbours, only a few next-nearest neighbours contribute to the propagation of the excitation wave, because distant neighbours take much longer to be reached directly by the cell originally excited than to be reached by relayed propagation involving nearest and next-nearest neighbours.

The use of a power-law ensures that as the occupancy of the lattice is varied, the size of the effective neighbourhood of an automaton, i.e. the distance over which direct excitation between automata is relevant to the propagation of the wave front, is automatically adjusted so that it always contains approximately the same number of cells. The number of cells in the neighbourhood is thus independent of the lattice occupancy; however it can be varied by changing the exponent in the power-law (decreasing the exponent will increase the size of the effective neighbourhood). Furthermore, the selection of the exponent in the power law (1) not only simulates diffusive coupling very well, but also ensures that neighbours are reached only at discrete times that are multiples of the time necessary to reach the nearest lattice point. This is very convenient for an automaton description.

4. Our simulations

For our simulations we used the following automaton rules: if one automaton k is excited, it will fire immediately. The automata that are at a distance of one lattice constant from automaton k will be excited in the next time step. Automata at a distance of n lattice constants from automaton k will be excited after n^2 time steps. In our investigations, we focused on the geometric properties of the wave front. We were thus only interested in the time at which a given cell was reached for the first time by an excitation and consequently had fired. After firing, our automata entered the refractory period and would not fire again.

We started with simulations of 10^4 excitable elements which we distributed on lattices of varying sizes, thus obtaining different lattice occupancies. As an example, Fig. 2 shows a typical result of a simulation of wave fronts at equally spaced instances for a lattice with 50 per cent of the sites occupied. The wave fronts presented, except for a scatter due to random occupation of lattice sites, exhibit clearly isotropic propagation. We consider this a key result of this contribution. This isotropic propagation has been achieved at small computational expense.

In order to analyse how isotropic propagation arises for a sufficiently low lattice occupancy we need to consider the convergence of propagation velocities in the axial and diagonal directions of the lattice. To monitor

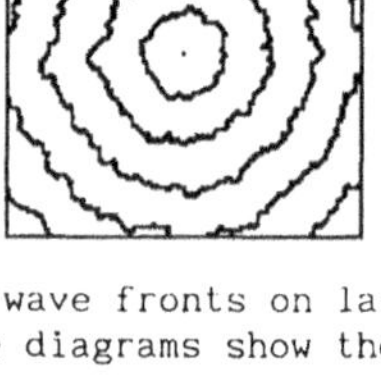

Figure 2. Simulation of wave fronts on lattices with 50 per cent occupancy. The diagrams show the wave front at equally space instances.

this convergence we consider mean propagation velocities in different lattice directions. Accurate values for these velocities can be obtained by averaging. For this purpose we have averaged wave front propagation over 150 runs like the one represented in Fig. 2. We also averaged over the eight equivalent octants in order to obtain smooth polar plots. The results are presented in Figs. 3 and 4. We would like to point out that the averaging procedure has only been adopted for the following mathematical analysis, but is not needed when our method is used for a simulation of isotropic propagation.

Figure 3 compares some averaged wave fronts obtained by simulations on lattices with 100 per cent, 70.7 per cent and 50 per cent occupancy. It can be observed that by reducing the occupancy of the lattice from 100 per cent to 50 per cent, the initially strong anistropy of the wave fronts virtually disappears. To detect a small remaining anistropy, Fig. 4 presents the velocities with which the automata surrounding the centre are reached. Below occupancies of 25 per cent, even the propagation velocities do not reveal any significant anisotropy.

For simulations of waves travelling **repeatedly** through an excitable medium, one would have to introduce a finite refractory period τ_{ref} into the automaton rules. Simultaneously, the time span during which one automaton affects neighbouring automata would have to be limited to an even shorter period than τ_{ref} to avoid artificial self-excitation in the trail of the wave front. As discussed previously, excitations which involve distant neighbours and which are slow, do not contribute to the propagation of wave fronts. Therefore, the above extensions of the automaton rules would not alter the propagation of the wave front.

Figure 4 also demonstrates a relationship between the curvature of the propagation front and propagation velocity. For a lattice occupancy of 25 per cent, this relationship can be described by the heuristic formula

$$v(r) = c - \frac{n}{r + m} \qquad (2)$$

where $c = 0.5$, $n = 3.5$ and $m = 16.5$, and where the radius r, expressed in units of the lattice constant, measures the curvature of the wave front. $v(r)$ is the velocity of the wave front in units of lattice constants per time step. c is the **asymptotic wave propagation velocity** for planar wave fronts. Equation 2 reproduces qualitatively the relations discussed in [1] for the curvature dependence of the propagation velocity of a wave front in an excitable medium. The constant m in our formula is needed to account for the discrete nature of our excitable medium.

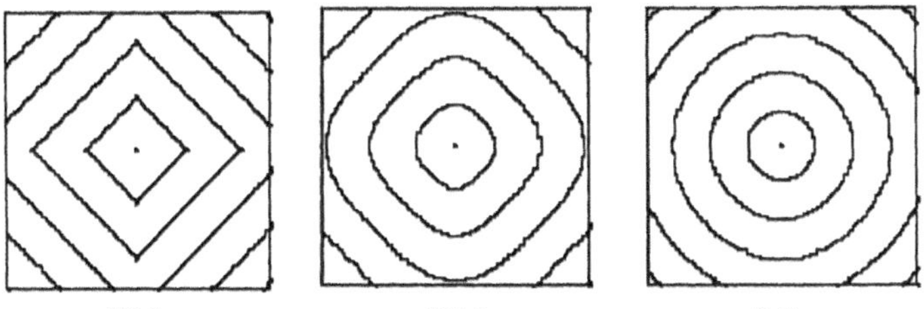

Figure 3. Simulation of averaged wave fronts on lattices with different occupancies. The diagrams show the wave fronts at equally space instances.

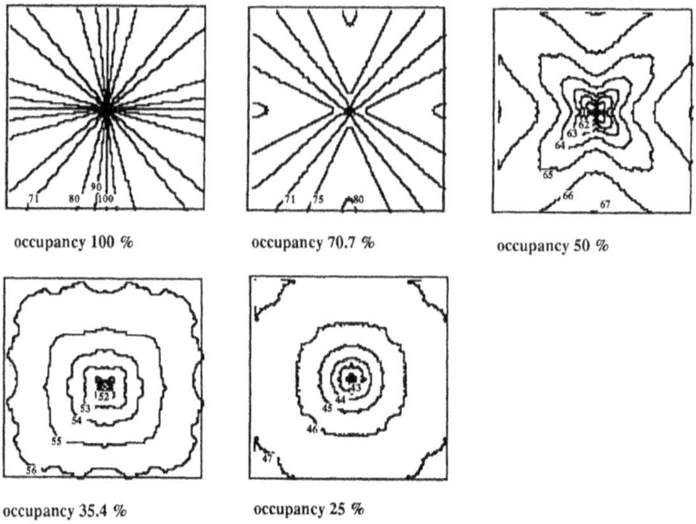

Figure 4. Simulation of averaged wave velocities on lattices with different occupancies. The contour lines denote positions that are reached with the same velocity, which is given as a percentage of the maximal propagation velocity in a fully occupied lattice.

5. Analysis of the transition to isotropic wave fronts

Figure 3 shows that wave fronts in an almost completely occupied lattice propagate fastest along the lattice axes and slowest along the lattice diagonals. Figure 5 presents and compares the asymptotic wave propagation velocities fór axial and diagonal propagation as a function of lattice occupancy. The following properties emerge:

- The velocity of axial wave propagation varies with the square root of the lattice occupancy. This dependence is expected for two-dimensional media with random distributions of excitable elements. (This expectation is based on the fact that the mean velocity for diffusion is inversely proportional to the average distance between elements and that, for two-dimensional media, this distance is inversely proportional to the square-root of the occupancy.)
- The velocity of diagonal wave propagation assumes a value of $\sqrt{1/2}$ lattice units per time step for lattice occupancies between 100 and 60 per cent.
- For occupancies below 50 per cent, the difference between the velocities for propagation along the diagonal and the axial directions is very small, i.e. one can speak of virtually isotropic wave fronts.

The fact that the velocity of wave propagation along the diagonal remains constant over a wide range of occupancies plays a key role in the fast convergence of axial and diagonal propagation velocities for occupancies below 50 per cent. We want to investigate in the following how this convergence arises. In particular, we will investigate the behaviour near the occupancy of 60 per cent at which the diagonal propagation velocity starts to decrease. We will show that this behaviour corresponds to that of a percolation transition [8].

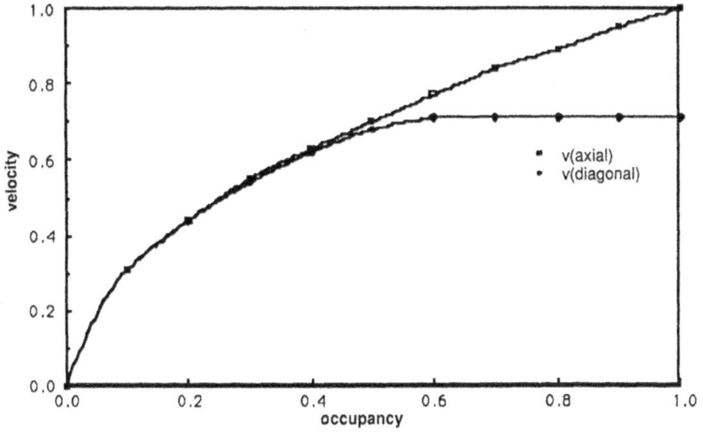

Figure 5. Occupancy dependency of asymptotic wave velocities for a quadratic lattice.

6. **Percolation theory approach**

Figures 6 and 7 show the propagation of planar waves in axial and diagonal lattice directions. Let us first discuss the propagation in the axial direction, which is schematically illustrated in Fig. 6. On the left side of this figure, the wave front propagates one lattice step forward for every time step. The wave front itself propagates in this situation with maximal velocity. If some of the excitable elements are missing, as on the right side, some parts of the wave front beyond the unoccupied sites will only be reached with some delay, because they take more time to be reached by more distant neighbours or through longer paths. The more empty sites the lattice of excitable elements has, i.e. the lower the occupancy of the lattice is, the more often such delay will occur, which translates into a decreasing propagation velocity.

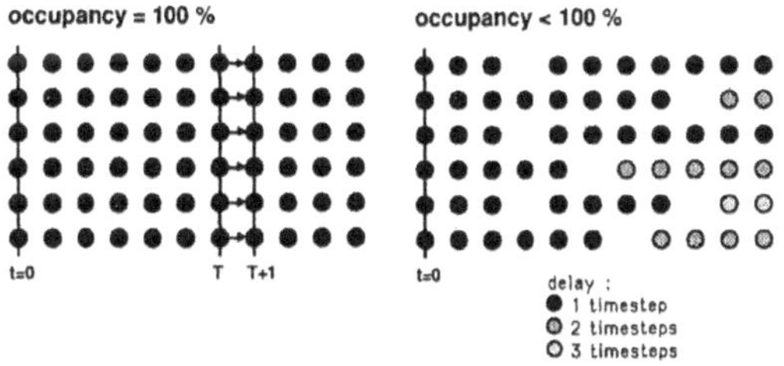

Figure 6. Schematic representation of the wave propagation in axial direction for a fully and partially occupied lattice. The left diagram shows the wave front at different times t = 0, t = T and t = T +1, and the direction from where the cells in the in the column eight will be excited at time T. The right diagram illustrates how unoccupied sites lead to a gradually increasing delay of the wave front, the latter being coded by increasigly lighter shaded cells.

The situation is different for the case of wave propagation along diagonals, which is represented in Fig. 7. In the case of a fully occupied lattice, the wave front advances $\sqrt{1/2}$ lattice constants per time step. However, along the diagonal direction, every cell can be reached by three of its neighbours. When the lattice occupancy is decreased slightly, there will still be enough connected paths left to each lattice site such that the wave front can propagate without delay, as shown on the right of Fig. 7. In the situation presented in Fig. 7, only two cells are reached with delay.

We have seen that along the diagaonal direction three lattice sites can be reached without delay from every lattice site that is occupied with an excitable element. In a figurative way one could say that one element is connected with three other sites by a **fast** link. If one of these three sites is occupied, it will again be connected with three further sites by **fast** links. In this way, clusters of cells are formed that are all connected by successive **fast** links to the original cell.

At low lattice occupancies, only a few cells might be reached by one specific cell through fast links. In this case the wave front often has to propagate directly to the next-nearest neighbours, a propagation mode that causes a delay of the wave front. At higher occupancies, the average number of cells that can be reached from one specific cell will steadily increase so that delays of the wave fronts will become more and more infrequent. This translates into an increasing mean propagation velocity. As long as a lattice contains an infinite cluster of cells connected by fast links, the wave front will propagate with maximal speed. Therefore, a knowledge of the dependence of the average size of clusters connected by fast links on the lattice occupancy will provide important clues to the understanding of the dependence of the wave velocity on the lattice occupancy: at occupancies for which infinite clusters exist, the diagonal propagation velocity should be at a maximum, at occupancies for which clusters are only finite, wave fronts should slow down in the diagonal direction.

The theory that deals with the size of clusters in partially connected lattices is percolation theory [8]. For our purposes, we define a cluster of cells as the set of all those cells that can be reached through fast links from a given set of initial cells. Percolation theory predicts that the mean size cs of a cluster, i.e. its linear dimension in the direction of the wave propagation, diverges with a power law behaviour

$$cs = const_1 \cdot (p_c - p)^{\gamma_1}, \qquad (3)$$

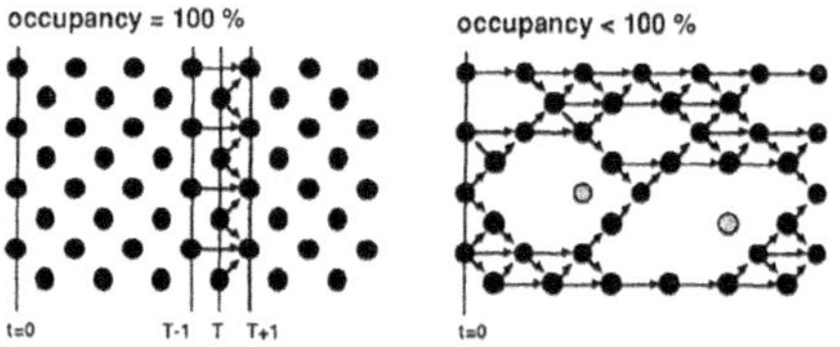

Figure 7. Wave propagation in a diagonal direction for a fully and partially occupied lattice. The left picture shows the wave front at different time steps and the directions, from where the cells have been excited for the first time. In the right picture, one can see that a limited amount of unoccupied space does not lead to a delay of the wave front.

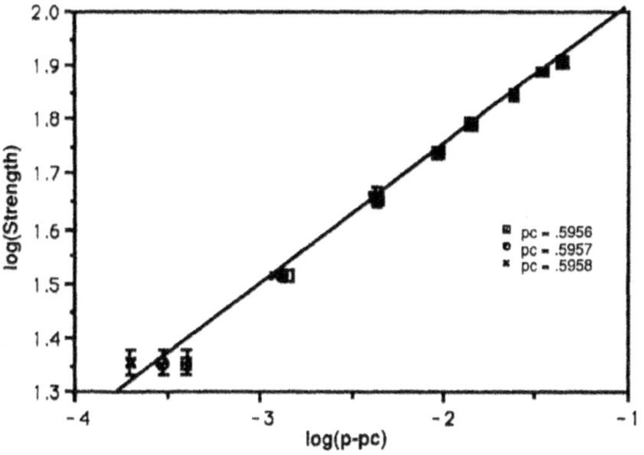

Figure 8. Strength st of the infinite cluster as a function of the lattice occupancy. The data are plotted for three different assumed percolation thresholds. A threshold p_c = .5957 and exponent γ_2 = .255 reproduces best the power-law (4).

as the lattice occupancy p approaches a critical value p_c, which is also called the percolation threshold. γ_1 is called a critical exponent. Above the percolation threshold, there will always be a cluster of infinite size. The strength st of the infinite cluster, which is defined as a ratio of cells that belong to the infinite cluster decreases from 1 down to 0 according to the power-law

$$st = const_1 \cdot (p - p_c)^{\gamma_2}, \qquad (4)$$

as the occupancy p is lowered from 1 down to p_c. Laws (3) and (4) can be used to extrapolate the percolation threshold from values of the cluster size or strength near the percolation threshold by matching a log-log-plot of the respective quantities to a straight line.

We performed simulations to determine the strength of the infinite cluster and mean cluster size for different lattice occupancies. Through resulting log-log-plots reproduced in Figs. 8 and 9 the percolation threshold and the critical exponents were determined to

$$p_c = 0.5957 \pm 0.0001$$

$$\gamma_1 = -1.50 \pm 0.05 \qquad (5)$$

$$\gamma_2 = 0.255 \pm 0.015$$

The value p_c = 0.5957 signifies that there will always be an infinite cluster in lattices with an occupancy higher than 59.57 per cent. For occupancies

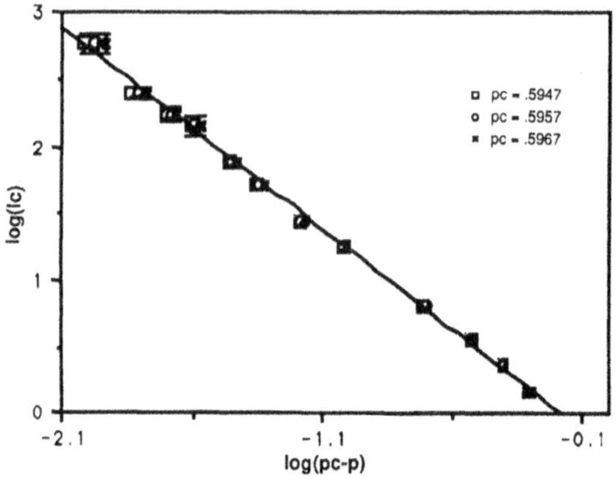

Figure 9. Mean cluster sice cs as a function of the lattice occupancy. The data are plotted for three different assumed percolation thresholds. The threshold p_c = .5957 and exponent γ_1 = -1.50 reproduces best the power-law (3).

higher than this critical value, the propagation velocity in diagonal lattice directions will always assume its maximum value. Put another way, lowering the occupancy of an initially fully occupied lattice will only decrease the faster axial wave propagation velocity until the percolation threshold p_c is reached (see Section 5 above). At this point, the anisotropy of the wave fronts has already been reduced significantly. As the occupancy is lowered further, both axial as well as diagonal wave propagation velocities will decrease. Therefore, as the occupancy is lowered below p_c, the anisotropy will decrease more slowly. From this behaviour one can derive the important practical conclusion that a lattice occupancy just below the percolation threshold will represent the best compromise to minimize anisotropy as well as minimize computational effort.

7. Comparison with the hexagonal lattice

We have carried out the above investigations also for the hexagonal lattice. Figure 10 shows the results of a simulation of the asymptotic propagation velocities in axial (0°, 120° or 240°) and diagonal (60°, 180° or 300°) directions. For a fully occupied lattice, wave propagation is less anisotropic in a hexagonal lattice than in a square lattice, the diagonal propagation velocity

$$v_{diag,hex} = \sqrt{3/4} \cdot v_{axial,hex} = 0.866\ldots \cdot v_{axial,hex} \tag{6}$$

being closer to $v_{axial,hex}$ than in the case of the square lattice, for which

$$v_{diag,sq} = 0.707\ldots \cdot v_{axial,sq} \tag{7}$$

holds. Furthermore, the two extreme velocities $v_{diag,hex}$ and $v_{axial,hex}$ for a

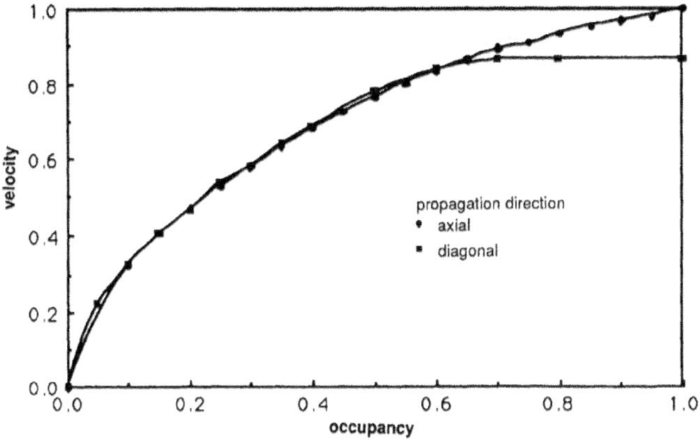

Figure 10. Simulation of the asymptotic wave velocities for different lattice occupancies for a hexagonal lattice.

hexagonal lattice converge already below occupancies of 70 per cent. This advantage of the hexagonal lattice over the quadratic lattice in producing isotropic propagation comes at the expense of a higher computational effort needed to code and handle a hexagonal lattice and at the expense of a larger number of lattice points per area for this lattice. From a computational point of view there is, therefore, no reason to prefer either one of these lattices.

The mechanism that leads to a rapid convergence of axial and diagonal propagation velocities below occupancies of 70 per cent is the same for the hexagonal lattice as for the quadratic lattice. This convergence can again be attributed to a percolation transition, i.e. there exists a critical occupancy above which the diagonal propagation velocity remains constant. For a hexagonal lattice and for the case of diagonal wave front propagation, a cell can be reached by only **two** cells through fast links, i.e. there are only two neighbouring sites that contribute to the fastest propagation of wave fronts in the diagonal direction. Therefore, one can expect that the percolation threshold, i.e. the occupancy below which an infinite cluster no longer exits, lies at a higher value than for the case of the quadratic lattice, for which every cell could be reached by **three** cells through fast links.

Numerically the percolation threshold and the critical exponents for the hexagonal lattice have been determined to be

$$p_{c,hexa} = 0.7055 \pm 0.0002$$

$$\gamma_1 = -1.377 \pm 0.020 \qquad (8)$$

$$\gamma_2 = 0.250 \pm 0.007$$

(The value for $p_{c,hexa}$ is in agreement with earlier calculations by Kinzel & Yeomans [3].) In hexagonal lattices, occupancies just below the value of $p_{c,hexa} = 0.7055$ thus will be best suited in terms of isotropy and computational effort for simulations of isotropic wave front propagation in a hexagonal lattice.

8. Conclusion

In this contribution, we have investigated an approach to obtain isotropic geometries in automata simulations of discrete excitable media. The approach involves lattices with randomly occupied sites. We have demonstrated that relatively few unoccupied lattice sites, i.e. around 50 per cent for a square lattice and around 33 per cent for a hexagonal lattice, suffice to obtain isotropic wave propagation. We have argued that the nature of the convergence to isotropic behaviour can be attributed to a percolation transition. We propose that lattices with randomly occupied lattice sites provide a computationally effective avenue to the study of spatial dynamical systems.

Acknowledgements

The authors would like to thank the University of Illinois at Urbana-Champaign for support. Some of the computations represented were carried out at the National Center for Supercomputing Applications in Urbana-Champaign, supported by the National Science Foundation.

References

[1] Foerster, P., Müller, S.C. & Hess, B. (1988). Curvature and propagation velocity of chemical waves. *Science* **241(4866)**, 685-687.
[2] Greenberg, J.M. & Hastings, S.P. (1978). Spatial patterns for discrete models of diffusion in excitable media. *SIAM J. Appl. Math.* **34(3)**, 515-523.
[3] Kinzel, W. & Yeomans, J.M. (1981). Directed percolation: a finite-size renormalisation group approach. *J. Phys.* **A14**, L163.
[4] Mackay, S. (1978). Computer simulation of aggregation in *dyctostelium discoideum*. *J. Cell Sci.* **33**, 16.
[5] Markus, M. & Hess, B. See contribution in this volume.
[6] Meinhardt, H. (1982). *Models of Biological Pattern Formation.* Academic Press: London.
[7] Nicolis, G. & Prigogine, I. (1977). *Self-Organisation in Non-Equilibrium Systems.* Wiley: New York.
[8] Stauffer, D. (19895). *Introduction to Percolation Theory.* Taylor and Francis: London and Philadelphia.
[9] Turing, A. (1952). The chemical basis for morphogenesis. *Philos. Trans. Roy. Soc. London Ser. B* **237**, 37-72.
[10] Tyson, J.J., Alexander, K.A., Manoranjan, V.S. & Murray, J.D. (1989). Spiral waves of cyclic AMP in a model of slime mold aggregation. *Physica* **D34**, 193-207.

44. PHASE DYNAMICS AND SPATIAL PATTERNS IN OSCILLATING AND EXCITABLE MEDIA

P. Hanusse, V. Pérez-Muñuzuri* and C. Vidal

Centre de Recherche Paul Pascal/CNRS
Université de Bordeaux I, Avenue Schwietzer
33600 Pessac, France
*on leave from Dept. of Física de la Materia Condensada
University of Santiago de Compostela, Spain

1. Introduction

As far as pattern formation, front propagation and phase dynamics are concerned, a lot has been achieved in various, yet related, specific fields. Let us mention just some of these: in hydrodynamics, the equations of the Rayleigh-Bénard convection by Newell & Whitehead [1] and the instability of convective flows [2], the phase instability of oscillating systems by Kuramoto [3], the propagation of fronts in combustion by Sivashinski [4], the Eckhaus instability of steady structures [5], the topological turbulence in the Ginzburg-Landau equation [6]. Finally, to return to topics more directly related to this work, and with no pretention to exhaustivity, we should cite the works of Winfree [7], Tyson [8], Fife [9], Keener [10], Mikhailov & Krinskii [11] Meron & Pelcé [12]. These mostly deal with theory. Of course, much has been done in experiments, which we should only cite indirectly, through review articles or collective works [13,14,15].

Nonetheless, many of these works are still influenced by the specific language used in their context. In this contribution, however, we address the question of the universal properties of front propagation and pattern formation in excitable as well as oscillating media. We wish to construct a description of spatio- temporal patterns occurring in such systems, having the minimum complexity and dimensionality, independently of mechanistic and system details. In doing so, we try altogether to define and describe the salient features of experimental observations, while preserving a universal character of the underlying mathematical structure. In particular, this description should capture two main features, typically found in chemical excitable systems, namely the existence of a strongly attracting slow manifold (cycle or pseudo-cycle) and a strongly relaxing dynamics onto this attractor (widely separated time scales).

We would like to show, in a very general context, that this could be achieved through an extension of the Hopf bifurcation scheme for distributed systems, by taking into account global aspects of the behaviour related to the aforementioned dynamical features. As a result, the dynamics is described by a universal equation in a one-dimensional periodic local phase variable. In particular the presence of a non-uniform local phase rotation velocity, a property which is absent from the Kuramoto-Sivashinski equation, turns out to be essential to account for pattern formation.

To study the behaviour of this equation we propose a cellular automaton model which implements a discrete version of this equation. We present a number of simulation results in one and two dimensions. The observed behaviour is strikingly similar to experimental observations of front propagation, front collisions, and spiral waves in excitable and oscillating systems. Dispersion is shown to occur as well as front interaction. Disordered and turbulent patterns in connection with topological defects are also depicted. Finally, we show that there are circumstances when the local dynamics can be influenced by walls and other passive geometrical defects, leading to target pattern centres.

2. **Towards a normal form for relaxation behaviour**

To describe the universal properties of spatial patterns in excitable or oscillating systems, independently of mechanistic system details, we have to define a proper level of description, an important issue which is often overlooked.

We want to find out some minimal mathematical or topological structure that is common to all systems, where excitability and/or oscillation occur. Some behaviours can be accounted for by a minimal description, and others require a more complex view. A general procedure to achieve this is to look for structure, rather than facts. The structure is expressed by the hierarchy of observation description, which ranges from specific to generic. In that sense, modelling is part of pattern recognition [16,17]. Why use a specific model or language to describe a generic behaviour, as soon as it has been recognized as generic? Finally, we are facing the following question: in all we know about pattern formation in excitable or oscillating systems, what part is generic?

Let us consider a reaction-diffusion system as a general dynamical system. We would like to find a general model for wave propagation. Doing so, we clearly pretend to reduce an infinite dimensional real phase space, to a low dimensional effective one. So stated, this is a quite universal issue. Experience tells us that, to achieve this, we should preserve the topology of structurally stable behaviours or features, rather than stick to a specific modelling.

Our starting point, which should appear quite natural to those with experience in observing chemical or biological dynamical systems, relies on the two following basic features. First, typical oscillator or excitator fields possess a strongly attracting slow manifold, a natural dynamical way to reduce the dimensionality of the space in which the system lives. This reduction is based on the existence of widely separated time scales. It also expresses the natural stability or robustness of such behaviours. Secondly, the dynamics onto this attractor is very relaxing. It involves at least two distinct time scales. Indeed, relaxation oscillators seem to be more common than sinusoidal ones in such dynamical systems. To sum up, we note that the slow manifold has the topology of a stable cycle, for excitators as well as oscillators, and that the evolution along this cycle is very non-uniform. These are the two main features that we want to consider.

Speaking of a cycle as a trajectory, a natural and generic starting point is the Hopf bifurcation. This is an example of the type we are precisely looking for, in which the genericity of behaviour can be established on firm mathematical grounds. What we stressed in the preceding paragraph indicates without ambiguity that we are not considering situations close to Hopf bifurcation. In general, it is not possible to use the information that can be rigorously derived **at** bifurcation, for finite cycle amplitude, i.e. far from bifurcation, or for highly relaxing systems, or when a global bifurcation is ahead. Nevertheless, ideally, we would like to derive all properties from bifurcation theory. In its usual form, it is a local theory, which means that to enrich our description we should go over to high codimensional bifurcations. An alternative way, which we have chosen, consists of going to global bifurcation theory. Unfortunately, it has not the completeness and firm status of its local counterpart. Therefore, we shall consider a semi-local approach. We start from Hopf bifurcation to take advantage of its universal modelling character. Next, we inject some non-local information to preserve the topology of behaviour, namely, the relaxation property.

What is significant is topology. Morphology comes next. This is what makes Hopf bifurcation generic. The Hopf theorem states that close to the bifurcation, the cycle, whatever its shape, is topologically equivalent to a circle around which the system state is rotating at constant velocity. The Hopf theory tells us how to construct the nonlinear transformation that maps one trajectory onto the other. We keep that idea, but we relax those conditions which contradict our goals, provided they do not violate Hopf theory **at** bifurcation. More specifically, we accept the topological equivalence to a circle, since it fits the existence of a stable circular manifold, but we do not require that the rotation should be uniform. To match these views with Hopf theory, we have to verify that this nonuniformity disappears at bifurcation. Since the system will essentially live on that circular manifold, we can naturally parametrize it by a phase variable. Thus, our approach amounts to assuming that the amplitude can renormalize as in Hopf theory, but that the phase keeps as much information as possible from the original system.

We shall not give here a formal derivation of the equations. Nor shall we attempt any mathematical justification, which is beyond the scope of this article [18,19,20]. At this point we shall simply consider it as a general model, rather than a generic theory.

Using the methods of bifurcation theory, and injecting the physical features described above, any system undergoing a Hopf bifurcation can be nonlinearly transformed - an extension of the Hopf-Cole transformation - to the following dynamical equation, in complex notation.

$$\dot{z} = az + a_3 z^2\bar{z} + a_5\, z^3\bar{z}^2 + \cdots + z\,\mathrm{Im}(b_1\bar{z}) + z\,\mathrm{Im}(b_2\bar{z}^2) + \cdots \quad . \tag{1}$$

with $z = re^{i\theta}$ complex, and $a = \lambda + i\omega$.

We have considered the restriction of the complete dynamics to the centre manifold, which, for Hopf bifurcation, is of dimension two, hence the description by a complex variable. In polar coordinates (r,θ), equation 1 reads

$$\dot{r} = \lambda r + \mathrm{Re}(a_3)r^3 + \mathrm{Re}(a_5)r^5 + \cdots \tag{2}$$

$$\dot{\theta} = \omega + \mathrm{Im}(a_3)r^2 + \mathrm{Im}(a_5)r^4 + \cdots$$
$$+ \mathrm{Re}(b_1)r\cos\theta + \mathrm{Im}(b_1)r\sin\theta$$
$$+ \mathrm{Re}(b_2)r^2\cos(2\theta) + \mathrm{Im}(b_2)r^2\sin(2\theta) + \cdots \qquad (3)$$

where the real functions Re and Im extract the real and imaginary part of their argument.

The symmetry that has been retained through the choice of the nonlinear transformation is meant to keep the topology of the circle while allowing a phase dynamics with full complexity. The essential benefit of this approach is to decouple amplitude and phase dynamics. Assuming, as stated before, that the slow manifold is "very" stable, i.e. that the dynamics rapidly relaxes towards the cycle, we obtain the following structure for the dynamical equations

$$\varepsilon\dot{r} = f(r) \qquad (4)$$

$$\dot{\theta} = g(r,\theta) \qquad (5)$$

In the limit when ε goes to zero, the relaxation to the cycle is infinitely fast, the amplitude r can be determined by equation 4 and the dynamics reduces to a phase equation 5 (or in detailed form 3).

Several benefits result from this approach. It extends the distance from bifurcation at which the reduced form defined by equation 1 can be used. It contains some knowledge about the global behaviour of the system. It provides an (at least) qualitative model of the relaxation behaviour. It decouples amplitude and phase, thus allowing amplitude elimination, when time scale separation is achieved, which is often the case in experimental situations for reaction-diffusion systems.

Let us have a closer look at equations (2) and (3). The amplitude equation (eq. 2) is exactly that of the generalized Hopf normal form, which was imposed from the beginning. The phase equation (eq. 3) reduces to the Hopf form when the amplitude r goes to zero, which satisfies the overmentioned condition. In this case, the terms involving b_1 and b_2 cancel out. These are the two first phase dependent contributions that we have included here. In general there are higher order terms.

The general modelling features of this phase equation can be shown to relate to global properties of the dynamics. We can understand this by inspecting Fig. 1, which describes four typical cases. Figure 1a depicts the case of a relaxation oscillator, with two time scales for phase rotation velocity. Figure 1b describes an excitable system, after a saddle-node bifurcation has occurred on the limit cycle. This bifurcation has been known for a long time, in bifurcation theory [20], in model systems, like the Van der Pol oscillator, as well as in many experimental situations [21]. This bifurcation is essentially controlled by parameter b_1, whose amplitude determines the distance from Hopf bifurcation at which this global bifurcation occurs. For some systems, for instance the Brussellator model, it could be only present as a way of describing relaxation behaviour, yet leaving the saddle-node global bifurcation out of reach of the domain of validity of the normal form, when the amplitude is too large.

Let us mention here, that this description of excitability does not

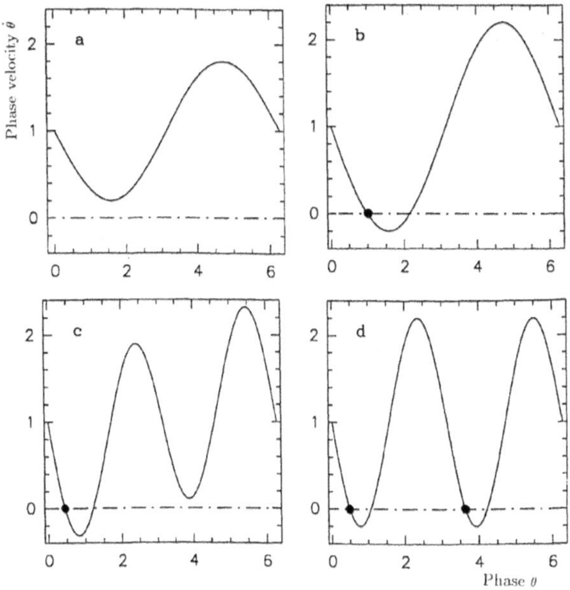

Figure 1. Schematic dependence of phase rotation velocity as a function of phase (see eq. (3)). (a) Relaxation oscillation. (b and c) Excitability. (d) Bistability. In (a) and (b) term b_1 is dominating. In (c) and (d) term b_2 is dominating.

contradict, nor rule out, another one used by many authors, sometimes referred to as the FitzHugh-Nagumo model [22]. Indeed, Fig. 1c presents another case of excitability, with four stages, equivalent to this model. In fact, as we have explained before, any specific model that leads to a topology of behaviour involving a circular manifold, on which a stable steady state exists, is equivalent to this generic form. The latter has been used as an ad hoc model by Winfree [7] and Kuramoto [23]. Our derivation replaces it in a general scheme for non-local or global effects.

Finally, in Fig. 1d two saddle-node bifurcations have occurred, leading to a bistable system. The connection between bistability, excitability and oscillations has been well established in experiments and in various models [24]. This effect is due to term b_2. Before any global bifurcation occurs, it describes a two-stage relaxation oscillator, involving two fast transitions between two slowly varying, or quasi-steady, states. This is also a situation commonly encountered in experimental situations.

3. **Adding diffusion**

Using the same approach, but including the diffusion of species, and making use of the same arguments of time scale separation between fast amplitude relaxation and slow phase dynamics, it is possible to treat diffusion as a small perturbation. In the limit of small gradients we have derived an explicit distributed phase equation similar to the Kuramoto-Sivashinski equation [3], but with a different effective content. We have used for that purpose the MAPLE computer algebra system. We shall neither give here any derivation, nor discuss the implications or conditions of applications [25]. The general form of this equation is

$$\frac{\partial\theta}{\partial t} = \Omega(\theta) + \alpha\nabla^2\theta + \beta|\nabla\theta|^2 - \gamma\nabla^4\theta \qquad (6)$$

The local term $\Omega(\theta)$ represents the phase dynamics described by equation (3). The nonuniformity of phase rotation introduces an important effect that turns out to be essential to account for the properties of pattern formation. It is a source of spiralling effects. It introduces information which is usually modelled by a complex Ginzburg-Landau equation [3]. To give an overview of some of the properties of such an equation, we shall later present simulation results obtained with a cellular automaton type of model, using the simplest possible form for the local phase term Ω

$$\Omega(\omega) = \omega(1 - \delta \sin\theta) \qquad (7)$$

The coefficient δ controls the relaxation behaviour (see Fig. 1). For δ = 0 a sinusoidal oscillation is obtained. For δ close to, but less than 1, the oscillation is of relaxation type. Its period is given by

$$T = \frac{2\pi}{\omega\sqrt{1 - \delta^2}} \qquad (8)$$

which goes to infinity as δ goes to 1. For δ = 1 a saddle-node bifurcation occurs on the cycle. For δ > 1 the system is excitable.

As a first simplification we leave out high order derivatives or non-linear space dependent terms. Most simulations will be performed using the resulting simple form for the dynamical phase equation

$$\frac{\partial\theta}{\partial t} = \omega(1 - \delta\sin\theta) + \alpha\nabla^2\theta \qquad (9)$$

Under these circumstances, there is only one free parameter, δ, since ω simply defines the time unit and α the space unit. We shall see that much can already be obtained using this quite simple and and general equation.

4. Cellular automaton

We could have tried to solve the distributed phase equation by usual PDE techniques. Along the same lines on which we conducted our search for genericity, we have considered that any numerical technique that would qualitatively provide a solution for this equation would be acceptable. This is what cellular automata do [26]. They have proved to be very efficient in various fields, although they may present some difficulties, for instance when one is to establish their qualitative or even quantitative validity when required, which is the case in hydrodynamics applications [27]. Also, it is well known [28] that one of the drawbacks resides in the influence of lattice local anisotropy on large scale behaviour. For this reason we have considered a class of automata that we call "medium dimensional automata" (MDA), in which the number of states is large, in our case 256, for coding and computation purposes. Traditional cellular automata have a very small number of local states. At the other extreme, the original PDE has an infinity of possible states. MDA present the advantage that they can be easily derived as a massively discretized version of the original PDE. The qualitative as well as quantitative interpretation is thus made easier. Moreover, the existence of a large enough number of local states provides a local averaging effect that smears out the lattice anisotropy. They still retain the simplicity and efficiency of low dimensional automata (a typical 128 by 128 array iteration time is of the order of 0.2 sec on a VAX 8600 computer).

This is neither the place to detail any further the formal background of such tools, nor to analyze the domain of validity of their formulation. Details will be found elsewhere [29]. To simulate eq. (6), we consider a square lattice, usually with periodic boundary conditions. Each site is coding a phase value in a 256-periodic integer space with values from 0 to 255. A lookup table gives the local state transformation $\theta_{i,n+1} = f(\theta_{i,n})$, where n represents the iteration counter (time) and i the lattice coordinates (space). First neighbour interaction is implemented by linear elastic coupling in the local periodic state space. For the simulation of eq. (9) this leads to the following general form

$$\theta_{i,n+1} = f(\theta_{i,n}) + a\theta_{i+1,n} + a\theta_{i-1,n} \tag{10}$$

with $f(\theta)$ of the form $k\omega(1 - \delta\sin\theta) - 2k\alpha\theta$ and $a = k\alpha$.

5. Simulation of the distributed phase equation in one and two dimensions

Let us first consider a one-dimensional system with periodic boundary conditions. In Fig. 2 is given a space-time picture of the behaviour of an excitable system ($\delta = 1.1$). The phase value is coded in grey tones. After a transient, four very localized fronts propagate in the system at constant velocity. The velocity is given by the inverse of the slope of the front (space is horizontal, time is vertical). Note that, due to system finite size and front interaction, a locked asymptotic regime is reached, satisfying the

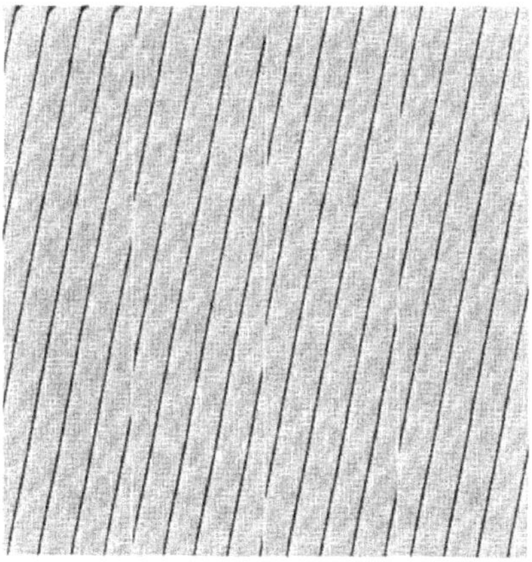

Figure 2. Space-time diagram of a one dimension excitable system. After a short transient there are four fronts travelling to the left (space is horizontal) at constant velocity (time is vertical, in four columns).

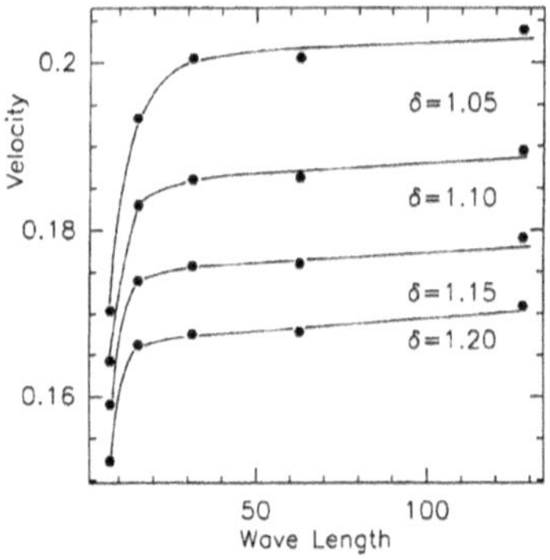

Figure 3. Dispersion curves of an excitable system for various values of parameter δ (see eq. 9).

relationship $vT = \lambda = L/4$, where v is the velocity, T the period, λ the wavelength and L the system length. This means that the front profile changes its shape until its velocity satisfies this equation.

In the previous calculation, the number of fronts was determined by the choice of initial conditions. We can perform the same calculations for various wavelengths by increasing, or decreasing the number of fronts. In Fig. 3 the measured velocity is plotted against the wavelength, for several values of parameter δ. This is the dispersion curve of the system. It presents a usual shape, as observed in experiments and models [15].

Figure 4 presents the same type of calculation as in Fig. 2, but in an oscillating system ($\delta < 1$). There are initially two propagating fronts. They

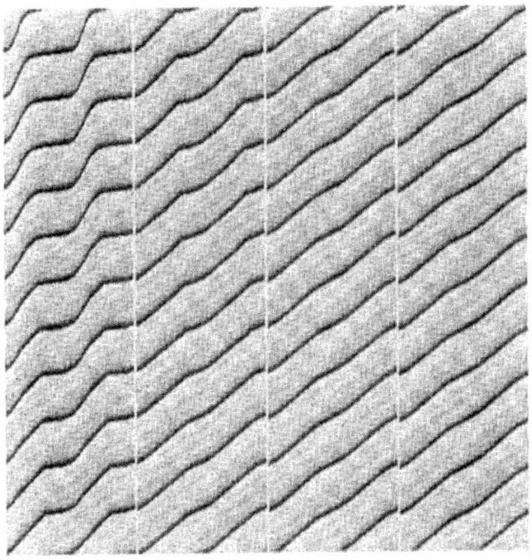

Figure 4. Same as Fig. 2 but for an oscillating system. There are two fronts travelling to the left at constant velocity, except when bulk oscillation (horizontal parts) sweeps across the system. Asymptotic behaviour is the same as for an excitable system.

Figure 5. A pair of counter-rotating spiral waves in a two dimension excitable system, whose morphology and behaviour is strikingly similar to experimental observations.

are only observable when steep straight segments are present in the picture, which indicates a rather slow intrinsic velocity. Almost horizontal sections correspond to the sweeping of the bulk oscillation (phase wave). As time goes on (down and right) the propagation regions extend, with a lower slope (higher velocity). The asymptotic regime is the same as in the excitable state. Large scale locking has occurred, where the front velocity is controlled by system size. Note that in zero flux boundary conditions no front could be sustained.

Figure 6. Pulsating behaviour of the core of a three arm spiral (4 different times), a phenomenon also observed experimentally. The pulsation is not symmetric, tips interact by pair.

Figure 7. Example of a disordered pattern in an excitable system. Several phase singularities interact in a complex way.

We now turn to two-dimensional systems. Figure 5 presents a typical pair of counter-rotating spirals. The aspect and behaviour are very similar to experimental observations. The fronts are sharp, corresponding to trigger waves, which is a result of the relaxation character of local dynamics (δ close to one). This is drastically different from the phase wave spirals observable with the Ginzburg-Landau eqution, which are isophase lines rather than fronts. The long time aspect does not differ much in excitable or oscillating conditions.

Spirals with two or more arms have been studied. In Fig. 6 we present the evolution of a three-arm spiral. The pulsating behaviour of the core is easily seen. As already reported in experimental observations, this core pulsation is not synchronous nor symmetric. Arm tips collide two by two. It can be periodic or more irregular. In the limit of high diffusion locking occurs, with arms oscillating symmetrically.

Disordered patterns of turbulent type can also be observed in this system (Fig. 7). They have been obtained by perturbation of the pair of spirals presented in Fig. 5. In most of the situations that we have considered, i.e. with $\beta = 0$ and $\gamma = 0$ (see eq. (6)), phase singularities seem to stay at a fixed location. Their number is therefore constant, determined by initial conditions. They interact only when very close to each other, e.g. as in the three-arm spiral (Fig. 6). The existence of a short length scale, due to the phase fast time scale, seems to induce some screening that shortens the interaction distance. Furthermore, the introduction of term γ seems to increase instability. This effect is still to be understood.

The fourth order spatial derivative controlled by γ seems also to play an important role in the stability of an oscillating phase field near boundaries or walls. In Fig. 8 we present an example of a local spatio-temporal instability due to the presence of a passive small cavity. It has no catalytic activity, but is a mere obstacle to diffusion. Nothing occurs when the system is initially uniformly oscillating. If we start the system with a uniform phase gradient, a phase wave first develops, then a front is formed and propagates through the system (as in one dimension as shown in Fig. 4). When

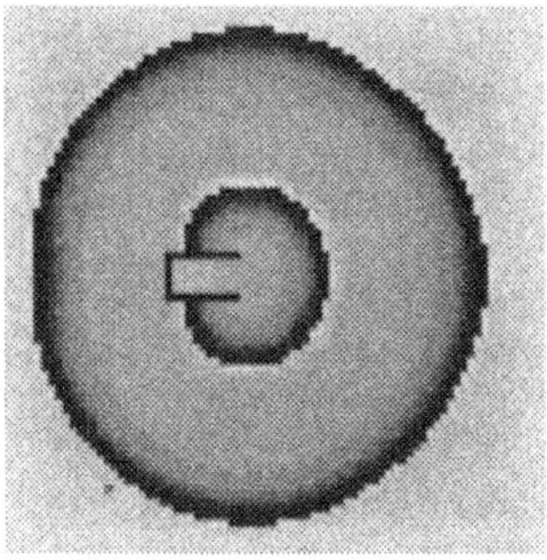

Figure 8. Model of target pattern due to the presence of a local passive cavity in a large scale phase gradient. It develops a local phase shift and local frequency increase, leading to a permanent triggering centre.

the front reaches the obstacle, it goes around it and passes through, with only a local transient deformation. But it leaves a trailing phase shift within the cavity which induces a local oscillation with a frequency higher than that of the bulk. This results in a sustained target pattern. Although we have gathered significant results on this effect, we have not yet studied all the parameters controlling the behaviour of such obstacles. We shall not go into more details for now. However we think that it provides a valuable direction of investigation for target pattern centres, at least those which seem to be related to the presence of physical heterogeneities.

6. Conclusions

The simulation results of the very simple phase equation (6) reported above are essentially preliminary, but we think that they already provide a clarification of the role of the relaxation behaviour of local dynamics, which is common to excitable and oscillating systems studied in chemical and biological media. We can definitely say that most of the qualitative behaviour is controlled by the parameter δ. It accounts for front formation, which supports the notion of trigger waves, in contrast to phase waves where no matter or information transport is involved. We are now studying extensively all the consequences that can be derived from this unique relaxation character. We are also considering the influence of higher order effects, as we may call them, controlled by parameters γ and β. In this respect, the stability properties near boundaries seem to open new interesting questions.

In the introduction we were calling for genericity. This is the foundation of the theory that leads to the phase equation. We are presently developing a general approach, based on symmetry considerations, that would extend the scope and validity of the distributed phase equation. As we stand now, at least with only modelling ambitions, we have already progressed towards a description of behaviour that is independent of system specifities. Surely, much can still be achieved in this direction.

Acknowledgements

One of us (V.P.M.) wants to acknowledge the support of a research grant from the Caixa Galicia, La Coruña, Spain.

References

[1] Newell, A.C. & Whitehead, J.C. (1969). *J. Fluid Mech.* **38**, 279.
[2] Pomeau, Y. & Manneville, P. (1979). *J. Phys. Lett.* **40**, L610.
[3] Kuramoto, Y. (1984). *Chemical Oscillations, Waves and Turbulence.* Springer.
[4] Sivashinski, G.I. (1977). *Acta Astron.* **4**, 1177.
[5] Eckhaus, W. (1965). *Studies in Nonlinear Stability Theory.* Springer.
[6] Coullet, P., Gil, L. & Lega, J. (1989). *Phys. Rev. Lett.* **62(14)**, 1619.
[7] Winfree, A.T. (1980). *The geometry of biological time.* Springer.
[8] Tyson, J. (1987). *J. Chim. Phys.* **84**, 1359-1365.
[9] Fife, P.C. (1984). In *Non-equilibrium Dynamics in Chemical systems,* Vidal, C. & Pacault, A. (eds.), pp. 76-88. Springer.
[10] Keener, J. (1989). *Physica* **D34**, 378-390.
[11] Mikhailov, A.C. & Krinskii, V.I. (1983). *Physica* **D9**, 346-371.
[12] Meron, E. & Pelcé, P. (1988). *Phys. Rev. Lett.* **60**, 1880-1883.
[13] Vidal, C. & Hanusse, P. (1986). *Int. rev. Phys. Chem.* **5**, 1-155.
[14] Field, R.J. & Burger, M. (eds.) (1985). *Oscillations and travelling waves in chemical systems.* Wiley: New York.
[15] Tyson, J. & Keener, J. (1988). *Physica* **D32**, 327-361.
[16] Hanusse, P. (1987). *J. Chim. Phys.* **84**, 1315-1327.
[17] Hanusse, P. & Guillataud, P. *Object detection and identification by hierarchical segmentation,* to be published.
[18] Marsden, J.E. & McCracken, M. (1976). *The Hopf bifurcation and its applications.* Springer.
[19] Guckenheimer, J. & Holmes, P. (1983). *Nonlinear oscillations, Dynamical sytems, Bifurcations of vector fields*, Vol. I. Springer.
[20] Golubitsky, M. & Schaeffer, D.G. (1985). *Singularities and groups in bifurcation theory.* Springer.
[21] de Kepper, P., Pacault, A. & Rossi, A. (1976). *C.R. Acad. Sci. Paris* **282C**, 199.
[22] FitzHugh, R. (1960). *J. Gen. Physiol.* **43**, 867-896.
[23] Kuramoto, Y. (1976). *Prog. Theor. Phys.* **56**, 724-740.
[24] Boissonade, J. & de Kepper, P. (1981). *J. Chem. Phys.* **75**, 189-195.
[25] Hanusse, P. & Pérez-Muñuzuri, V., to be published.
[26] Winfree, A.T., Winfree, E.M. & Seifert, H. (1985). *Physica* **D17**, 109.
[27] Wolfram, S. (1986). *Theory and applications of Cellular Automata.* World Scientific: Singapore.
[28] Markus, M., see contribution in this volume.
[29] Hanusse, P. & Pérez-Muñuzuri, V., in preparation.

45. EXCITABLE MEDIA AS COMPUTATIONAL SYSTEMS

A.V.Holden*, J.V.Tucker and B.C.Thompson**

*Centre for Nonlinear Studies
University of Leeds LS2 9JT, UK
**Department of Mathematics and Computer Science
University College, Swansea SA2 8PP Wales UK

1. Introduction

A number of mathematical approaches may be used to model a given excitable system. For an excitable system that is not spatially extensive a *map*, or a system of *nonlinear ordinary differential equations* may be appropriate. For a spatially extensive excitable medium a system of *partial differential equations*, or a *coupled map lattice*, or a *cellular automaton*, might be an appropriate model. These different types of model are all nonlinear, and are often intractable, and so their behaviour is usually investigated by numerical methods on a digital computer, using some appropriate algorithms.

In this paper we consider the inverse problem - whether an excitable medium can be considered and used as a computational device that carries out some set of algorithms. In particular, we are interested in the possible use of two or three dimensional excitable media (photo-chemical or opto-electronic) as computing devices, in which a spatially continuous function (image or field) is processed by the nonlinear dynamics and nonlinear wave properties of the medium.

The idea that nonlinear reaction-diffusion systems, such as actual excitable media, could be used to process patterns is not new. It is implicit in Kuhnert's use of the light sensitive Belousov-Zhabotinsky reaction to process images [1], and is explicit in Kirby and Conrad's [2] distinction between structurally intelligent systems, that perform intelligent actions without the use of symbols, and symbolically intelligent systems.

A spatially extensive excitable medium used as a computing device is an *analogue-field computer* and is of interest

[1] from the viewpoint of theoretical computer science as it represents the simplest form of infinite parallelism, and so could be used to investigate the asymptotic properties of parallel computer systems;

[2] if such devices were engineered, a likely application would be as special purpose elements (say for extracting global properties of a field) or preprocessors (say in image processing) in a hybrid computing system.

In both these cases it is necessary to specify the computational behaviour, programming and program verification of the analogue field computer within the same theoretical framework that can be applied to von Neumann and finite parallel digital architectures. This requires the construction of a theory of analogue computation, where computation is carried out on analogue signals (continuous variables) rather than on sets of discrete symbols. We bypass this problem by considering various discrete space, discrete time approximations to excitable media, whose dynamical behaviour provides a simulation of the behaviour of excitable media.

The theory of *synchronous concurrent algorithms* (SCA) provides a framework for investigating the computational behaviour of an excitable medium, if we take a discrete space, discrete time approximation to the excitable medium. An SCA is an algorithm based on a network of modules, channels, sources and sinks that compute and communicate in parallel and are synchronised by a global clock. SCAs process infinite streams of input data, and return infinite streams of output data, and as a general concept SCAs characterise the essential features of many classes of computer architectures. A general mathematical theory of SCAs has been developed using simultaneous recursive functions defined over classes of (many sorted) algebras [3]. The recursive functions represent the architecture and the algebra represent the data and given processing channels. These tools allow very general formulations of algorithms and architectures that process any kind of data.

Application of the theory of SCAs to discretised models of excitable media will allow us to determine the limitations of excitable media as computational systems.

2. Field equations for excitable media

An actual excitable medium (say a thin layer of Belousov-Zhabotinsky reagent) is a concrete object that can be idealized by a partial differential equation. The behaviour of excitable media and their models is surveyed in [4]. A general isotropic excitable medium may be represented by a nonlinear reaction diffusion:

$$\partial \mathbf{u}(\mathbf{x},t)\partial t = \mathbf{K}(\mathbf{u}(\mathbf{x},t)) + \mathbf{D}\,\nabla^2\mathbf{u}(\mathbf{x},t) \tag{1}$$

where $\mathbf{u} \in \mathbb{R}^n$ are the variables in state space, $\mathbf{x} \in \mathbb{R}^m$, m = 1, 2, or 3 forms the physical space and $\mathbf{D} \in \mathbb{R}^{n\times n}$ is a diagonal matrix of real-valued, positive diffusion coefficients.

For nerve and muscle all the "diffusion coefficients" D_{ij} are zero except for D_{11}: in this single diffusion case only the voltage interacts diffusively. In chemical excitable media all D_{ii} are positive and often approximately equal, while all D_{ij}, $i \neq j$, are zero.

3. Discretised excitable media

An actual excitable medium, which has a continuous state space, continuous physical space and operates in continuous time, may be modelled by a system in discrete space and discrete time. This may be done by discretising the field equations (1), or by considering models defined on discrete physical space and in discrete time, such as cellular automata [5] or coupled map lattice models [6] for actual excitable media

3.1 Numerical solutions of partial differential equations

The nonlinear partial differential equations that represent excitable media are almost invariably intractable, and so are usually approached numerically. Although a number of explicit and implicit methods are in current use the simplest finite-difference approximation

$$\delta t\left(\frac{\partial f_r}{\partial t}\right)_{t_0} = f_{r,t_0+\delta t} - f_{r,t_0}$$

$$f_{r,t_0+\delta t} = f_{r,t_0} + \frac{\delta t}{(\delta x)^2}\left(f_{r-1} - 2f_r + f_{r+1}\right)_{t_0}, \quad (r = 1,2,\ldots,n-1)$$

for a 1-dimensional parabolic partial differential equation illustrates the discretisation of space and time with a fixed space step and time step. For 2- and 3-dimensional systems a regular spatial grid may be used, or more sophisticated multigrid methods used.

The essential point is that the partial differential equation is replaced by a regular lattice, and the next value at a node is computed from the preceding values at the same and neighbouring nodes. The numerical solution of partial differential equations, although usually carried out on computers with a conventional, von Neuman architecture, is ideally suited for parallel machines such as an array processor.

3.2 Cellular automata

Cellular automata (CA) are dynamical systems with discrete time, discrete space and discrete state. CA were invented to model the development and organisation of biological systems [7] and recently many applications of CA have been discovered in physics and computer science [8]. For further information on the modelling of systems by CA the reader is referred to Soulié *et al.* [5] in which the basic ideas underlying the account of CA given here can be found.

Most generally a (deterministic) cellular automaton is an infinite dynamical system of processing elements or cells that occupy discrete n-dimensional space and compute on a finite state space Q in discrete time t = $0,1,2,\ldots,$.

Specifically we consider a cellular automaton on a finite subset $S \subseteq \mathbb{Z}^n$ such that cells are indexed by elements of S, and connections between a cell $x \in S$ and other cells are given by a neighbourhood $N(x) = \{y_1,\ldots,y_k\} \subseteq S^k$ when x has a direct connection from cells $y_1,\ldots,y_k$. In general both the number of neighbours k and the neighbours themselves may vary from cell to cell.

Each cell x has an associated *local transition function* $f_x : Q^k \to Q$ when x has k neighbours and the computation performed by a complete cellular automaton is the parallel execution of these transition functions in the following way:

Initially, that is at time t = 0, each cell x is in some given state $q^0(x)$. The system then evolves in discrete time t according to the iteration:

$$q^{t+1}(x) = f_x(q^t(y_1),\dots,q^t(y_k)) \qquad (2)$$

when $N(x) = \{y_1,\dots,y_k\}$.

Two further properties of CA are as follows. Firstly, it is usually the case that Q involves a distinguished state denoted by 0 called the *quiescent* state and each f_x must satisfy

$$f_x(0,\dots,0) = 0.$$

Thus local quiescence cannot generate non-quiescence.

Secondly, we may allow a cell's computation to vary with time. In such a case the local transition function for a cell x has the form $f_x: T \times Q^k \to Q$ where $T = \{0,1,2,\dots\}$ measures time t and the iteration (2) becomes

$$q^{t+1}(x) = f_x(t,q^t(y_1),\dots,q^t(y_k))$$

when $N(x) = \{y_1,\dots,y_k\}$. CA with such cells are called *flexible* CA [9].

Example

For any $m \geq 1$ consider the subset S of $\mathbb{Z}^2$ defined by

$$x = (i,j) \in S \Leftrightarrow 0 \leq i \leq m-1 \text{ and } 0 \leq j \leq m-1$$

This S is simply a square comprising m^2 cells lying in the first quadrant of the plane with one corner of the square at the origin (0,0).

We can make a cellular automaton on S by defining

$$N(x) = \{ (i,j-1),\ (i,j+1),\ (i-1,j),\ (i+1,j)\}$$

for every cell $x = (i,j)$. To ensure $N(x) \subseteq S$ we can assume the arithmetic on indices i and j is taken modulo m, and this imposes the topology of a torus on the automaton.

Alternatively we can define

$$N(x) = \begin{cases} \{(i,j-1),\ (i,j+1)\} & \text{if } i = 0 \text{ or } i = m-1 \\ \{(i,j-1),\ (i,j+1),\ (i-1,j),\ (i+1,j)\} & \text{otherwise} \end{cases}$$

for each $x = (i,j)$ and this defines a cylindrical structure (again taking arithmetic modulo m where necessary).

For a specific cellular automaton we choose the automaton of D.Griffeath described in [10]. We begin by chosing

$$N(x) = \{(i,j),\ (i,j-1),\ (i,j+1),\ (i-1,j),\ (i+1,j)\}$$

which gives a toroidal topology in which each cell is directly connected to itself in addition to its four orthogonal neighbours.

Next we take $Q = \{0,1,n-1\}$ for some $n \geq 1$ and define $f_x: T \times Q^5 \to Q$ by.

$$f_x(t,q,q_i,\ldots,q_5) = \begin{cases} q+1 \text{ if } q_1=q+1 \text{ or } q_2=q+1 \text{ or } q_3=q+1 \text{ or } q_4=q+1 \\ q \text{ otherwise} \end{cases} \quad (3)$$

and thus a cell changes state if at least one of its orthogonal neighbours has state one greater than the current state of x. (Here the arithmetic is considered modulo n so that $q = 0$ is considered one greater than $n - 1$).

If we visualise the behaviour of this automaton and execute the automaton on initial random states the system evolves in approximately 200 time steps into a spatially coherent system composed of a number of spirals.

3.3 Coupled Map Lattices

A coupled map lattice (CML) is a dynamical system with discrete time, discrete space and continuous state. At each lattice site i (i = 1,..., N, N = number of elements in the lattice), the activity $x(i)$ evolves as a nonlinear mapping f of its preceding value and some function of the the preceding activities at different sites $x(j)$. If there are no interactions between the activities at different sites the whole system is simply N independent mappings f , if there are interactions, these can be local, global, or in between these extremes and specified by some specific connectivity matrix.

If the connections are local, an element i interacts with some elements in its vicinity; these could be the nearest neighbours, or a larger vicinity. For a one-dimensional lattice, the nearest neigbours of i are simply ($i \pm 1$); in two- and higher dimensional lattices the neigbourhood needs to be specified (*e.g.* a von Neuman 4- or Moore 8- neighbour neighbourhood for a rectangular lattice in the plane). The connections can be unidirectional or symmetrical, and different weights can be associated with the different connections. An example of a one-dimensional, diffusively coupled CML is

$$x_{n+1}(i) = (1 - \varepsilon)\, f(x_n(i)) + \varepsilon/2[f(x_n(i + 1)) + f(x_n(i - 1))] \quad (4)$$

where n is a discrete time step. The behaviours of locally coupled map lattices are described in [11].

4. Synchronous Concurrent Algorithms.

A *synchronous concurrent algorithm* (SCA) is an algorithm based on a network of modules, channels, sources and sinks, that compute and communicate in parallel and are synchronised by a global clock T = {0,1,2,....}. The SCA processes *data* taken from a set A as a sequence $a(0)$, $a(1)$, $a(2)$, ... of clocked data: this sequence is a function a: $T \to A$. Each *module* is a unitary computational device that executes a (time-dependent) operation that is specified by a function $f_m : T \times A^n \to A$, when the module m has n input channels. Communication between modules occurs along *channels* that can transmit only a single datum $a \in A$ at any time; channels can branch but not merge. A *source* reads data into the network: it has no input channels, and a single output. A network with n sources will process n streams a_1, a_2, ..., a_n that form the vector-valued stream $a : T \to A^n$. A sink has a single input channel and no output channel; data is read out of the network. The *architecture* consists of a finite network of modules connected by channels. Two modules are neighbours if the output channel of one is an input channel of the other.

Let N be a SCA over a data set A with a clock T, with n > 0 sources. The input to N is a stream $a : T \to A^n$. If N has $k > 0$ modules the initial state of the network is a vector $x = (x_1, \ldots, x_k) \in A^k$, where x_i denotes the value output from the ith module at time zero. At each time $t \in T$ there is a single value output that can be determined from the t, a, and x. The *value functions* of the network $V(t,a,x)$, for $i = 1, \ldots, k$ are total functions

$$V_i : T \times [T \to A^n] \times A^k \to A$$

that denote the value output from the ith module at time t when the network is executed on input a and initial data x. The state of the channels is then given by *the* value function for the network

$$V_N(t,a,x) = \{V_1(t,a,x), \ldots, V_k(t,a,x)\}$$

for each $t \in A$, $x \in A^k$.

To obtain the value functions V_i for each $t \in T$ we first need $V_i(0,a,x)$, and then obtain $V_i(t+1,a,x)$ from $V_i(t,a,x)$. In the case $t = 0$:

$$V_i(0,a,x) = x_i.$$

If the ith module m_i has $n(i) > 0$ inputs and a functional specification f_i, then if at time t the input is $b_1, \ldots, b_{n(i)}$ then the value output at time $(t+1)$ is $f_i(t, b_1, \ldots, b_{n(i)})$. However, for $j = 1, \ldots, n(i)$, the jth input channel is the output of either a source $\lambda \in \{1, \ldots, n\}$, and so

$$b_j = a_\lambda(t), \qquad (5a)$$

or a module $\mu \in \{1, \ldots, k\}$, and so

$$b_j = V_\mu(t,a,x). \qquad (5b)$$

Thus the value functions of the network are

$$V_i(t+1, a, x) = f_i(t, b_1, \ldots, b_{n(i)}). \qquad (6)$$

The indices λ and μ are independent of t, a and x and are determined by the architecture; in programming terms they are syntactic quantities. For a specified architecture equations 5 and 6 collapse into a single equation.

Example .
Consider a linear array of n identical modules with nearest neighbour connections.
For $t = 0$, $\quad V_i(0,a,x) = x_i$,
and for $t > 0$

$$V_1(t+1, a, x) = f_1(t, V_1(t,a,x), V_2(t,a,x)),$$

For $i = 2, \ldots, n-1$, (7)

$$V_i(t+1, a, x) = f_i(t, V_{i-1}(t,a,x), V_i(t,a,x), V_{i+1}(t,a,x));$$

and $\quad V_n(t+1, a, x) = f_n(t, V_{n-1}(t,a,x), V_n(t,a,x)),$

where for $i = 1$ and $i = n$, $f_i : T \times A^2 \to A$, and for $i = 2, \ldots, n-1$, $f_i : T \times A^3 \to A$ are any given operations. (If $f_1, \ldots, f_n$ are all to have the same functionality then we need a ring structure.)

Thus we have a general formalisation that allows us to specify the output from a network of computing modules and any architecture if we know the architecture, functionalities of the modules and the initial inputs.

5. CA as SCAs.

Clearly the terminology of CA maps easily onto that of SCAs: a cellular automaton is an SCA over $A = Q$ whose modules are the cells of the automaton. For example, value functions for the Griffeath's automaton are

$$V_{00}, \ldots, V_{m-1,m-1} : T \times Q^{(m^2)} \to Q$$

$$V_{ij}(0) = x_{ij} \qquad (x = (x_{00}, \ldots, x_{m-1,m-1})$$

$$V_{ij}(t+1,x) = f_{ij}(t, V_{ij}(t,x), V_{i,j-1}(t,x), V_{i,j+1}(t,x), V_{i-1,j}(t,x), V_{i+1,j}(t,x))$$

where $f_{ij}: Q^5 \to Q$ is given by (3) above.

6. CMLs as SCAs.

The terminology of CMLs maps easily onto that of SCAs: a CML is an SCA over $A = \mathbb{R}$ whose modules are the elements of the CML. For example, value functions for the one dimensional diffusively coupled CML given by (4) are

$$V_1, \ldots, V_N : T \times \mathbb{R}^N \to \mathbb{R}$$

where, for $i + 1, \ldots, N$

$$V_i(0,x) = x_i$$

$$V_i(t+1,x) = f_i(t, V_{i-1}(t,x), \; V_i(t,x), V_{i+1}(t,x))$$

where $f_i: T \times \mathbb{R}^3 \to \mathbb{R}$ is defined by

$$f_i(t,u,v,w) = (1-\varepsilon)f(V) + \frac{\varepsilon}{2} \; (f(u) + f(w))$$

with $f: \mathbb{R} \to \mathbb{R}$ given, for example, by a piecewise linear map with a threshold [8].

7. Concluding remarks on the theory of SCAs.

The theory of synchronous concurrent algorithms has been developed in the context of theoretical computer science, to provide a neutral mathematical formalism for the analysis of architectures of processing elements that compute and communicate in parallel in a deterministic way. However, an SCA is a general notion that encompasses all discrete time, discrete physical space, continuous or discrete space dynamical systems; in fact, an SCA *is* a discrete time, discrete space dynamical system.

The theory of SCAs is developed in [13] and applied to coupled networks of excitable elements (biological and various formal neural networks) in [14]. Its application to discretised representations of excitable media emphasises the computational poverty of excitable media: the architecture of the SCA that is implemented by the representation of the excitable medium is simply nearest neighbour, diffusive coupling. The structure of this architecture is not changed by allowing the diffusion coefficients to be adjustable. The only way to introduce a richness into the computational capacity of an excitable medium is by radical anisotropy.

References

[1] Kuhnert, L. (1986) A new photochemical memory device in a light sensitive active medium. *Nature* **319** 393

[2] Kirby, K.G. and Conrad, M. (1996) Intraneuronal dynamics as a substrate for evolutionary learning. *Physica-D* **22** 205-215.

[3] Thompson, B.C. (1987) *A mathematical theory of synchronous concurrent algorithms.* PhD thesis, School of Computing Studies, University of Leeds.

[4] Zykov, V.S. (1987) *Simulation of wave processes in excitable media.* Manchester University Press.

[5] Fogelman-Soulié, F., Robert, Y. and Tchuente, M., eds. (1987) *Automata networks in computer science.* Manchester University Press.

[6] Crutchfield, J.P. and Kaneko, K. (1988) Phenomenology of spatio-temporal chaos. in *Directions in chaos*, Hao Bai-Lin (ed.) World Scientific, Singapore.

[7] Neumann, J. von (1966) in *Theory of self reproducing automata.* A.W.Burks, ed. University of Illinois Press, Urbana.

[8] Farmer, D., Tofoli, T and Wolfram, S., eds. (1984) *Cellular automata. Physica-D* **10** 1-248

[9] Tchuente, M (1987) Computation in automata networks. in [5].

[10] Dewdney, A.K. (1989) Computer recreations. *Scientific American* 88-91

[11] Kaneko, K. (1989). Pattern dynamics in spatio-temporal chaos. *Physica-D* **34** 1-41.

[12] Labos, E. (1987) Spike generating dynamical systems and networks. In *Dynamical systems: proceedings of IIASA workshop on mathematics of dynamical processes.* A.B.Kurzhanski and K.Sigmund (eds) Lecture Notes in Economics and Mathematical Systems. Springer-Verlag: Berlin.

[13] Thompson, B.C. and Tucker, J.V. (1990) Synchronous concurrent algorithms, Computer Science Division, University College of Swansea Research Report *(in preparation)*

[14] Holden, A.V., Tucker, J.V. and Thompson, B.C. (1990) The computational structure of neural systems. In: *Neurocomputers and attention: I: Neurobiology, synchronisation and chaos.* ed. A.V.Holden and V.I Kryukov. Manchester University Press.

INDEX

GPSR Compliance

The European Union's (EU) General Product Safety Regulation (GPSR) is a set of rules that requires consumer products to be safe and our obligations to ensure this.

If you have any concerns about our products, you can contact us on ProductSafety@springernature.com

In case Publisher is established outside the EU, the EU authorized representative is:

Springer Nature Customer Service Center GmbH
Europaplatz 3
69115 Heidelberg, Germany

Batch number: 10761599

Printed by Printforce, the Netherlands